Antifungals

From Genomics to Resistance and the Development of Novel Agents

Edited by

Alix T. Coste

Institute of Microbiology
University Hospital Lausanne; and
University Hospital Center
Lausanne
Switzerland

and

Patrick Vandeputte

Host–Pathogen Interactions Study Group
UPRES-EA 3142
L'UNAM Angers University; and
Parasitology–Mycology Laboratory
University Hospital Center
Angers
France

Caister Academic Press

Caister Academic Press
Norfolk, UK

www.caister.com

British Library Cataloguing-in-Publication Data
A catalogue record for this book is available from the British Library

ISBN (hardback): 978-1-910190-01-2
ISBN (ebook): 978-1-910190-02-9

Cover design adapted from images courtesy of Jean-Philippe Bouchara (*Aspergillus fumigatus*, *Scedosporium apiospermum*, *Trychophyton rubrum*, *Microsporum canis* and *Cryptococcus neoformans*), Patrick Vandeputte (*Candida albicans* and disk diffusion) and Amandine Gastebois (E-test) of Groupe d'Etude des Interactions Hôte-Pathogène, L'UNAM Université d'Angers, Angers, France

Contents

Contributors

Sara Amorim-Vaz
Institute of Microbiology
University of Lausanne; and
University Hospital Center
Lausanne
Switzerland

Sara.Martins-Amorim-Vaz@chuv.ch

Agostinho Carvalho
Life and Health Sciences Research Institute (ICVS)
School of Health Sciences
University of Minho; and
ICVS/3B's PT Government Associate Laboratory
Guimarães
Portugal

agostinhocarvalho@ecsaude.uminho.pt

Alix T. Coste
Institute of Microbiology
University of Lausanne; and
University Hospital Center
Lausanne
Switzerland

alix.coste@chuv.ch

Flavie Courjol
Inserm U995
University of Lille
Lille
France

courjolflavie@gmail.com

Leah Cowen
Department of Molecular Genetics
University of Toronto
Toronto, ON
Canada

leah.cowen@utoronto.ca

Cristina Cunha
Life and Health Sciences Research Institute (ICVS)
School of Health Sciences
University of Minho; and
ICVS/3B's PT Government Associate Laboratory
Guimarães
Portugal

cristinacunha@ecsaude.uminho.pt

Christophe d'Enfert
Institut Pasteur
Unité Biologie et Pathogénicité Fongiques
Département Génomes et Génétique; and
INRA USC2019
Paris
France

christophe.denfert@pasteur.fr

Thomas Edlind
MicrobiType LLC
Plymouth Meeting, PA
USA

tom@microbitype.com

Adeline Feri
Institut Pasteur
Unité Biologie et Pathogénicité Fongiques
Département Génomes et Génétique; and
INRA USC2019; and
University Paris Diderot
Magistère Européen de Génétique
Sorbonne Paris Cité
Cellule Pasteur
Paris
France

aferi@pasteur.fr

Emily P. Fox
Department of Microbiology and Immunology; and
Tetrad Program, Department of Biochemistry and Biophysics
University of California
San Francisco, CA
USA

Emily.Fox@ucsf.edu

Chantal Fradin
Inserm U995
University of Lille
Lille
France

chantal.fradin@univ-lille2.fr

Guillermo Garcia-Effron
Laboratorio de Micología y Diagnóstico Molecular – CONICET
Facultad de Bioquímica y Ciencias Biológicas (Planta Baja)
Universidad Nacional del Litoral – Ciudad Universitaria
Paraje el Pozo
Santa Fe
Argentina

ggarcia@unl.edu.ar; guillermo_garciaeffron@yahoo.com.ar

Claudia Galosi
Department of Experimental Medicine
University of Perugia
Perugia
Italy

claudia.gls@hotmail.it

Rebecca A. Hall
Institute of Microbiology and Infection
School of Biosciences
University of Birmingham
Birmingham
UK

r.a.hall@bham.ac.uk

Nairi Hartooni
Department of Microbiology and Immunology
University of California
San Francisco, CA
USA

Nairi.Hartooni@ucsf.edu

Thierry Jouault
Inserm U995
University of Lille
Lille
France

thierry.jouault@univ-lille2.fr

Santosh K. Katiyar
Department of Microbiology and Immunology
Drexel University College of Medicine
Philadelphia, PA
USA

santosh.katiyar@drexelmed.edu

Mélanie Legrand
Institut Pasteur
Unité Biologie et Pathogénicité Fongiques
Département Génomes et Génétique; and
INRA USC2019
Paris
France

mlegrand@pasteur.fr

Raphaël Loll-Krippleber
Institut Pasteur
Unité Biologie et Pathogénicité Fongiques
Département Génomes et Génétique; and
INRA USC2019; and
University Paris Diderot
Magistère Européen de Génétique
Sorbonne Paris Cité
Cellule Pasteur
Paris
France; and
Donnelly Centre for Cellular and Biomolecular Research
University of Toronto
Toronto, ON
Canada

raphael.loll.krippleber@utoronto.ca

Robin C. May
Institute of Microbiology and Infection
School of Biosciences
University of Birmingham
Birmingham
UK

r.c.may@bham.ac.uk

Clarissa J. Nobile
Department of Microbiology and Immunology
University of California
San Francisco, CA; and
Department of Molecular and Cell Biology
School of Natural Sciences
University of California
Merced, CA
USA

cnobile@ucmerced.edu

Adepemi O. Ogundeji
Department of Microbial, Biochemical, and Food Biotechnology
University of the Free State
Bloemfontein
South Africa

ogundejiao@ufs.ac.za

Luigina Romani
Department of Experimental Medicine
University of Perugia
Perugia
Italy

luigina.romani@unipg.it

Dominique Sanglard
Institute of Microbiology
University of Lausanne; and
University Hospital Center
Lausanne
Switzerland

dominique.sanglard@chuv.ch

Olihile M. Sebolai
Department of Microbial, Biochemical, and Food Biotechnology
University of the Free State
Bloemfontein
South Africa

sebolaiom@ufs.ac.za

Sheena D. Singh-Babak
Department of Microbiology and Immunology
University of California
San Francisco, CA
USA

Sheena.Singh@ucsf.edu

Luís A. Vale-Silva
Institute of Microbiology
University of Lausanne; and
University Hospital Center
Lausanne
Switzerland

luisvalesilva@gmail.com

Patrick Vandeputte
Host–Pathogen Interactions Study Group
UPRES-EA 3142
L'UNAM Angers University; and
Parasitology–Mycology Laboratory
University Hospital Center
Angers
France

pvandepu@gmail.com

Sadri Znaidi
Institut Pasteur
Genomes & Genetics Department
Fungal Biology and Pathogenicity Unit
Paris
France

sadri.znaidi@gmail.com

Preface

Over the last 30 years, the incidence of fungal infections in immuno-compromised patients has increased considerably. Systemic infections still have mortality rates of up to 50%, revealing the limitations of available antifungal therapies. These limitations are due partly to acquisition of antifungal resistance by pathogenic fungi and partly to the increasing incidence of intrinsically poorly susceptible pathogenic fungal species.

This book focuses on the most recent advances made to decipher the molecular mechanisms responsible for antifungal resistance to classic molecules that are azoles, polyenes, and echinocandins in the major fungal genera involved in human pathology. A large part is also dedicated to the contributions made to deepen the knowledge on fungal biology, always with the aim of defining new antifungal strategies. Thus, a special emphasis is given to the recent findings on a particular lifestyle that is biofilms, on fungal-specific biological pathways that constitute potential new targets, as well as on the research undertaken to enhance the efficacy of existing treatments. Likewise, high-throughput capabilities allowed the exponential increase in the number of screenings of chemical compound collections for antifungal activity, with the underlying decryption of the mode of action of the candidate molecules. Furthermore, the most recent technologies allowed a tremendous increase in the comprehension of the mechanisms settled by the host in response to fungi, resulting in major advances in the field of antifungal vaccine development as well as in moderating the infection effects by modulation of the human immune response. Finally, a chapter is dedicated to the description of animals models most recently developed to specifically study fungal infections, without which all the data from *in vitro* research would not be transposable to humans, which is the basis for improving antifungal treatments and consequently patients' quality of life.

Science has never been so deeply renewed since the recent advent of high-throughput technologies, which allow researchers to consider the whole organism's response to any particular condition. In the field of antifungal therapy, this unprecedented level of analyses has not only clarified prior knowledge but also opened the door to entirely new disciplines.

Patrick Vandeputte and Alix T. Coste

Molecular Mechanisms of Resistance of *Candida* spp. to Membrane-targeting Antifungals

1

Luís A. Vale-Silva

Abstract

Membrane-targeting antifungal drugs represent the most numerous group of available antifungals. Systemically available membrane-targeting drugs act either by inhibiting ergosterol biosynthesis or by binding ergosterol directly in the fungal cell membrane. Ergosterol biosynthesis inhibitors include the large group of azole drugs, as well as a few other drugs from three different chemical families: allylamines, morpholines and thiocarbamates. The sole representatives of ergosterol binding drugs are the polyenes, including two active compounds in clinical use: amphotericin B and nystatin. Various different mechanisms of resistance to membrane-targeting drugs have been described. Resistance mechanisms may be grouped as alterations of the intracellular accumulation of the drug, of the drug target (target enzyme sequence alterations or up-regulation), or of the sterol biosynthesis pathway. Often, different mechanisms combine in the same *Candida* spp. isolate, conferring stepwise increase in drug resistance. This chapter presents a review of the known mechanisms of resistance of *Candida* spp. to membrane-targeting antifungals currently in use to treat candidiasis.

Introduction

Antifungal drugs

The number of available antifungal drugs has increased over the last years. However, compared with antibacterial antibiotics available antifungal drug structural classes, as well as individual drugs within those classes, remain rather limited (Brown *et al.*, 2012; Pierce and Lopez-Ribot, 2013). Being eukaryotic organisms, fungal pathogens are phylogenetically much closer to the human host than bacterial pathogens. This sets serious limits to the range of exploitable fungal-specific drug targets. Currently available antifungal drugs have one of three general targets: RNA synthesis, the cell wall or the cell membrane. Overall, there are currently about 10 drugs available to treat disseminated fungal infections and a few more to treat superficial infections. Flucytosine, or 5-fluorocytosine, is the only available RNA synthesis inhibitor, or antimetabolite, while the group of cell wall targeting drugs includes the relatively recently developed echinocandins (caspofungin, anidulafungin and micafungin). These drugs are the subject of later chapters in this book. The present chapter focuses on membrane-targeting drugs, by far the most numerous group. Membrane-targeting drugs act either by inhibiting ergosterol biosynthesis or by binding ergosterol directly in the fungal cell membrane (Table 1.1). Ergosterol is the main fungal cell membrane sterol, as opposed to cholesterol in mammalian cells. Ergosterol biosynthesis inhibitors include the large group of azole drugs, as well as a few other drugs from three different chemical families: allylamines, morpholines and thiocarbamates. The sole family of ergosterol binding drugs is the polyenes, including two active compounds: amphotericin B and nystatin.

Definition of resistance to antifungal drugs

Antifungal resistance can be defined as clinical or microbiological resistance, or as a composite of the two. Clinical resistance is the situation

Table 1.1 Simplified ergosterol biosynthetic pathway and the targets of membrane-acting antifungal agents in *Candida* spp.

Enzyme	Gene	Sterol intermediate	Antifungal inhibitor class	Molecule
		Squalene		
Squalene epoxidase	*ERG1*	↓	Allylamine	Terbinafine
			Thiocarbamate	Tolnaftate
		2,3-Oxidosqualene		
Lanosterol synthase	*ERG7*	↓		
		Lanosterol		
C-14 Sterol demethylase	*ERG11*	↓	Triazoles	Fluconazole
				Itraconazole
				Voriconazole
				Posaconazole
C-14 Sterol reductase	*ERG24*	↓	Morpholine	Amorolfine
...	...	↓		
		Zymosterol		
C-24 Sterol methyltransferase	*ERG6*	↓		
		Fecosterol		
C-8 Sterol isomerase	*ERG2*	↓	Morpholine	Amorolfine
		Episterol		
C5-Sterol desaturase	*ERG3*	↓		
C22-Sterol desaturase	*ERG5*	↓	Triazoles (?)	
C-24 Sterol reductase	*ERG4*	↓		
		Ergosterol	Polyenes	Amphotericin B
				Nystatin

in which the antimicrobial drug concentration required to resolve an infection is higher than the concentration that could be safely achieved with normal therapeutic dosing. Microbiological resistance refers to the ability of the infecting microorganism to grow *in vitro* in the presence of concentrations of an antimicrobial drug higher than those sufficient to inhibit growth of wild-type strains. Microbiological drug resistance is detected using standardized *in vitro* drug susceptibility testing methods, which produce quantitative or semi-quantitative measurements of the minimum drug concentrations required to inhibit growth (minimum inhibitory concentration, MIC). Two standard broth microdilution protocols are available to test *Candida* spp.: one published in Europe by the Subcommittee on Antifungal Susceptibility Testing of the ESCMID European Committee for Antimicrobial Susceptibility Testing (AFST-EUCAST), protocol Edef. 7.1 (Rodriguez-Tudela, 2008); and one published in the US by the Clinical Laboratory Standards Institute (CLSI), protocol M27-A3 (Clinical and Laboratory Standards Institute, 2008, 2012). The methods present a few differences but both rely on measurements of yeast growth in the presence of a range of antifungal drug concentrations over a defined period of time. Additionally, a simpler agar disk diffusion method has also been standardized (Clinical and Laboratory Standards Institute, 2009; Matuschek *et al.*, 2014).

In practice, a composite definition is more often employed, according to which resistance is present when isolates are not inhibited *in vitro* by the clinically achievable concentrations of the agent with normal drug dosage schedules. Based on this premise, MIC breakpoints are defined by the above-referred authorized committees to guide therapy according to likelihood of treatment success (Pfaller, 2012). This process takes into account available microbiological data on *in vitro* susceptibility and MIC distributions, as well as drug pharmacokinetics/pharmacodynamics (PK/PD) and clinical experience (including cases of treatment success and failure). In this context, drug resistance is defined as the situation in which the MIC of an antimicrobial agent against an infecting organism is higher than the defined resistance breakpoint. In practical terms, this means that the use of the antimicrobial agent to treat that specific infection is associated with a low likelihood of therapeutic success and, therefore, not recommended (Pfaller and Diekema, 2012). MIC breakpoints in use with both broth microdilution standard methods have been under constant revision and the latest values have been published recently (Clinical and Laboratory Standards Institute, 2012; European Society of Clinical Microbiology and Infectious Diseases, 2013).

Even if microbiological resistance is the most common cause of refractory infection, several additional factors affect therapeutic success. Everything from problems with the diagnosis of the infection, the host's immune function, the fungal burden at initiation of treatment, the anatomical site of infection, alterations in virulence of the fungus, treatment length and/or compliance, and patient-related differences in the drug's bioavailability and metabolism may determine clinical resistance (Kanafani and Perfect, 2008). This way, the predictive power of the determined MIC is limited and the correlation between microbiological susceptibility data and clinical success is imperfect. In general, this *in vitro-in vivo* correlation can be well modelled by the so-called '90–60 rule'. The '90–60 rule' states that infections caused by susceptible isolates will respond to therapy in about 90% of the cases, while resistant isolates will respond in about 60% of cases (Rex and Pfaller, 2002). Like most of the above information, this rule was originally described for bacterial infections and has later been found to appropriately translate to fungal infections.

The focus of the present book will be on mechanisms of resistance as defined from the microbiological rather than the clinical point of view. The term 'resistance' will be used interchangeably with the term 'microbiological resistance'. Naturally, however, the importance of microbiological resistance ultimately relies primarily on the ability to predict clinical resistance.

The problem of resistance

Besides being rather limited in number, available antifungal drugs are further restricted by important problems of toxicity, undesirable drug interactions and resistance. Resistance can be either intrinsic, also known as primary resistance, or acquired, also called secondary resistance. The former refers to a shift towards colonization with intrinsically less susceptible or fully resistant organisms. In the case of primary resistance, all or almost all isolates of the species are inherently resistant even prior to any contact with the drug. Acquired resistance, on the other hand, involves the emergence of cell-specific resistance mechanisms in normally susceptible strains during exposure to the drug. Furthermore, acquired resistance can be stable or transient, meaning that the resistance phenotype can be either genetically stable or lost in the absence of the antifungal drug. Transient resistance can be the result of chromatin organization changes or of gene expression changes elicited by environment cue-responsive signal transduction pathways. Additionally, it may be the result of genetic alterations that cause fitness defects and are, thus, outcompeted by the wild type genotype once the drug pressure is removed. Yet another situation is the one described by the concept of drug tolerance, referring to the ability of the fungus to survive exposure to concentrations of the drug above the MIC.

Although less serious than the analogous problem in bacterial infections, the extensive use of antifungal drugs has raised concern about the emergence of resistant *Candida* spp. strains (Hof, 2008; Pfaller, 2012). The first major drug resistance problem arose with azole resistance in the context of oropharyngeal candidiasis in AIDS patients. This trend was successfully reverted before the end of the 1990s with the introduction of highly active antiretroviral therapy (HAART) to manage AIDS (Martins *et al.*, 1998; Revankar *et al.*, 1998). However, epidemiological surveillance programmes have since uncovered the continued development of antifungal drug resistance. Rather than the emergence of drug resistance within the most commonly isolated species, *C. albicans*, contemporary resistance appears to be due to an epidemiological drift towards the emergence of non-*albicans Candida* species with lower susceptibility to antifungals. The main concern is *C. glabrata*, which has emerged as the second most commonly encountered *Candida* species in the clinical setting. The ARTEMIS Global Antifungal Surveillance programme has identified the increase of *C. glabrata* invasive candidiasis as a driver for a significant increase in fluconazole resistance both in the United States (Pfaller *et al.*, 2009) and worldwide (Pfaller *et al.*, 2010). Similar data has been produced by the SENTRY Antimicrobial Surveillance Program (Pfaller *et al.*, 2011a,b,c). Moreover, besides rapid development of fluconazole resistance, *C. glabrata* shows the disturbing ability to develop multidrug resistance and can become progressively resistant to multiple unrelated antifungals during treatment (Pfaller, 2012).

Many different molecular mechanisms of antifungal drug resistance have been described. Regarding resistance to membrane-targeting drugs specifically, mechanisms may be grouped as alterations of the intracellular accumulation of the drug, drug target alterations (target enzyme sequence alterations or up-regulation) or compensatory adaptations of the sterol biosynthesis pathway (Fig. 1.1). An additional mechanism, the formation of fungal biofilms, is the subject of a later chapter (Chapter 4). Individual mechanisms are discussed in detail below in the context of the affected drug classes.

Azoles

Available drugs and mechanism of action

Azole antifungals are the most commonly used class of drugs to treat invasive fungal infections. The first therapeutically useful azoles were imidazole drugs, introduced in the late 1960s (Fromtling, 1988). Imidazoles are characterized chemically by the presence of two nitrogen atoms in their five-membered azole ring. A further and improved generation of drugs was later introduced with the development of the triazoles fluconazole and itraconazole, bearing three nitrogen atoms in their azole ring. Triazole drugs brought about

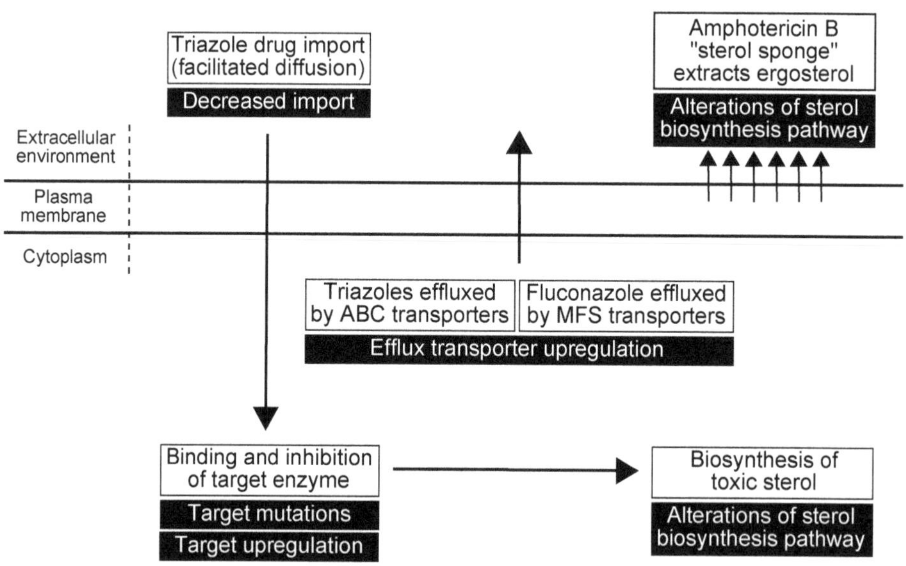

Figure 1.1 Mechanisms of action of azole and polyene drugs (white background) vis-à-vis mechanisms of resistance (dark background). See Table 1.2 for information on the involved genes. ABC, ATP-binding cassette; MFS, major facilitator superfamily.

significant improvement in both spectrum of antifungal activity and target specificity (Maertens, 2004). More recently, a second generation of triazoles has been introduced, with voriconazole, structurally related to fluconazole, and posaconazole, similar to itraconazole (Fig. 1.2). This meant further improvement in antifungal potency and spectrum of activity (Maertens, 2004). Azole drugs act by inhibiting the biosynthesis of ergosterol, the main fungal cell membrane sterol, equivalent to cholesterol in mammalian cells. Azoles target specifically the cytochrome P-450 enzyme 14α-sterol demethylase, encoded by the gene *ERG11* in *Candida* spp. (Table 1.1). The inhibition occurs through binding of the iron atom of a haem group, present at the active site of the enzyme, by a nitrogen atom of the azole ring. The extent of the inhibitory effect is affected by the structure of the specific azole molecule. Additionally, it is also affected by differences in the amino acid sequence and conformation of the active site of the enzyme between fungal species, thus accounting for inter-species differences in potency of the drugs. Inhibition of Erg11 leads to the accumulation of 14α-methylated sterols replacing ergosterol in the fungal cell membrane (Fig. 1.3). Ergosterol depletion, together with the accumulation of the toxic 14α-methylated sterol 14-methylergosta-8,24,(28)-dien-3β,6α-diol (Fig. 1.3), disrupts the normal structure and function of the cell membrane (Kelly *et al.*, 1995). The yeast cell then becomes more sensitive to further damage. In addition, the activity of membrane-bound enzymes, such as those involved in nutrient metabolism and transport or biosynthesis of cell wall components, is disturbed. Moreover, it has been suggested that azole treatment may render fungal cells more sensitive to attack by cells of the

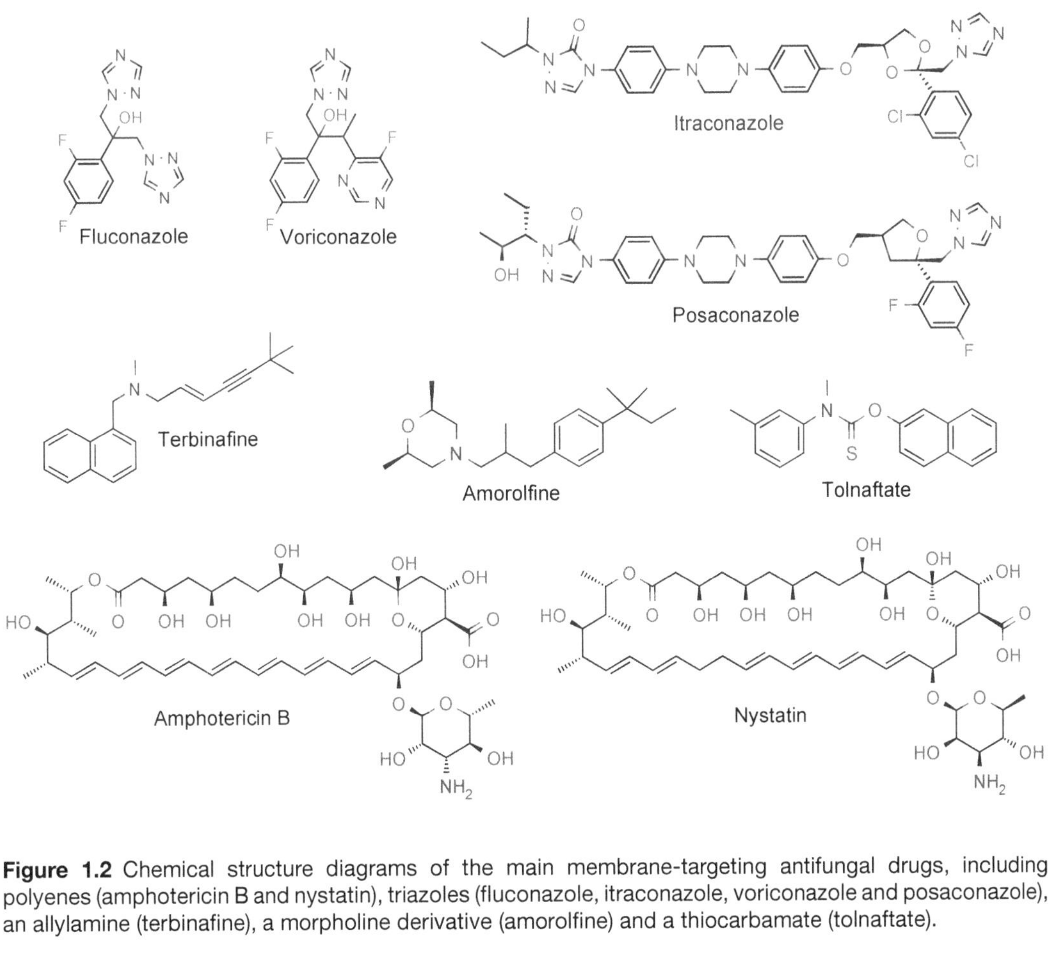

Figure 1.2 Chemical structure diagrams of the main membrane-targeting antifungal drugs, including polyenes (amphotericin B and nystatin), triazoles (fluconazole, itraconazole, voriconazole and posaconazole), an allylamine (terbinafine), a morpholine derivative (amorolfine) and a thiocarbamate (tolnaftate).

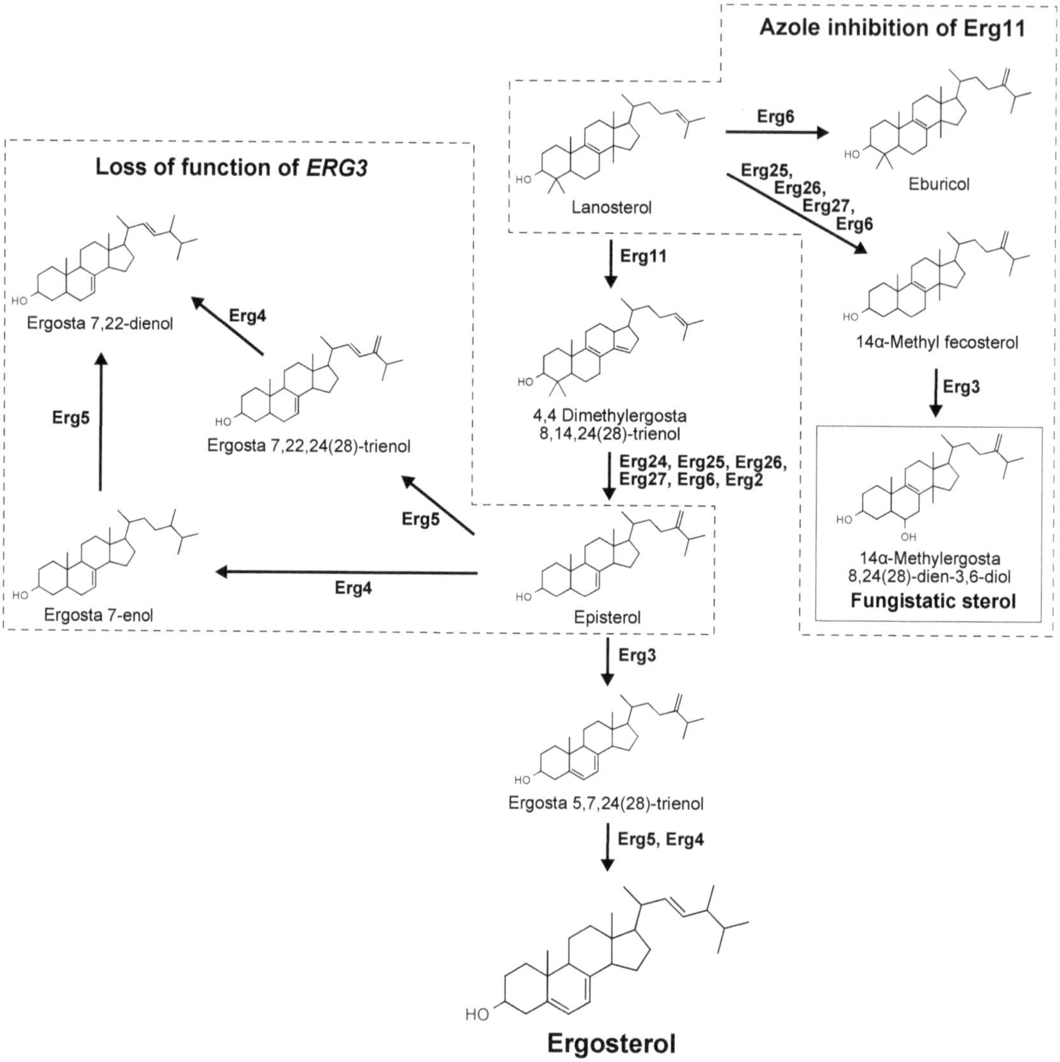

Figure 1.3 Chemical structure diagrams of post-cyclization sterols in the ergosterol biosynthesis pathway of *C. albicans*. The consequence of azole drug inhibition of Erg11 (right-hand side box) is the Erg3-dependent accumulation of a fungistatic sterol. Loss of function of Erg3 results in the accumulation of C5-C6 saturated sterols that support growth (left-hand side box). In the context of exposure to azole drugs, erg3 mutants bypass the biosynthesis of the fungistatic sterol and accumulate eburicol and 14α-methyl fecosterol. The exact proportions of the accumulated sterols represented in the two boxes, as well as the presence of specific precursors, vary between different *C. albicans* strains.

host's innate immunity, for example concerning oxidative metabolites produced by macrophages (Shimokawa and Nakayama, 1992). Azole drugs may also target other enzymes of the ergosterol biosynthesis pathway, among which at least C-22 sterol desaturase (*ERG5*; Table 1.1) has been proposed (Kelly *et al.*, 1997a). For more detail on the elucidation of the mechanism of action of azole antifungals see Chapter 9. Ultimately, these effects combined result in broad-spectrum fungistatic or fungicidal antifungal activity, depending on the specific drug and fungal species.

Mechanisms of resistance

Until the late 1980s, acquired resistance of *Candida* spp. to azole antifungals was rare (Vanden Bossche *et al.*, 1994). Following the development of the first successful triazoles, fluconazole and itraconazole, an emergence of resistant yeast isolates displaying various mechanisms of azole resistance has been documented. This is likely due to the mechanism of action of azole drugs and especially to the fact that they have long been the most commonly used drugs in clinical practice. In fact, azoles, and particularly fluconazole, have traditionally been by far the most commonly used drugs both for treatment and for prophylaxis of infection in patients at risk, in some cases with prolonged exposure. Even though extensive use of these drugs over the last few decades is generally thought to have been the main driver of resistance, epidemiological causality has not been unambiguously demonstrated (Blot *et al.*, 2006). Regardless, azoles are the drug class gathering the highest number of reports of resistant clinical fungal isolates and of different resistance mechanisms. In fact, all above-referred resistance mechanisms to membrane-targeting drugs affect azole drugs (Fig. 1.1 and Table 1.2). This includes alterations of the intracellular accumulation of the drug and of the drug target, alterations of the sterol biosynthesis pathway, as well as chromosome aneuploidies or the formation of fungal biofilms. With the exception of the formation of biofilms, which is the subject of a later chapter in this book, a closer look into the different mechanisms is taken below.

Reduced intracellular accumulation

Reduced accumulation of azole drugs in the fungal cell can be caused by reduced drug penetration in the cell, due to changes in membrane composition. These changes may consist on low ergosterol levels or possibly a decreased ratio between phosphatidylcholine and phosphatidylethanolamine, which may change the barrier function of the membrane. Such changes have been associated to fluconazole resistance in *C. albicans* clinical isolates (Loffler *et al.*, 2000). Additionally, itraconazole resistance in a *C. krusei* isolate that showed reduced intracellular levels of the drug has been suggested to be due to reduced permeability as well, rather than active efflux of the drug (Venkateswarlu *et al.*, 1996). Indirect evidence of the existence of a fluconazole uptake transporter and the hypothesis that resistance was caused by decreased activity of that transporter was suggested in a study of *C. lusitaniae* clinical isolates (Noel *et al.*, 2003). More recently, a confirmation of this hypothesis came about with the elucidation of the mechanism of azole drug import into yeast cells, shown to rely on facilitated diffusion through a transporter rather than passive diffusion (Mansfield *et al.*, 2010). Alteration of this unknown transporter may in principle be a potential cause of azole resistance, but it has not

Table 1.2 Summary of the most significant genetic mechanisms of resistance to membrane-targeting antifungal agents observed in clinical isolates of *Candida* spp.

Antifungal agents	Mechanism of resistance		Genetic target	
Triazoles	Drug target mutations		*C. albicans*	*ERG11*
	Drug target up-regulation	Mutations in target regulator	*C. albicans*	*UPC2*, regulator of *ERG11*
		Chromosomal aneuploidy	*C. albicans*	Chromosome 5
	Alterations of the sterol biosynthesis pathway	Mutations in enzymes	*C. albicans*	*ERG3*
	Up-regulation of ABC transporters	Mutations in regulators	*C. albicans*	*TAC1*, regulator of *CDR1*, *CDR2*
			C. glabrata	*CgPDR1*, regulator of *CgCDR1*, *CgCDR2*, *CgSNQ2*
		Chromosomal aneuploidy	*C. albicans*	Chromosome 5
Fluconazole	Up-regulation of MFS transporters	Mutations in regulators	*C. albicans*	*MRR1*, regulator of *MDR1*
Polyenes	Alterations of the sterol biosynthesis pathway	Mutations in enzymes	*C. albicans*	*ERG11*, *ERG3*, *ERG2*, *ERG6*,...

yet been definitively demonstrated. Also, for the above referred *C. albicans* and *C. krusei* isolates a decrease in the activity of this transporter secondary to the described alterations of the lipid composition of the plasma membrane cannot at this point be ruled out as the actual mechanism of resistance.

Reduced azole drug import is of limited importance, however, compared to resistance mediated by increased drug efflux. Although drug resistance mediated by multidrug transporters had previously been described in baker's yeast, the laboratory model yeast *Saccharomyces cerevisiae* (Balzi and Goffeau, 1994), the mechanism was first reported in the clinical setting by Sanglard *et al.* in 1995 (Sanglard *et al.*, 1995). The study described sequential clinical *C. albicans* isolates from AIDS patients with oropharyngeal candidiasis developing drug resistance following prolonged treatment, mainly with fluconazole (Sanglard *et al.*, 1995). In retrospect, this report came about surprisingly late, seeing that the first descriptions of azole-resistant *C. albicans* and *C. glabrata* clinical isolates were published starting from the early 1980s, first to the imidazoles, miconazole and ketoconazole, and later also to fluconazole and itraconazole (Vanden Bossche *et al.*, 1994). The study by Sanglard *et al.* correlated drug resistance with a failure to accumulate azoles in the yeast cell. In order to explain that observation, the authors analysed known efflux transporters and found increased mRNA levels of two multidrug transporters: the ATP-binding cassette (ABC) transporter *CDR1* (*Candida* drug resistance 1) and the major facilitator superfamily (MFS) transporter *MDR1* (multidrug resistance 1; then known as *BEN*^{R}, for benomyl resistance) (Sanglard *et al.*, 1995). Both transporters had been previously cloned, *MDR1* in 1991 (Fling *et al.*, 1991) and *CDR1* earlier that same year (Prasad *et al.*, 1995). Fluconazole was found to be a substrate for both transporters, while *CDR1* additionally accepted ketoconazole and itraconazole as well (Sanglard *et al.*, 1995). Soon after this report, a second multidrug ABC transporter, called *CDR2*, was cloned and implicated in azole resistance (Sanglard *et al.*, 1997). The study showed that Cdr2 could efflux azole drugs, but also terbinafine, amorolfine, and several unrelated drugs. *CDR2* was found to be overexpressed in azole-resistant clinical *C. albicans* isolates and contributed to resistance, together with *CDR1*. After these first reports of the involvement of *MDR1*, *CDR1* and *CDR2* in resistance to azole drugs, multiple clinical isolates have been found to display increased expression of these efflux pumps. In fact, overexpression of efflux transporters is the most common mechanism of azole drug resistance in *C. albicans*, even though combinations of two or more mechanisms are generally encountered in most resistant clinical isolates. A quantification of the prevalence of resistance mechanisms performed over a decade ago found up-regulation of efflux pumps in 85% of azole-resistant isolates, with 75% showing combined mechanisms (Perea *et al.*, 2001).

Building on the data obtained with *C. albicans*, parallels were made concerning azole resistance in non-*albicans Candida* species. An *MDR1* orthologue has been identified in *C. dubliniensis* and named *CdMDR1* (Moran *et al.*, 1998). This transporter was shown to mediate a very similar drug resistance profile to that mediated by *CaMDR1* in *C. albicans* (Moran *et al.*, 1998; Wirsching *et al.*, 2001). Similarly, experimentally induced azole resistance has been shown to be mediated by *MDR1* orthologues in *C. tropicalis* (Barchiesi *et al.*, 2000) and in *C. parapsilosis* (Silva *et al.*, 2011). This suggests *MDR1* up-regulation as a possible azole resistance mechanism in these species in the clinical setting. Besides the MFS transporter gene *MDR1*, ABC transporter orthologues have also been characterized in several non-*albicans Candida* species. Functional orthologues of *CDR1* and/or *CDR2* have long been known to be present in *C. glabrata* (Sanglard *et al.*, 1999, 2001), *C. dubliniensis* (Moran *et al.*, 1998) and *C. tropicalis* (Barchiesi *et al.*, 2000).

At the molecular level, the mechanism behind up-regulation of multidrug efflux pumps most commonly consists on selection for mutations in *trans*, on specific transcriptional regulators. This was described for the first time when the orthologue of *S. cerevisiae*'s transcription factor gene *NDT80*, named *CaNDT80*, was found to positively regulate *CDR1*, including the antifungal-mediated induction of *CDR1* expression (Chen *et al.*, 2004). Around the same time, following the description of a conserved *cis* regulatory element (drug-responsive element 1, DREI) on the promoters of *CDR1* and *CDR2* (de Micheli *et al.*, 2002), the major *trans* regulator of both these ABC transporter genes was discovered: a transcription factor containing the fungal-specific Zn(2)-Cys(6) (zinc cluster) DNA-binding domain and named *TAC1*, for transcriptional activator of *CDR* genes (Coste *et al.*, 2004). Tac1 was shown to bind the DREI and the deletion of the *TAC1* gene eliminated drug-induced up-regulation of *CDR1* and *CDR2*. Moreover, a mutant *TAC1* allele recovered from an azole-resistant *C. albicans* strain was found to mediate constitutive *CDR1* and *CDR2* up-regulation, thus configuring a gain-of-function (GOF) allele (Coste *et al.*, 2004). In clinical isolates, loss of heterozygosity at the *TAC1* locus, producing homozygous GOF alleles, often mediates increased azole resistance (discussed in more detail below in the subsection 'Chromosome aneuploidies').

The description of the main transcriptional activator of *MDR1* followed, with the identification of the transcription factor gene *MRR1* (multidrug resistance regulator 1) (Morschhauser *et al.*, 2007). The consequences of *MDR1* up-regulation had been dissected in an earlier study, using forced gene overexpression with a strong constitutive promoter (Hiller *et al.*, 2006). Natural *MDR1* overexpression was found to be most often caused by selection for homozygous GOF mutations on *MRR1*, a phenomenon that may occur not only *in vitro* but also *in vivo* during infection, under selective pressure from azole therapy (Dunkel *et al.*, 2008a; Morschhauser *et al.*, 2007). Even if Mrr1 is considered to be the most important *MDR1* regulator, other transcription factors are known to affect its expression. Most notably, the bZip transcription factor gene *CAP1*, which is implicated in responses to oxidative stress, also targets Mdr1 and allows efflux-mediated multidrug resistance (Alarco and Raymond, 1999; Rognon *et al.*, 2006).

In *C. glabrata* azole drug efflux plays a central role in the resistance to azole drugs. In fact, compared to *C. albicans* and other *Candida* species, *C. glabrata* generally displays reduced susceptibility to azole drugs (Jandric and Schuller, 2011). Additionally, *C. glabrata* more readily develops drug efflux-mediated secondary resistance upon exposure to azole antifungals such as fluconazole (Bennett *et al.*, 2004). This is known to be mainly mediated by development of GOF mutations on *C. glabrata*'s own *trans* regulator of ABC transporters, the Zn(2)-Cys(6) transcription factor *CgPDR1* (pleiotropic drug resistance 1). The characterization of *CgPDR1* was based on the analysis of the pleiotropic drug resistance network in the genetic model yeast *S. cerevisiae*, to which *C. glabrata* is actually phylogenetically closer than to *C. albicans*. Two related transcription factors, Pdr1 and Pdr3, are known to regulate efflux-mediated drug resistance in *S. cerevisiae* through a myriad of ABC transporters (Cannon *et al.*, 2009). In *C. glabrata*, *PDR1* orthologue *CgPDR1* was characterized in 2004 (Vermitsky and Edlind, 2004) and further confirmed two years later (Tsai *et al.*, 2006; Vermitsky *et al.*, 2006). The initial confusion concerning the fact that no *PDR3* orthologue could be found in *C. glabrata* was clarified more recently when *CgPDR1* was shown to combine in *C. glabrata* the regulatory activity of *PDR1* and *PDR3* of *S. cerevisiae* (Paul *et al.*, 2010). Interestingly, in addition to its previously known regulatory activity, *CgPDR1* can also be activated by direct binding to xenobiotics like the azole drugs. This configures a nuclear receptor-like pathway responding to drug stress by inducing ABC transporter expression and consequently causing drug resistance (Thakur *et al.*, 2008).

Constitutive up-regulation of ABC transporters and azole drug resistance in *C. glabrata* can also be caused by an additional *CgPDR1*-dependent mechanism known as high-frequency acquired resistance (HFAR). Wild type *C. glabrata* strains have been shown to develop azole resistance when cultured *in vitro* in the presence of the triazole fluconazole at frequencies higher than 10^{-4} (Sanglard *et al.*, 2001). The phenotype was shown to be accompanied by mutation or complete loss of mitochondrial DNA, closely resembling the 'petite' phenotype previously described in *S. cerevisiae* (Sanglard *et al.*, 2001). Such mutants show slow growth *in vitro*, producing smaller colonies in agar-containing medium, due to a respiration-deficiency that renders them unable to completely metabolize carbon sources. Since mitochondrial function participates in a wide range of metabolic processes, in order to cope with mitochondrial dysfunction cells undergo transcriptional reprogramming, within the regulatory connection between mitochondria and the nucleus referred to as retrograde regulation (Moye-Rowley, 2005). Among many other changes, this results in constitutive up-regulation of ABC transporters, thus establishing the connection between respiratory deficiency and drug resistance (Moye-Rowley, 2005).

Drug target alterations

Resistance mechanisms affecting azole drug target enzyme Erg11 include both the overexpression and the development of mutations in *ERG11*. The former mechanism relies on a gene dosage effect, while the latter works through a reduction of the binding affinity to azoles and/or an increase in the binding affinity to the sterol substrate. In either case, Erg11 ultimately evades the inhibitory interaction with the drugs without significantly affecting its function. Inactivation of *ERG11* is a third mechanism relying on alteration of the azole drug target. This causes alterations of the ergosterol biosynthesis pathway and will thus be discussed in the corresponding section below. The Zn(2)-Cys(6) transcription factor *UPC2* (uptake control 2) has been described as a *trans* regulator of *ERG11*, among other genes implicated in sterol metabolism, mediating *ERG11* overexpression in *C. albicans* (MacPherson *et al.*, 2005; Silver *et al.*, 2004; White and Silver, 2005). Following the establishment of its role in the control of *ERG11* and sterol biosynthesis in general, a few GOF mutations in *UPC2* have been identified and linked to azole resistance in clinical isolates of *C. albicans* (Dunkel *et al.*, 2008b; Heilmann *et al.*, 2010; Hoot *et al.*, 2011). The described mutations are closely located near the C-terminus of Upc2, which suggests this region may be a hot spot for mutations with functional implications (Vandeputte *et al.*, 2012). However, point mutations in *ERG11* are much more commonly encountered. The first non-synonymous point mutation leading to decreased azole affinity to Erg11 was reported as early as 1990 (Vanden Bossche *et al.*, 1990). A later study looked in detail at this mechanism and suggested that different mutations, as well as other mechanisms of resistance, may combine in the same *C. albicans* strain to provide stepwise increase in azole resistance levels (Sanglard *et al.*, 1998). After that, the collection of known mutations has grown impressively over the years. A review published in 2010 listed over 140 different amino acid substitutions in Erg11 (Morio *et al.*, 2010) and additional reports have accumulated since then. Even if published amino acid substitutions in Erg11 have not all been experimentally proven to mediate azole resistance, it seems clear that Erg11 displays great genetic polymorphism and structural plasticity. Specific amino acid substitutions affect azole drug binding and, consequently, azole drug resistance to different degrees. The understanding of this process has recently benefited from a major breakthrough, with the publication of the crystal structure of full-length Erg11 from *S. cerevisiae* (Monk *et al.*, 2014). The data shed light on how several previously published Erg11 mutations conferring azole resistance result in substitutions of amino acids located within the active site and substrate channel of the enzyme. Modelling of the interaction with azoles reveals how specific mutations affect the interaction of the enzyme with azole drugs (Monk *et al.*, 2014). On the other hand, the decrease in affinity to Erg11 may differ between different azole drugs. Quantitatively, the impact of such amino acid substitutions on the affinity of azoles with long side chains (itraconazole and posaconazole) is often milder than on smaller drugs like fluconazole and voriconazole, admittedly because longer hydrophobic side chains allow more points of contact and thus tighter binding to the enzyme (Xiao *et al.*, 2004). Also, mutations in the substrate channel located farther away from the haem-containing active site can be expected to affect susceptibility to the long-tailed triazoles itraconazole and posaconazole but not short-tailed fluconazole and voriconazole (Monk *et al.*, 2014). Even though mutations in Erg11 causing azole resistance are usually assumed to do so through a reduction in the enzyme's affinity to the drug, previous work shows that an increased affinity to the sterol substrate in the presence of azoles may also contribute to the overall resistance (Warrilow *et al.*, 2010). This way, this mechanism should also be kept in mind when studying the impact of target enzyme alterations on resistance.

Concerning the most relevant non-*albicans* *Candida* species, there is a striking scarcity of reports on target alteration-mediated azole resistance. Reported cases include *CtERG11* overexpression associated with missense mutations in *C. tropicalis* (Forastiero *et al.*, 2013; Jiang *et al.*, 2013; Vandeputte *et al.*, 2005). Additionally, *ERG11* polymorphisms and/or up-regulation have been found in fluconazole-resistant *C. dubliniensis* clinical isolates (Perea *et al.*, 2002). However, these strains combined different resistance mechanisms, including ABC transporter up-regulation, and the exact contribution of target alterations to the overall azole resistance was not studied in more detail (Perea *et al.*, 2002). In general, Erg11 alterations cause only minor to moderate decreases in azole susceptibility and frequently appear combined with additional mechanisms in fully resistant *Candida* spp. clinical isolates. This has been elegantly exemplified in a report dissecting the contribution of different resistance mechanisms in a clinical *C. albicans* isolate (MacCallum *et al.*, 2010). The exception is *C. krusei*, a species displaying intrinsic resistance to fluconazole due to target alterations. In fact, compared to other *Candida* spp. *C. krusei* displays Erg11 sequence changes that render the enzyme constitutively less susceptible to inhibition by fluconazole (Orozco *et al.*, 1998). However, drug efflux seems to play a role in *C. krusei* as well, through the constitutive expression of the multidrug efflux pump Abc1 (Lamping *et al.*, 2009).

Chromosomal aneuploidies

In *C. albicans*, chromosome 5 bears both the gene expressing the target enzyme of azole drugs (*ERG11*) and the gene expressing the transcriptional activator of ABC transporters (*TAC1*). Through a gene dosage effect, aneuploidies such as chromosome 5 trisomy may result in increased expression of drug target and efflux genes. This way, although treated separately here, this mechanism is directly related to the two previous ones, since it consists really in an alternative mechanism leading to up-regulation of ABC transporters and the azole drug target enzyme Erg11 (Kwon-Chung and Chang, 2012). The mechanism was first reported by Coste *et al.* (2006) when they found that the mechanism behind loss of *TAC1* heterozygosis, to generate an homozygous hyperactive *TAC1* allele that caused overexpression of the targets ABC transporters, could occur by recombination between portions of chromosome 5 or by chromosome 5 duplication. This was reminiscent of previous work showing the exact same mechanism behind spontaneous mating-type like (*MTL*) locus homozygosis, a locus actually present in the same chromosome (Wu *et al.*, 2005). A strong association was found between *MTL* locus homozygosis and azole drug resistance (Rustad *et al.*, 2002). Further investigation of that suggestion revealed that drug resistance was not directly affected by *MTL* locus zygosity, meaning that associated homozygosis of a drug resistance gene linked to the *MTL* locus might be the actual cause (Pujol *et al.*, 2003). With the discovery of *TAC1*, located about 14 kb upstream of the *MTL* locus in *C. albicans*, this transcription factor was suggested to be that gene. Even though this has so far not been definitively demonstrated, the mechanism is thought to consist on codominant hyperactive *TAC1* alleles requiring homozygosis in order to confer full ABC transporter-mediated azole resistance (Coste *et al.*, 2004). Building on the published notion that *C. albicans* displays surprisingly high tolerance of aneuploidies (Selmecki *et al.*, 2005), a new report showed that azole drug resistance could be caused by a specific segmental aneuploidy consisting on isochromosome formation composed of the left arm of chromosome 5, the genomic region bearing *ERG11* and *TAC1* (Selmecki *et al.*, 2006). Isochromosomes consisted either of a duplication of the left arm of chromosome 5 around a single centromere flanked by an inverted repeat, which was found as an independent chromosome, or fused at the telomere of a full-length chromosome 5 (Selmecki *et al.*, 2006). Another report finding a strong association between chromosome 5 aneuploidies and azole drug resistance followed (Coste *et al.*, 2007), while the definitive proof that the implication of chromosome 5 aneuploidies in azole drug resistance consisted on the up-regulation of *ERG11* and *TAC1* was published soon after by Selmecki *et al.* (Selmecki *et al.*, 2008). More recently, an in-depth study established a strong link between chromosome 5 aneuploidies, as well as other smaller chromosome aneuploidies, and increased fitness in the context of evolution under drug pressure, namely in the presence of the triazole drug fluconazole (Selmecki *et al.*, 2009).

Even though it has not been studied in similar detail, the association between chromosome aneuploidies and azole drug resistance has been described in *C. glabrata* as well. In a clinical isolate showing fluconazole and itraconazole resistance associated to the production of about twice the normal levels of ergosterol the whole chromosome harbouring *ERG11* was found to be duplicated (Marichal *et al.*, 1997; vanden Bossche *et al.*, 1992). More recently, a study of 40 clinical isolates from Danish hospitals identified wide chromosomal size variation, associated with chromosomal translocations or segmental duplications (Polakova *et al.*, 2009). Novel chromosomes were found to carry duplicated genes, including the ABC transporter family genes, and could thus result in efflux-mediated azole drug resistance. The occurrence of genomic rearrangements (which may contribute not only to drug resistance but also to virulence) appears to be quite common in *C. glabrata*, but its real clinical relevance remains under active discussion (Ahmad *et al.*, 2014). This particular mechanism of drug resistance in pathogenic *Candida* spp. is addressed in further detail in Chapter 10.

Alterations of the sterol biosynthesis pathway

Alteration of the sterol biosynthesis pathway, leading to the replacement of ergosterol by alternative sterols in the fungal cell membrane, is a rather uncommon mechanism of azole drug resistance. This mechanism often results in cross-resistance to the polyene amphotericin B (Fig. 1.1 and Table 1.2), since this drug acts by binding ergosterol directly (discussed in more detail below in the section on polyenes). The mechanism typically consists of loss of function of C-5 sterol desaturase, encoded by *ERG3*, an enzyme responsible for a late step in the ergosterol biosynthesis pathway. As referred in the section concerning the mechanism of action of azole drugs above, azole toxicity to yeast cells results not only from ergosterol depletion but also from the accumulation of the toxic sterol 14α-methylergosta-8,24,(28)-dien-3β,6α-diol. Biosynthesis of this toxic sterol arises from Erg3-catalysed modification of 14α-methyl fecosterol, which accumulates following azole inhibition of Erg11. In the context of *ERG3* inactivation the pathway produces growth-supporting episterol and 14α-methyl fecosterol, thus bypassing the accumulation of the toxic sterol even in the presence of azole drugs (Fig. 1.3) and conferring high-level azole resistance. This mechanism was first proposed in the late 1980s, through work done on the model yeast *S. cerevisiae* (Watson *et al.*, 1989). Several years later the first *C. albicans* clinical *erg3* mutants were identified. They consisted on two isolates from AIDS patients who had received prolonged fluconazole therapy (Kelly *et al.*, 1996, 1997b) and two strains isolated from leukaemia patients who had received short-term fluconazole prophylaxis plus amphotericin B therapy (Nolte *et al.*, 1997). Several other cases have been reported since then, generally concerning strains isolated from patients undergoing prolonged exposure to azole drugs (Chau *et al.*, 2005; Martel *et al.*, 2010a; Miyazaki *et al.*, 1999; Vale-Silva *et al.*, 2012). However, the real impact of *ERG3* inactivation on drug resistance is often confounded by the presence of additional mutations. This has been well studied in one of the above referred isolates, the so-called Darlington strain (Miyazaki *et al.*, 1999). Owing to the presence of point mutations on *ERG11*, restoration of *ERG3* activity and, consequently, of normal ergosterol content in the Darlington strain did not restore full azole susceptibility (Kakeya *et al.*, 2000).

The inactivation of *ERG3* opens the way for the development of loss-of-function mutations in *ERG11*, which mimics azole-mediated inhibition of Erg11, since it bypasses the biosynthesis of the toxic 14α-methylated sterol resulting from Erg11 inhibition (Fig. 1.3) (Bard et al., 1993; Kelly *et al.*, 1995). This way, clinical *C. albicans erg3* mutants often bear *erg11* mutations contributing to the drug resistance mechanism. The paradigm that, at least aerobically, *ERG11* inactivation can only occur in the context of previous *erg3* mutation has been challenged by the isolation of several *erg11* mutants in wild type backgrounds with functional *ERG3* alleles. This was first found in *C. glabrata* (Geber *et al.*, 1995) and then in *C. albicans* as well, when Sanglard *et al.* were able to recover an *erg11/*

erg11 mutant by exposure of an *ERG11/erg11* heterozygous strain to amphotericin B (Sanglard *et al.*, 2003). Similar *erg11* inactivation was recently described for the first time in clinical isolates, specifically in *C. glabrata* (Hull *et al.*, 2012b) and in *C. tropicalis* (Eddouzi *et al.*, 2013b; Forastiero *et al.*, 2013). Together, these reports show that the development of compensatory mutations allowing viability of drug-resistant *erg11* mutants with no accompanying *erg3* mutation is possible both *in vitro* and *in vivo*. This way, besides *ERG3* inactivation another adaptation allowing yeasts to bypass the lethal effect of *ERG11* gene disruption clearly exists. The exact mechanism has yet to be identified.

Interestingly, an association has been reported between *ERG3* inactivation and attenuation of *C. albicans* virulence (Chau *et al.*, 2005; Miyazaki *et al.*, 2006; Morio *et al.*, 2012; Nolte *et al.*, 1997). This is most likely because the analysed isolates all appeared locked in the yeast form and were unable to form hyphae, an important virulence factor in *C. albicans* (Mitchell, 1998). One recent study reports two clinical *erg3* mutant isolates that retain the ability to filament, although showing a reduced capacity for hyphal growth compared to the wild type isolate. The authors report statistically significant increase in mouse survival in a model of disseminated infection, translating into attenuated virulence (Morio *et al.*, 2012). Careful inspection of the data reveals that the differences in mouse survival are rather modest and certainly smaller than those reported previously for non-filamenting isolates (Chau *et al.*, 2005; Miyazaki *et al.*, 2006; Nolte *et al.*, 1997). Surprisingly, work done in laboratory-generated *erg3* mutants suggested that such mutants might actually respond to fluconazole treatment *in vivo*, in spite of their high-level *in vitro* azole resistance (Miyazaki *et al.*, 2006). However, this finding has been recently challenged by the analysis of a clinical *erg3* mutant which retained the ability to filament and not only displayed wild type virulence but also fluconazole resistance *in vivo* (Vale-Silva *et al.*, 2012). A good correlation between the ability to form hyphae and virulence appears to emerge from these reports in *C. albicans*: yeast-locked isolates show dramatic attenuation of virulence (Chau *et al.*, 2005; Miyazaki *et al.*, 2006; Nolte *et al.*, 1997), slowly filamenting isolates show less striking virulence defects (Morio *et al.*, 2012), while fully filamenting isolates may retain wild-type virulence (Vale-Silva *et al.*, 2012). Overall, published data also suggest that *ERG3* inactivation may be accompanied by compensatory mutations when developed *in vivo*, allowing preservation of hyphae formation and full virulence.

Azole drug resistance resulting from *ERG3* inactivation has been reported in *C. dubliniensis* as well. This consisted on laboratory selection of mutants able to grow on culture medium containing a high concentration of itraconazole. Seven mutants were isolated in that fashion, generally showing increased *CdCDR1* and *CdERG11* expression, but in six of them itraconazole resistance was primarily caused by inactivation of *CdERG3* (Pinjon *et al.*, 2003). Even if nothing is known about fitness of these mutants *in vivo*, the authors suggest that *CdERG3* inactivation may contribute to clinical azole resistance (Pinjon *et al.*, 2003). More recently, two separate reports described *ERG3* inactivation in *C. tropicalis* as well (Eddouzi *et al.*, 2013b; Forastiero *et al.*, 2013). The situation may differ in *C. glabrata*, however, since a laboratory generated *erg3* mutant remained susceptible to antifungal drugs (Geber *et al.*, 1995).

Non-azole ergosterol biosynthesis inhibitors

Available drugs and mechanism of action

Aside from azoles, several classes of drugs in medical use act by inhibiting ergosterol biosynthesis in fungi (Table 1.1). These include the thiocarbamate tolnaftate, the morpholine amorolfine and allylamines, of which the main representative is terbinafine (Fig. 1.2). Tolnaftate acts by inhibiting the *ERG1*-encoded enzyme squalene epoxidase (Ryder *et al.*, 1986), the first step in the biosynthesis of ergosterol from squalene (Table 1.1). However, it has been suggested that its penetration into *C. albicans* cells, and thus the interaction with its target enzyme, may be poor (Barrett-Bee *et al.*, 1986). Amorolfine acts by inhibiting two different enzymes of the pathway, the C-8 sterol isomerase, encoded by *ERG2*, and C-14 sterol reductase, encoded by *ERG24* (Table 1.1) (Haria and Bryson, 1995). These drugs have a wide spectrum of activity, but they present serious side-effects and are available for topical application only. They are thus indicated for the treatment of superficial fungal infections exclusively. Just like tolnaftate, the allylamine terbinafine targets the enzyme squalene epoxidase, encoded by *ERG1* (Table 1.1) (Petranyi *et al.*, 1984). Inhibition of Erg1 leads to accumulation of squalene and depletion of ergosterol in the cell, resulting in increased membrane permeability and eventually cell death. Terbinafine is available as a systemic agent (for oral administration), but it is generally considered to have limited activity against invasive fungal infections. This may, at least in part, be caused by reduced bioavailability due to association to serum components (Ryder and Frank, 1992). Terbinafine is thus currently not indicated to treat invasive candidiasis. Moreover, following systemic administration terbinafine accumulates in hair, nails and skin (Faergemann *et al.*, 1993, 1994), making it particularly suitable to treat superficial fungal infections. Synergistic activity of terbinafine in combination with azole antifungals has been repeatedly reported *in vitro*, including against problematic *C. albicans* isolates less susceptible to azoles (Barchiesi *et al.*, 1997). This has recently revived interest in terbinafine's pharmacokinetics and the development of its use for combination treatment of refractory invasive fungal infections (Dolton *et al.*, 2014). For the moment this interest remains mainly academic and it is not clear whether it will ever reach clinical practice. The general issue of combination therapy is discussed in detail in Chapter 5.

Mechanisms of resistance

There is limited evidence of resistance to terbinafine in *Candida* spp., which comes together with its limited implication in the treatment of invasive fungal infections. Consequently, not much is known about the molecular mechanisms of resistance of *Candida* spp. to terbinafine. In fact, no terbinafine-resistant clinical *Candida* spp. strain has thus far been isolated. However, a few reports have been published describing laboratory yeasts displaying alterations of the intracellular accumulation of terbinafine or of its target enzyme, raising the question of whether clinical terbinafine resistance may actually be going undetected due to lack of rigorous research.

Drug target alterations

Reports addressing terbinafine resistance in *S. cerevisiae* have provided important insight concerning resistance by alteration of the drug target. Laboratory generation of *S. cerevisiae* mutants allowed the identification of a few single point mutations in *ERG1* conferring resistance to terbinafine (Klobucnikova *et al.*, 2003; Leber *et al.*, 2003). These mutations localize to a hot-spot cluster, which likely constitutes an important functional domain playing a role in the interaction of terbinafine with the enzyme. Similar to susceptible strains, these mutants still seem to accumulate squalene upon terbinafine exposure. However, unlike susceptible strains, they do allow the production of adequate ergosterol levels. This suggests that the inhibitory activity of terbinafine on wild type cells may rely primarily on ergosterol depletion, rather than direct toxicity of squalene accumulation (Akins, 2005).

In a rather surprising observation, disruption of *ERG1* in *C. albicans* was found to lead to increased susceptibility to terbinafine, rather than the decreased susceptibility expected from the elimination of the drug's target (Pasrija *et al.*, 2005). A similar observation had previously been made in *C. glabrata* (Tsai *et al.*, 2004). This highlights the overall importance of a functional Erg1 to the fungal cell, ruling out *ERG1* inactivation as a potential mechanism of drug resistance. It further suggests the existence of additional, so far unidentified, molecular targets of terbinafine. Disruption of *UPC2*, encoding the transcriptional regulator of *ERG* genes, has also been shown to lead to increased terbinafine susceptibility (Silver *et al.*, 2004; White and Silver, 2005). This suggests a role for Upc2 in the up-regulation of ergosterol biosynthesis enzymes, including terbinafine's target Erg1, in response to terbinafine exposure. Again, this mechanism has never been observed in clinical *Candida* spp. isolates and at this point this hypothesis remains purely academic.

Reduced intracellular accumulation

Terbinafine is a substrate to ABC transporters Cdr1 and Cdr2 (Cannon *et al.*, 2009). ABC transporter up-regulation thus has the potential to mediate terbinafine resistance in *Candida* spp., even if it has never been reported in clinical isolates. Additionally, terbinafine exposure *in vitro* is also known to lead to *CDR1* up-regulation, among other transporters (Zeng *et al.*, 2007). Interestingly, low terbinafine susceptibility has been found to be associated to *MTL* locus homozygosis in *C. albicans* (Odds, 2009), much like what had previously been described for azoles (Rustad *et al.*, 2002), as discussed above. In the case of azoles, *TAC1* hyperactivity-mediated up-regulation of ABC transporters is the mechanism behind this observation. The same could be true for terbinafine, adding another piece of evidence to support the involvement of drug efflux in terbinafine resistance.

Polyenes

Mechanism of action

Polyenes were the first class of antifungal agents developed for clinical use. Clinically available polyenes include nystatin and amphotericin B (Fig. 1.2), with the latter being the only polyene currently in use to treat systemic infections. Polyenes have an unusual mechanism of action, since rather than inhibiting an enzyme and a specific fungal cell pathway they bind directly ergosterol (Table 1.1), the main sterol in the fungal cell membrane. The exact mechanism of action has been studied for amphotericin B in most detail (see Chapter 9). Amphotericin B is a macrolide amphoteric and amphiphilic molecule. Amphotericin B's amphiphilic nature allows the interaction with cell membrane ergosterol. The classical understanding of the mechanism of action suggests incorporation of the drug in the membrane bilayer, leading to the formation of aqueous channels that allow leakage of potassium ion and other cellular contents (Brajtburg *et al.*, 1990). Additionally, amphotericin B may also have an intracellular mechanism of action, resulting in oxidative damage to the cell (Ghannoum and Rice, 1999; Sokol-Anderson *et al.*, 1986). Amphotericin B activity ultimately results in plasma membrane damage and cell death. The classical model of the aqueous channel consists on an annulus of eight amphotericin B molecules hydrophobically linked to eight sterol molecules (Baginski *et al.*, 1997). The formation of the channel involves interaction of amphotericin B's lipophilic heptaene moiety with ergosterol through nonspecific van der Waals forces. The hydroxyl groups in the polar moiety allow the establishment of hydrogen bonds, with the participation of water molecules (Brajtburg *et al.*, 1990). This model has now been challenged by research showing that amphotericin B forms large extramembranous aggregates, rather than insertions in the lipid bilayer (Anderson *et al.*, 2014). This hypothesis builds on previous data showing that binding to ergosterol is required for the fungicidal activity of amphotericin B, while, surprisingly, pore formation is not (Gray *et al.*, 2012; Palacios *et al.*, 2007, 2011). In the new so-called 'sterol sponge model', amphotericin B's activity relies on the formation of large drug aggregates that extract and sequester sterols from the membrane (Fig. 1.1) (Anderson *et al.*, 2014). Besides ergosterol, amphotericin B is able to bind cholesterol as well, the major sterol in human cell membranes. This probably explains the serious adverse effects seen in the context of its clinical use (most commonly infusion-related reactions and nephrotoxicity). However, it does show higher binding affinity to ergosterol and ergosterol-containing membranes than cholesterol or cholesterol-containing membranes (Brajtburg *et al.*, 1990). The reason for this is thought to be the different tridimensional structure of the two sterols. Unlike the sigmoid cholesterol, ergosterol has a cylindrical structure that allows a stronger interaction with amphotericin B's rigid heptaene moiety (Fig. 1.4) (Kotler-Brajtburg *et al.*, 1974; Odds *et al.*, 2003). Additionally, there is a higher ratio of ergosterol to phospholipids in fungi (Kotler-Brajtburg *et al.*, 1974; Odds *et al.*, 2003), which may also contribute to the specificity of amphotericin B to the fungal target. An intracellular mechanism of action, resulting in oxidative damage to the cell, has also been proposed (Ghannoum and Rice, 1999; Sokol-Anderson *et al.*, 1986).

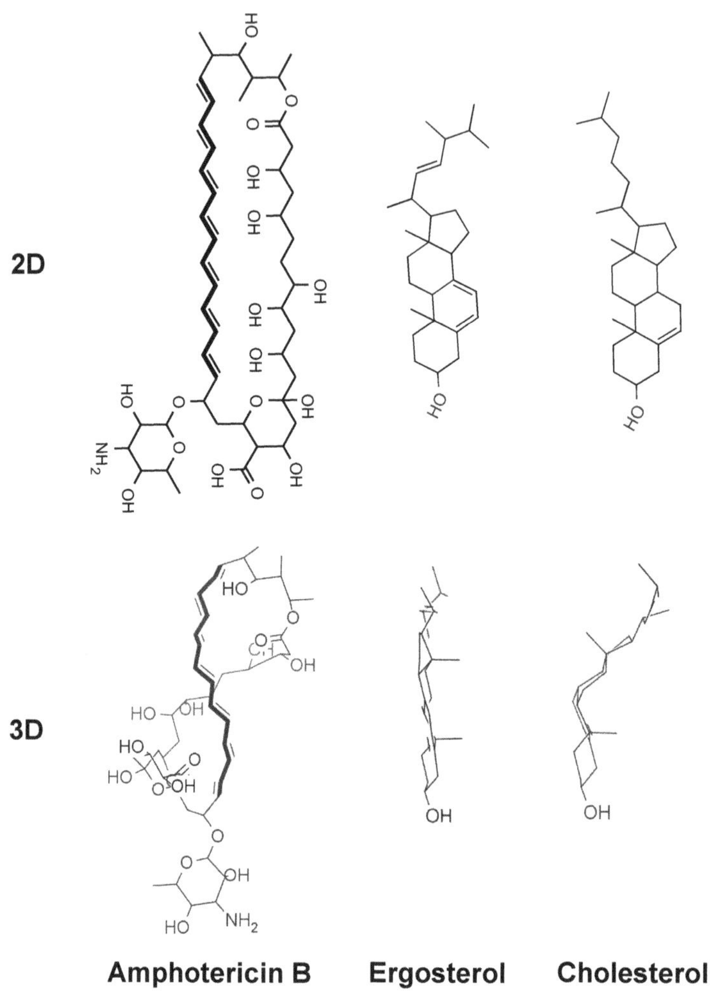

Figure 1.4 Chemical structure diagrams of the polyene antifungal agent amphotericin B and of the main fungal and mammalian membrane sterols, ergosterol and cholesterol, visualized in two (2D) or three (3D) dimensions. Amphotericin B's seven conjugated double bonds (in bold in the figure) provide a rigid hydrophobic chain that allows interaction with hydrophobic sterols. The double bond between carbon atoms C7 and C8 of ergosterol, absent in cholesterol, determines a cylindrical 3D conformation in all rotations. This allows a stronger binding by amphotericin B of ergosterol than the sigmoidal cholesterol.

Amphotericin B's mode of action results in potent, wide-spectrum antifungal activity. For decades, until the advent of azole drugs, amphotericin B deoxycholate was the only available drug for systemic administration in the treatment of disseminated fungal infections. Despite the above-referred serious toxic side-effects of its classical deoxycholate formulation, amphotericin B's high activity granted it the eminent position of 'gold standard' of antifungal therapy (Gallis, 1996). To this day it is sometimes thought of as the most reliable option in some difficult-to-treat invasive fungal infections (Klepser, 2011). Recently, in an effort to limit amphotericin B's toxicity three lipid formulations have been developed: lipid complex, colloidal dispersion and liposomal amphotericin B (Hamill, 2013). These formulations present different pharmacological properties and reduced nephrotoxicity, while not affecting the overall antifungal activity (Hamill, 2013; Wingard, 2002). Classical deoxycholate amphotericin B has since been losing popularity in the clinical management of invasive candidiasis, replaced by the newer lipid-based formulations. In any case, even in its lipid-based formulations, amphotericin B's clinical popularity has been in decline, as safer and effective alternatives become available (Pound *et al.*, 2011). This compound now tends to be relegated for second-line therapy (Kullberg *et al.*, 2011; Pappas *et al.*, 2009).

Mechanisms of resistance

Reports of amphotericin B-resistant *Candida* spp. isolates are uncommon. However, some confounding factors need to be considered. Detection of resistance to amphotericin B is not completely reliable (Cuenca-Estrella, 2013). Both available standardized methods for broth microdilution testing admit suboptimal performance, as amphotericin B MICs of tested strains tend to be clustered within a narrow range of values (Clinical and Laboratory Standards Institute, 2009; Rodriguez-Tudela, 2008). In fact, agar-based antifungal susceptibility testing seems to be more appropriate than the standard broth microdilution methods (Rex and Pfaller, 2002). To some degree, this situation may have resulted in an underestimation of the real extent of amphotericin B resistance. This being said, it is generally accepted that resistance to amphotericin B is truly low among the most commonly isolated *Candida* spp. Some unusual isolates with lower susceptibility to amphotericin B may emerge instead, under the pressure of drug therapy. This may the case of *C. lipolytica*, which has been reported to be refractory to amphotericin B treatment due to microbiological resistance (Belet *et al.*, 2006; Walsh *et al.*, 1989). Another good example may be a recently described new species displaying multidrug resistance, including resistance to amphotericin B (Eddouzi *et al.*, 2013a). However, the molecular mechanism of drug resistance has not been elucidated in these isolates. A further example is *C. lusitaniae*, discussed in more detail below.

Concerning the most common species, led by *C. albicans*, the reason for the low occurrence of amphotericin B resistance is thought to be the fact it is likely associated with a significant fitness cost. Clinical *Candida* spp. isolates for which the molecular mechanism behind resistance to amphotericin B is known all show alterations of the lipid composition in the plasma membrane. Such alterations seem to change membrane fluidity and permeability, having as an additional consequence hypersusceptibility to other growth inhibitory molecules and stress conditions. In fact, recent work showed that the development of mutations allowing amphotericin B resistance was accompanied by the generation of severe internal stresses to the yeast cell and increased susceptibility to external stresses, including those typically created by the human host's response to infection (Vincent *et al.*, 2013). A body of data is, however, available on the alteration of the plasma membrane and resistance to polyenes. More precisely, these alterations result from changes in the sterol biosynthesis pathway leading to reduction of the ergosterol content, which in turn limits binding by the drug (Fig. 1.1). This is the most significant mechanism of amphotericin B resistance reported so far and it is discussed in more detail below.

Alterations of the sterol biosynthesis pathway

Alterations of different enzymes involved in ergosterol biosynthesis have been implicated in the development of resistance to polyenes. That is the case, for example, for loss-of-function of *ERG2* in *C. albicans* (Vincent *et al.*, 2013), *ERG6* in both *S. cerevisiae* (Gaber *et al.*, 1989) and *C. albicans* (Jensen-Pergakes *et al.*, 1998) or the double loss-of-function of *ERG11* and *ERG5* in *C. albicans* (Martel *et al.*, 2010b). In *C. glabrata* alone, reduced amphotericin B susceptibility has been found to be caused by mutations in *ERG1* (Tsai *et al.*, 2004), *ERG2* (Hull *et al.*, 2012a), *ERG6* (Vandeputte *et al.*, 2007, 2008) and *ERG11* (Hull *et al.*, 2012b). Several additional reports of amphotericin B-resistant clinical *C. albicans* isolates are available in the literature, typically isolated from severely ill patients (Conly *et al.*, 1992; Krcmery *et al.*, 1998; Powderly *et al.*, 1988). Unfortunately the molecular mechanism behind resistance in these isolates has not been studied. All known molecular alterations in polyene-resistant *Candida* spp. isolates result in replacement of ergosterol in the membrane by sterols with rather distant tridimensional structures, thus precluding binding by polyenes. They are also probably accompanied by compensatory transcriptional alterations of other genes in the ergosterol biosynthesis pathway, as described for the up-regulation of *ERG5*, *ERG6* and *ERG25* in a *C. albicans* laboratory strain displaying high-level amphotericin B resistance (Barker *et al.*, 2004). In some instances, the accumulated sterols may still allow binding and the net result may even be increased susceptibility, on the contrary. That is the case for a reported laboratory study of *ERG4* inactivation in *S. cerevisiae* (Zweytick *et al.*, 2000).

Amphotericin B-resistant clinical *C. albicans* isolates often display loss-of-function mutations in *ERG3*. This mechanism may in fact be primarily selected for in patients during prolonged azole therapy or prophylaxis, since it mediates high-level azole resistance (see detailed description above in section 'Azoles', subsection 'Alterations of the sterol biosynthesis pathway'). It does, however, lead to replacement of ergosterol by alternative sterols in the fungal plasma membrane (Fig. 1.3) and can thus result in reductions of polyene susceptibility. Reported reductions in susceptibility range from slight increases in MIC (possibly due to the above-referred technical limitations of antifungal susceptibility testing methods) to full cross-resistance to amphotericin B. The implication of *ERG3* was originally described with the characterization of this gene in *S. cerevisiae* (Arthington *et al.*, 1991; Watson *et al.*, 1989) and subsequently identified in several clinical *C. albicans* isolates (Chau *et al.*, 2005; Kelly *et al.*, 1996, 1997b; Martel *et al.*, 2010a; Morio *et al.*, 2012; Nolte *et al.*, 1997). The situation seems to be rather complex, however, and at least one laboratory-generated *C. albicans* *erg3* mutant displayed no change in amphotericin B susceptibility (Sanglard *et al.*, 2003). That seems to be the case in *C. glabrata* as well, although in this species an *erg3* mutant has been shown to remain susceptible not only to amphotericin B but triazoles as well (Geber *et al.*, 1995). In turn, an *erg11* mutant in the same background did develop resistance to both triazoles and amphotericin B, an observation recapitulated by the double *erg11* and *erg3* mutant (Geber *et al.*, 1995). Similarly, in a recent first-time report of a *C. tropicalis* double *erg11* and *erg3* mutant, complementation of the *Cterg11* mutation, yielding a *Cterg3* mutant with a functional *CtERG11* gene, reversed amphotericin B resistance without affecting azole resistance (Eddouzi *et al.*, 2013b). A subsequent report described another case of *CtERG11* and *CtERG3* inactivation coupled with amphotericin B resistance in *C. tropicalis*, but it did not address this specific question (Forastiero *et al.*, 2013). Taken together, these data seem to suggest *ERG11* inactivation as the main mediator of amphotericin B resistance. *ERG3* inactivation would thus cause azole drug resistance and, subsequently, serve as a facilitator of *ERG11* inactivation to cause amphotericin B resistance. This is speculative at this point, however, and the exact mechanisms remain elusive.

A recent report of another *C. glabrata erg11* mutant sheds light on the influence of sterol uptake on amphotericin B resistance (Hull *et al.*, 2012b). Interestingly, the authors show that this *erg11* mutant, which does not normally contain any ergosterol in its membrane, shows increased susceptibility to amphotericin B when grown in medium with ergosterol and decreased susceptibility when grown in medium with cholesterol (Hull *et al.*, 2012b). This establishes a direct correlation between sterol uptake, the available membrane sterols and amphotericin B binding (and inhibitory activity). This correlation raises interesting questions concerning growth in the human host and resistance to amphotericin B.

C. lusitaniae is an emerging pathogen, which unlike other *Candida* spp. is often associated with amphotericin B resistance (Blinkhorn *et al.*, 1989). This species seems to be either inherently resistant to amphotericin B or to rapidly acquire secondary resistance. The reasons for this are not completely understood. In an attempt to elucidate this situation, Young *et al.* looked at transcriptional regulation of several ergosterol biosynthesis genes in clinical *C. lusitaniae* isolates. Reduced *ERG3* expression was found in amphotericin B-resistant clinical isolates, while all other analysed *ERG* genes were overexpressed (Young *et al.*, 2003). The exact mechanism of amphotericin B resistance could not be identified, but the authors suggest that a mutation either in *ERG3* or in a gene that controls *ERG3* expression might be the cause (Young *et al.*, 2003). Further work is necessary to establish the exact mechanism. In any case, even if it turns out that *ERG3* is not directly implicated, the common association of *C. lusitaniae* with amphotericin B resistance will still most likely be a result of alterations of the ergosterol biosynthetic pathway.

Concluding remarks

Several different molecular mechanisms of resistance of *Candida* spp. to membrane-targeting antifungals have been described over the years. Individual mechanisms can often mediate high-level drug resistance resulting in treatment failure and serious complications in patients presenting disseminated candidiasis. In addition, during prolonged therapy or prophylaxis regimens in individual patients, different resistance mechanisms often combine in *Candida* spp. This may cause a stepwise increase in drug resistance to an originally effective drug or the development of resistance to multiple drugs employed in the clinical management of the infection sequentially. In parallel, emergence of antifungal resistance has also been the result of epidemiological trends towards less susceptible non-*albicans Candida* species. This latter case raises the fear of emergence of a wider range of resistance mechanisms. Particularly, fluconazole-resistant, but also polyene-resistant, *C. albicans* strain and non-*albicans Candida* species isolation seems to be on the rise. It is also worth noting that known molecular mechanisms of drug resistance do not explain all currently observed cases of drug resistance. This means that additional, previously unidentified mechanisms of clinical significance are certainly at play in *Candida* spp. Knowledge of resistance mechanisms, not only of *Candida* spp. but other pathogenic fungi as well, helps drive rational therapeutic management and is particularly important in cases of life-threatening infection. Investigation of resistance to clinically useful membrane-targeting antifungals thus holds the highest importance today.

References

Ahmad, K.M., Kokosar, J., Guo, X., Gu, Z., Ishchuk, O.P., and Piskur, J. (2014). Genome structure and dynamics of the yeast pathogen *Candida glabrata*. FEMS Yeast Res. *14*, 529–535.

Akins, R.A. (2005). An update on antifungal targets and mechanisms of resistance in *Candida albicans*. Med. Mycol. *43*, 285–318.

Alarco, A.M., and Raymond, M. (1999). The bZip transcription factor Cap1p is involved in multidrug resistance and oxidative stress response in *Candida albicans*. J. Bacteriol. *181*, 700–708.

Anderson, T.M., Clay, M.C., Cioffi, A.G., Diaz, K.A., Hisao, G.S., Tuttle, M.D., Nieuwkoop, A.J., Comellas, G., Maryum, N., Wang, S., *et al.* (2014). Amphotericin forms an extramembranous and fungicidal sterol sponge. Nat. Chem. Biol. *10*, 400–406.

Arthington, B., Bennett, L., Skatrud, P., Guynn, C., Barbuch, R., Ulbright, C., and Bard, M. (1991). Cloning, disruption and sequence of the gene encoding yeast C-5 sterol desaturase. Gene *102*, 39–44.

Baginski, M., Resat, H., and McCammon, J.A. (1997). Molecular properties of amphotericin B membrane channel: a molecular dynamics simulation. Mol. Pharmacol. *52*, 560–570.

Balzi, E., and Goffeau, A. (1994). Genetics and biochemistry of yeast multidrug resistance. Biochimica Biophysica Acta *1187*, 152–162.

Barchiesi, F., Falconi Di Francesco, L., and Scalise, G. (1997). *In vitro* activities of terbinafine in combination with fluconazole and itraconazole against isolates of *Candida albicans* with reduced susceptibility to azoles. Antimicrob. Agents Chemother. *41*, 1812–1814.

Barchiesi, F., Calabrese, D., Sanglard, D., Falconi Di Francesco, L., Caselli, F., Giannini, D., Giacometti, A., Gavaudan, S., and Scalise, G. (2000). Experimental induction of fluconazole resistance in *Candida tropicalis* ATCC 750. Antimicrob. Agents Chemother. *44*, 1578–1584.

Bard, M., Lees, N., Turi, T., Craft, D., Cofrin, L., Barbuch, R., Koegel, C., and Loper, J. (1993). Sterol synthesis and viability of *ERG11* (cytochrome P450 lanosterol demethylase) mutations in *Saccharomyces cerevisiae* and *Candida albicans*. Lipids *28*, 963–967.

Barker, K.S., Crisp, S., Wiederhold, N., Lewis, R.E., Bareither, B., Eckstein, J., Barbuch, R., Bard, M., and Rogers, P.D. (2004). Genome-wide expression profiling reveals genes associated with amphotericin B and fluconazole resistance in experimentally induced antifungal resistant isolates of *Candida albicans*. J. Antimicrob. Chemother. *54*, 376–385.

Barrett-Bee, K.J., Lane, A.C., and Turner, R.W. (1986). The mode of antifungal action of tolnaftate. J. Med. Vet. Mycol. *24*, 155–160.

Belet, N., Ciftci, E., Ince, E., Dalgic, N., Oncel, S., Guriz, H., Yagmurlu, A., Dindar, H., and Dogru, U. (2006). Caspofungin treatment in two infants with persistent fungaemia due to *Candida lipolytica*. Scand. J. Infect. Dis. *38*, 559–562.

Bennett, J.E., Izumikawa, K., and Marr, K.A. (2004). Mechanism of increased fluconazole resistance in *Candida glabrata* during prophylaxis. Antimicrob. Agents Chemother. *48*, 1773–1777.

Blinkhorn, R.J., Adelstein, D., and Spagnuolo, P.J. (1989). Emergence of a new opportunistic pathogen, *Candida lusitaniae*. J. Clin. Microbiol. *27*, 236–240.

Blot, S., Janssens, R., Claeys, G., Hoste, E., Buyle, F., De Waele, J.J., Peleman, R., Vogelaers, D., and Vandewoude, K. (2006). Effect of fluconazole consumption on long-term trends in candidal ecology. J. Antimicrob. Chemother. *58*, 474–477.

Brajtburg, J., Powderly, W.G., Kobayashi, G.S., and Medoff, G. (1990). Amphotericin B: current understanding of mechanisms of action. Antimicrob. Agents Chemother. *34*, 183–188.

Brown, G.D., Denning, D.W., Gow, N.A., Levitz, S.M., Netea, M.G., and White, T.C. (2012). Hidden killers: human fungal infections. Sci. Transl. Med. *4*, 165rv113.

Cannon, R.D., Lamping, E., Holmes, A.R., Niimi, K., Baret, P.V., Keniya, M.V., Tanabe, K., Niimi, M., Goffeau, A., and Monk, B.C. (2009). Efflux-mediated antifungal drug resistance. Clin. Microbiol. Rev. *22*, 291–321.

Chau, A.S., Gurnani, M., Hawkinson, R., Laverdiere, M., Cacciapuoti, A., and McNicholas, P.M. (2005). Inactivation of sterol Delta5,6-desaturase attenuates virulence in *Candida albicans*. Antimicrob. Agents Chemother. *49*, 3646–3651.

Chen, C.G., Yang, Y.L., Shih, H.I., Su, C.L., and Lo, H.J. (2004). CaNdt80 is involved in drug resistance in *Candida albicans* by regulating *CDR1*. Antimicrob. Agents Chemother. *48*, 4505–4512.

Clinical and Laboratory Standards Institute (2008). Reference Method for Broth Dilution Antifungal Susceptibility Testing of Yeasts; Approved Standard – Third Edition, Document M27-A3 (Wayne, PA, USA).

Clinical and Laboratory Standards Institute (2009). Method for Antifungal Disk Diffusion Susceptibility Testing of Yeasts; Approved Guideline – Second Edition, Document M44-A2 (Wayne, PA, USA).

Clinical and Laboratory Standards Institute (2012). Reference Method for Broth Dilution Antifungal Susceptibility Testing of Yeasts: Fourth Informational Supplement M27-S4 (Wayne, PA, USA).

Conly, J., Rennie, R., Johnson, J., Farah, S., and Hellman, L. (1992). Disseminated candidiasis due to amphotericin B-resistant *Candida albicans*. J. Infect. Dis. *165*, 761–764.

Coste, A.T., Karababa, M., Ischer, F., Bille, J., and Sanglard, D. (2004). *TAC1*, transcriptional activator of *CDR* genes, is a new transcription factor involved in the regulation of *Candida albicans* ABC transporters *CDR1* and *CDR2*. Eukaryot. Cell 3, 1639–1652.

Coste, A., Turner, V., Ischer, F., Morschhauser, J., Forche, A., Selmecki, A., Berman, J., Bille, J., and Sanglard, D. (2006). A mutation in Tac1p, a transcription factor regulating *CDR1* and *CDR2*, is coupled with loss of heterozygosity at chromosome 5 to mediate antifungal resistance in *Candida albicans*. Genetics *172*, 2139–2156.

Coste, A., Selmecki, A., Forche, A., Diogo, D., Bougnoux, M.E., d'Enfert, C., Berman, J., and Sanglard, D. (2007). Genotypic evolution of azole resistance mechanisms in sequential *Candida albicans* isolates. Eukaryotic Cell *6*, 1889–1904.

Cuenca-Estrella, M. (2013). Antifungal drug resistance mechanisms in pathogenic fungi: from bench to bedside. Clin. Microbiol. Infect. *20* (Suppl. 6), 54–59.

Dolton, M.J., Perera, V., Pont, L.G., and McLachlan, A.J. (2014). Terbinafine in combination with other antifungal agents for treatment of resistant or refractory mycoses: investigating optimal dosing regimens using a physiologically based pharmacokinetic model. Antimicrob. Agents Chemother. *58*, 48–54.

Dunkel, N., Blass, J., Rogers, P.D., and Morschhauser, J. (2008a). Mutations in the multi-drug resistance regulator *MRR1*, followed by loss of heterozygosity, are the main cause of *MDR1* overexpression in fluconazole-resistant *Candida albicans* strains. Mol. Microbiol. *69*, 827–840.

Dunkel, N., Liu, T.T., Barker, K.S., Homayouni, R., Morschhauser, J., and Rogers, P.D. (2008b). A gain-of-function mutation in the transcription factor Upc2p causes upregulation of ergosterol biosynthesis genes and increased fluconazole resistance in a clinical *Candida albicans* isolate. Eukaryot. Cell *7*, 1180–1190.

Eddouzi, J., Lohberger, A., Vogne, C., Manai, M., and Sanglard, D. (2013a). Identification and antifungal susceptibility of a large collection of yeast strains isolated in Tunisian hospitals. Med. Mycol. *51*, 737–746.

Eddouzi, J., Parker, J.E., Vale-Silva, L.A., Coste, A., Ischer, F., Kelly, S., Manai, M., and Sanglard, D. (2013b). Molecular mechanisms of drug resistance in clinical *Candida* species isolated from Tunisian hospitals. Antimicrob. Agents Chemother. *57*, 3182–3193.

European Society of Clinical Microbiology and Infectious Diseases (2013). Clinical Breakpoints-Fungi (v6.1) and Definitions of Clinical Breakpoints and Epidemiological Cut Off Values. Available online: http://www.eucast.org/clinical_breakpoints/ (accessed 6 June 2014).

Faergemann, J., Zehender, H., Denouel, J., and Millerioux, L. (1993). Levels of terbinafine in plasma, stratum corneum, dermis-epidermis (without stratum corneum), sebum, hair and nails during and after 250 mg terbinafine orally once per day for four weeks. Acta Derm. Venereol. *73*, 305–309.

Faergemann, J., Zehender, H., and Millerioux, L. (1994). Levels of terbinafine in plasma, stratum corneum, dermis-epidermis (without stratum corneum), sebum, hair and nails during and after 250 mg terbinafine orally once daily for 7 and 14 days. Clin. Exp. Dermatol. *19*, 121–126.

Fling, M., Kopf, J., Tamarkin, A., Gorman, J., Smith, H., and Koltin, Y. (1991). Analysis of a *Candida albicans* gene that encodes a novel mechanism for resistance to benomyl and methotrexate. Mol. Gen. Genet. *227*, 318–329.

Forastiero, A., Mesa-Arango, A., Alastruey-Izquierdo, A., Alcazar-Fuoli, L., Bernal-Martinez, L., Pelaez, T., Lopez, J., Grimalt, J., Gomez-Lopez, A., Cuesta, I., *et al.* (2013). *Candida tropicalis* antifungal cross-resistance is related to different azole target (Erg11p) modifications. Antimicrob. Agents Chemother. *57*, 4769–4781.

Fromtling, R.A. (1988). Overview of medically important antifungal azole derivatives. Clin. Microbiol. Rev. *1*, 187–217.

Gaber, R.F., Copple, D.M., Kennedy, B.K., Vidal, M., and Bard, M. (1989). The yeast gene *ERG6* is required for normal membrane function but is not essential for biosynthesis of the cell-cycle-sparking sterol. Mol. Cell Biol. *9*, 3447–3456.

Gallis, H.A. (1996). Amphotericin B: a commentary on its role as an antifungal agent and as a comparative agent in clinical trials. Clin. Infect. Dis. *22* (Suppl. 2), S145–147.

Geber, A., Hitchcock, C.A., Swartz, J.E., Pullen, F.S., Marsden, K.E., Kwon-Chung, K.J., and Bennett, J.E. (1995). Deletion of the *Candida glabrata ERG3* and *ERG11* genes: effect on cell viability, cell growth, sterol composition, and antifungal susceptibility. Antimicrob. Agents Chemother. *39*, 2708–2717.

Ghannoum, M.A., and Rice, L.B. (1999). Antifungal agents: mode of action, mechanisms of resistance, and correlation of these mechanisms with bacterial resistance. Clin. Microbiol. Rev. *12*, 501–517.

Gray, K.C., Palacios, D.S., Dailey, I., Endo, M.M., Uno, B.E., Wilcock, B.C., and Burke, M.D. (2012). Amphotericin primarily kills yeast by simply binding ergosterol. Proc. Natl. Acad. Sci. U.S.A. *109*, 2234–2239.

Hamill, R.J. (2013). Amphotericin B formulations: a comparative review of efficacy and toxicity. Drugs *73*, 919–934.

Haria, M., and Bryson, H.M. (1995). Amorolfine. A review of its pharmacological properties and therapeutic potential in the treatment of onychomycosis and other superficial fungal infections. Drugs *49*, 103–120.

Heilmann, C.J., Schneider, S., Barker, K.S., Rogers, P.D., and Morschhauser, J. (2010). An A643T mutation in the transcription factor Upc2p causes constitutive *ERG11* upregulation and increased fluconazole resistance in *Candida albicans*. Antimicrob. Agents Chemother. *54*, 353–359.

Hiller, D., Sanglard, D., and Morschhauser, J. (2006). Overexpression of the *MDR1* gene is sufficient to confer increased resistance to toxic compounds in *Candida albicans*. Antimicrob. Agents Chemother. *50*, 1365–1371.

Hof, H. (2008). Will resistance in fungi emerge on a scale similar to that seen in bacteria? Eur. J. Clin. Microbiol. Infect. Dis. *27*, 327–334.

Hoot, S.J., Smith, A.R., Brown, R.P., and White, T.C. (2011). An A643V amino acid substitution in Upc2p contributes to azole resistance in well-characterized clinical isolates of *Candida albicans*. Antimicrob. Agents Chemother. *55*, 940–942.

Hull, C.M., Bader, O., Parker, J.E., Weig, M., Gross, U., Warrilow, A.G.S., Kelly, D.E., and Kelly, S.L. (2012a). Two clinical isolates of *Candida glabrata* exhibiting reduced sensitivity to amphotericin B both harbor mutations in *ERG2*. Antimicrob. Agents Chemother. *56*, 6417–6421.

Hull, C.M., Parker, J.E., Bader, O., Weig, M., Gross, U., Warrilow, A.G.S., Kelly, D.E., and Kelly, S.L. (2012b). Facultative sterol uptake in an ergosterol-deficient clinical isolate of *Candida glabrata* harboring a missense mutation in *ERG11* and exhibiting cross-resistance to azoles and amphotericin B. Antimicrob. Agents Chemother. *56*, 4223–4232.

Jandric, Z., and Schuller, C. (2011). Stress response in *Candida glabrata*: pieces of a fragmented picture. Future Microbiol. *6*, 1475–1484.

Jensen-Pergakes, K.L., Kennedy, M.A., Lees, N.D., Barbuch, R., Koegel, C., and Bard, M. (1998). Sequencing, disruption, and characterization of the *Candida albicans* sterol methyltransferase (*ERG6*) gene: drug susceptibility studies in *erg6* mutants. Antimicrob. Agents Chemother. *42*, 1160–1167.

Jiang, C., Dong, D., Yu, B., Cai, G., Wang, X., Ji, Y., and Peng, Y. (2013). Mechanisms of azole resistance in 52 clinical isolates of *Candida tropicalis* in China. J. Antimicrob. Chemother. *68*, 778–785.

Kakeya, H., Miyazaki, Y., Miyazaki, H., Nyswaner, K., Grimberg, B., and Bennett, J.E. (2000). Genetic analysis of azole resistance in the Darlington strain of *Candida albicans*. Antimicrob. Agents Chemother. *44*, 2985–2990.

Kanafani, Z.A., and Perfect, J.R. (2008). Resistance to antifungal agents: mechanisms and clinical impact. Clin. Infect. Dis. *46*, 120–128.

Kelly, S.L., Lamb, D.C., Corran, A.J., Baldwin, B.C., and Kelly, D.E. (1995). Mode of action and resistance to azole antifungals associated with the formation of 14 alpha-methylergosta-8,24(28)-dien-3 beta,6 alpha-diol. Biochem. Biophys. Res. Commun. *207*, 910–915.

Kelly, S.L., Lamb, D.C., Kelly, D.E., Loeffler, J., and Einsele, H. (1996). Resistance to fluconazole and amphotericin in *Candida albicans* from AIDS patients. Lancet *348*, 1523–1524.

Kelly, S.L., Lamb, D.C., Baldwin, B.C., Corran, A.J., and Kelly, D.E. (1997a). Characterization of *Saccharomyces cerevisiae CYP61*, sterol delta22-desaturase, and inhibition by azole antifungal agents. J. Biol. Chem. *272*, 9986–9988.

Kelly, S.L., Lamb, D.C., Kelly, D.E., Manning, N.J., Loeffler, J., Hebart, H., Schumacher, U., and Einsele, H. (1997b). Resistance to fluconazole and cross-resistance to amphotericin B in *Candida albicans* from AIDS patients caused by defective sterol delta5,6-desaturation. FEBS Lett. *400*, 80–82.

Klepser, M. (2011). The value of amphotericin B in the treatment of invasive fungal infections. J. Crit. Care *26*, 225.

Klobucnikova, V., Kohut, P., Leber, R., Fuchsbichler, S., Schweighofer, N., Turnowsky, F., and Hapala, I. (2003). Terbinafine resistance in a pleiotropic yeast mutant is caused by a single point mutation in the *ERG1* gene. Biochem. Biophys. Res. Commun. *309*, 666–671.

Kotler-Brajtburg, J., Price, H.D., Medoff, G., Schlessinger, D., and Kobayashi, G.S. (1974). Molecular basis for the selective toxicity of amphotericin B for yeast and filipin for animal cells. Antimicrob. Agents Chemother. *5*, 377–382.

Krcmery, V., Jr., Oravcova, E., Spanik, S., Mrazova-Studena, M., Trupl, J., Kunova, A., Stopkova-Grey, K., Kukuckova, E., Krupova, I., Demitrovicova, A., *et al.* (1998). Nosocomial breakthrough fungaemia during antifungal prophylaxis or empirical antifungal therapy in 41 cancer patients receiving antineoplastic chemotherapy: analysis of aetiology risk factors and outcome. J. Antimicrob. Chemother. *41*, 373–380.

Kullberg, B.J., Verweij, P.E., Akova, M., Arendrup, M.C., Bille, J., Calandra, T., Cuenca-Estrella, M., Herbrecht, R., Jacobs, F., Kalin, M., *et al.* (2011). European expert opinion on the management of invasive candidiasis in adults. Clin. Microbiol. Infect. *17* (Suppl. 5), 1–12.

Kwon-Chung, K.J., and Chang, Y.C. (2012). Aneuploidy and drug resistance in pathogenic fungi. PLoS Pathog. *8*, e1003022.

Lamping, E., Ranchod, A., Nakamura, K., Tyndall, J.D., Niimi, K., Holmes, A.R., Niimi, M., and Cannon, R.D. (2009). Abc1p is a multidrug efflux transporter that tips the balance in favor of innate azole resistance in *Candida krusei*. Antimicrob. Agents Chemother. *53*, 354–369.

Leber, R., Fuchsbichler, S., Klobucnikova, V., Schweighofer, N., Pitters, E., Wohlfarter, K., Lederer, M., Landl, K., Ruckenstuhl, C., Hapala, I., *et al.* (2003). Molecular mechanism of terbinafine resistance in *Saccharomyces cerevisiae*. Antimicrob. Agents Chemother. *47*, 3890–3900.

Loffler, J., Einsele, H., Hebart, H., Schumacher, U., Hrastnik, C., and Daum, G. (2000). Phospholipid and sterol analysis of plasma membranes of azole-resistant *Candida albicans* strains. FEMS Microbiol. Lett. *185*, 59–63.

MacCallum, D.M., Coste, A., Ischer, F., Jacobsen, M.D., Odds, F.C., and Sanglard, D. (2010). Genetic dissection of azole resistance mechanisms in *Candida albicans* and their validation in a mouse model of disseminated infection. Antimicrob. Agents Chemother. *54*, 1476–1483.

MacPherson, S., Akache, B., Weber, S., De Deken, X., Raymond, M., and Turcotte, B. (2005). *Candida albicans* zinc cluster protein Upc2p confers resistance to antifungal drugs and is an activator of ergosterol biosynthetic genes. Antimicrob. Agents Chemother. *49*, 1745–1752.

Maertens, J.A. (2004). History of the development of azole derivatives. Clin. Microbiol. Infect. *10* (Suppl. 1), 1–10.

Mansfield, B., Oltean, H., Oliver, B., Hoot, S., Leyde, S., Hedstrom, L., and White, T. (2010). Azole drugs are imported by facilitated diffusion in *Candida albicans* and other pathogenic fungi. PLoS Pathog. *6*, e1001126.

Marichal, P., Vanden Bossche, H., Odds, F.C., Nobels, G., Warnock, D.W., Timmerman, V., Van Broeckhoven, C., Fay, S., and Mose-Larsen, P. (1997). Molecular biological characterization of an azole-resistant *Candida glabrata* isolate. Antimicrob. Agents Chemother. *41*, 2229–2237.

Martel, C.M., Parker, J.E., Bader, O., Weig, M., Gross, U., Warrilow, A.G., Rolley, N., Kelly, D.E., and Kelly, S.L. (2010a). Identification and characterization of four azole-resistant *erg3* mutants of *Candida albicans*. Antimicrob. Agents Chemother. *54*, 4527–4533.

Martel, C.M., Parker, J.E., Bader, O., Weig, M., Gross, U., Warrilow, A.G.S., Kelly, D.E., and Kelly, S.L. (2010b). A clinical isolate of *Candida albicans* with mutations in *ERG11* (encoding sterol 14alpha-demethylase) and *ERG5* (encoding C22 desaturase) is cross resistant to azoles and amphotericin B. Antimicrob. Agents Chemother. *54*, 3578–3583.

Martins, M.D., Lozano-Chiu, M., and Rex, J.H. (1998). Declining rates of oropharyngeal candidiasis and carriage of *Candida albicans* associated with trends toward reduced rates of carriage of fluconazole-resistant *C. albicans* in human immunodeficiency virus-infected patients. Clin. Infect. Dis. *27*, 1291–1294.

Matuschek, E., Brown, D.F., and Kahlmeter, G. (2014). Development of the EUCAST disk diffusion antimicrobial susceptibility testing method and its implementation in routine microbiology laboratories. Clin. Microbiol. Infect. *20*, O255–266.

de Micheli, M., Bille, J., Schueller, C., and Sanglard, D. (2002). A common drug-responsive element mediates the upregulation of the *Candida albicans* ABC transporters *CDR1* and *CDR2*, two genes involved in antifungal drug resistance. Mol. Microbiol. *43*, 1197–1214.

Mitchell, A.P. (1998). Dimorphism and virulence in *Candida albicans*. Curr. Opin. Microbiol. *1*, 687–692.

Miyazaki, T., Miyazaki, Y., Izumikawa, K., Kakeya, H., Miyakoshi, S., Bennett, J.E., and Kohno, S. (2006). Fluconazole treatment is effective against a *Candida albicans erg3/erg3* mutant *in vivo* despite *in vitro* resistance. Antimicrob. Agents Chemother. *50*, 580–586.

Miyazaki, Y., Geber, A., Miyazaki, H., Falconer, D., Parkinson, T., Hitchcock, C., Grimberg, B., Nyswaner, K., and Bennett, J.E. (1999). Cloning, sequencing, expression and allelic sequence diversity of *ERG3* (C-5 sterol desaturase gene) in *Candida albicans*. Gene *236*, 43–51.

Monk, B.C., Tomasiak, T.M., Keniya, M.V., Huschmann, F.U., Tyndall, J.D., O'Connell, J.D., 3rd, Cannon, R.D., McDonald, J.G., Rodriguez, A., Finer-Moore, J.S., *et al.* (2014). Architecture of a single membrane spanning cytochrome P450 suggests constraints that orient the catalytic domain relative to a bilayer. Proc. Natl. Acad. Sci. U.S.A. *111*, 3865–3870.

Moran, G.P., Sanglard, D., Donnelly, S.M., Shanley, D.B., Sullivan, D.J., and Coleman, D.C. (1998). Identification and expression of multidrug transporters responsible for fluconazole resistance in *Candida dubliniensis*. Antimicrob. Agents Chemother. *42*, 1819–1830.

Morio, F., Loge, C., Besse, B., Hennequin, C., and Le Pape, P. (2010). Screening for amino acid substitutions in the *Candida albicans* Erg11 protein of azole-susceptible and azole-resistant clinical isolates: new substitutions and a review of the literature. Diagn. Microbiol. Infect. Dis. *66*, 373–384.

Morio, F., Pagniez, F., Lacroix, C., Miegeville, M., and Le Pape, P. (2012). Amino acid substitutions in the *Candida albicans* sterol delta5,6-desaturase (Erg3p) confer azole resistance: characterization of two novel mutants with impaired virulence. J. Antimicrob. Chemother. *67*, 2131–2138.

Morschhauser, J., Barker, K.S., Liu, T.T., Bla, B.W.J., Homayouni, R., and Rogers, P.D. (2007). The transcription factor Mrr1p controls expression of the *MDR1* efflux pump and mediates multidrug resistance in *Candida albicans*. PLoS Pathog. *3*, e164.

Moye-Rowley, W.S. (2005). Retrograde regulation of multidrug resistance in *Saccharomyces cerevisiae*. Gene *354*, 15–21.

Noel, T., Francois, F., Paumard, P., Chastin, C., Brethes, D., and Villard, J. (2003). Flucytosine–fluconazole cross-resistance in purine-cytosine permease-deficient *Candida lusitaniae* clinical isolates: indirect evidence of a fluconazole uptake transporter. Antimicrob. Agents Chemother. *47*, 1275–1284.

Nolte, F.S., Parkinson, T., Falconer, D.J., Dix, S., Williams, J., Gilmore, C., Geller, R., and Wingard, J.R. (1997). Isolation and characterization of fluconazole- and amphotericin B-resistant *Candida albicans* from blood of two patients with leukemia. Antimicrob. Agents Chemother. *41*, 196–199.

Odds, F.C. (2009). In *Candida albicans*, resistance to flucytosine and terbinafine is linked to MAT locus homozygosity and multilocus sequence typing clade 1. FEMS Yeast Res. *9*, 1091–1101.

Odds, F.C., Brown, A.J., and Gow, N.A. (2003). Antifungal agents: mechanisms of action. Trends Microbiol. *11*, 272–279.

Orozco, A.S., Higginbotham, L.M., Hitchcock, C.A., Parkinson, T., Falconer, D., Ibrahim, A.S., Ghannoum, M.A., and Filler, S.G. (1998). Mechanism of fluconazole resistance in *Candida krusei*. Antimicrob. Agents Chemother. *42*, 2645–2649.

Palacios, D.S., Anderson, T.M., and Burke, M.D. (2007). A post-PKS oxidation of the amphotericin B skeleton predicted to be critical for channel formation is not required for potent antifungal activity. J. Am. Chem. Soc. *129*, 13804–13805.

Palacios, D.S., Dailey, I., Siebert, D.M., Wilcock, B.C., and Burke, M.D. (2011). Synthesis-enabled functional group deletions reveal key underpinnings of amphotericin B ion channel and antifungal activities. Proc. Natl. Acad. Sci. U.S.A. *108*, 6733–6738.

Pappas, P.G., Kauffman, C.A., Andes, D., Benjamin, D.K., Jr., Calandra, T.F., Edwards, J.E., Jr., Filler, S.G., Fisher, J.F., Kullberg, B.J., Ostrosky-Zeichner, L., *et al.* (2009). Clinical practice guidelines for the management of candidiasis: 2009 update by the Infectious Diseases Society of America. Clin. Infect. Dis. *48*, 503–535.

Pasrija, R., Krishnamurthy, S., Prasad, T., Ernst, J.F., and Prasad, R. (2005). Squalene epoxidase encoded by *ERG1* affects morphogenesis and drug susceptibilities of *Candida albicans*. J. Antimicrob. Chemother. *55*, 905–913.

Paul, S., Schmidt, J.A., and Moye-Rowley, W.S. (2010). Regulation of the CgPdr1 transcription factor from the pathogen *Candida glabrata*. Eukaryot. Cell *10*, 187–197.

Perea, S., Lopez-Ribot, J.L., Kirkpatrick, W.R., McAtee, R.K., Santillan, R.A., Martinez, M., Calabrese, D., Sanglard, D., and Patterson, T.F. (2001). Prevalence of molecular mechanisms of resistance to azole antifungal agents in *Candida albicans* strains displaying high-level fluconazole resistance isolated from human immunodeficiency virus-infected patients. Antimicrob. Agents Chemother. *45*, 2676–2684.

Perea, S., Lopez-Ribot, J.L., Wickes, B.L., Kirkpatrick, W.R., Dib, O.P., Bachmann, S.P., Keller, S.M., Martinez, M., and Patterson, T.F. (2002). Molecular mechanisms of fluconazole resistance in *Candida dubliniensis* isolates from human immunodeficiency virus-infected patients with oropharyngeal candidiasis. Antimicrob. Agents Chemother. *46*, 1695–1703.

Petranyi, G., Ryder, N.S., and Stutz, A. (1984). Allylamine derivatives: new class of synthetic antifungal agents inhibiting fungal squalene epoxidase. Science *224*, 1239–1241.

Pfaller, M.A. (2012). Antifungal drug resistance: mechanisms, epidemiology, and consequences for treatment. Am. J. Med. *125*, S3–13.

Pfaller, M.A., and Diekema, D.J. (2012). Progress in antifungal susceptibility testing of *Candida* spp. by use of Clinical and Laboratory Standards Institute broth microdilution methods, 2010 to 2012. J. Clin. Microbiol. *50*, 2846–2856.

Pfaller, M.A., Messer, S.A., Hollis, R.J., Boyken, L., Tendolkar, S., Kroeger, J., and Diekema, D.J. (2009). Variation in susceptibility of bloodstream isolates of *Candida glabrata* to fluconazole according to patient age and geographic location in the United States in 2001 to 2007. J. Clin. Microbiol. *47*, 3185–3190.

Pfaller, M.A., Diekema, D.J., Gibbs, D.L., Newell, V.A., Ellis, D., Tullio, V., Rodloff, A., Fu, W., Ling, T.A., and Global Antifungal Surveillance, G. (2010). Results from the ARTEMIS DISK Global Antifungal Surveillance Study, 1997 to 2007: a 10.5-year analysis of susceptibilities of *Candida* species to fluconazole and voriconazole as determined by CLSI standardized disk diffusion. J. Clin. Microbiol. *48*, 1366–1377.

Pfaller, M.A., Messer, S.A., Moet, G.J., Jones, R.N., and Castanheira, M. (2011a). *Candida* bloodstream infections: comparison of species distribution and resistance to echinocandin and azole antifungal agents in Intensive Care Unit (ICU) and non-ICU settings in the SENTRY Antimicrobial Surveillance Program (2008–2009). Int. J. Antimicrob. Agents *38*, 65–69.

Pfaller, M.A., Moet, G.J., Messer, S.A., Jones, R.N., and Castanheira, M. (2011b). *Candida* bloodstream infections: comparison of species distributions and antifungal resistance patterns in community-onset and nosocomial isolates in the SENTRY Antimicrobial Surveillance Program, 2008–2009. Antimicrob. Agents Chemother. *55*, 561–566.

Pfaller, M.A., Moet, G.J., Messer, S.A., Jones, R.N., and Castanheira, M. (2011c). Geographic variations in species distribution and echinocandin and azole antifungal resistance rates among *Candida* bloodstream infection isolates: report from the SENTRY Antimicrobial Surveillance Program (2008 to 2009). J. Clin. Microbiol. *49*, 396–399.

Pierce, C.G., and Lopez-Ribot, J.L. (2013). Candidiasis drug discovery and development: new approaches targeting virulence for discovering and identifying new drugs. Expert Opin. Drug Discov. *8*, 1117–1126.

Pinjon, E., Moran, G.P., Jackson, C.J., Kelly, S.L., Sanglard, D., Coleman, D.C., and Sullivan, D.J. (2003). Molecular mechanisms of itraconazole resistance in *Candida dubliniensis*. Antimicrob. Agents Chemother. *47*, 2424–2437.

Polakova, S., Blume, C., Zarate, J.A., Mentel, M., Jorck-Ramberg, D., Stenderup, J., and Piskur, J. (2009). Formation of new chromosomes as a virulence mechanism in yeast *Candida glabrata*. Proc. Natl. Acad. Sci. U.S.A. *106*, 2688–2693.

Pound, M.W., Townsend, M.L., Dimondi, V., Wilson, D., and Drew, R.H. (2011). Overview of treatment options for invasive fungal infections. Med. Mycol. *49*, 561–580.

Powderly, W.G., Kobayashi, G.S., Herzig, G.P., and Medoff, G. (1988). Amphotericin B-resistant yeast infection in severely immunocompromised patients. Am. J. Med. *84*, 826–832.

Prasad, R., De Wergifosse, P., Goffeau, A., and Balzi, E. (1995). Molecular cloning and characterization of a novel gene of *Candida albicans, CDR1*, conferring multiple resistance to drugs and antifungals. Curr. Genet. *27*, 320–329.

Pujol, C., Messer, S., Pfaller, M., and Soll, D. (2003). Drug resistance is not directly affected by mating type locus zygosity in *Candida albicans*. Antimicrob. Agents Chemother. *47*, 1207–1212.

Revankar, S.G., Sanche, S.E., Dib, O.P., Caceres, M., and Patterson, T.F. (1998). Effect of highly active antiretroviral therapy on recurrent oropharyngeal candidiasis in HIV-infected patients. AIDS *12*, 2511–2513.

Rex, J.H., and Pfaller, M.A. (2002). Has antifungal susceptibility testing come of age? Clin. Infect. Dis. *35*, 982–989.

Rodriguez-Tudela, J.L.A., Arendrup, M.C., Barchiesi, F., Bille, J., Chryssanthou, E., Cuenca-Estrella, M., Dannaoui, E., Denning, D.W., Donnelly, J.P., Dromer, F., *et al.* (2008). EUCAST definitive document EDef 7.1: method for the determination of broth dilution MICs of antifungal agents for fermentative yeasts. Clin. Microbiol. Infect. *14*, 398–405.

Rognon, B., Kozovska, Z., Coste, A.T., Pardini, G., and Sanglard, D. (2006). Identification of promoter elements responsible for the regulation of *MDR1* from *Candida albicans*, a major facilitator transporter involved in azole resistance. Microbiology *152*, 3701–3722.

Rustad, T.R., Stevens, D.A., Pfaller, M.A., and White, T.C. (2002). Homozygosity at the *Candida albicans* MTL locus associated with azole resistance. Microbiology *148*, 1061–1072.

Ryder, N.S., and Frank, I. (1992). Interaction of terbinafine with human serum and serum proteins. J. Med. Vet. Mycol. *30*, 451–460.

Ryder, N.S., Frank, I., and Dupont, M.C. (1986). Ergosterol biosynthesis inhibition by the thiocarbamate antifungal agents tolnaftate and tolciclate. Antimicrob. Agents Chemother. *29*, 858–860.

Sanglard, D., Kuchler, K., Ischer, F., Pagani, J.L., Monod, M., and Bille, J. (1995). Mechanisms of resistance to azole antifungal agents in *Candida albicans* isolates from AIDS patients involve specific multidrug transporters. Antimicrob. Agents Chemother. *39*, 2378–2386.

Sanglard, D., Ischer, F., Monod, M., and Bille, J. (1997). Cloning of *Candida albicans* genes conferring resistance to azole antifungal agents: characterization of *CDR2*, a new multidrug ABC transporter gene. Microbiology *143 (Pt 2)*, 405–416.

Sanglard, D., Ischer, F., Koymans, L., and Bille, J. (1998). Amino acid substitutions in the cytochrome P-450 lanosterol 14alpha-demethylase (*CYP51A1*) from azole-resistant *Candida albicans* clinical isolates contribute to resistance to azole antifungal agents. Antimicrob. Agents Chemother. *42*, 241–253.

Sanglard, D., Ischer, F., Calabrese, D., Majcherczyk, P.A., and Bille, J. (1999). The ATP binding cassette transporter gene *CgCDR1* from *Candida glabrata* is involved in the resistance of clinical isolates to azole antifungal agents. Antimicrob. Agents Chemother. *43*, 2753–2765.

Sanglard, D., Ischer, F., and Bille, J. (2001). Role of ATP-binding-cassette transporter genes in high-frequency acquisition of resistance to azole antifungals in *Candida glabrata*. Antimicrob. Agents Chemother. *45*, 1174–1183.

Sanglard, D., Ischer, F., Parkinson, T., Falconer, D., and Bille, J. (2003). *Candida albicans* mutations in the ergosterol biosynthetic pathway and resistance to several antifungal agents. Antimicrob. Agents Chemother. *47*, 2404–2412.

Selmecki, A., Bergmann, S., and Berman, J. (2005). Comparative genome hybridization reveals widespread aneuploidy in *Candida albicans* laboratory strains. Mol. Microbiol. *55*, 1553–1565.

Selmecki, A., Forche, A., and Berman, J. (2006). Aneuploidy and isochromosome formation in drug-resistant *Candida albicans*. Science *313*, 367–370.

Selmecki, A., Gerami-Nejad, M., Paulson, C., Forche, A., and Berman, J. (2008). An isochromosome confers drug resistance *in vivo* by amplification of two genes, *ERG11* and *TAC1*. Mol. Microbiol. *68*, 624–641.

Selmecki, A.M., Dulmage, K., Cowen, L.E., Anderson, J.B., and Berman, J. (2009). Acquisition of aneuploidy provides increased fitness during the evolution of antifungal drug resistance. PLoS Genet. *5*, e1000705.

Shimokawa, O., and Nakayama, H. (1992). Increased sensitivity of *Candida albicans* cells accumulating 14 alpha-methylated sterols to active oxygen: possible relevance to *in vivo* efficacies of azole antifungal agents. Antimicrob. Agents Chemother. *36*, 1626–1629.

Silva, A.P., Miranda, I.M., Guida, A., Synnott, J., Rocha, R., Silva, R., Amorim, A., Pina-Vaz, C., Butler, G., and Rodrigues, A.G. (2011). Transcriptional profiling of azole-resistant *Candida parapsilosis* strains. Antimicrob Agents Chemother. *55*, 3546–3556.

Silver, P.M., Oliver, B.G., and White, T.C. (2004). Role of *Candida albicans* transcription factor Upc2p in drug resistance and sterol metabolism. Eukaryot. Cell *3*, 1391–1397.

Sokol-Anderson, M.L., Brajtburg, J., and Medoff, G. (1986). Amphotericin B-induced oxidative damage and killing of *Candida albicans*. J. Infect. Dis. *154*, 76–83.

Thakur, J.K., Arthanari, H., Yang, F., Pan, S.J., Fan, X., Breger, J., Frueh, D.P., Gulshan, K., Li, D.K., Mylonakis, E., *et al.* (2008). A nuclear receptor-like pathway regulating multidrug resistance in fungi. Nature *452*, 604–609.

Tsai, H.F., Bard, M., Izumikawa, K., Krol, A.A., Sturm, A.M., Culbertson, N.T., Pierson, C.A., and Bennett, J.E. (2004). *Candida glabrata erg1* mutant with increased sensitivity to azoles and to low oxygen tension. Antimicrob. Agents Chemother. *48*, 2483–2489.

Tsai, H.F., Krol, A.A., Sarti, K.E., and Bennett, J.E. (2006). *Candida glabrata PDR1*, a transcriptional regulator of a pleiotropic drug resistance network, mediates azole resistance in clinical isolates and petite mutants. Antimicrob. Agents Chemother. *50*, 1384–1392.

Vale-Silva, L.A., Coste, A.T., Ischer, F., Parker, J.E., Kelly, S.L., Pinto, E., and Sanglard, D. (2012). Azole resistance by loss of function of the sterol delta5,6-desaturase gene (*ERG3*) in *Candida albicans* does not necessarily decrease virulence. Antimicrob. Agents Chemother. *56*, 1960–1968.

Vanden Bossche, H., Marichal, P., Gorrens, J., Bellens, D., Moereels, H., and Janssen, P.A. (1990). Mutation in cytochrome P-450-dependent 14 alpha-demethylase results in decreased affinity for azole antifungals. Biochem. Soc. Trans. *18*, 56–59.

Vanden Bossche, H., Marichal, P., Odds, F.C., Le Jeune, L., and Coene, M.C. (1992). Characterization of an azole-resistant *Candida glabrata* isolate. Antimicrob. Agents Chemother. 36, 2602–2610.

Vanden Bossche, H., Warnock, D.W., Dupont, B., Kerridge, D., Sen Gupta, S., Improvisi, L., Marichal, P., Odds, F.C., Provost, F., and Ronin, O. (1994). Mechanisms and clinical impact of antifungal drug resistance. J. Med. Vet. Mycol. *32* (Suppl. 1), 189–202.

Vandeputte, P., Larcher, G., Berges, T., Renier, G., Chabasse, D., and Bouchara, J.P. (2005). Mechanisms of azole resistance in a clinical isolate of *Candida tropicalis*. Antimicrob. Agents Chemother. *49*, 4608–4615.

Vandeputte, P., Tronchin, G., Berges, T., Hennequin, C., Chabasse, D., and Bouchara, J. (2007). Reduced susceptibility to polyenes associated with a missense mutation in the *ERG6* gene in a clinical isolate of *Candida glabrata* with pseudohyphal growth. Antimicrob. Agents Chemother. *51*, 982–990.

Vandeputte, P., Tronchin, G., Larcher, G., Ernoult, E., Berges, T., Chabasse, D., and Bouchara, J.P. (2008). A nonsense mutation in the *ERG6* gene leads to reduced susceptibility to polyenes in a clinical isolate of *Candida glabrata*. Antimicrob. Agents Chemother. *52*, 3701–3709.

Vandeputte, P., Ferrari, S., and Coste, A.T. (2012). Antifungal resistance and new strategies to control fungal infections. Int. J. Microbiol. *2012*, 713687.

Venkateswarlu, K., Denning, D.W., Manning, N.J., and Kelly, S.L. (1996). Reduced accumulation of drug in *Candida krusei* accounts for itraconazole resistance. Antimicrob. Agents Chemother. *40*, 2443–2446.

Vermitsky, J.P., and Edlind, T.D. (2004). Azole resistance in *Candida glabrata*: coordinate upregulation of multidrug transporters and evidence for a Pdr1-like transcription factor. Antimicrob. Agents Chemother. *48*, 3773–3781.

Vermitsky, J.P., Earhart, K.D., Smith, W.L., Homayouni, R., Edlind, T.D., and Rogers, P.D. (2006). Pdr1 regulates multidrug resistance in *Candida glabrata*: gene disruption and genome-wide expression studies. Mol. Microbiol. *61*, 704–722.

Vincent, B.M., Lancaster, A.K., Scherz-Shouval, R., Whitesell, L., and Lindquist, S. (2013). Fitness trade-offs restrict the evolution of resistance to amphotericin B. PLoS Biol. *11*, e1001692.

Walsh, T.J., Salkin, I.F., Dixon, D.M., and Hurd, N.J. (1989). Clinical, microbiological, and experimental animal studies of *Candida lipolytica*. J. Clin. Microbiol. *27*, 927–931.

Warrilow, A.G., Martel, C.M., Parker, J.E., Melo, N., Lamb, D.C., Nes, W.D., Kelly, D.E., and Kelly, S.L. (2010). Azole binding properties of *Candida albicans* sterol 14-alpha demethylase (*CaCYP51*). Antimicrob. Agents Chemother. *54*, 4235–4245.

Watson, P.F., Rose, M.E., Ellis, S.W., England, H., and Kelly, S.L. (1989). Defective sterol C5-6 desaturation and azole resistance: a new hypothesis for the mode of action of azole antifungals. Biochem. Biophys. Res. Commun. *164*, 1170–1175.

White, T.C., and Silver, P.M. (2005). Regulation of sterol metabolism in *Candida albicans* by the *UPC2* gene. Biochem. Soc. Trans. 33, 1215–1218.

Wingard, J.R. (2002). Lipid formulations of amphotericins: are you a lumper or a splitter? Clin. Infect. Dis. *35*, 891–895.

Wirsching, S., Moran, G.P., Sullivan, D.J., Coleman, D.C., and Morschhauser, J. (2001). *MDR1*-mediated drug resistance in *Candida dubliniensis*. Antimicrob. Agents Chemother. *45*, 3416–3421.

Wu, W., Pujol, C., Lockhart, S.R., and Soll, D.R. (2005). Chromosome loss followed by duplication is the major mechanism of spontaneous mating-type locus homozygosis in *Candida albicans*. Genetics *169*, 1311–1327.

Xiao, L., Madison, V., Chau, A.S., Loebenberg, D., Palermo, R.E., and McNicholas, P.M. (2004). Three-dimensional models of wild-type and mutated forms of cytochrome P450 14alpha-sterol demethylases from *Aspergillus fumigatus* and *Candida albicans* provide insights into posaconazole binding. Antimicrob. Agents Chemother. *48*, 568–574.

Young, L.Y., Hull, C.M., and Heitman, J. (2003). Disruption of ergosterol biosynthesis confers resistance to amphotericin B in *Candida lusitaniae*. Antimicrob. Agents Chemother. *47*, 2717–2724.

Zeng, Y.B., Qian, Y.S., Ma, L., and Gu, H.N. (2007). Genome-wide expression profiling of the response to terbinafine in *Candida albicans* using a cDNA microarray analysis. Chin. Med. J. (Engl.) *120*, 807–813.

Zweytick, D., Hrastnik, C., Kohlwein, S.D., and Daum, G. (2000). Biochemical characterization and subcellular localization of the sterol C-24(28) reductase, Erg4p, from the yeast *Saccharomyces cerevisiae*. FEBS Lett. *470*, 83–87.

Point Mutations and Membrane-targeting Antifungal Resistance in *Aspergillus fumigatus* and Other Non-*Candida* Species

2

Guillermo Garcia-Effron

Abstract

Invasive fungal infections are occurring with an increasing frequency. The management of these infections is challenging and the mortality is high. These facts make clinicians to indiscriminately use antifungal agents to treat and prevent fungal infections. The increased use of these drugs has resulted in the development of resistance. Antifungal resistance is linked with different molecular mechanisms including alteration in drug target, reduction of the intracellular concentration of the drug, overexpression of the drug target, etc. The aim of this chapter is to summarize the mechanisms of intrinsic and acquired resistance to ergosterol biosynthesis inhibitors and polyenes in *Aspergillus fumigatus* and other non-*Candida* species due to point mutations in enzymes participating in ergosterol biosynthesis pathways. This chapter will focus mainly in azole drugs since azole resistance is one of the main problems that modern Mycology is facing. A comparison and correlation between yeasts and moulds mechanism of resistance will be described and analysed although it is challenged by the inherent differences existing between these groups of fungi.

Introduction

Antifungal agents can be grouped based on their site of action. The main antifungal classes target fungal membrane sterols. Azoles and allylamines inhibit ergosterol synthesis while polyenes physicochemically interact with it. As any microorganisms, fungi demonstrate two types of antimicrobial resistance: intrinsic or natural and secondary or acquired. Intrinsic antifungal resistance is defined as the innate ability of an organism to resist the activity of an antifungal agent through its inherent structural or functional characteristics that are encoded in its wild type genotype. Secondary or acquired resistance is defined as the loss of activity of an antifungal agent that was once effective against the organism due to a genotypic change.

Antifungal resistance is linked with different molecular mechanisms including: point mutations altering enzyme target affinity to drugs, reduction of drug intracellular concentration (overexpression of efflux pump genes), overexpression of the target enzyme and the existence of secondary metabolic routes able to bypass the inhibition of the main pathway. The aim of this chapter is to summarize the molecular mechanisms of intrinsic and acquired resistance to ergosterol biosynthesis inhibitors and polyenes in *A. fumigatus* and other non-*Candida* species due to point mutations in enzymes participating in ergosterol biosynthesis pathway.

The first part of the following chapter will focus in antifungal resistance in clinical settings; especially for *A. fumigatus* and azole antifungal agents. The second part of the chapter will refer to the ergosterol synthesis pathways and the primary and three-dimensional structure of the main enzymes targeted by antifungals. Ergosterol biosynthesis pathways differ between fungal species, not in the final product (ergosterol) but in the enzymatic steps leading to its synthesis. Those biochemical divergences could explain the differences in antifungal activity of different drugs between species and particular mechanisms of resistance. This information is essential as a theoretical basis to understand the mechanism of azole and allylamine resistance associated with enzyme modifications. The third part of this chapter will describe azole resistance mechanism in *A. fumigatus* and other non-*Candida* species. The final part of the chapter will focus in the mechanism of resistance to other non-azole antifungal drugs targeting fungal membrane.

Aspergillus fumigatus azole resistance in clinical settings

The clinical impact of antifungal resistance is not the main goal of this chapter. However, it is important to briefly describe how azole antifungal resistance in *A. fumigatus* is emerging to understand its importance. *Aspergillus* spp. is intrinsically resistant to fluconazole and normally susceptible to the other triazoles (itraconazole, voriconazole, posaconazole, etc.). Azole acquired resistance started to be seen in 1997 in UK were the first three *A. fumigatus* itraconazole-resistant strains were isolated (Denning *et al.*, 1997). Since then several reports demonstrate that azole resistance in *A. fumigatus* is emerging (Verweij *et al.*, 2007; Baddley *et al.*, 2009; Tashiro *et al.*, 2012). However, the prevalence of *A. fumigatus* azole resistance is not accurately established due that most laboratories do not routinely perform susceptibility testing. The most accurate available data about *A. fumigatus* azole resistance incidence is the obtained from three independent reports published by the SCARE-Network, the ARTEMIS worldwide surveillance program and a meta-analysis performed by Drs. Howard and Arendrup which studied 3249 strains from 20 countries, 497 strains isolated mostly from China and more than 5000 strains isolated between 1990 and 2010, respectively (Howard and Arendrup, 2011; Lockhart *et al.*, 2011; van der Linden *et al.*, 2011a). These reports show that azole resistance incidence in *A. fumigatus* clinical strains is low (around 3%) and that almost 100% of the azole-resistant strains were resistant to itraconazole while around 60% of those were also resistant to voriconazole and/or posaconazole (Howard and Arendrup, 2011; Lockhart *et al.*, 2011; van der Linden *et al.*, 2011a).

The incidence of azole resistance in UK (Manchester) and the Netherlands is significantly higher than the worldwide incidence described above. In Britain, the azole resistance rate jumped from 1% to 8% in 2004 and to 20% in 2009. The majority of those strains (97%) were itraconazole resistant and 78% were multi-azole resistant (Howard *et al.*, 2009; Bueid *et al.*, 2010). In the Netherlands, *A. fumigatus* azole resistance prevalence is steadily increasing from 1.7% in 1997 to 6% in 2006 (Snelders *et al.*, 2008; Verweij *et al.*, 2002; van der Linden *et al.*, 2011b). It has to be noted that there are important differences between the resistant isolates from UK and The Netherlands. The British resistant strains were mainly isolated during or after azole treatment while the Dutch ones were obtained from azole naive patients (Bueid *et al.*, 2010; Howard and Arendrup, 2011; van der Linden *et al.*, 2011b; Howard *et al.*, 2006). Also, the British strains showed multiple mechanisms of resistance involving a wide range of mutations in azole antifungal target enzyme (60% to 70% of the isolates) and other less studied mechanisms, mainly involving efflux pump overexpression (Bueid *et al.*, 2010). Oppositely, the Dutch resistant isolates showed mainly one particular mechanism involving two modifications in the target enzyme (Snelders *et al.*, 2008; Verweij *et al.*, 2002; van der Linden *et al.*, 2011b). This last mechanism was linked with the use of azole antifungals in agriculture while the UK resistance is due to long-term azole therapy in the context of chronic aspergillosis (Howard *et al.*, 2006; Snelders *et al.*, 2008).

The resistance frequency in *A. fumigatus* not only depends on the country but also on the affected population. *A. fumigatus* strains isolated from patient suffering cystic fibrosis and chronic pneumonia showed higher azole resistance prevalence than strains infecting patient with oncohematologic diseases (up to 8% versus 0.85% prevalence, respectively) (Alanio *et al.*, 2011; Arabatzis *et al.*, 2011; Burgel *et al.*, 2012; Dannaoui *et al.*, 1999; Howard and Arendrup, 2011; Lockhart *et al.*, 2011; Morio *et al.*, 2012; Mortensen *et al.*, 2011; Tashiro *et al.*, 2012; van der Linden *et al.*, 2011a,b).

The molecular target of azoles, allylamines and polyenes: ergosterol and ergosterol biosynthesis pathway

Ergosterol (ergosta-5,7,22-trien-3β-ol) is the most abundant membrane sterol in fungi. Its main function is to maintain the fungal membrane integrity by regulating membrane fluidity and asymmetry (Bard *et al.*, 1993; Rodriguez *et al.*, 1985). Ergosterol is found exclusively in fungi, making it an excellent target for antifungal agents. Three of the major groups of antifungal agents in clinical use, polyenes, azoles and allylamines owe their antifungal activity to a direct interaction with ergosterol or to an inhibition of its synthesis (White *et al.*, 1998; Sanglard *et al.*, 2003; Ghannoum and Rice, 1999). Polyenes mechanism of action is still not totally understood. What is clear is that there is a strict association between the presence of membrane sterols and polyenes susceptibility (Anderson *et al.*, 2014; Dick *et al.*, 1980; Gray *et al.*, 2012; Kelly *et al.*, 1994). On the other hand, allylamines and azoles reduce ergosterol content by inhibiting fungal squalene epoxidases or squalene monooxigenases (EC 1.14.13.132) and 14-α sterol demethylases (EC 1.14.13.70), respectively (Ghannoum and Rice, 1999; Howard *et al.*, 2009; Leber *et al.*, 2001; Sakakibara *et al.*, 1995).

Ergosterol biosynthesis pathway is relatively well studied in *Saccharomyces cerevisiae* and is used as a model of study for other fungi. In this yeast, the sterol biosynthesis can be divided in five stages (Alcazar-Fuoli *et al.*, 2008; Lai *et al.*, 1994) (Fig. 2.1). The first stage involves linear molecules and is common for Plantae, Animalia and Fungi kingdoms. It starts with the production of acetoacetyl-CoA using two acetyl-CoA molecules and finishes with the production of squalene (Dimster-Denk and Rine, 1996).

At the beginning of the second stage, lanosterol (5α-lanosta-o, 24-en-3β-ol), the first cyclic molecule of the ergosterol pathway, is produced. These reactions are catalysed by squalene epoxidase (Erg1p) and squalene cyclase (Erg7p). Erg1p is the molecular target of allylamine class of antifungals. From lanosterol to zymosterol (the final product of this stage) three demethylation steps are described (Parks *et al.*, 1985). The first one is the 14-α methyl group removal from lanosterol catalysed by the 14-α sterol demethylase (SDM) or lanosterol demethylase (Lai *et al.*, 1994). Fungal SDMs are codified by *ERG11* and *CYP51* genes in yeasts and filamentous fungi, respectively. SDMs are the target for azole antifungal agents. The other reactions of this stage are carried out by four different enzymes encoded by *ERG24* (target of the morpholine antifungal drugs), *ERG25*, *ERG26* and *ERG27*. The final product of stage two is zymosterol (IUPAC name: cholesta-8,24-dien-3β-ol) (Aaron *et al.*, 2001; Bard *et al.*, 1996; Gachotte *et al.*, 1998, 1999; Jia *et al.*, 2002; Lai *et al.*, 1994; Lorenz and Parks, 1992; Mo *et al.*, 2002; Morris and Richards, 1991; Vanden Bossche *et al.*, 1993). The third stage involves the methylation of zymosterol to get fecosterol (24 methyl-cholesta-8,24(28)-dien-3β-ol). This reaction is catalysed by the C-24 sterol methyl transferase (codified by *ERG6*). Fecosterol is the first sterol in the ergosterol pathway with 28 carbons, which is the main differential characteristic between ergosterol and cholesterol (Aaron *et al.*, 2001; Bard *et al.*, 1996; Gaber *et al.*, 1989; Jensen-Pergakes *et al.*, 1998; Lees *et al.*, 1995). The fourth stage is the desaturation of the sterol cycle B (C-8 and C-5) by the C-8 sterol isomerase (encoded by *ERG2*) and the C-5 sterol desaturase (Erg3p), respectively (Ashman *et al.*, 1991; Keon *et al.*, 1994). The last stage in ergosterol biosynthesis pathway is the lateral chain modification including a desaturation in C-22 by the C-22 sterol desaturase (Erg5p) and a reduction in C-24 carried out by the C-24 sterol reductase (codified by *ERG4* gene) (Lees *et al.*, 1995; Lamb *et al.*, 1999; Skaggs *et al.*, 1996).

Differences in ergosterol pathway between *Aspergillus* spp. and yeasts

As stated before, *S. cerevisiae* ergosterol biosynthesis pathway is a complex enzymatic route that was considered for many years as the model route in fungi. However, as early as 1973, alternative steps in fungal and plant sterol pathways were described (Fryberg *et al.*, 1973; Nes *et al.*, 1989, 2002).

These alternative steps seem to be fungal taxa-specific and involve changes in the order in which enzymes act, differences in the growth conditions

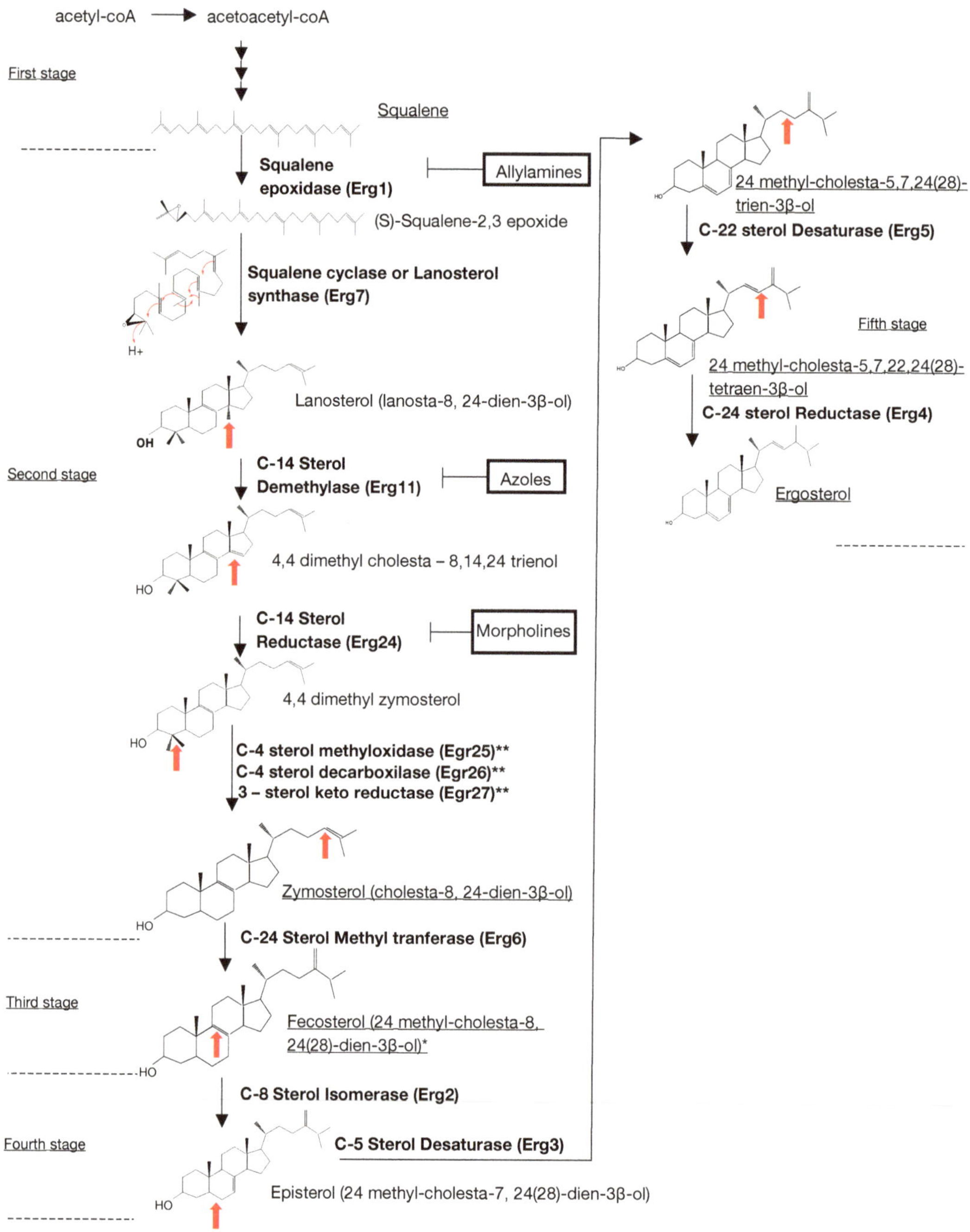

Figure 2.1 Ergosterol biosynthesis in yeasts (model organism: *Saccharomyces cerevisiae*). Red arrows indicate where the chemical modification takes place in the following reaction. Chemical names in parenthesis are IUPAC suggested names. Enzymes participating in each of the steps of the pathway are named in bold letters (in parenthesis and bold are the genes encoding the enzymes in each of the steps). Underline sterol precursors indicate the end of one of the described ergosterol biosynthesis stages. Compiled from Aaron *et al.* (2001), Bard *et al.* (1996), Gachotte *et al.* (1998), Gachotte *et al.* (1999), Jia *et al.* (2002), Lai *et al.* (1994), Lorenz and Parks (1992), Mo *et al.* (2002), Pierson *et al.* (2004) and Vanden Bossche *et al.* (1993). *Fecosterol is the first molecule with 28 carbon atoms and is characteristic of the Fungi kingdom. **Double demethylation in C-4 performed sequentially by the three listed enzymes.

where a particular metabolic route is taken, and the participation of 'new' (not described for *S. cerevisiae*) precursors. All the newly described steps or pathway changes include genes and enzymes upstream lanosterol synthesis, demonstrating its key function as ergosterol precursor (Lees *et al.*, 1995; Ruan *et al.*, 2002; Harmouch *et al.*, 1995).

Ergosterol biosynthesis pathway is more complex in filamentous fungi than in yeasts. In *Aspergillus* spp., two homologous SDMs (Cyp51Ap and Cyp51Bp) and three homologous C-5 sterol desaturases (Erg3Ap, Erg3Bp and Erg3Cp) were described (Mellado *et al.*, 2001; Alcazar-Fuoli *et al.*, 2006). Moreover, in *A. flavus* three SDMs (Cyp51Ap, Cyp51Bp and Cyp51Cp) were reported (Liu *et al.*, 2012). These data suggest that *Aspergillus* spp. ergosterol biosynthesis pathway should be reconsidered as a branched pathway on the basis of the existence of multiple enzymes for the same reaction. Alcazar-Fuoli *et al.* confirmed the 'branched idea' by studying the membrane sterol composition of *A. fumigatus* strains defective in different ergosterol pathway enzymes (Alcazar-Fuoli *et al.*, 2008).

The first stage of *A. fumigatus* ergosterol biosynthesis pathway starts with the production of acetoacetyl-CoA and finishes with the production of squalene, as described for *S. cerevisiae*. The second stage of *A. fumigatus* ergosterol biosynthesis is where the first branched pathway was described (Fig. 2.2) (Alcazar-Fuoli *et al.*, 2008). At the beginning of this stage, lanosterol is methylated in C-24 position (catalysed by Erg6p) to give eburicol (4α,4β,14,24-tetramethylcholesta-8,24(28)-dien-3β-ol) which is a 31-C sterol and one of the *A. fumigatus* Cyp51p's substrates together with lanosterol (the only Erg11p substrate in yeast) (Alcazar-Fuoli *et al.*, 2008; Warrilow *et al.*, 2010). The ability of *A. fumigatus* Cyp51ps to catalyse the demethylation of two different substrates divides the second stage of ergosterol synthesis pathway into two branches. This stage starts with lanosterol and finishes with episterol and/or 4,4,24-trimethylcholesta-8,24(28)-dien-3β-ol depending on the used branch (Fig. 2.2). The first branch reaches zymosterol utilizing the *S. cerevisiae* sequence of enzymatic reactions (Cyp51p, Erg24p, Erg25p, Erg26p, Erg27p, Erg6p and Erg2p). The other branch use eburicol as precursor to give 4,4,24-trimethylcholesta-8,24(28)-dien-3β-ol (zymosterol is not produced), using the following enzymes C-31 methyl transferase, Cyp51p, Erg25p, Erg26p, Erg27p and Erg2p (Alcazar-Fuoli *et al.*, 2008). The existence of these branches was described in other fungal species and the choice of one, the other or both depends on the species (Fryberg *et al.*, 1973; Harmouch *et al.*, 1995; Nes *et al.*, 1989). In *A. fumigatus*, there were no detection of any of the intermediates of the branch that used zymosterol (Alcazar-Fuoli *et al.*, 2008). Moreover, Warrilow *et al.* demonstrated that Cyp51s have a lower affinity for lanosterol than for eburicol (Warrilow *et al.*, 2010). Thus, both reports suggest that the synthesis of episterol is done mainly by the eburicol branch, at least in regular growing conditions. However, the existence in *A. fumigatus* genome of all the described genes (*ERG24*, *ERG25*, *ERG26*, *ERG27* and *ERG6*) and the Cyp51ps lanosterol affinity (low but existing) suggest that both branches could be used but in yet unknown situations (e.g. particular growing conditions or inhibition of the main pathway by antifungal agents).

The third and fourth stages of ergosterol biosynthesis (fecosterol and episterol synthesis) have no differences when compared with *S. cerevisiae*.

The last stage in *A. fumigatus* ergosterol biosynthesis starts with episterol (5α-ergosta-8,24(28)-dien-3β-ol) and finishes in ergosterol. This stage is the second branching point of the pathway (Nes *et al.*, 2002) (Fig. 2.3). These branches were reported in other fungi and may coexist (Alcazar-Fuoli *et al.*, 2006, 2008; Benveniste, 2004; Mejanelle *et al.*, 2001; Zhou *et al.*, 2002; Fryberg *et al.*, 1973; Nes *et al.*, 1989). However, using *A. fumigatus* defective strains in *ERG3* genes, Alcazar-Fuoli *et al.* demonstrated that Erg3Bp is the only *A. fumigatus* enzyme capable to catalyse the transformation of 24-methylcholesta-7-22-dien-3β-ol to ergosterol and that the branches involving 24-methylcholesta-5,7,22,24(28)-tetraen-3β-ol as intermediate seem to be the way ergosterol is synthesized in this fungi. The branch that involves 24-methylcholesta-7-en-3β-ol (dotted lines in Fig. 2.3), essential for *Candida* spp. and melanized fungi (Fryberg *et al.*, 1973; Mejanelle *et al.*, 2001), is not important for

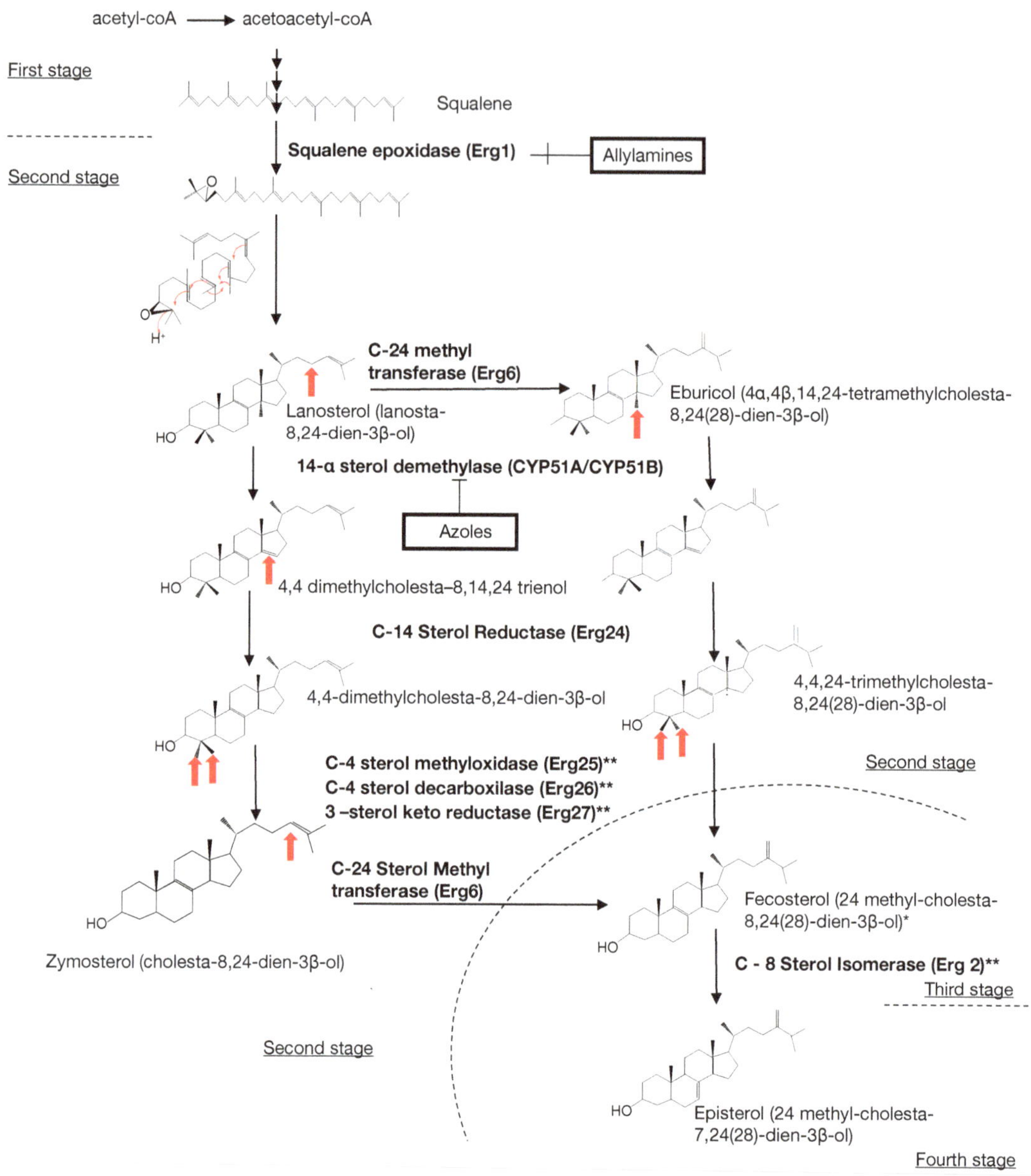

Figure 2.2 Ergosterol biosynthesis in *Aspergillus fumigatus* (stages 1 to 4). Red arrows indicate where the chemical modification takes place in the following reaction. Chemical names in parenthesis are IUPAC suggested names. Enzymes participating in each of the steps of the pathway are named in bold letters (in parenthesis and bold are the genes encoding the enzymes in each of the steps). Underline sterol precursors indicate the end of one of the described ergosterol biosynthesis stages (Alcazar-Fuoli *et al.*, 2008b; Warrilow *et al.*, 2010). *Fecosterol is the first molecule with 28 carbon atoms and is characteristic of the Fungi kingdom. **Double demethylation in C-4 performed sequentially by the three listed enzymes.

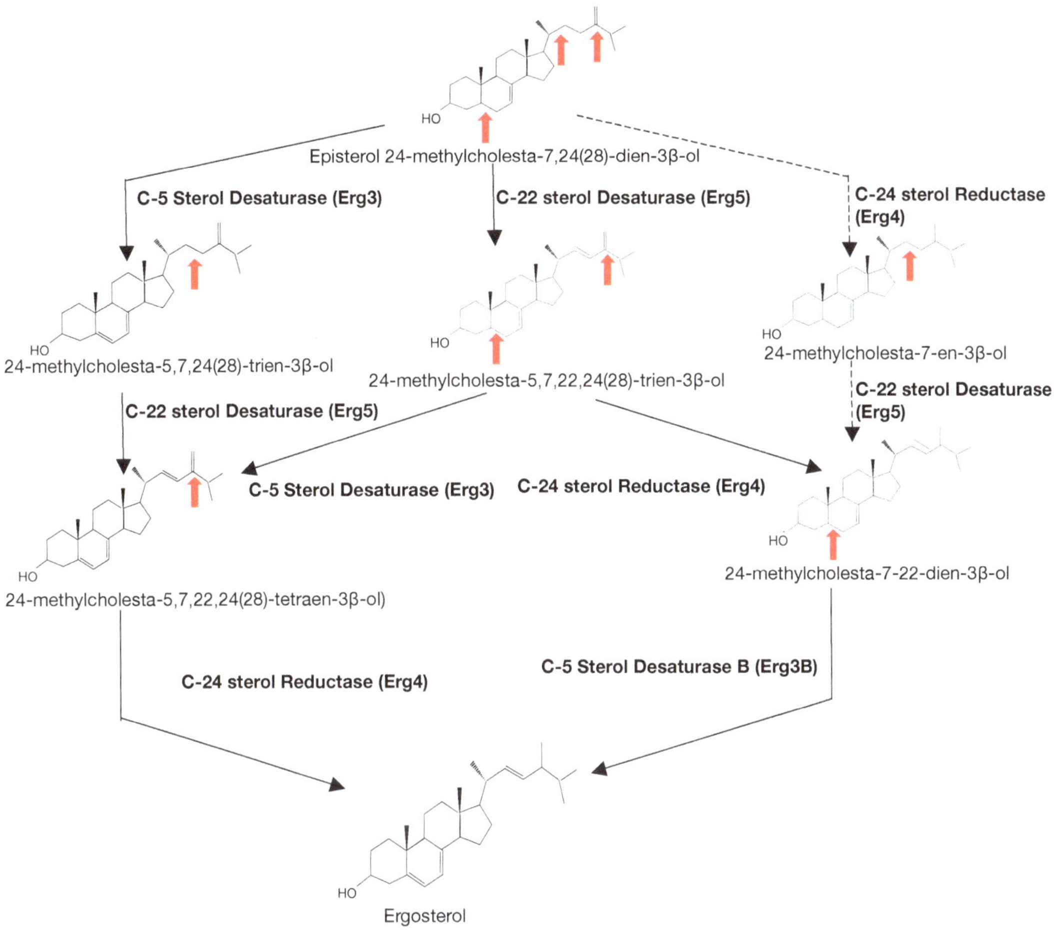

Figure 2.3 *Aspergillus fumigatus* last stage of ergosterol biosynthesis (episterol to ergosterol). Red arrows show where the following enzyme will modify the precursor. IUPAC systematical names are in parenthesis. Names in bold letters refer to enzymes of the biosynthesis pathways (in parenthesis and bold are the genes encoding the enzymes in each of the steps). Dotted arrows show common enzyme reactions in *Candida* spp. and other fungi but not observed in *A. fumigatus* (Alcazar-Fuoli *et al.*, 2008).

ergosterol synthesis in *A. fumigatus* (Alcazar-Fuoli *et al.*, 2008). Altogether, it is clear that *A. fumigatus* is able to synthesize ergosterol using different pathways bypassing chemically or genetically suppressed enzymes. Subsequently, the mechanism of antifungal resistance associated with ergosterol pathway enzymes would be more complex in this filamentous fungus than in yeasts.

Fungal 14-a sterol demethylases (SDMs): the main target for azole antifungals

SDMs (Erg11p and Cyp51p) are cytochrome P450s superfamily members and have monooxigenase activity. SDMs active site comprises a haem group (van den Brink *et al.*, 1998; Waterman and Lepesheva, 2005). During their catalytic activity, SDMs remove the 14-α-methyl group from lanosterol and eburicol (in *A. fumigatus*) by a triple successive monooxigenation reaction. During the process, a formic acid is eliminated, oxygen and NADPH are consumed and a 14,15 double bond is introduced to give 4,4,24-trimethylcholesta-8,14,24-trienol (from eburicol) and 4,4-dimethylcholesta-8,14,24-trienol (from lanosterol) (Fig. 2.4) (Aoyama *et al.*, 1989; Trzaskos *et al.*, 1986).

The fungal SDMs are integral membrane proteins located in the endoplasmic reticulum

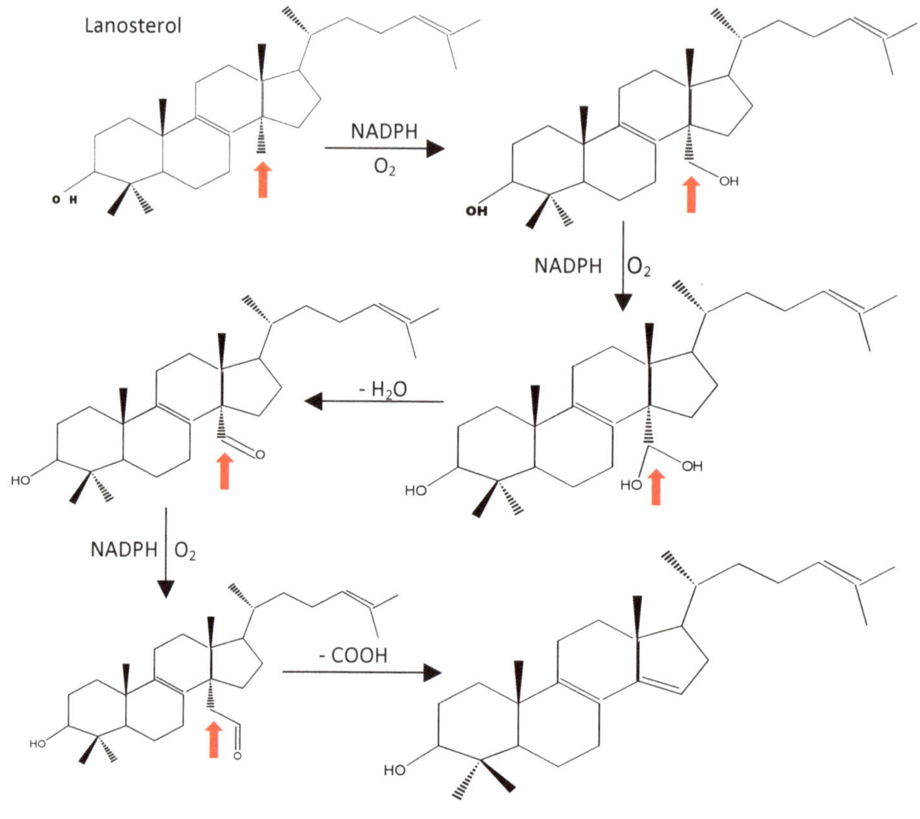

Figure 2.4 Ergosterol precursor demethylation by three successive monooxigenation reactions at C-14 catalysed by fungal 14-α sterol demethylase.

with an N-terminal membrane anchor (Kalb *et al.*, 1987; Warrilow *et al.*, 2010; Xiao *et al.*, 2004), which renders their structural characterization more complex than for soluble proteins. The first fungal SDMs primary structure was studied in *S. cerevisiae* (Kalb *et al.*, 1987) and more recently, several *ERG11/CYP51* have been cloned and characterized including several species of *Candida, Aspergillus, Cryptococcus* and other fungal human and plant pathogens (Mellado *et al.*, 2001; Chen *et al.*, 1988; Delye *et al.*, 1998; Geber *et al.*, 1995; Hargreaves and Keon, 1996; Revankar *et al.*, 2004; van Nistelrooy *et al.*, 1996).

As described before, in *Aspergillus* spp. two SDMs homologues named Cyp51Ap and Cyp51Bp (encoded by *CYP51A* and *CYP51B* genes) were described. *A. fumigatus CYP51A* (GenBank accession number: EU626229) has 1617 base pairs (bp) and one 72 bp intron leading to a 515 amino acids (aa) protein. *A. fumigatus CYP51B* (GenBank: AF338660) is a 1731 bp gene that has three introns (of 59, 46 and 54 bp) and encodes a 524 amino acids protein (Mellado *et al.*, 2001). *A. fumigatus* Cyp51Ap shows a high homology (59.4%) with Cyp51Bp. However, each of the proteins shows different homologies with other fungal SDMs. *A. fumigatus* Cyp51Ap shows the strongest homologies with other *Aspergillus* spp. Cyp51Ap (higher than 70%) and *Penicillium italicum* Cyp51p (63.3%) while *A. fumigatus* Cyp51Bp shows high percentages of identity with *Uncinula necator* Cyp51p (58.1%), *Ustilago maydis* Cyp51p (43%) and yeast's Erg11ps (higher than 45% homology with *Candida* spp., *S. cerevisiae* and *Cryptococcus neoformans*).

A. fumigatus Cyp51ps harbour all the characteristic of other fungal, bacterial and plant SDMs. Following the proposed nomenclature for *P.*

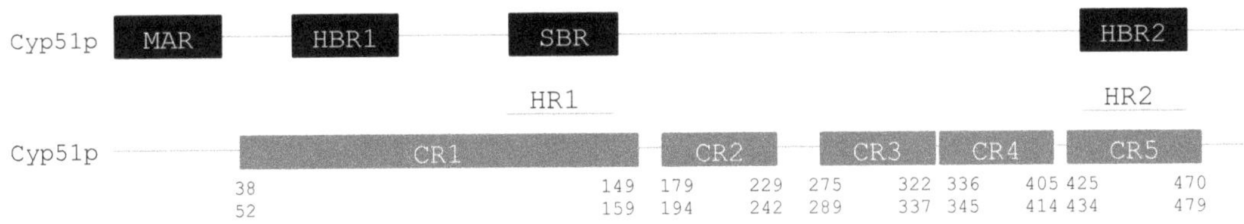

Figure 2.5 Cyp51p primary structure conserved regions. Black squares represent the Cyp51p regions following Podust *et al.* (2001a,b) nomenclature: MAR, membrane anchor region; HBR, haem binding regions; SBR, substrate binding region (Podust *et al.*, 2001a,b; van Nistelrooy *et al.*, 1996). Grey squares represent the clusters of conserved amino acids (CR regions from CR1 to CR5) and lines represent the functional subregions (HR) of Cyp51p using van Nistelrooy's nomenclature (Podust *et al.*, 2001a,b; van Nistelrooy *et al.*, 1996). Numbers beneath CR regions represent the amino acid number where each region starts and finishes for Cyp51Ap (upper row) and for Cyp51Bp (bottom row).

italicum and *Mycobacterium tuberculosis* Cyp51ps, four regions and five clusters (with subregions) can be distinguished (Podust *et al.*, 2001a,b; van Nistelrooy *et al.*, 1996) (Fig. 2.5). The names of the regions describe their putative function. The first region is named the membrane anchor region (MAR). It includes the first 29 and 46 aa of *A. fumigatus* Cyp51Ap and Cyp51Bp, respectively and it is a transmembrane region. The second and fourth region are named haem binding regions (HBR) while the third is the substrate binding region (SBR) (Xiao *et al.*, 2004). Another way to study the SDMs primary sequence is to divide it in clusters of conserved aa sequences. Van Nistelrooy *et al.* (van Nistelrooy *et al.*, 1996) described five clusters for *P. italicum* Cyp51p named CR1 to CR5 (CR stands for Conserved Regions) with subregions with putative functions named HR (HR hypothetical regions). All the described CRs and HRs are conserved in *A. fumigatus* Cyp51ps (Fig. 2.5). For example, HR1 subregion (aa 110–133 and 115–140 for Cyp51Ap and Cyp51Bp, respectively) included in the cluster CR1 was linked with the substrate recognition (and coincide with the aa included in Podust's SBR). On the other hand, the subregion HR2 has a cysteine residue (residue 454 and 463 for Cyp51Ap and Cyp51Bp, respectively) that is conserved among Basidiomycetes, Mucormycetes and Ascomycetes. This cysteine is essential for the binding of the protein to the haem group. Thus, HR2 is part of one of the HBRs (HBR2) (van den Brink *et al.*, 1998; van Nistelrooy *et al.*, 1996).

Most of the actual knowledge about fungal SDMs secondary structure is based on the soluble *M. tuberculosis* SDM (Podust *et al.*, 2001a). This enzyme was the first SDM crystallized and studied by X-ray crystallography (Aoyama *et al.*, 1998; Podust *et al.*, 2001a,b, 2007; Bellamine *et al.*, 1999). This crystal structure was used as a template to obtain the homology models of *A. fumigatus* Cyp51p enzymes. From that, twelve α-helices and the same number of β-pleated sheets were described (Gollapudy *et al.*, 2004; Xiao *et al.*, 2004) (Fig. 2.6). In addition, two substrates and antifungal drugs entrance channels were proposed. Both channels are able to accommodate azole drugs with long lateral chains such as itraconazole and posaconazole. One of the channels (channel 1) is parallel to the haem group and is formed by the curved part of the I, B and G α helices and the loop between both (loop BC). Channel 2 is perpendicular to the haem group plane and its internal surface is formed by the FG loop, the A′ helix and the amino acid chain between β4-1 and β4-2 pleated sheets (Gollapudy *et al.*, 2004; Macchiarulo *et al.*, 2002; Wester *et al.*, 2003; Xiao *et al.*, 2004). This last channel is where azoles interact with SDMs based on 3D models. Confirming this interaction, amino acid substitutions in *A. fumigatus* Cyp51Ap A′ helix (aa S49 to G54) and FG loop (T215 and L226) are linked with itraconazole and posaconazole cross-resistance (Diaz-Guerra *et al.*, 2003; Mann *et al.*, 2003; Xiao *et al.*, 2004) (Fig. 2.7). Despite these data, Podust *et al.* (2001a) described that *M. tuberculosis* Cyp51p exhibits a site-access channel (I helix and BC loop) parallel to the haem plane and open to the surface while the substrate entry channel is

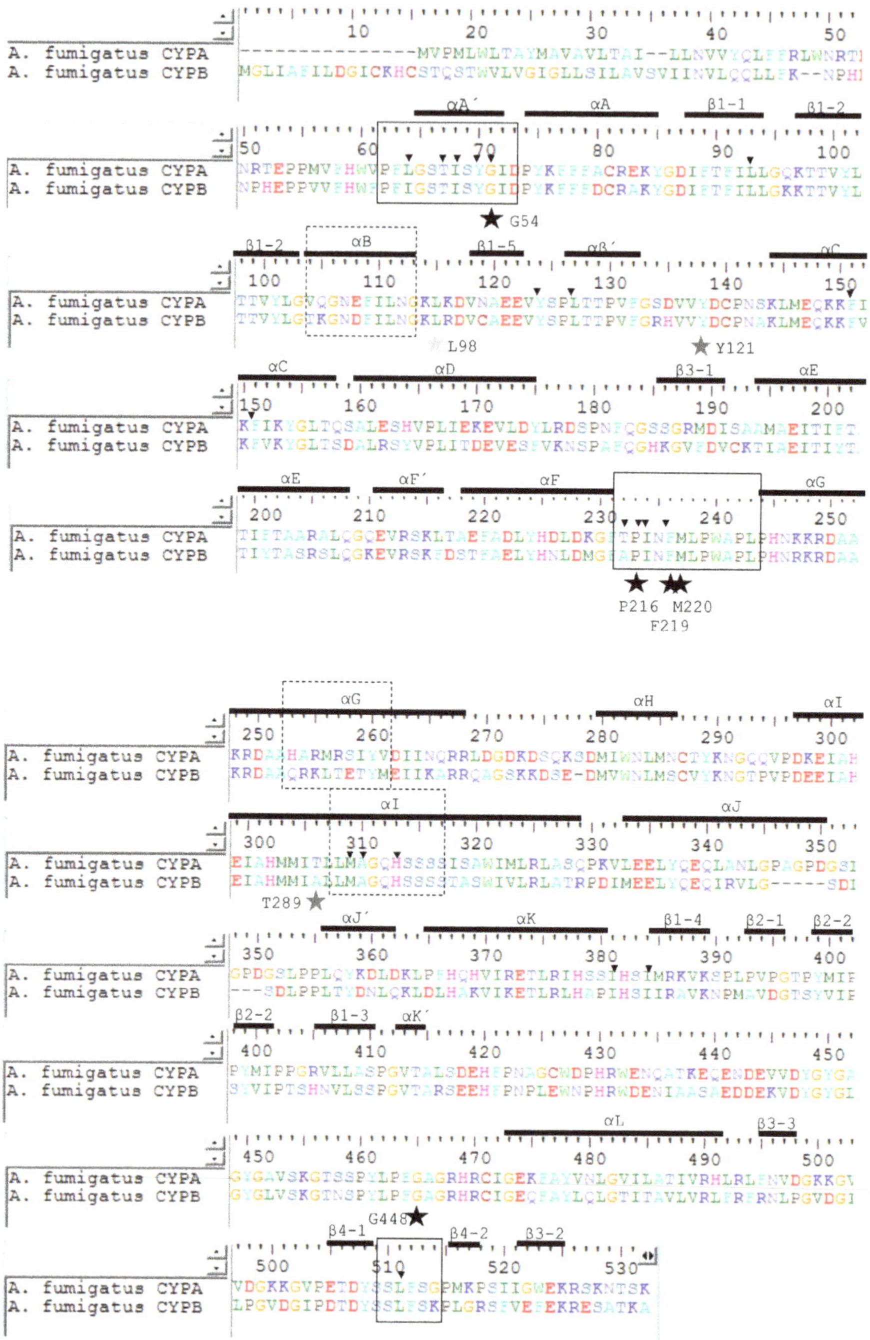

Figure 2.6 Clustal alignment of the *A. fumigatus* Cyp51ps sequences. The secondary structure elements are marked with black lines and named as described by Xiao *et al.* (2004). Arrow heads label the amino acid residues involved in posaconazole binding by *A. fumigatus* Cyp51Ap (Xiao *et al.*, 2004). Dotted lined squares show the amino acid wich form channel 1. Full lined squares show the FG loop, the A′ helix and the chain between β4-1 and β4-2 sheets which form the internal surface of Cyp51Ap channel 2. Stars show the amino acids of the Cyp51Ap implicated in azole resistance and confirmed by gene replacement. Grey stars represent amino acid linked with resistance only if its substitution is coupled with a modification in the promoter of the *Cyp51A* gene (dark grey stars shows Y121 and T289 amino acid that needs a repetition of 46 nucleotides in the *CYP51A* promoter (TR_{46}/Y121F/T289A) and the light grey star show the amino acid L98 that needs to be coupled with a repetition of 34 nucleotides in the *CYP51A* promoter (TR_{34}/L98H)).

A´ Helix	60 70 80
A. fumigatus CYP51A	TEPPMVFHWVPFLGSTISYGIDPYKFFFACRE
A. fumigatus CYP51B	H...V....F..I...............D..A
C. albicans ERG11	DRA.L..Y.I.WF..AA...QQ..E..ES..Q
C. glabrata ERG11	DR..L..Y.I.WV..A.P..TK..E..ED.QK
S. cerevisiae ERG11	DR..L..Y.I.WV..AVV..MK..E..EE.QK

FG Loop	220 230 240 250
A. fumigatus CYP51A	RSKLTAEFADLYHDLDKGFTPINFMLPWAPLPH
A. fumigatus CYP51B	...FDST..E...N..M..A.............
C. albicans ERG11	.RIFDRS..Q..S...........VF.NL....
C. glabrata ERG11	.D..DTD..Y..S...........VF.NL..E.
S. cerevisiae ERG11	.A..DTD..Y..S...........VF.NL..E.

β4-1 and β4-2 sheets	220 230 240 250
A. fumigatus CYP51A	RLFNVDGKKGVPETDYSSLFSGPMKPSIIGWEK
A. fumigatus CYP51B	.FR.LP.VD.I.D........K.LGR.FVEF..
C. albicans ERG11	.WTID--GYK..DP....MVVL.TE.AE.I...
C. glabrata ERG11	KWRYPTEGET..PS.FT.MVTL.TA.AK.Y...
S. cerevisiae ERG11	KWHYP-EG.T..PP.FT.MVTL.TG.AK.I...

Figure 2.7 Clustal alignment of the amino acid residues of the FG loop (aa F214 to P225 in *A. fumigatus* Cyp51Ap), A′ helix (aa S49 to G54) and β4-1 and β4-2 pleated sheets (aa S492 G498) that form the internal surface of SDM channel 2. The aa of the described regions are highlighted in black. Red arrows indicate the residues involved in binding to posaconazole (Gollapudy *et al.*, 2004; Macchiarulo *et al.*, 2002; Wester *et al.*, 2003; Xiao *et al.*, 2004).

closed. The existence of two channels (one open and the other closed) suggests a conformational regulated substrate-in/product-out opening in SDMs (Podust *et al.*, 2001a). Very recently, full-length (with the transmembrane domain) *S. cerevisiae* SDM (Erg11p) was crystalized and co-crystallized with lanosterol and three azole drugs including: itraconazole, fluconazole and voriconazole (Monk *et al.*, 2014). This work shows that the N-terminal transmembrane helix limits the orientation of the putative substrate channel by contacting with catalytic domain. Overall, these data described by Monk *et al.* (2014) for *S. cerevisiae* Erg11p confirms what Xiao *et al.* (2004) and Gollapudy *et al.* (2004) described 10 years before for *A. fumigatus* Cyp51Ap using *M. tuberculosis* SDM as a model. Briefly, *S. cerevisiae* Erg11p has two channels (named primary and secondary vestibules by Monk *et al.*, 2014) where azole drugs and substrate interact with the enzyme. One of the channels is perpendicular while the other is parallel to the haem group. Also, Monk *et al.* (2014) coincide with the other two authors in (i) the way triazole groups interact with the iron of the haem group by a coordination bond occupying the space used by the substrate and the oxygen and (ii) that amino acid residues of channel 2 (secondary vestibule mouth for Monk *et al.*, 2014) interact with azole drugs (Gollapudy *et al.*, 2004; Monk *et al.*, 2014; Xiao *et al.*, 2004).

A. fumigatus azole susceptibility patterns and mechanisms of azole resistance linked with *CYP51* point mutations

A. fumigatus is intrinsically resistant to fluconazole and ketoconazole and normally susceptible to the rest of the available azole antifungal agents (e.g. itraconazole, posaconazole, voriconazole, ravuconazole, isovuconazole) (Garcia-Effron *et al.*, 2005b; Mellado *et al.*, 2005; Gregson *et al.*, 2013; Pfaller *et al.*, 2009, 2011; Rodriguez-Tudela *et al.*, 2008) (Table 2.1).

Table 2.1 Published amino acid substitutions in *A. fumigatus* Cyp51Ap resulting in azole resistance

Cyp51Ap aa substitution	Minimal inhibitory concentration (μg/ml)				Reference
	Itraconazole	Posaconazole	Voriconazole	Ravuconazole	
WT	0.12–0.50	0.03–0.12	0.06–0.50	0.12–0.50	
F46Y/M172V/E427K[a,c]	>8.00	0.12–0.50	2.00–4.00	NI	Howard *et al.* (2009)
F46Y/M172V/E427K/N248T/D255[a,d]	>8.00	0.12–0.50	2.00–4.00	NI	Howard *et al.* (2009)
E427G[a]	>8.00	0.125–0.50	2.00–4.00	NI	Howard *et al.* (2009)
G138C[a]	>8.00	2->8.00	8.00->8.00	8.00	Howard *et al.* (2006, 2009), Garcia-Effron *et al.* (2008)
H147Y[a,e]	>8.00	>8.00	0.50	NI	Howard *et al.* (2009)
H147Y/G448S[a,e]	>8.00	>8.00	0.50	NI	Howard *et al.* (2009)
G54E/V/R	>8.00	0.25–0.50	0.25–0.50	0.50–1.00	da Silva Ferreira *et al.* (2004), Diaz-Guerra *et al.* (2003), Garcia-Effron *et al.* (2005c, 2008); Mellado *et al.* (2005b), Nascimento *et al.* (2003)
G54W	>8.00	>8.00	0.25–0.50	0.50–1.00	Garcia-Effron *et al.* (2008), Mann *et al.* (2003)
M220K/V/T/I	>8.00	0.25–0.50	0.50–1.00	1.00–2.00	Chen *et al.* (2005b), Garcia-Effron *et al.* (2008), Mellado *et al.* (2004)
TR34-L98H	>8.00	0.50–1.00	2.00–4.00	4.00–8.00	Garcia-Effron *et al.* (2008), Mellado *et al.* (2007), Verweij *et al.* (2007, 2009)
F219I[b]	>8.00	0.50 – >8.00	1.00- 8.00	NI	Camps *et al.* (2012)
P216L	>8.00	1.00	1.00	NI	Camps *et al.* (2012)
TR_{46}/Y121F/T289A	2.00->8.00	0.25–2.00	>8.00	NI	Camps *et al.* (2012), van der Linden *et al.* (2013), Vermeulen *et al.* (2012)
Y431C[a,d]	>8.00	4.00	1.00	NI	Howard *et al.* (2009)
G434C[a,d]	>8.00	>8.00	0.50–1.00	NI	Howard *et al.* (2009)
G448S	0.50	0.125	8.00	NI	Bellete *et al.* (2010), Manavathu *et al.* (2000), Pelaez *et al.* (2012)

[a]The implication on azole resistance was not confirmed by gene replacement or other molecular methodology. [b]The strains that showed MIC values for PSC >8.00 μg/ml where isolated during or after PSC treatment (see text for more detail). [c]Found along with 3 silent mutations (4 isolates). [d]Only 1 isolate described. [e]The H147Y substitution is probably not related with azole resistance since the G448S substitution alone showed the same phenotype (Bellete *et al.*, 2010; Manavathu *et al.*, 2000; Pelaez *et al.*, 2012).
The minimal inhibitory concentration was obtained following the CLSI or EUCAST protocol.

A. fumigatus intrinsic fluconazole resistance

The *A. fumigatus* fluconazole intrinsic resistance is not yet molecularly explained. However, a potential mechanism was described by Edlind *et al.* (2001) and would involve a naturally occurring aa substitution in *A. fumigatus* Cyp51Ap. This idea is based on what is known for *C. albicans* fluconazole-resistant isolates. In this yeast, several Erg11p aa residues were implicated in fluconazole resistance. The most important and studied are the tyrosine 132 (Y132), threonine 315 (T315), serine 405 (S405), glycine 464 (G464), and arginine 467 (R467) (Lamb *et al.*, 1997). In *A. fumigatus* Cyp51Bp all five residues are conserved while in *A. fumigatus* Cyp51Ap, the T315 equivalent residue is naturally replaced by an isoleucine (codon I301 for *A. fumigatus*) (Fig. 2.8). The *C. albicans* Erg11p T315 plays a key role in both enzyme-substrate and enzyme–drug interactions since it binds the 3-OH of the substrate (lanosterol) near the haem group (Marichal *et al.*, 1999; Sanglard *et al.*, 1998). Similarly, *A. fumigatus* Cyp51Ap I301 is located in the centre of a conserved domain (the CR3 region: aa 275 to 322) part of the α I loop (D280 to Q312), linked with enzyme–substrate (SBR) and enzyme–drug interactions (van Nistelrooy *et al.*, 1996) (Fig. 2.6). In *C. albicans*, the substitution of the polar threonine for the non-polar alanine (T315A) was linked with fluconazole resistance [4- to 5-fold minimum inhibitory concentration (MIC) increase], and coupled with a half reduction of the demethylase activity of the Erg11p (V_{max} 50% of the wild type enzyme) (Lamb *et al.*, 1997). In *A. fumigatus* Cyp51Ap, the threonine is naturally replaced by isoleucine, another non-polar amino acid. Thus, analogous changes would be expected in the mould SDM's enzymatic properties. Accordingly, similar changes in fluconazole susceptibilities were observed by Dr Mellado's group in *A. fumigatus Cyp51A* defective strains (Mellado *et al.*, 2005; Garcia-Effron *et al.*, 2005a,b). Those strains showed a 4- to 8-fold decrease in fluconazole MICs. These results demonstrated that Cyp51Bp (the only active SDM in *CYP51A* defective strains) is more susceptible to fluconazole and that Cyp51Ap would be responsible for fluconazole intrinsic resistance.

A. fumigatus secondary or acquired triazole resistance

The first three itraconazole-resistant *A. fumigatus* strains were reported by Dr David Denning *et al.* in the UK in 1997 (Denning *et al.*, 1997). Two of them were genetically identical (same RAPDs patterns) and all three were isolated from patients who failed azole therapy. The resistance phenotype was confirmed using a murine model of aspergillosis. The intracellular concentration

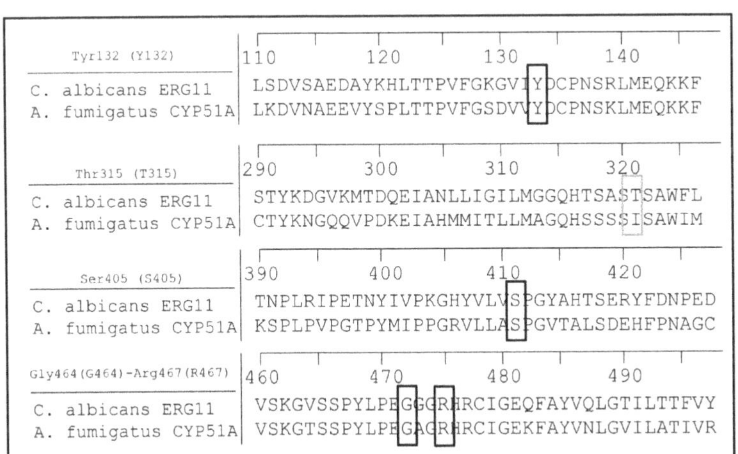

Figure 2.8 Clustal alignment of *C. albicans* Erg11p and *A. fumigatus* Cyp51Ap. Boxes indicate amino acid linked with fluconazole resistance in *C. albicans*. Grey box shows the amino acid T315 of *C. albicans* Erg11p wich is naturally substituted by an isoleucine in *A. fumigatus* Cyp51Ap.

of [^{3}H]itraconazole was measured and one of the strains (named AF-72) showed reduced drug cytoplasmic concentration compared with wild type and the other resistant (named AF-91) strains. Thus, Denning *et al.* proposed that the possible mechanism of itraconazole resistance would be the overexpression of efflux pumps (Denning *et al.*, 1997). Five years later, the same group cloned an *A. fumigatus* putative ABC transporter gene (*AtrF*) that was overexpressed upon itraconazole induction in the strain AF-72 (Slaven *et al.*, 2002). Based on these results, different groups studied the efflux pumps overexpression in *A. fumigatus* as azole resistance mechanism (da Silva Ferreira *et al.*, 2004; Manavathu *et al.*, 2000; Nascimento *et al.*, 2003). However, no other clinical strain showed this mechanism of resistance until recently (Fraczek *et al.*, 2013). In 2003, Dr Mellado's group studied the SDMs genes of a collection of azole-resistant strains including the first resistant isolates (AF-72 and AF-91) (Mosquera and Denning, 2002). The open reading frames of the *CYP51A* and *CYP51B* genes were sequenced and point mutations were found in both genes. Mutations in *CYP51B* genes were silent or have no impact in azole susceptibility. Six of the azole-resistant isolates (including AF-72) showed an aa substitution at the glycine 54 (G54) while five strains (including AF-91) showed an amino acid change at the methionine 220 (M220). In this work and in another published one year later, the mutated *CYP51A* genes from AF-72 and AF-91 strains were PCR amplified and transformed into a wild-type azole susceptible *A. fumigatus* strain by electroporation. The transformants showed the same exact resistance phenotype as AF-72 and AF-91 strains confirming that the efflux pumps has little or no implication in azole resistance, at least for these particular strains (Diaz-Guerra *et al.*, 2003; Mellado *et al.*, 2004). In 2005, two reports from Dr Mellado's group reinforced the idea that the main azole resistance mechanisms in *A. fumigatus* clinical strains are point mutations in *CYP51A*. In those reports, *A. fumigatus CYP51A* and *CYP51B* defective strains were obtained. All *CYP51A* knockout strains showed 20- to 500-fold reduction of azoles MIC (Mellado *et al.*, 2005). On the other hand, *CYP51B* knockout strains showed no changes in azoles susceptibility as compared with isogenic wild-type strains (Garcia-Effron *et al.*, 2005a).

Since then, different molecular mechanisms of azole resistance associated with mutations in *CYP51A* gene were described. In some cases, the implications of these aa changes were molecularly confirmed by gene replacements and knockouts. That is the case of the mutations at codons: 54 (G54), 220 (M220), 216 (P216), 219 (F219), 98 combined with a tandem 34 bp repetition in the *CYP51A* promoter (TR34-L98H), 448 (G448), and double mutations at codons 121 and 289 combined with a 46 bp tandem duplication in the *CYP51A* promoter (TR46-Y121F-T289A) were described from 2003 to 2013 (Zhao *et al.*, 2013; Vermeulen *et al.*, 2012; van der Linden *et al.*, 2013; Snelders *et al.*, 2008; Pelaez *et al.*, 2012; Nascimento *et al.*, 2003; Mosquera and Denning, 2002; Mellado *et al.*, 2004, 2007; Mann *et al.*, 2003; Manavathu *et al.*, 2000; Garcia-Effron *et al.*, 2005b, 2008; Diaz-Guerra *et al.*, 2003; Denning *et al.*, 1997; da Silva Ferreira *et al.*, 2004; Chowdhary *et al.*, 2012, 2013; Chen *et al.*, 2005a; Camps *et al.*, 2012; Bellete *et al.*, 2010). Other *CYP51A* mutations were described but not molecularly confirmed as the only azole resistance mechanism. These mechanisms included unique and multiple aa substitutions in Cyp51Ap such as G138C, H147Y, E427G, Y431C and G434C and F46Y/M172V/E427K, F46Y/M172V/E427K/N248T/D255 and H147Y/G448S (Howard *et al.*, 2006, 2009; Manavathu *et al.*, 2000). Whether or not the mutations at the *A. fumigatus* Cyp51Ap are molecularly confirmed as responsible for resistance, it is clear that the position and the nature of the aa substitution greatly influence the pattern of azole resistance. All the published *A. fumigatus CYP51A* mutations associated with triazole resistance are described in Table 2.1. Of them, the most prevalent resistance profiles are (i) resistance to itraconazole associated with substitutions at Cyp51p glycine 54 (G54E/V/R) (da Silva Ferreira *et al.*, 2004; Diaz-Guerra *et al.*, 2003; Garcia-Effron *et al.*, 2005b; Garcia-Effron *et al.*, 2008; Mellado *et al.*, 2005b; Nascimento *et al.*, 2003); (ii) itraconazole and posaconazole cross-resistance observed when the G54 substituted by a tryptophan (G54W) (Garcia-Effron *et al.*, 2008; Mann *et al.*, 2003); (iii) itraconazole resistance and high voriconazole, ravuconazole and posaconazole MICs when methionine 220 is substituted (M220/K/V/T/I) (Chen *et al.*, 2005b; Garcia-Effron *et al.*, 2008; Mellado *et al.*, 2004), and (iv) azole cross-resistance associated with a *CYP51A* overexpression produced by a tandem repeat of a 34-bp sequence located in the gene promoter in combination with an amino acid substitution at the Cyp51Ap leucine 98 named TR-L98H (prevalent in the Netherlands) (Garcia-Effron *et al.*, 2008; Mellado *et al.*, 2007; Verweij *et al.*, 2007, 2009).

Structure–function relationship of *A. fumigatus* Cyp51Ap amino acid substitutions in azole resistance

The way Cyp51Ap substitutions impact on azole resistance is controversial. Two different possibilities were postulated, both based on crystallographic studies of the prokaryotic Cyp51p and *S. cerevisiae* Erg11p. The first option is a direct interaction between the amino acids of the channel 2 of the fungal SDMs with azole side chains (Gollapudy *et al.*, 2004; Monk *et al.*, 2014; Xiao *et al.*, 2004). Any amino acid substitution in residues of channel 2 (or its access entry) alters azole–SDM interaction and the enzyme become resistant to inhibition. The second option is based on modifications in protein regions which participate in the passage of the SDM through different conformational stages during catalysis (Podust *et al.*, 2001a). In this case, azole resistance due to amino acid substitutions would be mediated by the alteration of SDMs protein regions involved in the regulation of the enzyme conformation (Podust *et al.*, 2001a; Joseph-Horne and Hollomon, 1997). In the case of the mutations at codon 220 (M220) both options could explain the observed phenotype. M220 is part of the *A. fumigatus* Cyp51Ap F-G loop (T215 to P230) and CR-2 cluster (aa 179 to 229). This loop forms the internal part of the SDM channel 1 which is mainly lined by aromatic and hydrophobic residues (F214, I217, F219, M220 and L221). The M220 substitution alters the hydrophobicity of the channel and subsequently the conformational stages of the SDM (Fig. 2.9) (Mellado *et al.*, 2004; Podust *et al.*, 2001a; Joseph-Horne and Hollomon, 1997). However, itraconazole resistance and voriconazole susceptibility phenotype mediated by M220 substitutions could be also explained by an alteration in the azole–SDM interaction. Docking of both triazoles was explored and itraconazole may interacts through its long side chain with some of the residues of the SDM F-G loop (T215 to L226 – part of the channel 1 entry) while voriconazole may not (Mellado *et al.*, 2004; Xiao *et al.*, 2004).

The precise manner in which other substitutions at Cyp51Ap produce azole resistance is better explained by alteration in the docking of azoles in the SDMs channels. That is the case of the G54 residue that is located at channel 2 entry (A′ α helix). This residue interacts with the long side chain of itraconazole and posaconazole but not with voriconazole. The replacement of the glycine (G54) with aa with side chains may create van der Waals conflicts between the azole drug and the aa of the pocket formed by the alpha A′ helix, the FG loop and the chain between the β4–1 and β4–2 strands, thus explaining the observed phenotype in *A. fumigatus* strains with G54 substitutions (Fig. 2.10) (Diaz-Guerra *et al.*, 2003; Garcia-Effron *et al.*, 2008; Mann *et al.*, 2003; Xiao *et al.*, 2004). Similarly, G138 is part of the C α helix of the Cyp51Ap which interacts with the haem group (HBR) and is located in the bottom of channel 2. The G138 mutants show cross-triazole resistance since all triazoles (independently of the length of their side chain) interact with aa located in this region of the channel (Diaz-Guerra *et al.*, 2003; Garcia-Effron *et al.*, 2008; Gollapudy *et al.*, 2004; Howard *et al.*, 2006; Monk *et al.*, 2014; Snelders *et al.*, 2008; Xiao *et al.*, 2004) (Fig. 2.10).

The mechanism how L98H substitutions at *A. fumigatus* Cyp51Ap produce triazole resistance phenotype is better explained by the alteration of the SDMs protein regions involved in the regulation of the conformation of the enzyme during the catalytic cycle. L98 residue is not near any of ligand access channels or in the vicinity of the haem group. It is located in a small loop (formed by three aa: K97, L98 and K99) connecting the α helix B (aa V87 to G96) with the 1–5 β sheet (aa D100 to A103) itself connected with other α helix named B′ (aa P109 to G117) (Fig. 2.6) (Gollapudy *et al.*, 2004; Snelders *et al.*, 2008; Xiao *et al.*, 2004; Podust *et al.*, 2001a; Snelders *et*

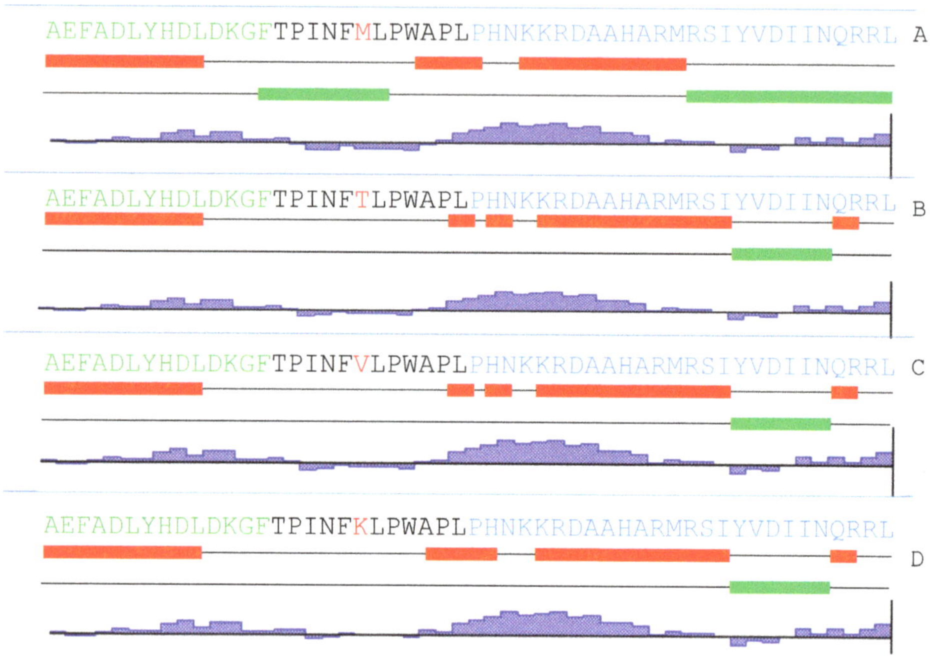

Figure 2.9 Analysis by Granier-Robson (α and β chains prediction) and by Kyte-Doolitle (hydrophobicity prediction) of the amino acid sequences of *A. fumigatus* Cyp51Ap from A201 to L252 (αF and αG helices) wild type and M220 substituted. Amino acid of the αF and αG helices are highlighted in green and blue, respectively. Amino acids in red show M220 (A) and the substitutions by T (B), V (C) and K (D). Red and green rectangles show the predicted α and β chains and the blue histograms represent hydrophobicity (under the middle line) or hydrophilicity (over the middle line).

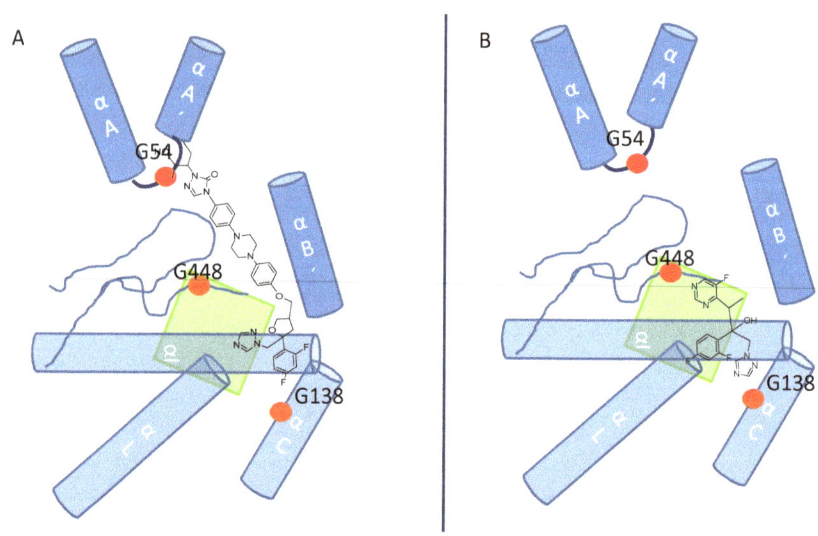

Figure 2.10 Schematic representation of the pocket formed by helix A′, the FG loop and the chain between the β4–1 and β4–2 strands where azole drugs interact with Cyp51Ap. Figure shows the interaction between posaconazole (A) and voriconazole (B) and *A. fumigatus* Cyp51Ap. The locations of substitutions in Cyp51Ap resulting in azole resistance are represented with red circles, α helices are symbolized by light-blue cylinders (transparent cylinders are nearer the viewer than the solid ones), the haem cofactor is denoted as a green square, β strands are shown as light-blue lines and the amino acid chain between the α helices A and A′ (channel 2 entrance) is represented as a dark-blue line (modified from Xiao *et al.* (2004)).

al., 2011). L98 is the only hydrophilic residue at the surface of the region formed by the B and B′ helices. This region interacts with the F helix and the BC loop forming a gate-like structure that is open to the haem group. This 3D configuration is conserved among the Cyp51p family of proteins. In wild type SDMs, L98 residue interacts with the hydrophobic P124 residue of the BC loop. This interaction is no longer possible when the L98 is substituted by a polar histidine. Molecular dynamics simulations performed with the L98H mutated protein showed that the BC loop is more flexible than in the wild type Cyp51Ap resulting in a closure of the first ligand entry channel (channel 1) (Gollapudy *et al.*, 2004; Podust *et al.*, 2001a; Snelders *et al.*, 2011). Slenders *et al.* described another possible consequence of the L98H substitution that is the disruption of the interaction between the Y121 residue and the haem group, and a rapprochement of the histidine towards the carboxylic groups of the haem. These modifications alter the local environment of the Cyp51Ap. The side chains of the Y107 and Y121 residues are relocated into a hydrophobic pocket (A102, V106, L110, T111, M368, L494 and F495) producing the channel-2 closure (Snelders *et al.*, 2011). The modification of the 3D structure of the Cyp51Ap by the L98H mutation is shown in Fig. 2.11.

Even if the conformational regulated openings of the Cyp51Ap altered by the L98H mutation explains the azole resistance phenotype, it is possible (as a consequence of the local environment change) that direct interaction between amino

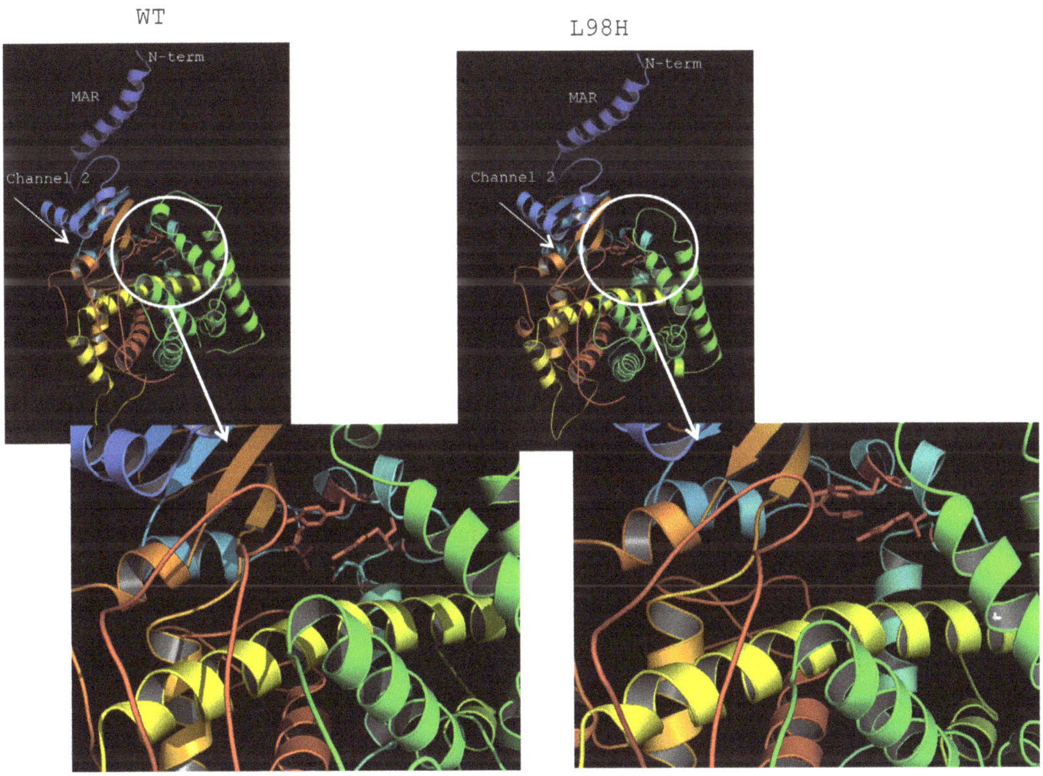

Figure 2.11 Structure of the homology model of the wild type (WT) and L98H mutant *A. fumigatus* Cyp51Ap obtained using the 3D jigsaw comparative modelling software and the P MOL Molecular Graphics System, version 1.5.0.4 (Schrödinger, LLC). The haem group was not included to reduce the complexity of the graphic. White circles show the pocket where azole drugs interact with Cyp51Ap. The pockets are magnified to show how the L98H mutation produces a dilatation of the pocket caused by the flexibility augment of the BC loop and the disruption of the interaction between Y121 residue and the haem group. Residues L98 (and H98 in the mutant), Y107 and Y121 are highlighted in red.

acids and triazoles may also play a role in azole resistance. Podust *et al.* (2001a) suggested that both tyrosine residues (Y107 and Y121) may directly interact with triazoles. Thus, the described modifications, especially those involving Y107 and Y121 could play a key role in triazole resistance linked with L98 substitutions. An issue not totally explained about these substitutions is the fact that the L98H substitution is necessary but not sufficient to display a complete resistance phenotype. In 2007, Mellado *et al.* demonstrated by homologous recombination experiments that a 34 bp tandem repetition (TR34) in the *CYP51A* gene promoter coupled with the L98H substitution is necessary to significantly increase the triazole MIC values. In those experiments, the TR34 repetition and the L98H ORF mutation were introduced in a wild type *A. fumigatus* separately and combined. Both individual modifications produce slight triazole MIC increase (1- to 2-fold). The TR34 promoter modification alone is responsible of a 8-fold increase in *CYP51A* expression but mutants harbouring the TR34 together with the L98H substitution showed a 60-fold increase in itraconazole MIC and a 8- to 16-fold increase in posaconazole and voriconazole MICs (Mellado *et al.*, 2007). Despite these efforts using either classic homologous recombination experiments or molecular dynamics studies, the necessity of this combination of mechanisms (overexpression plus mutation) was not explained. One possibility is that the L98H mutation alters the enzyme kinetic reducing its maximum reaction rate (V_{max}) and the overexpression is needed to compensate it by increasing the number of available enzyme molecules.

Voriconazole resistance as a sole phenotype is rare but was described in some laboratory and clinical strains (Bellete *et al.*, 2010; Howard and Arendrup, 2011; Manavathu *et al.*, 2000; Pelaez *et al.*, 2012). Those strains have G448 substitutions in Cyp51Ap, sometimes coupled with H147 substitutions (Bellete *et al.*, 2010; Howard and Arendrup, 2011; Manavathu *et al.*, 2000; Pelaez *et al.*, 2012). The implication of the G448S substitution on voriconazole resistance was molecularly confirmed by gene replacement by Pelaez *et al.* (2012). On the other hand, H147Y substitution is probably not related with this phenotype since the G448S and the H147Y/G448S substitutions resulted in the same resistance level. The G448 residue forms part of the HBR and CR5 conserved regions. The G448S substitution in *A. fumigatus* corresponds to the G484S and G464S amino acid change in the *Cryptococcus neoformans* and *Candida albicans* Erg11ps, respectively implicated in fluconazole resistance (Kelly *et al.*, 1999; Rodero *et al.*, 2003). In yeasts, G464 substitutions induce the reduction of both catalytic activity and azole binding capacity of Erg11p. Thus, it was suggested that a similar effect would account for voriconazole resistance in *A. fumigatus* (Pelaez *et al.*, 2012). Docking of azoles in the SDMs channels explains why G448 mutants show voriconazole resistance but itraconazole and posaconazole susceptibility. Effectively, G448 being located in one of the pocket sides, would only interact with voriconazole but not with the two other azole drugs (Fig. 2.10) (Manavathu *et al.*, 2000).

Azole resistance in non-fumigatus *Aspergillus*

As described for *A. fumigatus* all *Aspergillus* spp. are intrinsically resistant to fluconazole and almost all the susceptible to other azole agents. The most important species intrinsically resistant to other azole drugs than fluconazole are included in the sections *Fumigati* (*A. lentulus, Neosartorya pseudofischeri* and *A. viridinutans*) (Alcazar-Fuoli *et al.*, 2008a; Samson *et al.*, 2006), *Nigri* (*A. awamori, A. tubingensis, A. niger* and *A. acidus*) (Alcazar-Fuoli *et al.*, 2009; Howard *et al.*, 2011), *Usti* (*A. ustus* and *A. calidoustus*) (Alastruey-Izquierdo *et al.*, 2010; Samson *et al.*, 2011; Varga *et al.*, 2008) and *Circumdati* (*A. tanneri*) (Sugui *et al.*, 2012). The molecular mechanism of azole resistance was studied in *A. lentulus* by disrupting its *Cyp51A* gene. The deletion mutant showed no morphologic or virulence changes as compared with wild-type strain but became azole susceptible. The *A. lentulus* Cyp51Ap showed differences in 26 amino acid residues when compared with *A. fumigatus* Cyp51Ap (Mellado *et al.*, 2011). Most of these amino acid substitutions which could be found are located in the CR-4 region. They correspond to residues linked with posaconazole docking in the K α helix and 1–4 β sheet (I354V,

I360L and I367L) and also in the putative channel 2 entry (S49N and E488D) (Mellado *et al.*, 2011; Xiao *et al.*, 2004). This suggests that azole resistance in this species could be due to alteration in the docking of azoles in the SDMs channels.

Clinical isolates of *Aspergillus* belonging to the section *Nigri* showed itraconazole resistance in 36%, 90%, 33% and 100% of the strains of the *A. awamori*, *A. tubingensis*, *A. niger* and *A. acidus* groups, respectively (Alcazar-Fuoli *et al.*, 2009; Howard *et al.*, 2011). These phenotypes were not directly linked with mutations in *CYP51A* gene although some amino acid substitutions (e.g. G427S) described in azole-resistant *A. fumigatus* were present in some of the itraconazole-resistant *Aspergillus* section *Nigri* strains (Howard *et al.*, 2011).

Secondary azole resistance is rare in non-*fumigatus Aspergillus* species. *A. terreus* species complex and *A. flavus* are the second and third most common *Aspergillus* spp. causing invasive aspergillosis. However, there is only one description of azole resistance in each of the species. In both cases, resistance phenotype was related with amino acid substitutions in Cyp51p. In the *A. terreus* azole-resistant isolate, an M217I amino acid substitution was described in Cyp51Ap (Arendrup *et al.*, 2012). This residue is equivalent to the *A. fumigatus* Cyp51Ap M220 linked with azole resistance (see above) (Mellado *et al.*, 2004).

The voriconazole-resistant *A. flavus* isolate showed a substitution at Cyp51Cp (S240A). The implication of this mutation in azole resistance was confirmed by gene replacement (Liu *et al.*, 2012).

Azole secondary resistance linked with point mutations in the 14-α sterol demethylase enzyme in non-*Candida* and non-*Aspergillus* species

Candida spp. and *Aspergillus* spp. account for more than 70% of the invasive fungal infections (Pappas *et al.*, 2010). The third most common fungal human pathogen is *Cryptococcus* spp. followed by different non-*Aspergillus* moulds, endemic fungi and Mucormycetes (Pappas *et al.*, 2010). Acquired resistance in the less common fungal species is an even rarer phenomenon than in non-fumigatus *Aspergillus* species. However, a high percentage of the *Cryptococcus neoformans* (both var. *neoformans* and var. *grubii*) and *C. gattii* isolates showed fluconazole reduced susceptibility (Cordoba *et al.*, 2011; Espinel-Ingroff *et al.*, 2012). However, fluconazole treatment failure in patients with cryptococcosis cannot be fully attributed to point mutations in the *C. neoformans* 14- α sterol demethylase. In *C. neoformans* and *C. gattii*, hetero-resistance was suggested as an alternative mechanism of azole resistance. Hetero-resistance is a kind of intrinsic resistance where one single isolate shows a heterogeneous genetic composition. Most of the cells are susceptible to fluconazole but a subpopulation shows a high level of resistance (around 1% of the cell can be selected on 64 μg/ml fluconazole) (Cheong and McCormack, 2013; Mondon *et al.*, 1999; Varma and Kwon-Chung, 2010; Yamazumi *et al.*, 2003). Hetero-resistance was linked with disomy of the chromosome 1 (where *ERG11* and the major ABC-transporter associated with azole efflux in *C. neoformans* named *AFR1* are located) and chromosome 4 (where genes responsible for endoplasmic reticulum integrity as *YOP1*, *SEY1*, *GLO3*, and *GCS2* are placed). This mechanism of fluconazole resistance is induced by fluconazole exposure and is the most prevalent in clinical *C. neoformans* azole-resistant isolates (Sionov *et al.*, 2009, 2010; Ngamskulrungroj *et al.*, 2012).

Point mutations in *C. neoformans* Erg11p were also described as molecular mechanism of azole resistance. That is the case of a *C. neoformans* var. *grubii* strain isolated in Argentina from a patient with recurrent cryptococcosis (Rodero *et al.*, 2003). During a 15-month period, five episodes of cryptococcal meningitis were detected. The first episode was treated with amphotericin B and the other four with fluconazole. The five isolates were genetically identical and showed the same voriconazole, itraconazole and amphotericin B MIC. On the other hand, the fifth strain showed a 4-fold increase in fluconazole MIC (from 2 μg/ml to 16 μg/ml). This MIC increase was linked with a G484S amino acid substitution in a conserved region of the Erg11p (Rodero *et al.*, 2003). The described glycine residue corresponds to the G448 in *A. fumigatus* Cyp51Ap confirmed to be responsible for voriconazole resistance in *A. fumigatus* (Pelaez *et al.*, 2012). The Y145F substitution in Erg11p was also linked with azole resistance in a clinical *C. neoformans* var. *grubii* strain. This strain displays a fluconazole and voriconazole cross-resistance (128 µg/ml and 2 µg/ml, respectively) and an hypersensitivity to itraconazole and posaconazole. This strain carried five substitutions in Erg11p, but the introduction of the Y145F substitution in a wild-type strain was sufficient to reproduce the phenotype of the clinical isolate (Sionov *et al.*, 2012). The Y145 residue forms part of the putative channel 1 of the *C. neoformans* 14-α-sterol demethylase and is located in the loop that link the B' and C α helixes of the SDMs.

In the amphotericin B-resistant human pathogen *Trichosporon asahii*, a recent report showed that G453R substitution in Erg11p is responsible for an increase in fluconazole MIC (Kushima *et al.*, 2012). This glycine residue is equivalent to the G434 residue of the *A. fumigatus* Cyp51Ap described as a possible cause of azole resistance (Howard *et al.*, 2006, 2009; Manavathu *et al.*, 2000).

Taking all together, amino acid changes in the Erg11p seem to be a universal mechanism of resistance in fungi even in less commonly isolated pathogens.

Polyene resistance

Resistance to polyenes is rare. It was described in only some *Candida* spp., *Cryptococcus* spp., *A. fumigatus* and *A. flavus* strains (Dick *et al.*, 1980; Fryberg *et al.*, 1974; Kelly *et al.*, 1994; Manavathu *et al.*, 1998; Seo *et al.*, 1999; Sokol-Anderson *et al.*, 1986; Vincent *et al.*, 2013; Guinea *et al.*, 2013; Sharma *et al.*, 2014; Ren *et al.*, 2014). The best known mechanisms of polyene resistance involves the alteration of membrane sterol composition by the double loss of function in Erg3p (C-5 sterol desaturase) and Erg11p (14-α sterol demethylase) in *C. albicans*, by the alteration of the function of Erg2p (C-8 sterol isomerase) in *Cryptococcus neoformans* or by the loss of the activity of Erg2p and Erg6p (C-24 Sterol Methyl transferase) in *C. glabrata* and *C. lusitaniae* (Kelly *et al.*, 1994, 1997; Martel *et al.*, 2010; Sanglard *et al.*, 2003b; Hull *et al.*, 2012; Vandeputte *et al.*, 2008; Young *et al.*, 2003). Turning to *Aspergillus* spp., little is known about polyene acquired resistance. *A. fumigatus* laboratory mutants isolated by UV mutagenesis showed amphotericin B reduced susceptibility but no molecular study were performed to explain it (Manavathu *et al.*, 1998). Similarly, an *A. flavus* mutant showing amphotericin B resistance was isolated by serial transfers on agar plates containing polyene drugs. These strain's spheroplasts were susceptible to amphotericin B. Thus, the resistance phenotype was attributed to a peculiar fungal cell wall. In fact, cell wall chemically studies, showed a higher level of alkali-soluble and -insoluble glucans in the mutant than in the parental strain. This suggested that 1,3-α glucans and cell wall proteins were involved in the resistance phenotype (Seo *et al.*, 1999). The scarcity of polyene-resistant clinical strains was not understood until recently when Vincent *et al.* (2013) showed that amphotericin B resistance came at a great fitness cost to the cell. Amphotericin B-resistant mutants have a higher basal level of HSP90p (chaperone molecule required to tolerate cellular stresses) than polyene susceptible isolates. This fact leaves less HSP90 available in the mutant strains to tolerate external stresses as host immunity, oxidative stress, temperature variation, etc. explaining why these mutants are avirulent in animal.

Ergosterol synthesis pathway seems not the only pathway implicated in polyene resistance. Sharma *et al.* (2014) described the implication of sphingolipid biosynthetic pathway genes *FEN1* and *SUR4* in amphotericin B resistance in yeasts. Moreover, the mechanism of amphotericin B toxicity in *C. albicans* was explained as a consequence of the oxidative damage of the drug over the yeast. Cells could be protected from amphotericin B by exogenous catalase and/or superoxide dismutase (Sokol-Anderson *et al.*, 1986). This last mechanism was suggested to be responsible for the intrinsic reduced amphotericin B susceptibility of *A. terreus*. In this *Aspergillus* species, the resistance phenotype seems to be linked with the higher catalase activity of *A. terreus* when compared with *A. fumigatus*. Moreover, membrane sterol composition, cell-associated drug level, intracellular efflux and prooxidant effect were studied in two *A. terreus* strains differing in amphotericin B susceptibility. The resistant strain absorbed less antifungal and had better protection against oxidative stress. However, susceptible strain was more virulent in a murine model of infection (Blum *et al.*, 2008, 2013).

The limited numbers of amphotericin B-resistant isolates explain the scant attention that received the study of the genetic basis of polyene resistance. Nevertheless, we can conclude that amphotericin B resistance in *Aspergillus* spp. should be linked with the reduction of membrane ergosterol concentration, extracellular catalases activity and/or alteration in the fungal cell wall composition.

Resistance to allylamines

Allylamines, such as terbinafine, reduce ergosterol biosynthesis by inhibiting one of the earliest steps in the pathway named squalene epoxidase (SQE or Erg1p) (Ryder, 1985, 1992; Ryder and Dupont, 1985). Terbinafine is highly effective, both *in vivo* and *in vitro*, against dermatophytes and other agents of cutaneous mycoses (Ghannoum *et al.*, 2004). Also, it shows good *in vitro* activity against *Candida* spp., *Cryptococcus* spp., moulds and dimorphic fungi (Garcia-Effron *et al.*, 2004; Moore *et al.*, 2001; Ryder *et al.*, 1998). However, terbinafine has little clinical use in the treatment of deep mycosis because of its low volume of distribution in deep tissues such as brain, spleen and lungs (around 200-fold lower than in skin and adipose) (Hosseini-Yeganeh and McLachlan, 2002).

In *Saccharomyces cerevisiae*, SQE is encoded by *ERG1* (Jandrositz *et al.*, 1991). This flavoprotein shows two conserved domains, a N-terminal flavin adenine dinucleotide binding site (FAD-BS) and a monooxigenase domain (MOD). *S. cerevisiae* and *C. albicans* SQEs utilize NADP or NAD as redox cofactors, respectively for epoxidation of squalene (Ono *et al.*, 1982; Sakakibara *et al.*, 1995). Amino acid substitutions in the FAD-BS have been associated with hyper-susceptibility to terbinafine in *S. cerevisiae* (Leber *et al.*, 2003). On the other hand, resistance to terbinafine in this yeast was linked to modifications in the *ERG1* promoter (leading to over expression of Erg1p) and with mutation in the MOD and C-terminal part of the SQE (Leber *et al.*, 2001, 2003). Similarly, in *Trichophyton rubrum*, amino acid substitutions at L393F or F398L (C-term portion) of Erg1p leads to terbinafine resistance (Osborne *et al.*, 2005).

In *A. fumigatus*, Erg1p has 472 aa. In this species the FAD-BS comprises 27 amino acids (from the residue 30 to 57) while the MOD includes 139 aa (from the residue 187 to 326) (Liu *et al.*, 2004). In this species, *in vitro* terbinafine MICs shows a high range of values (from 0.06 to 16 mg/l) (Garcia-Effron *et al.*, 2004). However, MIC differences observed between different clinical strains seemed to have no direct relation with amino acid substitutions in Erg1p, *ERG1* extra copies, or gene overexpression (Garcia-Effron *et al.*, 2005a). However, the implication of Erg1p in *A. fumigatus* terbinafine resistance cannot be discarded. Using *A. fumigatus* laboratory mutants other authors demonstrated that mutations in Erg1p (F389L) and extra copies of this gene were associated with terbinafine resistance (Liu *et al.*, 2004b; Rocha *et al.*, 2006).

Terbinafine resistance in *A. nidulans* was analysed by Rocha and co-workers using a different approach (Rocha *et al.*, 2002). They established, through classical genetics, that at least two loci (named *tebA* and *tebB*) participate in a high level of terbinafine resistance phenotype. The same group demonstrated, in UV generated *A. nidulans* mutants, a causal relationship between overexpression of *SAL-A* gene, encoding a putative salicylate 1-monooxigenase, and antifungal resistance (Graminha *et al.*, 2004). Sal-Ap enzymes participate in the degradation pathway of salicylate-like aromatic compounds in prokaryotes (Bosch *et al.*, 1999). Terbinafine chemical structure harbours a naphthalene nucleus able to be degraded by Sap-Ap. All the presented results concern *in vitro* assessment of allylamine resistance and not true clinical resistance. Effectively, as terbinafine and other allylamines are currently essentially used for the treatment of superficial mycoses, mainly dermatophytosis, the only described molecular mechanism of resistance with clinical implications is the presence of mutations in *T. rubrum* Erg1p (Osborne *et al.*, 2005).

General conclusion

The emergence and spread of resistance to antifungal agents is the inevitable consequence of using these drugs for human, animal and plant fungal infections treatments. Antifungal resistance in non-*Candida* species is still a rare phenomenon but its prevalence is increasing over years. The study of the molecular mechanism underlying resistance to these agents will give us theoretical bases which will help in the design of diagnostic tools to rapidly and objectively detect resistant strains. This is crucial to develop new treatment strategies to prevent and control the spreading of antifungal resistance.

Concerning *Aspergillus* spp. and other non-*Candida* species, point mutations in the target enzymes appears clearly to be the main molecular mechanism of resistance to membrane-targeting antifungals. However, overexpression of efflux pumps genes is an emerging mechanism of azole resistance in *A. fumigatus*.

All these resistance mechanisms should be carefully studied, monitored and informed to delay worldwide dissemination of resistant strains since only few antifungal agents are available or in development.

References

Aaron, K.E., Pierson, C.A., Lees, N.D., and Bard,M. (2001). The *Candida albicans* ERG26 gene encoding the C-3 sterol dehydrogenase (C-4 decarboxylase) is essential for growth. FEMS Yeast Res. *1*, 93–101.

Alanio, A., Sitterle, E., Liance, M., Farrugia, C., Foulet, F., Botterel, F., Hicheri, Y., Cordonnier, C., Costa, J.M., and Bretagne, S. (2011). Low prevalence of resistance to azoles in *Aspergillus fumigatus* in a French cohort of patients treated for haematological malignancies. J. Antimicrob. Chemother. *66*, 371–374.

Alcazar-Fuoli, L., Mellado, E., Garcia-Effron, G., Buitrago, M.J., Lopez, J.F., Grimalt, J.O., Cuenca-Estrella, J.M., and Rodriguez-Tudela, J.L. (2006). *Aspergillus fumigatus* C-5 sterol desaturases Erg3A and Erg3B: role in sterol biosynthesis and antifungal drug susceptibility. Antimicrob. Agents Chemother. *50*, 453–460.

Alcazar-Fuoli, L., Mellado, E., Garcia-Effron, G., Lopez, J.F., Grimalt, J.O., Cuenca-Estrella, J.M., and Rodriguez-Tudela, J.L. (2008). Ergosterol biosynthesis pathway in *Aspergillus fumigatus*. Steroids 73, 339–347.

Anderson, T.M., Clay, M.C., Cioffi, A.G., Diaz, K.A., Hisao, G.S., Tuttle, M.D., Nieuwkoop, A.J., Comellas, G., Maryum, N., Wang, S., *et al.* (2014). Amphotericin forms an extramembranous and fungicidal sterol sponge. Nat. Chem. Biol. *10*, 400–406.

Aoyama, Y., Horiuchi, T., Gotoh, O., Noshiro, M., and Yoshida, Y. (1998). CYP51-like gene of *Mycobacterium tuberculosis* actually encodes a P450 similar to eukaryotic CYP51. J. Biochem. *124*, 694–696.

Aoyama, Y., Yoshida, Y., Sonoda, Y., and Sato, Y. (1989). Deformylation of 32-oxo-24,25-dihydrolanosterol by the purified cytochrome P-45014DM (lanosterol 14 alpha-demethylase) from yeast evidence confirming the intermediate step of lanosterol 14 alpha-demethylation. J. Biol. Chem. *264*, 18502–18505.

Arabatzis, M., Kambouris, M., Kyprianou, M., Chrysaki, A., Foustoukou, M., Kanellopoulou, M., Kondyli, L., Kouppari, G., Koutsia-Karouzou, C., Lebessi, E., *et al.* (2011). Polyphasic identification and susceptibility to seven antifungals of 102 Aspergillus isolates recovered from immunocompromised hosts in Greece. Antimicrob. Agents Chemother. *55*, 3025–3030.

Arendrup, M.C., Jensen, R.H., Grif, K., Skov, M., Pressler, T., Johansen, H.K., and Lass-Florl, C. (2012). *In vivo* emergence of *Aspergillus terreus* with reduced azole susceptibility and a Cyp51a M217I alteration. J. Infect. Dis. *206*, 981–985.

Ashman, W.H., Barbuch, R.J., Ulbright, C.E., Jarrett, H.W., and Bard, M. (1991). Cloning and disruption of the yeast C-8 sterol isomerase gene. Lipids *26*, 628–632.

Baddley, J.W., Marr, K.A., Andes, D.R., Walsh, T.J., Kauffman, C.A., Kontoyiannis, D.P., Ito, J.I., Balajee, S.A., Pappas, P.G., and Moser, S.A. (2009). Patterns of susceptibility of *Aspergillus* isolates recovered from patients enrolled in the Transplant-Associated Infection Surveillance Network. J. Clin. Microbiol. *47*, 3271–3275.

Bard, M., Bruner, D.A., Pierson, C.A., Lees, N.D., Biermann, B., Frye, L., Koegel, C., and Barbuch, R. (1996). Cloning and characterization of ERG25, the *Saccharomyces cerevisiae* gene encoding C-4 sterol methyl oxidase. Proc. Natl. Acad. Sci. U.S.A 93, 186–190.

Bard, M., Lees, N.D., Turi, T., Craft, D., Cofrin, L., Barbuch, R., Koegel, C., and Loper, J.C. (1993). Sterol synthesis and viability of erg11 (cytochrome P450 lanosterol demethylase) mutations in *Saccharomyces cerevisiae* and *Candida albicans*. Lipids *28*, 963–967.

Bellamine, A., Mangla, A.T., Nes, W.D., and Waterman, M.R. (1999). Characterization and catalytic properties of the sterol 14α-demethylase from *Mycobacterium tuberculosis*. Proc. Natl. Acad. Sci. U.S.A *96*, 8937–8942.

Benveniste, P. (2004). Biosynthesis and accumulation of sterols. Annu. Rev. Plant Biol. *55*, 429–457.

Blum, G., Hortnagl, C., Jukic, E., Erbeznik, T., Pumpel, T., Dietrich, H., Nagl, M., Speth, C., Rambach, G., and Lass-Florl, C. (2013). New insight into amphotericin B resistance in *Aspergillus terreus*. Antimicrob. Agents Chemother. *57*, 1583–1588.

Blum, G., Perknofer, S., Haas, H., Schrettl, M., Wurzner, R., Dierich, M.P., and Lass-Florl, C. (2008). Potential basis for amphotericin B resistance in *Aspergillus terreus*. Antimicrob. Agents Chemother. *52*,1553–1555.

Bosch, R., Moore, E.R., Garcia-Valdes, E., and Pieper, D.H. (1999). NahW, a novel, inducible salicylate hydroxylase involved in mineralization of naphthalene by *Pseudomonas stutzeri* AN10. J. Bacteriol. *181*, 2315–2322.

van den Brink, H.M., van Gorcom, R.F., van den Hondel, C.A., and Punt, P.J. (1998). Cytochrome P450 enzyme systems in fungi. Fungal. Genet. Biol. *23*, 1–17.

Bueid, A., Howard, S.J., Moore, C.B., Richardson, M.D., Harrison, E., Bowyer, P., and Denning, D.W. (2010). Azole antifungal resistance in *Aspergillus fumigatus*: 2008 and 2009. J. Antimicrob. Chemother. *65*, 2116–2118.

Burgel, P.R., Baixench, M.T., Amsellem, M., Audureau, E., Chapron, J., Kanaan, R., Honore, I., Dupouy-Camet, J., Dusser, D., Klaassen, C.H., *et al.* (2012). High prevalence of azole-resistant *Aspergillus fumigatus* in adults with cystic fibrosis exposed to itraconazole. Antimicrob. Agents Chemother. *56*, 869–874.

Chen, C., Kalb, V.F., Turi, T.G., and Loper, J.C. (1988). Primary structure of the cytochrome P450 lanosterol 14 alpha-demethylase gene from *Candida tropicalis*. DNA *7*, 617–626.

Chowdary, A., Kathuria, S., Randhawa, H.S., Gaur, S.N., Klaussen, C.H., and Meis, J.F. (2012). Isolation of multiple-triazole-resistant *Aspergillus fumigatus* strains carrying the TR/L98H mutations in the cyc51A gene in India. J. Antimicrob. Chemother. *67*, 362–366.

Chowdhary, A., Sharma, C., Kathuria, S., Hagen, F., and Meis, J.F. (2014). Azole-resistant *Aspergillus fumigatus* with the environmental TR46/Y121F/T298A mutation in India. J. Antimicrob. Chemother. *69*, 555–557.

Dannaoui, E., Persat, F., Monier, M.F., Borel, E., Piens, M.A., and Picot, S. (1999). *In vitro* susceptibility of *Aspergillus* spp. isolates to amphotericin B and itraconazole. J. Antimicrob. Chemother. *44*, 553–555.

Delye, C., Bousset, L., and Corio-Costet, M.F. (1998). PCR cloning and detection of point mutations in the eburicol 14α-demethylase (CYP51) gene from *Erysiphe graminis* f. sp. *hordei*, a "recalcitrant" fungus. Curr. Genet. *34*, 399–403.

Denning, D.W., Venkateswarlu, K., Oakley, K.L., Anderson, M.J., Manning, N.J., Stevens, D.A., Warnock, D.W., and Kelly, S.L. (1997). Itraconazole resistance in *Aspergillus fumigatus*. Antimicrob. Agents Chemother. *41*, 1364–1368.

Diaz-Guerra, T.M., Mellado, E., Cuenca-Estrella, M., and Rodriguez-Tudela, J.L. (2003). A point mutation in the 14α-sterol demethylase gene cyp51A contributes to itraconazole resistance in *Aspergillus fumigatus*. Antimicrob. Agents Chemother. *47*, 1120–1124.

Dick, J.D., Merz, W.G., and Saral, R. (1980). Incidence of polyene-resistant yeasts recovered from clinical specimens. Antimicrob. Agents Chemother. *18*, 158–163.

Dimster-Denk, D., and Rine, J. (1996). Transcriptional regulation of a sterol-biosynthetic enzyme by sterol levels in *Saccharomyces cerevisiae*. Mol. Cell Biol. *16*, 3981–3989.

Edlind, T.D., Henry, K.W., Metera, K.A., and Katiyar, S.K. (2001). *Aspergillus fumigatus* CYP51 sequence: potential basis for fluconazole resistance. Med. Mycol. *39*, 299–302.

Fraczek, M.G., Bromley, M., Buied, A., Moore, C.B., Rajendran, R., Rautemaa, R., Ramage, G., Denning, D.W., and Bowyer, P. (2013). The cdr1B efflux transporter is associated with non-cyp51a-mediated itraconazole resistance in *Aspergillus fumigatus*. J. Antimicrob. Chemother. *68*, 1486–1496.

Fryberg, M., Oehlschlager, A.C., and Unrau, A.M. (1973). Biosynthesis of ergosterol in yeast. Evidence for multiple pathways. J. Am. Chem. Soc. *95*, 5747–5757.

Gaber, R.F., Copple, D.M., Kennedy, B.K., Vidal, M., and Bard, M. (1989). The yeast gene ERG6 is required for normal membrane function but is not essential for biosynthesis of the cell-cycle-sparking sterol. Mol. Cell Biol. *9*, 3447–3456.

Gachotte, D., Barbuch, R., Gaylor, J., Nickel, E., and Bard, M. (1998). Characterization of the *Saccharomyces cerevisiae* ERG26 gene encoding the C-3 sterol dehydrogenase (C-4 decarboxylase) involved in sterol biosynthesis. Proc. Natl. Acad. Sci. U.S.A *95*, 13794–13799.

Gachotte, D., Sen, S.E., Eckstein, J., Barbuch, R., Krieger, M., Ray, B.D., and Bard, M. (1999). Characterization of the *Saccharomyces cerevisiae* ERG27 gene encoding the 3-keto reductase involved in C-4 sterol demethylation. Proc. Natl. Acad. Sci. U.S.A *96*, 12655–12660.

Garcia-Effron, G., Dilger, A., Alcazar-Fuoli, L., Park, S., Mellado, E., and Perlin, D.S. (2008). Rapid detection of triazole antifungal resistance in *Aspergillus fumigatus*. J. Clin. Microbiol. *46*, 1200–1206.

Garcia-Effron, G., Gomez-Lopez, A., Mellado, E., Monzon, A., Rodriguez-Tudela, J.L., and Cuenca-Estrella, M. (2004). *In vitro* activity of terbinafine against medically important non-dermatophyte species of filamentous fungi. J. Antimicrob. Chemother. *53*, 1086–1089.

Garcia-Effron, G., Mellado, E., Alcazar-Fuoli, L., Buitrago, M.J., Cuenca-Estrella, M., and Rodriguez-Tudela, J.L. (2005a). Role of *Aspergillus fumigatus* 14- alpha sterol demethylase (*cyp*51B) gene on cell growth and viability, antifungal susceptibility and sterol composition. 45th International Conference on Antimicrobial Agents and Chemotherapy (Washington DC, USA).

Garcia-Effron, G., Mellado, E., Gomez-Lopez, A., Alcazar-Fuoli, L., Cuenca-Estrella, M., and Rodriguez-Tudela, J.L. (2005b). Differences in interactions between azole drugs related to modifications in the 14-alpha sterol demethylase gene (cyp51A) of *Aspergillus fumigatus*. Antimicrob. Agents Chemother. *49*, 2119–2121.

Geber, A., Hitchcock, C.A., Swartz, J.E., Pullen, F.S., Marsden, K.E., Kwon-Chung, K.J., and Bennett, J.E. (1995). Deletion of the Candida glabrata ERG3 and ERG11 genes: effect on cell viability, cell growth, sterol composition, and antifungal susceptibility. Antimicrob. Agents Chemother. *39*, 2708–2717.

Ghannoum, M.A., Chaturvedi, V., Espinel-Ingroff, A., Pfaller, M.A., Rinaldi, M.G., Lee-Yang, W., and Warnock, D.W. (2004). Intra- and interlaboratory study of a method for testing the antifungal susceptibilities of dermatophytes. J. Clin. Microbiol. *42*, 2977–2979.

Ghannoum, M.A., and Rice, L.B. (1999). Antifungal agents: mode of action, mechanisms of resistance, and correlation of these mechanisms with bacterial resistance. Clin. Microbiol. Rev. *12*, 501–517.

Gollapudy, R., Ajmani, S., and Kulkarni, S.A. (2004). Modeling and interactions of *Aspergillus fumigatus* lanosterol 14-alpha demethylase 'A' with azole antifungals. Bioorg. Med. Chem. *12*, 2937–2950.

Gray, K.C., Palacios, D.S., Dailey, I., Endo, M.M., Uno, B.E., Wilcock, B.C., and Burke, M.D. (2012). Amphotericin primarily kills yeast by simply binding ergosterol. Proc. Natl. Acad. Sci. U.S.A. *109*, 2234–2239.

Gregson, L., Goodwin, J., Johnson, A., McEntee, L., Moore, C.B., Richardson, M., Hope, W.W., and Howard, S.J. (2013). *In vitro* susceptibility of *Aspergillus fumigatus* to isavuconazole: correlation with itraconazole, voriconazole, and posaconazole. Antimicrob. Agents Chemother. *57*, 5778–5780.

Hargreaves, J.A., and Keon, J.P. (1996). Isolation of an *Ustilago maydis* ERG11 gene and its expression in a mutant deficient in sterol 14 alpha-demethylase activity. FEMS Microbiol. Lett. *139*, 203–207.

Harmouch, N., Coulon, J., and Bonaly, R. (1995). Identification of 24-methylene-24,25-dihydrolanosterol as a precursor of ergosterol in the yeasts *Schizosaccharomyces pombe* and *Schizosaccharomyces octosporus*. FEMS Microbiol. Lett. *134*, 147–152.

Hosseini-Yeganeh, M., and McLachlan, A.J. (2002). Physiologically based pharmacokinetic model for terbinafine in rats and humans. Antimicrob. Agents Chemother. *46*, 2219–2228.

Howard, S.J., and Arendrup, M.C. (2011). Acquired antifungal drug resistance in Aspergillus fumigatus: epidemiology and detection. Med. Mycol. *49* (Suppl. 1), S90-S95.

Howard, S.J., Cerar, D., Anderson, M.J., Albarrag, A., Fisher, M.C., Pasqualotto, A.C., Laverdiere, M., Arendrup, M.C., Perlin, D.S., and Denning, D.W. (2009). Frequency and evolution of azole resistance in *Aspergillus fumigatus* associated with treatment failure. Emerg. Infect. Dis. *15*, 1068–1076.

Howard, S.J., Webster, I., Moore, C.B., Gardiner, R.E., Park, S., Perlin, D.S., and Denning, D.W. (2006). Multi-azole resistance in *Aspergillus fumigatus*. Int. J. Antimicrob. Agents *28*, 450–453.

Jandrositz, A., Turnowsky, F., and Hogenauer, G. (1991). The gene encoding squalene epoxidase from *Saccharomyces cerevisiae*: cloning and characterization. Gene *107*, 155–160.

Jensen-Pergakes, K.L., Kennedy, M.A., Lees, N.D., Barbuch, R., Koegel, C., and Bard, M. (1998). Sequencing, disruption, and characterization of the *Candida albicans* sterol methyltransferase (ERG6) gene: drug susceptibility studies in erg6 mutants. Antimicrob. Agents Chemother. *42*, 1160–1167.

Jia, N., Arthington-Skaggs, B., Lee, W., Pierson, C.A., Lees, N.D., Eckstein, J., Barbuch, R., and Bard, M. (2002). *Candida albicans* sterol C-14 reductase, encoded by the ERG24 gene, as a potential antifungal target site. Antimicrob. Agents Chemother. *46*, 947–957.

Kalb, V.F., Woods, C.W., Turi, T.G., Dey, C.R., Sutter, T.R., and Loper, J.C. (1987). Primary structure of the P450 lanosterol demethylase gene from *Saccharomyces cerevisiae*. DNA *6*, 529–537.

Kelly, S.L., Lamb, D.C., Taylor, M., Corran, A.J., Baldwin, B.C., and Powderly, W.G. (1994). Resistance to amphotericin B associated with defective sterol delta 8-->7 isomerase in a *Cryptococcus neoformans* strain from an AIDS patient. FEMS Microbiol. Lett. *122*, 39–42.

Kelly, S.L., Lamb, D.C., Kelly, D.E., Manning, N.J., Loeffler, J., Hebart, H., Schumacher, U., and Einsele, H. (1997). Resistance to fluconazole and cross-resistance to amphotericin B in *Candida albicans* from AIDS patients caused by defective sterol delta 5, 6-desaturation FEBS Lett. *400*, 80–82.

Keon, J.P., James, C.S., Court, S., Baden-Daintree, C., Bailey, A.M., Burden, R.S., Bard, M., and Hargreaves, J.A. (1994). Isolation of the ERG2 gene, encoding sterol delta 8-->delta 7 isomerase, from the rice blast fungus *Magnaporthe grisea* and its expression in the maize smut pathogen *Ustilago maydis*. Curr. Genet. *25*, 531–537.

Lai, M.H., Bard, M., Pierson, C.A., Alexander, J.F., Goebl, M., Carter, G.T., and Kirsch, D.R. (1994). The identification of a gene family in the *Saccharomyces cerevisiae* ergosterol biosynthesis pathway. Gene *140*, 41–49.

Lamb, D.C., Kelly, D.E., Schunck, W.H., Shyadehi, A.Z., Akhtar, M., Lowe, D.J., Baldwin, B.C., and Kelly, S.L. (1997). The mutation T315A in *Candida albicans* sterol 14alpha-demethylase causes reduced enzyme activity and fluconazole resistance through reduced affinity. J. Biol. Chem. 272, 5682–5688.

Lamb, D.C., Maspahy, S., Kelly, D.E., Manning, N.J., Geber, A., Bennett, J.E., and Kelly, S.L. (1999). Purification, reconstitution, and inhibition of cytochrome P-450 sterol delta22-desaturase from the pathogenic fungus *Candida glabrata*. Antimicrob. Agents Chemother. *43*, 1725–1728.

Leber, R., Zenz, R., Schrottner, K., Fuchsbichler, S., Puhringer, B., and Turnowsky, F. (2001). A novel sequence element is involved in the transcriptional regulation of expression of the ERG1 (squalene epoxidase) gene in *Saccharomyces cerevisiae*. Eur. J. Biochem. *268*, 914–924.

Leber, R., Fuchsbichler, S., Klobucnikova, V., Schweighofer, N., Pitters, E., Wohlfarter, K., Lederer, M., Landl, K., Ruckenstuhl, C., Hapala, I., and Turnowsky, F. (2003). Molecular mechanism of terabinafine resistance in *Saccharomyces cerevisiae*. Antimicrob. Agents Chemother. *47*, 3890–3900.

Lees, N.D., Skaggs, B., Kirsch, D.R., and Bard, M. (1995). Cloning of the late genes in the ergosterol biosynthetic pathway of *Saccharomyces cerevisiae*--a review. Lipids *30*, 221–226.

van der Linden, J.W., Arendrup, M.C., Verweij, P.E., and SCARE-Network. (2011a). Prospective international surveillance of azole resistance in *Aspergillus fumigatus* (SCARE-Network). Conference Proceeding at: International Conference on Antimicrobial Agents and Chemotherapy (ICAAC, Chicago, IL, USA).

van der Linden, J.W., Snelders, E., Kampinga, G.A., Rijnders, B.J., Mattsson, E., Debets-Ossenkopp, Y.J., Kuijper, E.J., van Tiel, F.H., Melchers, W.J., and Verweij, P.E. (2011b). Clinical implications of azole resistance in *Aspergillus fumigatus*, The Netherlands, 2007–2009. Emerg. Infect. Dis. *17*, 1846–1854.

Liu, W., May, G.S., Kontoyiannis, D.P., and Li, R.Y. (2004). [Cloning of the *Aspergillus fumigatus* squalene epoxidase gene]. Beijing Da. Xue. Xue. Bao. *36*, 476–482.

Liu, W., Sun, Y., Chen, W., Liu, W., Wan, Z., Bu, D., and Li, R. (2012). The T788G mutation in the cyp51C gene confers voriconazole resistance in *Aspergillus flavus* causing aspergillosis. Antimicrob. Agents Chemother. *56*, 2598–2603.

Lockhart, S.R., Frade, J.P., Etienne, K.A., Pfaller, M.A., Diekema, D.J., and Balajee, S.A. (2011). Azole resistance in *Aspergillus fumigatus* isolates from the ARTEMIS global surveillance study is primarily due to the TR/L98H mutation in the cyp51A gene. Antimicrob. Agents Chemother. *55*, 4465–4468.

Lorenz, R.T., and Parks, L.W. (1992). Cloning, sequencing, and disruption of the gene encoding sterol C-14 reductase in *Saccharomyces cerevisiae*. DNA Cell Biol. *11*, 685–692.

Macchiarulo, A., Costantino, G., Fringuelli, D., Vecchiarelli, A., Schiaffella, F., and Fringuelli, R. (2002). 1,4-Benzothiazine and 1,4-benzoxazine imidazole derivatives with antifungal activity: a docking study. Bioorg. Med. Chem. *10*, 3415–3423.

Manavathu, E.K., Alangaden, G.J., and Chandrasekar, P.H. (1998). *In vitro* isolation and antifungal susceptibility of amphotericin B-resistant mutants of *Aspergillus fumigatus*. J. Antimicrob. Chemother. *41*, 615–619.

Manavathu, E.K., Cutright, J.L., Loebenberg, D., and Chandrasekar, P.H. (2000). A comparative study of the *in vitro* susceptibilities of clinical and laboratory-selected resistant isolates of *Aspergillus* spp. to amphotericin B, itraconazole, voriconazole and posaconazole (SCH 56592). J. Antimicrob. Chemother. *46*, 229–234.

Mann, P.A., Parmegiani, R.M., Wei, S.Q., Mendrick, C.A., Li, X., Loebenberg, D., DiDomenico, B., Hare, R.S., Walker, S.S., and McNicholas, P.M. (2003). Mutations in *Aspergillus fumigatus* resulting in reduced susceptibility to posaconazole appear to be restricted to a single amino acid in the cytochrome P450 14alpha-demethylase. Antimicrob. Agents Chemother. *47*, 577–581.

Marichal, P., Koymans, L., Willemsens, S., Bellens, D., Verhasselt, P., Luyten, W., Borgers, M., Ramaekers, F.C., Odds, F.C., and Bossche, H.V. (1999). Contribution of mutations in the cytochrome P450 14alpha-demethylase (Erg11p, Cyp51p) to azole resistance in *Candida albicans*. Microbiology *145* (Pt 10), 2701–2713.

Mejanelle, L., Lopez, J.F., Gunde-Cimerman, N., and Grimalt, J.O. (2001). Ergosterol biosynthesis in novel melanized fungi from hypersaline environments. J. Lipid Res. *42*, 352–358.

Mellado, E., Diaz-Guerra, T.M., Cuenca-Estrella, M., and Rodriguez-Tudela, J.L. (2001). Identification of two different 14-alpha sterol demethylase-related genes (cyp51A and cyp51B) in *Aspergillus fumigatus* and other *Aspergillus* species. J. Clin. Microbiol. *39*, 2431–2438.

Mellado, E., Garcia-Effron, G., Alcazar-Fuoli, L., Cuenca-Estrella, M., and Rodriguez-Tudela, J.L. (2004). Substitutions at methionine 220 in the 14alpha-sterol demethylase (Cyp51A) of *Aspergillus fumigatus* are responsible for resistance *in vitro* to azole antifungal drugs. Antimicrob. Agents Chemother. *48*, 2747–2750.

Mellado, E., Garcia-Effron, G., Buitrago, M.J., Alcazar-Fuoli, L., Cuenca-Estrella, M., and Rodriguez-Tudela, J.L. (2005). Targeted gene disruption of the 14-alpha sterol demethylase (cyp51A) in *Aspergillus fumigatus* and its role in azole drug susceptibility. Antimicrob. Agents Chemother. *49*, 2536–2538.

Mellado, E., Garcia-Effron, G., Alcazar-Fuoli, L., Melchers, W.J., Verweij, P.E., Cuenca-Estrella, M., and Rodriguez-Tudela, J.L. (2007). A new *Aspergillus fumigatus* resistance mechanism conferring *in vitro* cross-resistance to azole antifungals involves a combination of cyp51A alterations. Antimicrob. Agents Chemother. *51*, 1897–1904.

Mo, C., Valachovic, M., Randall, S.K., Nickels, J.T., and Bard, M. (2002). Protein–protein interactions among C-4 demethylation enzymes involved in yeast sterol biosynthesis. Proc. Natl. Acad. Sci. U.S.A. *99*, 9739–9744.

Monk, B.C., Tomasiak, T.M., Keniya, M.V., Huschmann, F.U., Tyndall, J.D., O'Connell, J.D., III, Cannon, R.D., McDonald, J.G., Rodriguez, A., Finer-Moore, J.S., and Stroud, R.M. (2014). Architecture of a single membrane spanning cytochrome P450 suggests constraints that orient the catalytic domain relative to a bilayer. Proc. Natl. Acad. Sci. U.S.A. *111*, 3865–3870.

Morio, F., Aubin, G.G., Danner-Boucher, I., Haloun, A., Sacchetto, E., Garcia-Hermoso, D., Bretagne, S., Miegeville, M., and Le, P.P. (2012). High prevalence of triazole resistance in *Aspergillus fumigatus*, especially mediated by TR/L98H, in a French cohort of patients with cystic fibrosis. J. Antimicrob. Chemother. *67*, 1870–1873.

Morris, G.M., and Richards, W.G. (1991). Molecular modelling of the sterol C-14 demethylase of *Saccharomyces cerevisiae*. Biochem. Soc. Trans. *19*, 793–795.

Mortensen, K.L., Jensen, R.H., Johansen, H.K., Skov, M., Pressler, T., Howard, S.J., Leatherbarrow, H., Mellado, E., and Arendrup, M.C. (2011). *Aspergillus* species and other molds in respiratory samples from patients with cystic fibrosis: a laboratory-based study with focus on *Aspergillus fumigatus* azole resistance. J. Clin. Microbiol. *49*, 2243–2251.

Mosquera, J., and Denning, D.W. (2002). Azole cross-resistance in *Aspergillus fumigatus*. Antimicrob. Agents Chemother. *46*, 556–557.

Nascimento, A.M., Goldman, G.H., Park, S., Marras, S.A., Delmas, G., Oza, U., Lolans, K., Dudley, M.N., Mann, P.A., and Perlin, D.S. (2003). Multiple resistance mechanisms among *Aspergillus fumigatus* mutants with high-level resistance to itraconazole. Antimicrob. Agents Chemother. *47*, 1719–1726.

Nes, W.D., Xu, S.H., and Haddon, W.F. (1989). Evidence for similarities and differences in the biosynthesis of fungal sterols. Steroids *53*, 533–558.

Nes, W.D., Zhou, W., Dennis, A.L., Li, H., Jia, Z., Keith, R.A., Piser, T.M., and Furlong, S.T. (2002). Purification, characterization and catalytic properties of human sterol 8-isomerase. Biochem. J. *367*, 587–599.

van Nistelrooy, J.G., van den Brink, J.M., van Kan, J.A., van Gorcom, R.F., and de Waard, M.A. (1996). Isolation and molecular characterisation of the gene encoding eburicol 14 alpha-demethylase (cYP51) from *Penicillium italicum*. Mol. Gen. Genet. *250*, 725–733.

Parks, L.W., Bottema, C.D., Rodriguez, R.J., and Lewis, T.A. (1985). Yeast sterols: yeast mutants as tools for the study of sterol metabolism. Methods Enzymol. *111*, 333–346.

Pfaller, M., Boyken, L., Hollis, R., Kroeger, J., Messer, S., Tendolkar, S., and Diekema, D. (2011). Use of epidemiological cutoff values to examine 9-year trends in susceptibility of *Aspergillus* species to the triazoles. J. Clin. Microbiol. *49*, 586–590.

Pfaller, M.A., Diekema, D.J., Ghannoum, M.A., Rex, J.H., Alexander, B.D., Andes, D., Brown, S.D., Chaturvedi, V., Espinel-Ingroff, A., Fowler, C.L., *et al.* (2009). Wild-type MIC distribution and epidemiological cutoff values for *Aspergillus fumigatus* and three triazoles as determined by the Clinical and Laboratory Standards Institute broth microdilution methods. J. Clin. Microbiol. *47*, 3142–3146.

Podust, L.M., Poulos, T.L., and Waterman, M.R. (2001a). Crystal structure of cytochrome P450 14alpha-sterol demethylase (CYP51) from *Mycobacterium tuberculosis* in complex with azole inhibitors. Proc. Natl. Acad. Sci. U.S.A. *98*, 3068–3073.

Podust, L.M., Stojan, J., Poulos, T.L., and Waterman, M.R. (2001b). Substrate recognition sites in 14alpha-sterol demethylase from comparative analysis of amino acid sequences and X-ray structure of *Mycobacterium tuberculosis* CYP51. J. Inorg. Biochem. *87*, 227–235.

Podust, L.M., von Kries, J.P., Eddine, A.N., Kim, Y., Yermalitskaya, L.V., Kuehne, R., Ouellet, H., Warrier, T., Altekoster, M., Lee, J.S., *et al.* (2007). Small-molecule scaffolds for CYP51 inhibitors identified by high-throughput screening and defined by X-ray crystallography. Antimicrob. Agents Chemother. *51*, 3915–3923.

Revankar, S.G., Fu, J., Rinaldi, M.G., Kelly, S.L., Kelly, D.E., Lamb, D.C., Keller, S.M., and Wickes, B.L. (2004). Cloning and characterization of the lanosterol 14alpha-demethylase (ERG11) gene in *Cryptococcus neoformans*. Biochem. Biophys. Res. Commun. *324*, 719–728.

Rocha, E.M., Almeida, C.B., and Martinez-Rossi, N.M. (2002). Identification of genes involved in terbinafine resistance in *Aspergillus nidulans*. Lett. Appl. Microbiol. *35*, 228–232.

Rodriguez, R.J., Low, C., Bottema, C.D., and Parks, L.W. (1985). Multiple functions for sterols in *Saccharomyces cerevisiae*. Biochim. Biophys. Acta *837*, 336–343.

Rodriguez-Tudela, J.L., Alcazar-Fuoli, L., Mellado, E., Alastruey-Izquierdo, A., Monzon, A., and Cuenca-Estrella, M. (2008). Epidemiological cutoffs and cross-resistance to azole drugs in *Aspergillus fumigatus*. Antimicrob. Agents Chemother. *52*, 2468–2472.

Ruan, B., Lai, P.S., Yeh, C.W., Wilson, W.K., Pang, J., Xu, R., Matsuda, S.P., and Schroepfer, G.J., Jr. (2002). Alternative pathways of sterol synthesis in yeast. Use of C(27) sterol tracers to study aberrant double-bond migrations and evaluate their relative importance. Steroids *67*, 1109–1119.

Ryder, N.S. (1985). Effect of allylamine antimycotic agents on fungal sterol biosynthesis measured by sterol side-chain methylation. J. Gen. Microbiol. *131*, 1595–1602.

Ryder, N.S. (1992). Terbinafine: mode of action and properties of the squalene epoxidase inhibition. Br. J. Dermatol. *126* (Suppl. 39), 2–7.

Ryder, N.S., and Dupont, M.C. (1985). Inhibition of squalene epoxidase by allylamine antimycotic compounds. A comparative study of the fungal and mammalian enzymes. Biochem J. *230*, 765–770.

Sakakibara, J., Watanabe, R., Kanai, Y., and Ono, T. (1995). Molecular cloning and expression of rat squalene epoxidase. J. Biol. Chem. *270*, 17–20.

Sanglard, D., Ischer, F., Koymans, L., and Bille, J. (1998). Amino acid substitutions in the cytochrome P-450 lanosterol 14alpha-demethylase (CYP51A1) from azole-resistant *Candida albicans* clinical isolates contribute to resistance to azole antifungal agents. Antimicrob. Agents Chemother. *42*, 241–253.

Sanglard, D., Ischer, F., Parkinson, T., Falconer, D., and Bille, J. (2003). *Candida albicans* mutations in the ergosterol biosynthetic pathway and resistance to several antifungal agents. Antimicrob. Agents Chemother. *47*, 2404–2412.

Seo, K., Akiyoshi, H., and Ohnishi, Y. (1999). Alteration of cell wall composition leads to amphotericin B resistance in *Aspergillus flavus*. Microbiol. Immunol. *43*, 1017–1025.

Sharma, S., Alfatah, M., Bari, V.K., Rawal, Y., Paul, S., and Ganesan, K. (2014). Sphingolipid biosynthetic pathway genes FEN1 and SUR4 modulate amphotericin B resistance. Antimicrob. Agents Chemother.

da Silva Ferreira, M.E., Capellaro, J.L., dos Reis, M.E., Malavazi, I., Perlin, D., Park, S., Anderson, J.B., Colombo, A.L., Arthington-Skaggs, B.A., Goldman, M.H., and Goldman, G.H. (2004). *In vitro* evolution of itraconazole resistance in *Aspergillus fumigatus* involves multiple mechanisms of resistance. Antimicrob. Agents Chemother. *48*, 4405–4413.

Sionov, E., Chang, Y.C., Garraffo, H.M., and Kwon-Chung, K.J. (2009). Heteroresistance to fluconazole in *Cryptococcus neoformans* is intrinsic and associated with virulence. Antimicrob. Agents Chemother. *53*, 2804–2815.

Sionov, E., Lee, H., Chang, Y.C., and Kwon-Chung, K.J. (2010) *Cryptococcus neoformans* overcomes stress of azole drugs by formation of disomy in specific multiple chromosomes. PLoS Pathog. *6*, e1000848.

Sionov, E., Chang, Y.C., Garraffo, H.M., Dolan, M.A., Ghannoum, M.A., and Kwon-Chung, K.J. (2012). Identification of a *Cryptococcus neoformans* cytochrome P450 lanosterol 14 alpha-demethylase (Erg11) residue critical for differential susceptibility between fluconazole/voriconazole and itraconazole/posaconazole. Antimicrob. Agents Chemother. *56*, 1162–1169.

Skaggs, B.A., Alexander, J.F., Pierson, C.A., Schweitzer, K.S., Chun, K.T., Koegel, C., Barbuch, R., and Bard, M. (1996). Cloning and characterization of the *Saccharomyces cerevisiae* C-22 sterol desaturase gene, encoding a second cytochrome P-450 involved in ergosterol biosynthesis. Gene *169*, 105–109.

Slaven, J.W., Anderson, M.J., Sanglard, D., Dixon, G.K., Bille, J., Roberts, I.S., and Denning, D.W. (2002). Increased expression of a novel *Aspergillus fumigatus* ABC transporter gene, atrF, in the presence of itraconazole in an itraconazole resistant clinical isolate. Fungal. Genet. Biol. *36*, 199–206.

Snelders, E., van der Lee, H.A., Kuijpers, J., Rijs, A.J., Varga, J., Samson, R.A., Mellado, E., Donders, A.R., Melchers, W.J., and Verweij, P.E. (2008). Emergence of azole resistance in *Aspergillus fumigatus* and spread of a single resistance mechanism. PLoS Med. *5*, e219.

Sokol-Anderson, M.L., Brajtburg, J., and Medoff, G. (1986). Amphotericin B-induced oxidative damage and killing of *Candida albicans*. J. Infect. Dis. *154*, 76–83.

Tashiro, M., Izumikawa, K., Hirano, K., Ide, S., Mihara, T., Hosogaya, N., Takazono, T., Morinaga, Y., Nakamura, S., Kurihara, S., *et al.* (2012). Correlation between triazole treatment history and susceptibility in clinically isolated *Aspergillus fumigatus*. Antimicrob. Agents Chemother. *56*, 4870–4875.

Trzaskos, J.M., Fischer, R.T., and Favata, M.F. (1986). Mechanistic studies of lanosterol C-32 demethylation. Conditions which promote oxysterol intermediate accumulation during the demethylation process. J. Biol. Chem. *261*, 16937–16942.

Vanden Bossche, H., Marichal, P., Le, J.L., Coene, M.C., Gorrens, J., and Cools, W. (1993). Effects of itraconazole on cytochrome P-450-dependent sterol 14 alpha-demethylation and reduction of 3-ketosteroids in *Cryptococcus neoformans*. Antimicrob. Agents Chemother. *37*, 2101–2105.

Verweij, P.E., Mellado, E., and Melchers, W.J. (2007). Multiple-triazole-resistant aspergillosis. N. Engl. J. Med. *356*, 1481–1483.

Verweij, P.E., Te Dorsthorst, D.T., Rijs, A.J., De Vries-Hospers, H.G., and Meis, J.F. (2002). Nationwide survey of *in vitro* activities of itraconazole and voriconazole against clinical *Aspergillus fumigatus* isolates cultured between 1945 and 1998. J. Clin. Microbiol. *40*, 2648–2650.

Verweij, P.E., Howard, S.J., Melchers, W.J., and Denning, D.W. (2009). Azole-resistance in *Aspergillus*: proposed nomenclature and breakpoints. Drug Resist. Updat. *12*, 141–147.

Vincent, B.M., Lancaster, A.K., Scherz-Shouval, R., Whitesell, L., and Lindquist, S. (2013). Fitness trade-offs restrict the evolution of resistance to amphotericin B. PLoS Biol. *11*, e1001692.

Warrilow, A.G., Melo, N., Martel, C.M., Parker, J.E., Nes, W.D., Kelly, S.L., and Kelly, D.E. (2010). Expression, purification, and characterization of *Aspergillus fumigatus* sterol 14-alpha demethylase (CYP51) isoenzymes A and B. Antimicrob. Agents Chemother. *54*, 4225–4234.

Waterman, M.R., and Lepesheva, G.I. (2005). Sterol 14 alpha-demethylase, an abundant and essential mixed-function oxidase. Biochem. Biophys. Res. Commun. *338*, 418–422.

Wester, M.R., Johnson, E.F., Marques-Soares, C., Dansette, P.M., Mansuy, D., and Stout, C.D. (2003). Structure of a substrate complex of mammalian cytochrome P450 2C5 at 2.3 A resolution: evidence for multiple substrate binding modes. Biochemistry *42*, 6370–6379.

White, T.C., Marr, K.A., and Bowden, R.A. (1998). Clinical, cellular, and molecular factors that contribute to antifungal drug resistance. Clin. Microbiol. Rev. *11*, 382–402.

Xiao, L., Madison, V., Chau, A.S., Loebenberg, D., Palermo, R.E., and McNicholas, P.M. (2004). Three-dimensional models of wild-type and mutated forms of cytochrome P450 14alpha-sterol demethylases from *Aspergillus fumigatus* and *Candida albicans* provide insights into posaconazole binding. Antimicrob. Agents Chemother. *48*, 568–574.

Zhou, W., Nguyen, T.T., Collins, M.S., Cushion, M.T., and Nes, W.D. (2002). Evidence for multiple sterol methyl transferase pathways in *Pneumocystis carinii*. Lipids *37*, 1177–1186.

Echinocandins: Resistance Mechanisms

3

Santosh Katiyar and Thomas Edlind

Abstract

Echinocandins, including caspofungin, micafungin, and anidulafungin, are semi-synthetic lipopeptides that inhibit β-1,3-glucan synthase and hence fungal cell wall synthesis. They display excellent to moderate activity versus most *Candida* and *Aspergillus* species, although acquired resistance increasingly compromises their use versus *C. glabrata*. A primary limitation is the intrinsic resistance of all other fungal pathogens. Genetic analysis of acquired resistance in *Saccharomyces cerevisiae* identified the integral membrane protein Fks1 as the echinocandin target. Mutations cluster into hot spots 1 and, less commonly, 2 and 3 which share a similar topology: within or adjacent to the outer leaflet of the plasma membrane, and flanking a central cytoplasmic domain containing the likely substrate binding and catalytic sites. The differential resistance demonstrated by certain Fks1 mutations is consistent with direct interaction between the mutated residues and echinocandin-specific side chains. Differential resistance is also conferred by sphingolipid biosynthesis pathway mutations, suggesting a tripartite model for echinocandin–Fks1-membrane interaction. Analysis of Fks1 hot spot regions from fungi that exhibit intrinsically reduced susceptibility (*Candida parapsilosis*) or resistance (including *Fusarium* and *Scedosporium* species) identified substitutions that are likely to contribute to those phenotypes. Additional contributors likely include a reduced role for β-1,3-glucan in intrinsically resistant fungi, and increased cell wall chitin or melanin. Support for these Fks1-independent mechanisms is provided by dimorphic fungi such as *Histoplasma capsulatum* that encode a single Fks1 but alternate between echinocandin susceptible and resistant in their hyphal and yeast phases, respectively.

Introduction

Echinocandin antifungals have excellent activity versus most *Candida* species including *Candida albicans* and *Candida glabrata,* and moderate activity versus *Aspergillus* species. However, they have little or no activity versus most other fungal pathogens. Also, consistent with their complex lipopeptide structures, they have limited oral bioavailability and hence must be administered intravenously. Despite these serious limitations with respect to spectrum of activity and administration, echinocandins have risen to prominence in their brief history since *Candida* and *Aspergillus* species represent the major causes of life-threatening fungal infection. Furthermore, in contrast to azoles, echinocandins are generally fungicidal, and in contrast to amphotericin B, echinocandins are non-toxic. Finally, owing to their unique mechanism of action involving inhibition of cell wall synthesis, echinocandins retain activity versus fungi that develop resistance to other antifungals, particularly the ergosterol synthesis-inhibiting azoles. These advantages form the basis for recent guidelines specifying echinocandins as first line agents for *C. glabrata* infections, as second line agents for infections with other *Candida* and *Aspergillus* species, and as empiric therapy before the *Candida* species is identified (Pappas *et al.*, 2009). Echinocandins are also increasingly used prophylactically in high risk patients such as those undergoing haematopoietic stem cell transplants.

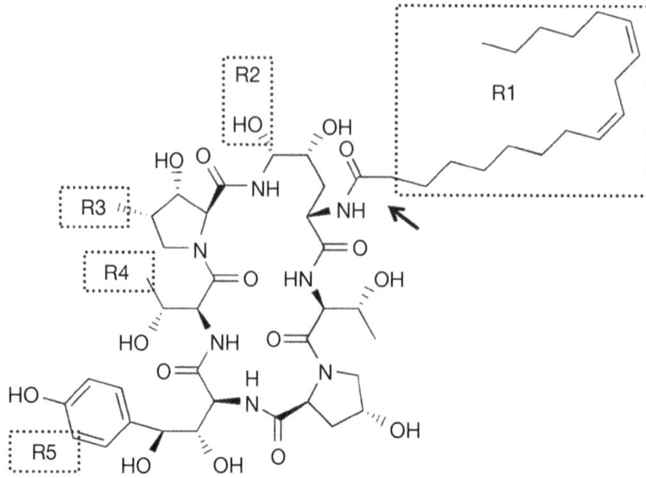

	caspofungin	micafungin	anidulafungin
R1	$-C_{15}H_{31}$	$-(C_6H_4)(C_4NO)(C_6H_4)OC_5H_{11}$	$-(C_6H_4)_3OC_5H_{11}$
R2	$-NHC_2H_4NH_2$	-OH	-OH
R3	-H	$-CH_3$	$-CH_3$
R4	$-C_2H_4NH_2$	$-CH_2CONH_2$	$-CH_3$
R5	-H	$-SO_3H$	-H

Figure 3.1 Structures of semi-synthetic echinocandins caspofungin, micafungin, and anidulafungin, relative to natural product echinocandin B. The arrow indicates the site of deacylase action used to generate free cyclic peptide for subsequent acylation with alternative lipids (R1).

The focus of this chapter is on *in vitro* studies – molecular, genetic, and genomic – of echinocandin mechanisms of resistance. For recent reviews with a greater focus on clinical aspects of echinocandin use, readers should consult Perlin (2011), Chen *et al.* (2011), Mukherjee *et al.* (2011), Beyda *et al.* (2012), and Pfaller (2012).

Structures

Caspofungin was the first echinocandin approved for human use (in 2001), followed by micafungin (2005) and anidulafungin (2006). All three are semisynthetic derivatives of fungal natural products; specifically, caspofungin derives from the *Glarea lozoyensis* product pneumocandin B_0, micafungin from the *Coleophama empedri* product FR901379, and anidulafungin from the *Aspergillus nidulans* var. *echinulatus* product echinocandin B. Additional investigational echinocandins include aculeacin A and aminocandin. All share a cyclic hexapeptide core, but differ in their side chain modifications to this core and to their lipid chains (Fig. 3.1). In particular, the lipid chains of anidulafungin and, especially, micafungin have aryl moieties, while the caspofungin lipid chain is strictly alkyl. Lipid chain modifications were facilitated by the use of a specific deacylase (Fig. 3.1); this enzyme was also used to demonstrate that removal of the lipid chain is essential to antifungal activity (Debono *et al.*, 1988). Other lipopeptide cell wall synthesis inhibitors include aerothricin 3 (Kondoh *et al.*, 2002) and the arbocandins (Ohyama *et al.*, 2000), with 12 or 10 residue cyclic peptide cores and one or two lipid chains, respectively.

The papulacandins represent a distinct group of natural products that also specifically inhibit fungal cell wall synthesis. In common with echinocandins, papulacandins have a lipid tail essential to their activity, but they substitute a glycoside core for the hexapeptide (Rommele *et al.*, 1983). Several additional lipoglycoside natural products have similar activity, and structure–activity relationships in this group are being explored (van der Kaaden *et al.*, 2012). As discussed below, papulacandin-resistant mutants demonstrate at least some cross-resistance to echinocandins, implying a common target despite their structural differences (Martins *et al.*, 2011). On the other hand, the enfumafungins which also target cell wall synthesis are lipid-free terpenoid glycosides (Peláez *et al.*, 2000), and do not exhibit cross-resistance with echinocandins (Pfaller *et al.*, 2013).

Echinocandin target: Fks1

Echinocandins inhibit cell wall synthesis in susceptible fungi, and in this sense they represent the penicillins of the antifungal world. Specifically, echinocandins inhibit the synthesis of β-1,3-glucan, a major cell wall polysaccharide in *Candida* species and related yeast including the genetic model *Saccharomyces cerevisiae* (Douglas, 2001). The responsible enzyme, β-1,3-glucan synthase, employs UDP-glucose as substrate, but has eluded definitive characterization since its activity is strictly membrane dependent. Nevertheless, partial purification, aided by a product entrapment protocol, strongly implicated a large (ca. 210 kd) integral membrane protein as the catalytic subunit, with the small GTPase Rho1 as a putative regulatory subunit (Kondoh *et al.*, 1997). Peptide sequencing demonstrated that this large protein is the product of the gene *FKS1* (a.k.a. *GSC1* in *C. albicans*) (Mio *et al.*, 1997). Since *FKS1* (or its paralogue *FKS2*) is also the the target of mutations conferring echinocandin resistance (see below), the case for Fks1 (or Fks2) equating to β-1,3-glucan synthase is strong. As detailed below, this correspondence is further strengthened by cross-linking studies with azido-modified UDP-glucose, and bioinformatic analyses that identified distant homologies to known UDP-glucosyltransferases.

On the other hand, an attempt to directly identify the echinocandin target in *C. albicans* by cross-linking to azido-modified anidulafungin failed to implicate Fks1 (Radding *et al.*, 1998). Rather, the cross-linked protein, Pil1, is a major component of eisosomes, which associate with membrane components during endocytosis (Edlind and Katiyar, 2004). Hence, Pil1 was presumably artifactually cross-linked to anidulafungin. Less direct but nevertheless convincing evidence that echinocandins do indeed target Fks1 comes from analysis of Fks1 mutations that confer differential resistance (Johnson *et al.*, 2011). Most mutations confer echinocandin cross-resistance; however, rare mutants were identified that confer micafungin resistance but unaltered caspofungin susceptibility, or conversely caspofungin resistance but unaltered micafungin susceptibility. These results are best explained by a direct, differential interaction of the mutated Fks1 residue (e.g. wild-type W695 vs. mutant L695) with an echinocandin-specific side chain (e.g. the aryl vs. alkyl regions of the micafungin versus caspofungin lipid chains).

Fks gene family

To understand the mechanisms behind echinocandin resistance, it is necessary to understand its target Fks1. The *FKS1* gene encoding this protein has been shown to be essential in the echinocandin-susceptible fungi *C. albicans* (Mio *et al.*, 1997; Douglas *et al.*, 1997) and *Aspergillus fumigatus* (Firon *et al.*, 2002) and, importantly, even in the non-susceptible fungi *Cryptococcus neoformans* (Thompson *et al.*, 1999), *Coccidioides posadasii* (Kellner *et al.*, 2005), and *Fusarium solani* (Ha *et al.*, 2006). By way of comparison, the *AGS* genes responsible for synthesizing the related cell wall polysaccharide α-1,3-glucan are non-essential. This is most obvious in *Candida* species and *S. cerevisiae* since they lack this gene and its product, but even in fungi such as *A. nidulans* and *C. neoformans* where α-1,3-glucan is a major cell wall polysaccharide, *AGS* deletion is tolerated (Henry *et al.*, 2011; Reese *et al.*, 2007).

Somewhat paradoxically, *FKS1* is non-essential in *S. cerevisiae* and *C. glabrata*, even though they are highly echinocandin susceptible. The explanation for this is redundancy: both yeast have paralogous *FKS2* genes that can fully complement *fks1* deletion or null mutation (Inoue *et al.*, 1995; Mazur *et al.*, 1995; Niimi *et al.*, 2012; Katiyar *et al.*, 2012). This is contingent, however, on *FKS2* expression, which in *S. cerevisiae* is strongly calcineurin dependent; calcineurin inhibitors such as FK506 effectively render FKS1 essential (the gene name derives from FK506 sensitivity). (In contrast to *S. cerevisiae*, in *C. glabrata FKS2* expression is largely constitutive, and hence the sensitivity of its *fks1* disruptant to calcineurin inhibitors implies that this phosphatase regulates

Fks2 activity at the protein level.) These two yeast encode an additional, more divergent paralogue, Fks3 (ca. 55% identity to Fks1 and Fks2), which has been shown in *S. cerevisiae* to play a more specialized role in spore wall formation (Ishihara *et al.*, 2007). Indeed, three *FKS* genes is the rule in ascomycetous yeast, including *C. albicans* and its close relatives *Candida parapsilosis* and *Candida tropicalis*, although roles for their Fks2 and Fks3 (a.k.a. Gsl1 and Gsl2) remain to be identified.

With respect to other clinically relevant fungi, multiple *FKS* genes are similarly detectable by BLAST analysis of the genome sequences of the ascomycete *F. solani* (two) and zygomycetes including *Rhizopus* species (three). In contrast, *Aspergillus* and *Cryptococcus* species encode a single essential *FKS1*, as do the dimorphic fungi such as *Histoplasma capsulatum* and the obligately parasitic *Pneumocystis* species. It should be noted here that *FKS1* orthologues are also present in the genomes of certain protists (e.g. *Toxoplasma gondii* and oomycete species) as well as many plants (where β-1,3-glucan is commonly referred to as callose).

Fks1 structure–function

Consistent with its large size and integral membrane location, deciphering structure–function relationships in Fks1 has been challenging. Initial sequence analysis by Douglas *et al.* (1994) predicted 16 transmembrane helices, divided into N-terminal and C-terminal clusters, and separated by a central region of ca. 600 residues that was hypothesized to represent the catalytic domain with cytosolic localization. In more recent work by Johnson *et al.* (2012), these topology predictions were refined using the TMHMM algorithm, and then experimentally tested in *S. cerevisiae* using panels of Fks1 fragments C-terminally fused to HA-Suc2-His4C. Depending on location of the Fks1 fusion, the HA-Suc2-His4 tag was either cytosolic or external. A model summarizing the results of Fks1 topological analysis is presented in Fig. 3.2.

Having confirmed its cytosolic localization, the central domain was further analysed in terms of its putative role in catalysis. This role was initially supported (albeit weakly given limited specificity) by cross-linking studies with azido-UDP-glucose and partially purified *Neurospora crassa* β-1,3-glucan synthase preparations which identified four peptides that clustered between the equivalent of *S. cerevisiae* Fks1 residues 1070–1273 (Schimoler-O'Rourke *et al.*, 2003). Also, multiple mutations yielding temperature sensitive Fks1 were mapped to the central domain (Okada *et al.*, 2010). Using the structure-based search algorithm HHpred, Johnson *et al.* (2012) identified previously undetected homologies to known UDP-glucosyltransferases, specifically to DXD-containing regions that coordinate Mn^{2+}

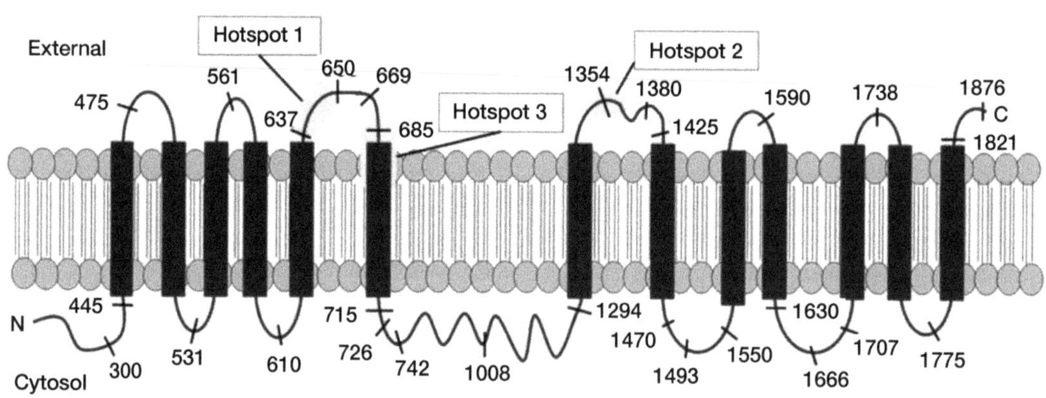

Figure 3.2 *S. cerevisiae* Fks1 membrane topology according to Johnson *et al.* (2012). Transmembrane helices (dark bars) were first predicted using TMHMM, then tested using HA–Suc2-His4C fusions to the C-termini of Fks1 fragments at the indicated residues. Note the external location of all three hot spots. Drawing is not to scale. (Modified from Johnson *et al.*, 2012.)

interaction with the substrate diphosphate. The corresponding Fks1 sequence is DAN (residues 1102–1104), and the conserved D1102 was shown by site-directed mutagenesis to play an essential role in Fks1 function. Based on its location relative to D1102 and its presence within a universally (fungi, protists, and plants) conserved six residue region, D1197 was subsequently identified as a candidate catalytic site residue. Again, site-directed mutagenesis demonstrated that D1197 is essential to Fks1 function. Together, these data support the hypothesis that D1102 and D1197 represent the substrate binding and catalytic sites of β-1,3-glucan synthase, although further genetic and biochemical studies to confirm this are clearly needed. Recently described piperazinyl-pyridazinones derivatives with β-1,3-glucan synthase inhibiting activity could prove useful in this regard; resistance-conferring mutations were mapped to Fks1 residues 1175 and 1297 (Walker *et al.*, 2011).

Cell-free β-1,3-glucan synthase systems

Crude cell-free systems, or more specifically partially purified membrane preparations, were developed as early as 1981 in order to study echinocandin mechanism of action. These systems demonstrated that echinocandins specifically inhibit synthesis of β-1,3-glucan from UDP-glucose substrate (Perez *et al.*, 1981; Sawistowska-Schroder *et al.*, 1984). Since UDP-glucose serves as substrate for multiple glycosyltransferases, it was essential to confirm that the product was indeed β-1,3-glucan (e.g. by digestion with β-1,3-glucanase). Most importantly, these systems showed that echinocandin inhibition was non-competitive with UDP-glucose substrate. These cell-free β-1,3-glucan synthase assays were subsequently improved through an additional purification step involving product entrapment (Inoue *et al.*, 1995). This purification permitted partial sequencing of Fks1, and identification of the putative regulatory subunit Rho1 (Kondoh *et al.*, 1997). It also enabled precise comparisons of the effects of different echinocandins, and different Fks1 mutations, on the V_{max} and Km of this complex enzyme (Garcia-Effron *et al.*, 2009a,b).

Echinocandin uptake and efflux

One rationale for using cell-free inhibition assays is that it eliminates the variables of drug uptake and efflux. With respect to echinocandins, however, there is no convincing evidence that uptake is required for activity, or relatedly that efflux mediates resistance. One study attempted to directly assess 'uptake' kinetics by incubating intact yeast with radiolabelled caspofungin, followed by separation of bound and unbound caspofungin by centrifugation (Paderu *et al.*, 2004). However, there was no evidence presented that the bound caspofungin was internalized. Alternative interpretations are that it was trapped in the cell wall or bound to the outer leaflet of the membrane, both not unlikely since caspofungin is relatively large and lipophilic. An independent argument against uptake is the unaltered echinocandin susceptibility in *Candida* species and *S. cerevisiae* mutants with altered expression of classic multidrug transporters in the ATP binding cassette and major facilitator families. This has been confirmed by multiple laboratories (Posteraro *et al.*, 2006; Niimi *et al.*, 2006; Richards *et al.*, 2008), although there is one report to the contrary that implicated *C. albicans CDR2* in caspofungin resistance (Schuetzer-Muehlbauer *et al.*, 2003). However, this study relied on high level overexpression from the constitutive *ACT1* promoter, and controls for non-specific effects (in particular, activation of cell wall stress pathways) were lacking.

Acquired echinocandin resistance: Fks hot spots 1 and 2

Mutationally acquired echinocandin resistance was first demonstrated in the genetic model *S. cerevisiae,* and subsequently in four of the five most clinically relevant *Candida* species; specifically, *C. albicans, C. glabrata, C. tropicalis,* and *C. krusei* (for references see Fig. 3.3 legend) The exception is *Candida parapsilosis,* which exhibits intrinsically low susceptibility as discussed below. Spontaneously acquired resistance, either clinical or laboratory, has not been characterized in any *Aspergillus* species, although an *A. fumigatus* mutant was generated by transformation (Rocha *et al.*, 2007). Note that 'resistance' as used here refers to any reproducible decrease in *in vitro* susceptibility; this may or may not translate to *in vivo* resistance, since mutations variably affect fitness and hence response to treatment.

The *S. cerevisiae* resistance-conferring gene was identified as *FKS1* by genetic complementation (exploiting the hypersensitivity of one mutant to chitin synthase inhibitor nikkomycin) and characterized at the sequence level by Merck researchers in 1994 (Douglas *et al.*, 1994). Characterization of the *C. albicans* orthologue followed three years later (Douglas *et al.*, 1997; Mio *et al.*, 1997). It was not until 2005, however, that the first resistance-conferring mutations in these genes were described in a peer-reviewed publication (Park *et al.*, 2005). This lengthy 11 year timeline can be attributed to the relative rarity of echinocandin resistance mutations combined with the large size and complexity of the Fks1 target, but to some extent it may also reflect the reluctance of pharmaceutical companies to publish studies relating to resistance of a drug in development.

The initial resistance-conferring mutations described by Park *et al.* (2005) in *S. cerevisiae*, *C. albicans*, and *C. krusei* totalled only seven, but together they defined the two 'hot spot' regions within which all mutations subsequently reported in clinical isolates have been shown to fall. Specifically, using *S. cerevisiae* Fks1 numbering, hot spot 1 spans residues F639 to P647, with 44 unique mutations reported to date in five different yeasts (Fig. 3.3). Hot spot 2 spans residues W1354 to S1361, with four unique mutations in three different yeasts. As suggested by these numbers, hot spot 1 mutations are at least 10-fold more likely than hot spot 2 mutations. Interestingly, mutations in hot spot 1 appear to cluster every four residues or so (639–643–647), similar to the periodicity of an α-helix.

The relevance of Fks1 hot spots 1 and 2 as primary targets for inhibition of β-1,3-glucan synthase is strengthened by studies with two structurally distinct inhibitors. The fission yeast *Schizosaccharomyces pombe* exhibits only moderate susceptibility to echinocandins, but its high susceptibility to the papulacandin facilitated the isolation of mutants resistant to this lipoglucoside (Martins *et al.*, 2011). Intriguingly, one of these involve *S. pombe* Bgs4 residue E700, which is equivalent in *S. cerevisiae* to Fks1 residue E635. This is only 4 residues upstream of hot spot 1 (Fig. 3.3), and indeed this *S. pombe* mutant exhibits cross-resistance to caspofungin. Aerothricin3 is a lipopeptide, like the echinocandins, but with a considerably larger (12 residue) cyclic peptide core. It exhibits differential activity versus *S. cerevisiae* Fks1 versus Fks2, and through the use of chimeric genes and site-directed mutagenesis the responsible residue was identified as K1336/I1336 (Kondoh *et al.*, 2002). This is 18 residues upstream of hot spot 2, within the same external loop (Fig. 3.2). It seems likely that aerothricin3 shares this target with echinocandins, but has a longer reach owing to its larger peptide.

Hot spots 1 and 2 have dissimilar sequences, but similar topologies (Fig. 3.2). Specifically, both have external locations, adjacent to or partly within transmembrane helices that are symmetrically located on either side of the central cytoplasmic domain.

The effects of Fks hot spot mutation on echinocandin susceptibility are largely comparable whether measured in terms of cell growth or cell-free β-1,3-glucan synthesis (Niimi *et al.*, 2010). In both *C. albicans* and *C. glabrata*, these mutations did not significantly affect Km, as expected since echinocandins are non-competitive with substrate (Garcia-Effron *et al.*, 2009a, 2009b). Also, in *C. glabrata* these mutations minimally affected V_{max}, implying that the mutated residues do not have critical role in catalysis. This was not the case in *C. albicans*, and relatedly Fks1 mutants in this yeast exhibit reduced fitness *in vitro* and virulence *in vivo* (Ben-Ami *et al.*, 2011).

Differential echinocandin resistance: discovery of hot spot 3

The clustering of numerous resistance-conferring mutations within Fks1 hot spots 1 (635–647) and 2 (1354–1361) implies but does not prove that these regions represent echinocandin binding sites. Their external locations (Fig. 3.2) are clearly distinct from the cytoplasmic central domain residues putatively involved in substrate binding (1102) and catalysis (1197); this separation is consistent with echinocandins as non-competitive

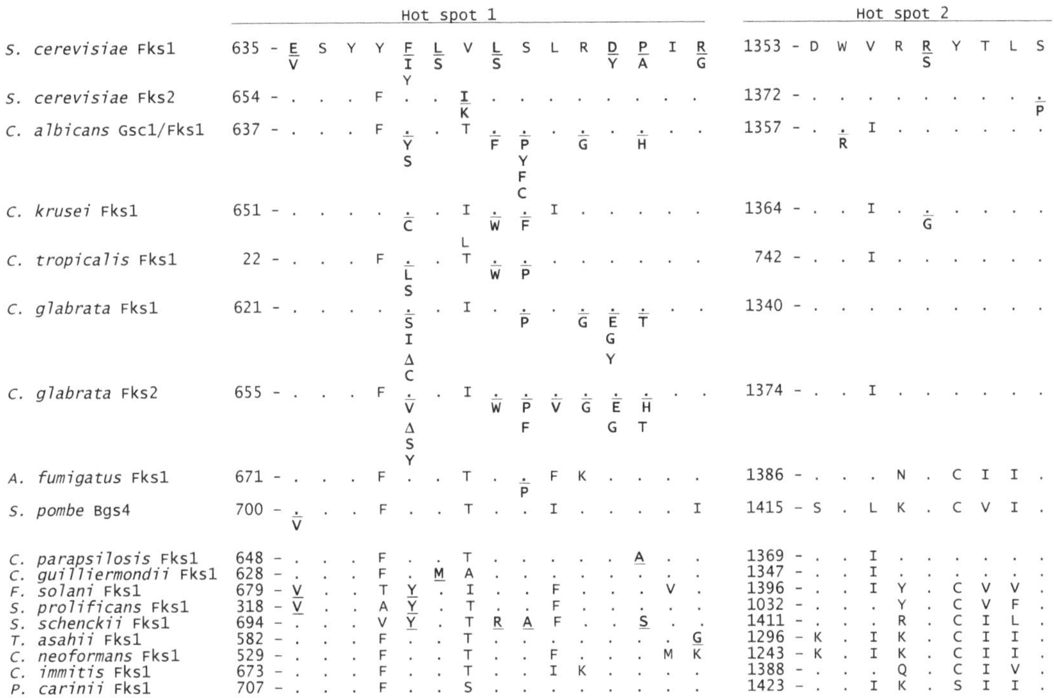

Figure 3.3 Fks1 and Fks2 hot spot 1 mutations associated with echinocandin resistance in clinical isolates of *C. albicans*, *C. krusei*, *C. tropicalis* (partial sequence), and *C. glabrata*, and laboratory mutants of *S. cerevisiae*, *C. glabrata*, *A. fumigatus*, and *S. pombe* (Balashov *et al.*, 2006; Castanheira *et al.*, 2010; Cleary *et al.*, 2008; Dannaoui *et al.*, 2012; Desnos-Ollivier *et al.*, 2008; Garcia-Effron *et al.*, 2008, 2009a; Kahn *et al.*, 2007; Katiyar *et al.*, 2006, 2012; Laverdiere *et al.*, 2006; Messer *et al.*, 2010; Niimi *et al.*, 2010, 2012; Ohyama *et al.*, 2004; Park *et al.*, 2005; Pfaller *et al.*, 2012; Pfeiffer *et al.*, 2010; Zimbeck *et al.*, 2010; Johnson *et al.*, 2011; Rocha *et al.*, 2007). Dots indicate identity at that position to the *S. cerevisiae* Fks1 sequence. Mutations are shown beneath the underlined wild-type residue. Δ, deletion. Residues that potentially contribute to the intrinsic echinocandin reduced susceptibility or resistance of *C. parapsilosis*, *C. guilliermondii*, *F. solani*, *S. prolificans*, *S. schenckii*, and *T. asahii* are underlined. *S. cerevisiae* sequences obtained from SGD (www.yeastgenome.org), *Candida* sequences from CGD (www.candidagenome.org), all others from NCBI (www.ncbi.nlm.nih.gov).

inhibitors. Nevertheless, it remains possible that hot spot 1 and 2 mutations confer resistance by indirect means; for example, by inducing conformational changes in Fks1 that disrupt echinocandin binding elsewhere. Evidence in support of a direct interaction between an Fks residue and an echinocandin would be differential resistance. That is, if a mutation confers resistance to one echinocandin but not another, the simplest explanation is that the mutated amino acid interacts with an echinocandin-specific side chain. For example, caspofungin and micafungin have alkyl and aryl-alkyl side chains, respectively (Fig. 3.1). If these side chains are in proximity to an aromatic Trp residue which mutates to an alkyl Leu, the interaction between this residue and micafungin would likely decrease while interaction with caspofungin would likely increase.

Using *S. cerevisiae* as model, Johnson *et al.* (2011) screened mutants selected on caspofungin or micafungin-containing media for differential resistance. As with most clinical isolates, the *S. cerevisiae* mutants predominantly exhibited echinocandin cross-resistance and mutations in hot spot 1. However, one mutant exhibited differential resistance to caspofungin. Sequencing identified Fks2 mutation S1380P; the equivalent residue in Fks1 is 4 residues downstream from the previously defined hot spot 2. These data provide support for a direct interaction between echinocandins and hot spot 2. A second mutant exhibited differential resistance to micafungin. Sequencing followed by site-directed mutagenesis implicated Fks2 mutation W714L, a location not previously associated with echinocandin resistance. Mutational analysis of the equivalent Fks1 residue, W695, and adjacent residues defined a new region located 50 residues downstream of hot spot 1 that is likely to interact directly with and hence determine susceptibility to echinocandins. Mutations in this 'hot spot 3' have also been identified in laboratory mutants of *C. glabrata* selected on micafungin (Johnson *et al.*, 2011) and *S. pombe* selected on papulacandin (Martins *et al.*, 2011), and its role in the intrinsic echinocandin resistance of *Fusarium* species has been proposed (see below). To date, however, there are no reports associating hot spot 3 mutations with acquired resistance in clinical isolates.

Similar to hot spots 1 and 2, topological analysis demonstrated that hot spot 3 is external, but in contrast to the former the latter is largely embedded in the outer leaflet of the plasma membrane (Fig. 3.2).

In subsequent studies (M. Johnson and S. Katiyar, unpublished data), an additional *S. cerevisiae* mutant exhibiting differential resistance was isolated, and sequencing identified a novel mutation at the C-terminus of hot spot 1, R649G (Fig. 3.3). The relevance of this mutation to intrinsic resistance is discussed below.

Impact of Fks heterozygosity and redundancy on acquired resistance

In the diploid species *C. albicans*, *C. tropicalis*, and *C. krusei*, the impact of Fks1 mutations on echinocandin susceptibility depends on their heterozygosity versus homozygosity. There have been several reports, in both laboratory mutants and clinical isolates, of mutants that shift from low-level to high-level resistance due to loss of heterozygosity (Douglas *et al.*, 1997; Park *et al.*, 2005; Niimi *et al.*, 2010; Jensen *et al.*, 2013). On the other hand, *C. glabrata* is haploid, and therefore might be expected to behave differently. In fact, in many *C. glabrata* strains *FKS1* and *FKS2* are expressed at variable but roughly comparable levels (Garcia-Effron *et al.*, 2009a; Katiyar *et al.*, 2012), making it functionally diploid with respect to the effects of echinocandin resistance mutation. This was demonstrated in laboratory mutants; for example, an Fks2-S663F mutation decreased susceptibility 4- to 8-fold in a wild-type strain versus 128- to 256-fold in an *fks1Δ* strain (Katiyar *et al.*, 2012). Similarly, clinical isolates have been characterized with hot spot mutations in one gene and null mutations in the other (Garcia-Effron *et al.*, 2009a; Niimi *et al.*, 2012). The impact of this *FKS1-FKS2* redundancy will vary depending on their relative expression levels (or relative activity of the gene products). Calcineurin inhibitors such as FK506 can influence this (Katiyar *et al.*, 2012), but it is likely that genetic or epigenetic changes will do this as well.

The Fks redundancy resulting from either diploid alleles as in *C. albicans* or comparable *FKS1–FKS2* expression as in *C. glabrata* has positive clinical implications with respect to echinocandin resistance, as recently noted (Niimi *et al.*, 2010; Katiyar *et al.*, 2012). Redundancy attenuates the degree of resistance (i.e. MIC) conferred by mutation in one Fks allele/gene, since the second allele continues to confer susceptibility. As a direct consequence, the frequency of resistance is also attenuated, provided that echinocandin concentration remains above the MIC conferred by the single allele/gene mutation. Note that these positive effects of Fks redundancy would be reduced in *C. albicans* strains that more readily undergo loss of heterozygosity (hypothetical) or in *C. glabrata* strains in which one *FKS* becomes recessive due to reduced expression or null mutation (Garcia-Effron *et al.*, 2009a; Niimi *et al.*, 2012).

Fks-independent acquired echinocandin resistance

The first report characterizing an *S. cerevisiae* gene whose mutation conferred resistance to an echinocandin (specifically, pneumocandin B_0 derivative and caspofungin precursor L-733,560) did not in fact identify *FKS1* but rather *GNS1* (el-Sherbeini and Clemas, 1995), which proved to be allelic to *FEN1* encoding a fatty acid elongase required for ceramide and hence sphingolipid synthesis. A *fen1/gns1* disruptant reproduced the mutant phenotype. More recently, Healey *et al.* (2011,

2012) similarly identified *fen1/gns1* disruptants and null mutants following *in vitro* selection for caspofungin reduced susceptibility (CRS) in both *S. cerevisiae* and *C. glabrata*. Further analysis implicated null mutants in additional genes critical to both arms of the sphingolipid biosynthesis pathway; i.e. phytosphingosine synthesis as well as fatty acid elongation. Paradoxically, the *C. glabrata* CRS mutants exhibited micafungin increased susceptibility (MIS). As a consequence of their sphingolipid pathway mutations, CRS-MIS mutants accumulate long chain bases such as phytosphingosine, and exogenous addition of these sphingolipid intermediates to wild-type cells induces a CRS-MIS phenotype. With respect to clinical implications, CRS-MIS mutants were also identified within a large collection of *C. glabrata* clinical isolates (some presumably derived from caspofungin-treated patients); nevertheless, they may play a minimal role in caspofungin treatment failure due to the variably deleterious effects of sphingolipid pathway mutation. One *C. albicans* strain (among 20 or so screened) also readily yielded CRS-MIS mutants following *in vitro* selection on caspofungin-containing medium, and sequencing implicated loss of heterozygosity in the sphingolipid biosynthesis pathway gene *TSC13* (K. Healey and S. Katiyar, unpublished data). The specific consequence on the protein level is conversion of Tsc13-H104/P104 to Tsc13-P104/P104, which is likely an inactive form of this enoyl reductase as evidenced by lipid analysis (the mutant accumulates long chain bases) and the conservation of H104 in the Tsc13 orthologues of widely diverse fungi. All other *C. albicans* strains analysed encode homozygous Tsc13-H104/H104, consistent with their inability to yield CRS-MIS mutants.

The *S. cerevisiae* component of the studies described above relied on the widely used genome-wide deletion libraries; specifically, the homozygous and heterozygous diploid collections. Two other laboratories employed the haploid deletion collection in screens for reduced caspofungin susceptibility, and failed to identify disruptants in sphingolipid biosynthesis pathway genes (Markovich *et al.*, 2004; Lesage *et al.*, 2004). The screening approach may have been biased against disruptants with subtle growth defects; alternatively, the haploid deletion collections suffer from more artefacts than previously appreciated. The latter is the most likely explanation for the identification by both laboratories of *SLG1* (alias *WSC1*) disruptants as caspofungin resistant, in contrast to other studies with independent disruptants that clearly demonstrated caspofungin hypersusceptibility (Reinoso-Martin *et al.*, 2003; Bermejo *et al.*, 2010). Slg1 is a membrane-localized sensor required for activation of the Pkc1-Mpk1 cell wall integrity pathway, which plays a key role in the cellular response to caspofungin treatment (Levin, 2011), and hence disruption of its gene should indeed confer hypersusceptibility, not resistance.

One group of genes whose expression is upregulated following activation of the cell wall integrity pathway are the chitin synthase genes, and it has been repeatedly observed that cell wall chitin levels increase in response to caspofungin treatment (Walker *et al.*, 2013; Lee *et al.*, 2012; Cota *et al.*, 2008). Indeed, this is the likely basis for the paradoxical (a.k.a. Eagle) effect, i.e. growth at echinocandin concentrations substantially above the MIC (Stevens *et al.*, 2006). One might expect, therefore, that a second class of Fks-independent mutants with reduced echinocandin susceptibility would result from mutations leading to increased cell wall chitin. Indeed, a laboratory mutant with decreased caspofungin susceptibility, increased chitin, reduced chitinase activity, and mutations in chitinase 2 and 3 genes has been described (Drakulovski *et al.*, 2011). The relevance of these mutations was not directly tested, however, and to date there are no reports linking acquired echinocandin resistance in clinical isolates with increased chitin.

Intrinsic echinocandin resistance: hot spot 1 substitutions

A major limitation of echinocandins is the intrinsic resistance (MIC ≥8 μg/ml) demonstrated by many pathogenic fungi, including ascomycetous *Fusarium*, *Scedosporium*, and *Sporothrix* species; basidiomycetous *Cryptococcus* and *Trichosporon* species; and zygomycetes such as *Rhizopus oryzae*. Other fungi exhibit reduced (intermediate) echinocandin susceptibility (MIC = 1–4 µg/ml), including *Candida* species *C. parapsilosis* and *C. guilliermondii*. A third category includes the endemic dimorphic fungi and *Pneumocystis* species that exhibit stage-specific resistance, as discussed below.

Multiple explanations for intrinsic resistance have been proposed, although few have been tested. One explanation is that β-1,3-glucan plays only a minor role in intrinsically resistant fungi compared with susceptible *Candida* species, since cell walls in the former include α-1,3-glucan which is lacking in the latter. This is inconsistent, however, with observations that the *FKS1* gene is essential to viability in all three of the intrinsically resistant fungi in which its disruption has been attempted (Thompson *et al.*, 1999; Kellner *et al.*, 2005; Ha *et al.*, 2006). Furthermore, fungi in which the non-essential α-1,3-glucan synthase genes (*AGS*) have been disrupted did not exhibit enhanced echinocandin susceptibility (Reese *et al.*, 2007; Yoshimi *et al.*, 2013). A second explanation invokes a protective role for melanin, an amorphous polymer with drug-binding properties that is present to varying extents in the cell walls of many different fungal species (Eisenman and Casadevall, 2012). This has received some experimental support (see below), but has yet to be rigorously tested, for example by conversion of echinocandin resistance to susceptibility following genetic disruption of melanin synthesis. Third, intrinsic resistance could result from echinocandin degradation, perhaps by a deacylase that removes the lipid tail or peptidase that cleaves the cyclic peptide. Since echinocandins appear to act externally, these degradative enzymes would need to secreted or cell wall localized. This mechanism also remains to be tested.

A mechanism for echinocandin intrinsic resistance that has, however, received experimental support returns to the role of Fks1 sequence, and specifically hot spots 1, 2, and 3 that represent the targets for acquired resistance. The model for this mechanism is *C. parapsilosis*, which exhibits intrinsically low susceptibility to all three echinocandins. Inspection of its Fks1 sequence revealed that, in contrast to nearly all other fungi, residue 660 (equivalent to *S. cerevisiae* Fks1 residue 647) within hot spot 1 is A rather than P. Mutations of this residue were previously implicated in the echinocandin resistance of *C. albicans* and *C. glabrata* clinical isolates (Fig. 3.3). As a direct test, a P647A mutation was engineered into *S. cerevisiae* Fks1; the result was a 16-fold increase in MIC and comparably increased glucan synthase IC_{50} (Garcia-Effron *et al.*, 2008).

Similar to *C. parapsilosis*, other pathogenic fungi have natural polymorphisms within hot spot 1 that most likely contribute to their intrinsic echinocandin resistance. Specifically, in *Fusarium* species such as *F. solani*, and in *Scedosporium prolificans*, the residues equivalent to *S. cerevisiae* Fks1 E635 and F639 are V and Y, respectively, substitutions previously shown to confer echinocandin resistance in *S. pombe* and two *Candida* species, respectively (Fig. 3.3). A role for these substitutions was supported by construction and testing of an *S. cerevisiae-F. solani* hybrid Fks1 (Katiyar and Edlind, 2009). As a further test, the double mutation E635V/F639Y was engineered into *S. cerevisiae* Fks1, and resistance confirmed (S. Katiyar, unpublished data). *Sporothrix schenckii* Fks1 (unpublished; GenBank accession ERS95454) also has the equivalent of Y639, along with non-conservative substitutions of three additional hot spot 1 residues that very likely contribute to its intrinsic echinocandin resistance (Fig. 3.3).

The *C. guilliermondii FKS* gene PGUG_01168 that is syntenic with *C. albicans FKS1/GSC1*, and hence encodes the likely echinocandin target, substitutes M for the residue equivalent to *S. cerevisiae* Fks1 L640. In the latter, an L640S mutation confers echinocandin resistance (Fig. 3.3; M. Johnson and S. Katiyar, unpublished data). L at this position is conserved in most if not all fungi (although curiously it is mutated to S in the initial description of the *C. albicans FKS1/GSC1* gene; Mio *et al.*, 1997). Thus, M633 of *C. guilliermondii* Fks1 may explain its intrinsically low echinocandin susceptibility; this could be tested by generating an L640M mutation in *S. cerevisiae* Fks1.

The basidiomycetous yeast *Trichosporon asahii*, a common skin commensal that can cause life-threatening invasive infection in the immunocompromised, substitutes G for the R at the C-terminus of hot spot 1 (Fig. 3.3). The equivalent mutation in *S. cerevisiae* Fks1, R649G, confers micafungin resistance with minimal effect on caspofungin susceptibility (M. Johnson and S. Katiyar, unpublished data). Thus this substitution can explain, but only in part, the broad spectrum echinocandin resistance exhibited by this yeast (Espinel-Ingroff, 2003).

Intrinsic echinocandin resistance: hot spot 3 substitutions

The *Scedosporium* species *S. prolificans* and *S. apiospermum* exhibit intrinsic resistance not only to echinocandins but also to amphotericin B; *S. prolificans* is furthermore resistant to azoles. Hence, infections with these ascomycetous moulds can be particularly difficult to treat. As reviewed above, mutations of *S. cerevisiae* Fks1 residues 691 to 700, including W695F and W695C, defined the new hot spot 3 for acquired echinocandin resistance (Johnson *et al.*, 2011). In *Scedosporium* species, substitutions within Fks1 hot spot 3 are present that may contribute to their intrinsic resistance (in the case of *S. prolificans,* in addition to its hot spot 1 substitutions). Specifically, both species have the equivalent of F695. The potential role of this substitution was more directly tested using a hybrid *S. cerevisiae-S. prolificans* Fks1, which exhibited 8- to 16-fold resistance to micafungin and anidulafungin, although caspofungin susceptibility was unchanged (Johnson *et al.*, 2011). The Fks1 of the dimorphic fungus *Blastomyces dermatitidis* has the equivalent of C695, associated with caspofungin-specific decreased susceptibility in *S. cerevisiae* (Johnson *et al.*, 2011). However, the related fungi *Histoplasma capsulatum* and *Coccidioides immitis* retain the equivalent of W695; see below for discussion of stage-specific susceptibility. As with *S. prolificans,* additional Fks1 substitutions, or non-Fks1 factors such as those proposed for *C. neoformans* (see below), must contribute to the broad-spectrum echinocandin resistance exhibited by these fungal pathogens.

Intrinsic resistance: mechanism to be determined

With respect to *Cryptococcus* species *C. neoformans* and *C. gattii,* no direct correlations can be made between their Fks1 sequences and specific resistance-conferring hot spot mutations described to date in other fungi (Fig. 3.3; and data not shown). It remains to be seen whether or not other Fks1 regions, such as the highly divergent external loop upstream of hotspot 2 (including the aerothricin3 susceptibility-determining residue described above), contribute to intrinsic echinocandin resistance in these basidiomycetous yeast. An alternative mechanism with some experimental support involves cell wall melanin, which is induced in *C. neoformans* during growth *in vivo* and has been shown to bind caspofungin *in vitro* (van Duin *et al.*, 2002). Indeed, based on cell-free glucan synthase assays, it has been suggested that *C. neoformans* Fks1 itself is intrinsically echinocandin susceptible (Maligie and Selitrennikoff, 2005), although unpublished studies cited by Thompson *et al.* (1999) indicate otherwise. The reliability of these assays may be low for *C. neoformans* given the technical challenge of distinguishing UDP-glucose incorporation specifically into β-1,3-glucan versus the large background of other polysaccharides synthesized by this encapsulated yeast. Attempts to resolve this issue by disrupting both signalling pathways (calcineurin and Pkc1-Mpk1) responsible for cell wall integrity were equivocal; i.e. the caspofungin susceptibility of these strains increased but remained in the resistant range (MIC $\geq 4\,\mu g/ml$) (Kraus *et al.*, 2003; Gerik *et al.*, 2005).

Aspergillus species present two paradoxes. First, they are relatively resistant to echinocandins in conventional growth inhibition assays, but highly susceptible in assays that use morphological change as the endpoint. Thus, their susceptibility is generally expressed in terms of MEC (minimum effective concentration) rather than MIC. The conventional wisdom is that β-1,3-glucan synthase activity is concentrated in the growing hyphal tip (Kurtz *et al.*, 1994; Beauvais *et al.*, 2001), while other hyphal regions undergo Fks1-independent growth, but it is unclear how the latter could be sustained given Fks1 essentiality. A second paradox is that the anidulafungin parent echinocandin B is produced by an anidulafungin susceptible *Aspergillus* species, *A. nidulans* (Hof and Dietz, 2009; Toth *et al.*, 2012). Similarly, the *Aspergillus aculeatus* product aculeacin A possesses potent activity versus at least some *Aspergillus* species, including *A. fumigatus* but not *A. niger* (Mizuno *et al.*, 1977). These two natural products differ from anidulafungin only in the structure of their lipid chains: alkyl versus arylalkyl, respectively. Evidence has been presented that *A. nidulans* tolerance to its product is due to chitin synthase up-regulation (Toth *et al.*, 2012). However, comparison of available Fks1 sequences from *Aspergillus* species suggests another potential contributor: the residue immediately upstream of mutationally defined hot spot 2. Specifically, the residue equivalent to *S. cerevisiae* Fks1 D1353 is D in *A. fumigatus* (Fig. 3.3), *A. flavus*, and *A. terreus*, but A in *A. niger* and N in the two echinocandin producers *A. nidulans* and *A. aculeatus*. The relevance of this residue warrants testing (e.g. by site-directed mutagenesis in *S. cerevisiae* Fks1), although its role in intrinsic resistance may be limited to the natural products.

Stage-specific intrinsic resistance

The dimorphic fungi including *Histoplasma capsulatum*, *Blastomyces dermatitidis*, and *Coccidioides* species *C. immitis* and *C. posadasii* represent a special case, since they are intrinsically resistant in their pathogenic yeast (or spherule) phase but susceptible in their environmental hyphal phase (Hage *et al.*, 2011; Nakai *et al.*, 2003). Inspection of the single Fks1 encoded by these fungi confirms that their hot spot sequences are consistent with echinocandin susceptibility (with the exception of the *B. dermatitidis* hot spot 3 substitution noted above which may contribute specifically to caspofungin resistance). Thus, the echinocandin resistance of their yeast/spherule phases must depend on factors other than Fks1. Analogous to other fungi discussed above, potential factors include a reduced role for β-1,3-glucan along with increased production of chitin and melanin in the yeast versus hyphal stages. There is some evidence to support the melanin hypothesis (van Duin *et al.*, 2002), and *FKS1* expression is down-regulated in mature spherules of *C. posadasii* (Kellner *et al.*, 2005), but further studies on the basis for this stage specific susceptibility are needed.

The human opportunist *Pneumocystis jirovecii* (formerly referred to as *Pneumocystis carinii*) latently infects the respiratory tract, but reactivation in the immunocompromised can result in life-threatening pneumonia. The vegetative, trophozoite stage has an amoeboid morphology with minimal cell wall, and consistent with this exhibits echinocandin resistance (Powles *et al.*, 1998). However, differentiation into the cyst stage with its pronounced cell wall is blocked by echinocandin treatment, and this susceptibility is consistent with *Pneumocystis* Fks1 hot spot sequences (Fig. 3.3).

Conclusions and future prospects

The 'first generation' of research on echinocandin resistance has focused on understanding mechanisms for acquired resistance in the experimentally tractable fungi *S. cerevisiae*, *C. albicans*, and *C. glabrata*. This research has built a strong foundation, including mapping of mutations conferring echinocandin resistance to three externally localized hot spots within the integral membrane protein Fks1. The second generation of research in this field will experimentally explore the basis for intrinsic reduced susceptibility or resistance in the diverse fungal pathogens that increasingly threaten immunocompromised patients, including *Aspergillus*, *Cryptococcus*, *Fusarium*, and *Scedosporium* species. These fungi can be studied initially through their genome sequences, and subsequently using heterologous expression systems in models such as *S. cerevisiae*. Ultimately, understanding the molecular basis for intrinsic resistance should facilitate the design of second generation echinocandins with extended spectrum activity.

References

Balashov, S.V., Park, S., and Perlin, D.S. (2006). Assessing resistance to the echinocandin antifungal drug caspofungin in *Candida albicans* by profiling mutations in *FKS1*. Antimicrob. Agents Chemother. *50*, 2058–2063.

Beauvais, A., Bruneau, J.M., Mol, P.C., Buitrago, M.J., Legrand, R., and Latgé, J.P. (2001). Glucan synthase complex of *Aspergillus fumigatus*. J. Bacteriol. *183*, 2273–2279.

Ben-Ami, R., Garcia-Effron, G., Lewis, R.E., Gamarra. S., Leventakos, K., Perlin, D.S., and Kontoyiannis, D.P. (2011). Fitness and virulence costs of Candida albicans FKS1 hot spot mutations associated with echinocandin resistance. J. Infect. Dis. *204*, 626–635.

Bermejo, C., García, R., Straede, A., Rodríguez-Peña, J.M., Nombela, C., Heinisch, J.J., and Arroyo, J. (2010). Characterization of sensor-specific stress response by transcriptional profiling of wsc1 and mid2 deletion strains and chimeric sensors in *Saccharomyces cerevisiae*. OMICS *14*, 679–688.

Beyda, N.D., Lewis, R.E., and Garey, K.W. (2012). Echinocandin resistance in Candida species: mechanisms of reduced susceptibility and therapeutic approaches. Ann. Pharmacother. *46*, 1086–1096.

Castanheira, M., Woosley, L.N., Diekema, D.J., Messer, S.A., Jones, R.N., and Pfaller, M.A. (2010). Low prevalence of fks1 hot spot 1 mutations in a worldwide collection of Candida strains. Antimicrob. Agents Chemother. *54*, 2655–2659.

Chen, S.C., Slavin, M.A., and Sorrell, T.C. (2011). Echinocandin antifungal drugs in fungal infections: a comparison. Drugs *71*, 11–41.

Cleary, J.D., Garcia-Effron, G., Chapman, S.W., and Perlin, D.S. (2008). Reduced Candida glabrata susceptibility secondary to an FKS1 mutation developed during candidemia treatment. Antimicrob. Agents Chemother. *52*, 2263–2265.

Cota, J.M., Grabinski, J.L., Talbert, R.L., Burgess, D.S., Rogers, P.D., Edlind, T.D., and Wiederhold, N.P. (2008). Increases in SLT2 expression and chitin content are associated with incomplete killing of Candida glabrata by caspofungin. Antimicrob. Agents Chemother. *52*, 1144–1146.

Dannaoui, E., Desnos-Ollivier, M., Garcia-Hermoso, D., Grenouillet, F., Cassaing, S., Baixench, M.T., Bretagne, S., Dromer, F., Lortholary, O., and French Mycoses Study Group (2012). Candida spp. with acquired echinocandin resistance, France, 2004–2010. Emerg Infect. Dis. *18*, 86–90.

Debono, M., Abbott, B.J., Turner, J.R., Howard, L.C., Gordee, R.S., Hunt, A.S., Barnhart, M., Molloy, R.M., Willard, K.E., Fukuda, D., *et al.* (1988). Synthesis and evaluation of LY121019, a member of a series of semisynthetic analogues of the antifungal lipopeptide echinocandin Ba. Ann. N.Y. Acad. Sci. *544*, 152–167.

Desnos-Ollivier, M., Bretagne, S., Raoux, D., Hoinard, D., Dromer, F., Dannaoui, E., and European Committee on Antibiotic Susceptibility Testing (2008). Mutations in the fks1 gene in *Candida albicans, C. tropicalis,* and *C. krusei* correlate with elevated caspofungin MICs uncovered in AM3 medium using the method of the European Committee on Antibiotic Susceptibility Testing. Antimicrob. Agents Chemother. *52*, 3092–3098.

Douglas, C.M. (2001) Fungal beta(1,3)-D-glucan synthesis. Med. Mycol. 39 (Suppl. 1), 55–66.

Douglas, C.M., Foor, F., Marrinan, J.A., Morin, N., Nielsen, J.B., Dahl, A.M., Mazur, P., Baginsky, W., Li, W., el-Sherbeini, M., *et al.* (1994). The *Saccharomyces cerevisiae* FKS1 (ETG1) gene encodes an integral membrane protein which is a subunit of 1,3-β-D-glucan synthase. Proc. Natl. Acad. Sci. U.S.A. *91*, 12907–12911.

Douglas, C.M., D'Ippolito, J.A., Shei, G.J., Meinz, M., Onishi, J., Marrinan, J.A., Li, W., Abruzzo, G.K., Flattery, A., Bartizal, K., *et al.* (1997). Identification of the FKS1 gene of *Candida albicans* as the essential target of 1,3-beta-D-glucan synthase inhibitors. Antimicrob. Agents Chemother. *41*, 2471–2479.

Douglas, C.M., Marrinan, J.A., Li, W., and Kurtz, M.B. (1994). A *Saccharomyces cerevisiae* mutant with echinocandin-resistant 1,3-beta-D-glucan synthase. J. Bacteriol. *176*, 5686–5696.

Drakulovski, P., Dunyach, C., Bertout, S., Reynes, J., and Mallié, M. (2011). A *Candida albicans* strain with high MIC for caspofungin and no FKS1 mutations exhibits a high chitin content and mutations in two chitinase genes. Med. Mycol. *49*, 467–474.

van Duin, D., Casadevall, A., and Nosanchuk, J.D. (2002). Melanization of *Cryptococcus neoformans* and *Histoplasma capsulatum* reduces their susceptibilities to amphotericin B and caspofungin. Antimicrob. Agents Chemother. *46*, 3394–3400.

Edlind, T.D., and Katiyar, S.K. (2004). The echinocandin "target" identified by cross-linking is a homolog of Pil1 and Lsp1, sphingolipid-dependent regulators of cell wall integrity signaling. Antimicrob. Agents Chemother. *48*, 4491.

Eisenman, H.C., and Casadevall, A. (2012). Synthesis and assembly of fungal melanin. Appl. Microbiol. Biotechnol. *93*, 931–940.

el-Sherbeini, M., and Clemas, J.A. (1995). Cloning and characterization of GNS1: a *Saccharomyces cerevisiae* gene involved in synthesis of 1,3-beta-glucan *in vitro*. J. Bacteriol. *177*, 3227–3234.

Espinel-Ingroff, A. (2003). *In vitro* antifungal activities of anidulafungin and micafungin, licensed agents and the investigational triazole posaconazole as determined by NCCLS methods for 12,052 fungal isolates: review of the literature. Rev. Iberoam. Micol. *20*, 121–136.

Firon, A., Beauvais, A., Latge, J.P., Couve, E., Grosjean-Cournoyer, M.C., and D'Enfert, C. 2002. Characterization of essential genes by parasexual genetics in the human fungal pathogen *Aspergillus fumigatus*: impact of genomic rearrangements associated with electroporation of DNA. Genetics *161*, 1077–1087.

Garcia-Effron, G., Kontoyiannis, D.P., Lewis, R.E., and Perlin, D.S. (2008). Caspofungin-resistant *Candida tropicalis* strains causing breakthrough fungemia in patients at high risk for hematologic malignancies. Antimicrob. Agents Chemother. *52*, 4181–4183.

Garcia-Effron, G., Lee, S., Park, S., Cleary, J.D., and Perlin, D.S. (2009a). Effect of *Candida glabrata* FKS1 and FKS2 mutations on echinocandin sensitivity and kinetics of 1,3-beta-D-glucan synthase: implication for the existing susceptibility breakpoint. Antimicrob. Agents Chemother. *53*, 3690–3699.

Garcia-Effron, G., Park, S., and Perlin, D.S. (2009b). Correlating echinocandin MIC and kinetic inhibition of fks1 mutant glucan synthases for *Candida albicans*:

implications for interpretive breakpoints. Antimicrob. Agents Chemother. 53, 112–122.

Gerik, K.J., Donlin, M.J., Soto, C.E., Banks, A.M., Banks, I.R., Maligie, M.A., Selitrennikoff, C.P., and Lodge, J.K. (2005). Cell wall integrity is dependent on the PKC1 signal transduction pathway in Cryptococcus neoformans. Mol. Microbiol. *58*, 393–408.

Ha, Y.S., Covert, S.F., and Momany, M. (2006). FsFKS1, the 1,3-beta-glucan synthase from the caspofungin-resistant fungus *Fusarium solani*. Eukaryot. Cell *5*, 1036–1042.

Hage, C.A., Connolly, P., Horan, D., Durkin, M., Smedema, M., Zarnowski, R., Smith, P., and Wheat, L.J. (2011). Investigation of the efficacy of micafungin in the treatment of histoplasmosis using two North American strains of *Histoplasma capsulatum*. Antimicrob. Agents Chemother. *55*, 4447–4450.

Healey, K.R., Katiyar, S.K., Castanheira, M., Pfaller, M.A., and Edlind, T.D. (2011). *Candida glabrata* mutants demonstrating paradoxical reduced caspofungin susceptibility but increased micafungin susceptibility. Antimicrob. Agents Chemother. *55*, 3947–3949.

Healey, K.R., Katiyar, S.K., Raj, S., and Edlind, T.D. (2012). CRS-MIS in *Candida glabrata*: sphingolipids modulate echinocandin–Fks interaction. Mol. Microbiol. *86*, 303–313.

Henry, C., Latgé, J.P., and Beauvais, A. (2011). αi,3-glucans are dispensable in *Aspergillus fumigatus*. Eukaryot. Cell *11*, 26–29.

Hof, H., and Dietz, A. (2009). Antifungal activity of anidulafungin, a product of *Aspergillus nidulans*, against *Aspergillus nidulans*. Int. J. Antimicrob. Agents 33, 285–286.

Inoue, S.B., Takewaki, N., Takasuka, T., Mio, T., Adachi, M., Fujii, Y., Miyamoto, C., Arisawa, M., Furuichi, Y., and Watanabe, T. (1995). Characterization and gene cloning of 1,3-beta-D-glucan synthase from *Saccharomyces cerevisiae*. Eur. J. Biochem. *231*(3), 845–54.

Ishihara, S., Hirata, A., Nogami, S., Beauvais, A., Latge, J.P., and Ohya, Y. (2007). Homologous subunits of 1,3-beta-glucan synthase are important for spore wall assembly in *Saccharomyces cerevisiae*. Eukaryot. Cell *6*, 143–156.

Jensen, R.H., Johansen, H.K., and Arendrup, M.C. (2013). Stepwise development of a homozygous S80P substitution in Fks1p, conferring echinocandin resistance in *Candida tropicalis*. Antimicrob. Agents Chemother. *57*, 614–617.

Johnson, M.E., and Edlind, T.D. (2012). Topological and mutational analysis of *Saccharomyces cerevisiae* Fks1. Eukaryot. Cell *11*, 952–960.

Johnson, M.E., Katiyar, S.K., and Edlind, T.D. (2011). New Fks hot spot for acquired echinocandin resistance in *Saccharomyces cerevisiae* and its contribution to intrinsic resistance of Scedosporium species. Antimicrob. Agents Chemother. *55*, 3774–3781.

van der Kaaden, M., Breukink, E., and Pieters, R.J. (2012). Synthesis and antifungal properties of papulacandin derivatives. Beilstein J. Org. Chem. *8*, 732–737.

Kahn, J.N., Garcia-Effron, G., Hsu, M.J., Park, S., Marr, K.A., and Perlin, D.S. (2007). Acquired echinocandin resistance in a *Candida krusei* isolate due to modification of glucan synthase. Antimicrob. Agents Chemother. *51*, 1876–1878.

Katiyar, S.K., Alastruey-Izquierdo, A., Healey, K.R., Johnson, M.E., Perlin, D.S., and Edlind, T.D. (2012). Fks1 and Fks2 are functionally redundant but differentially regulated in *Candida glabrata*: implications for echinocandin resistance. Antimicrob. Agents Chemother. *56*, 6304–6309.

Katiyar, S.K., and Edlind, T.D. (2009). Role for Fks1 in the intrinsic echinocandin resistance of *Fusarium solani* as evidenced by hybrid expression in *Saccharomyces cerevisiae*. Antimicrob. Agents Chemother. *53*, 1772–1778.

Katiyar, S., Pfaller, M., and Edlind, T. (2006). *Candida albicans* and *Candida glabrata* clinical isolates exhibiting reduced echinocandin susceptibility. Antimicrob. Agents Chemother. *50*, 2892–2894.

Kellner, E.M., Orsborn, K.I., Siegel, E.M., Mandel, M.A., Orbach, M.J., and Galgiani, J.N. (2005). *Coccidioides posadasii* contains a single 1,3-beta-glucan synthase gene that appears to be essential for growth. Eukaryot. Cell *4*, 111–120.

Kondoh, O., Tachibana, Y., Ohya, Y., Arisawa, M., and Watanabe, T. (1997). Cloning of the RHO1 gene from *Candida albicans* and its regulation of beta-1,3-glucan synthesis. J. Bacteriol. *179*, 7734–7741.

Kondoh, O., Takasuka, T., Arisawa, M., Aoki, Y., and Watanabe, T. (2002). Differential susceptibility between Fks1p and Fks2p against a novel β-1,3-glucan synthase inhibitor, aerothricin3. J. Biol. Chem. *277*, 41744–41749.

Kraus, P.R., Fox, D.S., Cox, G.M., and Heitman, J. (2003). The *Cryptococcus neoformans* MAP kinase Mpk1 regulates cell integrity in response to antifungal drugs and loss of calcineurin function. Mol. Microbiol. *48*, 1377–1387.

Kurtz, M.B., Heath, I.B., Marrinan, J., Dreikorn, S., Onishi, J., and Douglas, C. (1994). Morphological effects of lipopeptides against Aspergillus fumigatus correlate with activities against (1,3)-beta-D-glucan synthase. Antimicrob. Agents Chemother. *38*, 1480–1489.

Laverdiere, M., Lalonde, R.G., Baril. J.G., Sheppard. D.C., Park, S., and Perlin, D.S. (2006). Progressive loss of echinocandin activity following prolonged use for treatment of *Candida albicans* oesophagitis. J. Antimicrob. Chemother. *57*, 705–708.

Lee, K.K., MacCallum, D.M., Jacobsen, M.D., Walker, L.A., Odds, F.C., Gow, N.A.R., and Munro, C.A. (2012). Elevated cell wall chitin in *Candida albicans* confers echinocandin resistance *in vivo*. Antimicrob. Agents. Chemother. *56*, 208–217.

Lesage, G., Sdicu, A.M., Ménard, P., Shapiro, J., Hussein, S., and Bussey, H. (2004). Analysis of beta-1,3-glucan assembly in *Saccharomyces cerevisiae* using a synthetic interaction network and altered sensitivity to caspofungin. Genetics *167*, 35–49.

Levin, D.E. (2011). Regulation of cell wall biogenesis in *Saccharomyces cerevisiae*: the cell wall integrity signaling pathway. Genetics *189*, 1145–1175.

Maligie, M.A., and Selitrennikoff, C.P. (2005). *Cryptococcus neoformans* resistance to echinocandins: (1,3) beta-glucan synthase activity is sensitive to echinocandins. Antimicrob, Agents Chemother. *49*, 2851–2856.

Markovich, S., Yekutiel, A., Shalit, I., Shadkchan, Y., and Osherov, N. (2004). Genomic approach to identification of mutations affecting caspofungin susceptibility in *Saccharomyces cerevisiae*. Antimicrob. Agents Chemother. *48*, 3871–3876.

Martins, I.M., Cortés, J.C., Muñoz, J., Moreno, M.B., Ramos, M., Clemente-Ramos, J.A., Durán, A., and Ribas, J.C. (2011). Differential activities of three families of specific beta(1,3)glucan synthase inhibitors in wild-type and resistant strains of fission yeast. J. Biol. Chem. *286*, 3484–496.

Mazur, P., Morin, N., Baginsky, W., el-Sherbeini, M., Clemas, J.A., Nielsen, J.B., and Foor, F. (1995). Differential expression and function of two homologous subunits of yeast 1,3-beta-D-glucan synthase. Mol. Cell. Biol. *15*, 5671–5681.

Messer, S.A., Jones, R.N., Moet, G.J., Kirby, J.T., and Castanheira, M. (2010). Potencty of anidulafungin compared to nine other antifungal agents tested against *Candida* spp., *Cryptococcus* spp., and *Aspergillus* spp.,: results from the Global SENTRY Antimicrobial Surveillance Program (2008). J. Clin. Microbiol. *48*, 2984–2987.

Mio, T., Adachi-Shimizu, M., Tachibana, Y., Tabuchi, H., Inoue, S.B., Yabe, T., Yamada-Okabe, T., Arisawa, M., Watanabe, T., and Yamada-Okabe, H. (1997). Cloning of the *Candida albicans* homolog of *Saccharomyces cerevisiae* GSC1/FKS1 and its involvement in beta-1,3-glucan synthesis. J. Bacteriol. *179*, 4096–4105.

Mizuno, K., Yagi, A., Satoi, S., Takada, M., and Hayashi, M. (1977). Studies on aculeacin. I. Isolation and characterization of aculeacin A. J. Antibiot. (Tokyo) *30*, 297–302.

Mukherjee, P.K., Sheehan, D., Puzniak, L., Schlamm, H., and Ghannoum, M.A. (2011). Echinocandins: are they all the same? J. Chemother. *23*, 319–25.

Nakai, T., Uno, J., Ikeda, F., Tawara, S., Nishimura, K., and Miyaji, M. (2003). *In vitro* antifungal activity of Micafungin (FK463) against dimorphic fungi: comparison of yeast-like and mycelial forms. Antimicrob. Agents Chemother. *47*, 1376–1381.

Niimi, K., Maki, K., Ikeda, F., Holmes, A.R., Lamping, E., Niimi, M., Monk, B.C., and Cannon, R.D. (2006). Overexpression of *Candida albicans* CDR1, CDR2, or MDR1 does not produce significant changes in echinocandin susceptibility. Antimicrob, Agents Chemother. *50*, 1148–1155.

Niimi, K., Monk, B.C., Hirai, A., Hatakenaka, K., Umeyama, T., Lamping, E., Maki, K., Tanabe, K., Kamimura, T., Ikeda, F., *et al.* (2010). Clinically significant micafungin resistance in *Candida albicans* involves modification of a glucan synthase catalytic subunit GSC1 (FKS1) allele followed by loss of heterozygosity. J. Antimicrob. Chemother. *65*, 842–852.

Niimi, K., Woods, M.A., Maki, K., Nakayama, H., Hatakenaka, K., Chibana, H., Ikeda, F., Ueno, K., Niimi, M., Cannon, R.D., and Monk, B.C. (2012). Reconstitution of high-level micafungin resistance detected in a clinical isolate of *Candida glabrata* identifies functional homozygosity in glucan synthase gene expression. J. Antimicrob. Chemother. *67*, 1666–1676.

Ohyama T., Kurihara, Y., Ono, Y., Ishikawa, T., Miyakoshi, S., Hamano, K., Arai, M., Suzuki, T., Igari, H., Suzuki, Y., and Inukai, M., (2000). Arborcandins A, B, C, D, E and F, novel 1,3-beta-glucan synthase inhibitors: production and biological activity. J. Antibiot. *53*, 1108–1116.

Ohyama, T., Miyakoshi, S., and Isono, F. (2004). *FKS1* mutations responsible for selective resistance of *Saccharomyces cerevisiae* to the novel 1,3-β-glucan synthase inhibitor arborcandin C. Antimicrob. Agents Chemother. *48*, 319–322.

Okada, H., Abe, M., Asakawa-Minemura, A., Hirata, H., Qadota, K., Morishita, S., Ohnuki, S., Nogami, S., and Ohya, Y. (2010). Multiple functional domains of the yeast 1,3-β-glucan synthase subunit Fks1p revealed by quantitative phenotypic analysis of temperature-sensitive mutants. Genetics *184*, 1013–1024.

Paderu, P., Park, S., and Perlin, D.S. (2004). Caspofungin uptake is mediated by a high-affinity transporter in *Candida albicans*. Antimicrob. Agents Chemother. *48*, 3845–3849.

Pappas, P.G., Kauffman, C.A., Andes, D., Benjamin, D.K., Jr., Calandra, T.F., Edwards, J.E., Jr., Filler, S.G., Fisher, J.F., Kullberg, B.J., Zeichner, L.O., *et al.* (2009) Clinical practice guidelines for the management of candidiasis: 2009 update by the Infectious Diseases Society of America. Clin. Infect. Dis. *48*, 503–535.

Park, S., Kelly, R., Kahn, J.N., Robles, J., Hsu, M.J., Register, E., Li, W., Vyas, V., Fan, H., Abruzzo, G., *et al.* (2005). Specific substitutions in the echinocandin target Fks1p account for reduced susceptibility of rare laboratory and clinical *Candida* sp. isolates. Antimicrob. Agents Chemother. *49*, 3264–3273.

Peláez, F., Cabello, A., Platas, G., Díez, M.T., González del Val, A., Basilio, A., Martán, I., Vicente, F., Bills, G.E., Giacobbe, R.A., *et al.* (2000). The discovery of enfumafungin, a novel antifungal compound produced by an endophytic Hormonema species biological activity and taxonomy of the producing organisms. Syst. Appl. Microbiol. *23*, 333–343.

Perez, P., Varona, R., Garcia-acha, I., and Duran, A. (1981). Effect of papulacandin B and aculeacin A on β-(1,3)-glucan synthase from Geotrichum lactis. FEBS Letters *129*, 249–252.

Perlin, D.S. (2011). Current perspectives on echinocandin class drugs. Future Microbiol. *6*, 441–457.

Pfaller, M.A. (2012). Antifungal drug resistance: mechanisms, epidemiology, and consequences for treatment. Am. J. Med. *125*, S3–13.

Pfaller, M.A., Castanheira, M.A., Lockhart, S.R., Ahlquist, A.M., Messer, S.A., and Jones, R.N. (2012). Frequency of decreased susceptibility and resistance to echinocandins among fluconazole-resistant bloodstream isolates of *Candida glabrata*. J. Clin. Microbiol. *50*, 1199–1203.

Pfaller, M.A., Messer, S.A., Motyl, M.R., Jones, R.N., and Castanheira, M. (2013). Activity of MK-3118, a new oral glucan synthase inhibitor, tested against *Candida* spp. by two international methods (CLSI and EUCAST). J. Antimicrob. Chemother. *68*, 858–863.

Pfeiffer, C.D., Garcia-Effron, A.K., Zaas, R.N., Perfect, J.R., Perlin, D.S., and Alexander, B.D. (2010). Breakthrough invasive candidiasis in patients on micafungin. J. Clin. Microbiol. *48*, 2373–2380.

Posteraro, B., Sanguinetti, M., Fiori, B., La Sorda, M., Spanu, T., Sanglard, D., and Fadda, G. (2006). Caspofungin activity against clinical isolates of azole cross-resistant *Candida glabrata* overexpressing efflux pump genes. J. Antimicrob. Chemother. *58*, 458–461.

Powles, M.A., Liberator, P., Anderson, J., Karkhanis, Y., Dropinski, J.F., Bouffard, F.A., Balkovec, J.M., Fujioka, H., Aikawa, M., McFadden, D., and Schmatz, D. (1998). Efficacy of MK-991 (L-743,872), a semisynthetic pneumocandin, in murine models of *Pneumocystis carinii*. Antimicrob. Agents Chemother. *42*, 1985–1989.

Radding, J.A., Heidler, S.A., and Turner, W.W. (1998). Photoaffinity analog of the semisynthetic echinocandin LY303366: identification of echinocandin targets in *Candida albicans*. Antimicrob. Agents Chemother. *42*, 1187–1194.

Reese, A.J., Yoneda, A., Breger, J.A., Beauvais, A., Liu, H., Griffith, C.L., Bose, I., Kim, M.J., Skau, C., Yang, S., *et al.* (2007). Loss of cell wall alpha(1–3) glucan affects *Cryptococcus neoformans* from ultrastructure to virulence. Mol. Microbiol. *63*, 1385–1398.

Reinoso-Martín, C., Schüller, C., Schuetzer-Muehlbauer, M., and Kuchler, K. (2003). The yeast protein kinase C cell integrity pathway mediates tolerance to the antifungal drug caspofungin through activation of Slt2p mitogen-activated protein kinase signaling. Eukaryot. Cell 2, 1200–1210.

Richards, T.S., Oliver, B.G., and White, T.C. (2008). Micafungin activity against *Candida albicans* with diverse azole resistance phenotypes. J. Antimicrob. Chemother. *62*, 349–355.

Rocha, E.M., Garcia-Effron, G., Park, S., and Perlin, D.S. (2007). A Ser678Pro substitution in Fks1p confers resistance to echinocandin drugs in *Aspergillus fumigatus*. Antimicrob. Agents Chemother. *51*, 4174–4176.

Römmele, G., Traxler, P., and Wehrli, W. (1983). Papulacandins – the relationship between chemical structure and effect on glucan synthesis in yeast. J. Antibiot. (Tokyo) *36*, 1539–1542.

Sawistowska-Schröder, E.T., Kerridge, D., and Perry, H. (1984). Echinocandin inhibition of 1,3-beta-D-glucan synthase from *Candida albicans*. FEBS Lett. *173*, 134–138.

Schimoler-O'Rourke, R., Renault, S., Mo, W., and Selitrennikoff, C.P. (2003). *Neurospora crassa* FKS protein binds to the (1,3)-β-glucan synthase substrate, UDP-glucose. Curr. Microbiol. *46*, 0408–0412.

Schuetzer-Muehlbauer, M., Willinger, B., Krapf, G., Enzinger, S., Presterl, E., and Kuchler, K. (2003). The *Candida albicans* Cdr2p ATP-binding cassette (ABC) transporter confers resistance to caspofungin. Mol. Microbiol. *48*(1), 225–235.

Stevens, D.A., Ichinomiya, M., Koshi, Y., and Horiuchi, H. (2006). Escape of *Candida* from caspofungin inhibition at concentrations above the MIC (paradoxical effect) accomplished by increased cell wall chitin; evidence for beta-1,6-glucan synthesis inhibition by caspofungin. Antimicrob, Agents Chemother. *50*, 3160–3161.

Thompson, J.R., Douglas, C.M., Li, W., Jue, C.K., Pramanik, B., Yuan, X., Rude, T.H., Toffaletti, D.L., Perfect, J.R., and Kurtz, M. (1999). A glucan synthase FKS1 homolog in *Cryptococcus neoformans* is single copy and encodes an essential function. J. Bacteriol. *181*, 444–453.

Tóth, V., Nagy, C.T., Pócsi, I., and Emri, T. (2012). The echinocandin B producer fungus *Aspergillus nidulans* var. *roseus* ATCC 58397 does not possess innate resistance against its lipopeptide antimycotic. Appl. Microbiol. Biotechnol. 95, 113–122.

Walker, L.A., Gow, N.A.R., and Munro, C.A. (2010). Fungal echinocandin resistance. Fungal Genet. Biol. *47*, 117–126.

Walker, L.A., Gow, N.A., and Munro, C.A. (2013). Elevated chitin content reduces the susceptibility of Candida species to caspofungin. Antimicrob. Agents Chemother. *57*, 146–154.

Walker, S.S., Xu, Y., Triantafyllou, I., Waldman, M.F., Mendrick, C., Brown, N., Mann, P., Chau, A., Patel, R., Bauman, N., *et al.* (2011). Discovery of a novel class of orally active antifungal β-1,3-d-glucan synthase inhibitors. Antimicrob. Agents Chemother. *55*, 5099–5106.

Yoshimi, A., Sano, M., Inaba, A., Kokubun, Y., Fujioka, T., Mizutani, O., Hagiwara, D., Fujikawa, T., Nishimura, M., Yano, S., *et al.* (2013). Functional analysis of the α-1,3-glucan synthase genes agsA and agsB in *Aspergillus nidulans*: agsB is the major α-1,3-glucan synthase in this fungus. PLoS One *8*, e54893.

Zimbeck, A.J., Iqbal, N., Ahlquist, A.M., Farley, M.M., Harrison, L.H., Chiller, T., and Lockhart, S.R. (2010). FKS mutations and elevated echinocandin MIC values among *Candida glabrata* isolates from U.S. population-based surveillance. Antimicrob. Agents Chemother. *54*, 5042–5047.

Biofilms and Antifungal Resistance

4

Emily P. Fox, Sheena D. Singh-Babak, Nairi Hartooni and
Clarissa J. Nobile

Abstract

Infections caused by pathogenic fungi are a significant medical problem, as they are able to disseminate to nearly every organ of the human body and there are few classes of antifungal drugs available as therapeutic options. Fungal infections are even more difficult to manage when they are biofilm-associated due to the natural properties of the biofilm mode of growth. Like bacterial biofilms, fungal biofilms consist of adherent communities of cells that are attached to a substrate and to one another, and are enclosed in a protective extracellular matrix material. Biofilms in general are able to withstand much higher concentrations of antimicrobial agents compared with single free-floating (or planktonic) cells, making biofilm infections extremely challenging to treat. In this chapter, we review the current knowledge of biofilm formation in representative, pathogenic species from several phyla of fungi. We also discuss the molecular mechanisms of drug resistance in fungal biofilms, the current standards of care for treating these biofilm-associated infections, and strategies for overcoming challenges in developing new antifungal drugs with efficacies against biofilms.

Introduction to fungal biofilms

Microbial species have long been thought to exist as single, free-floating cells, in self-contained units that are self-reliant for metabolism, cell division, reproduction, and survival. While it is true that microbial cells have all of these characteristics, over the last two decades it has become apparent that the predominant lifestyle of microbes is to exist in a biofilm. Biofilms are adherent communities of cells that have distinct properties from their planktonic, or free-floating, counterparts (Costerton *et al.*, 1999; Donlan, 2001; López *et al.*, 2010; Nobile and Mitchell, 2007; Wolcott *et al.*, 2012). Biofilms were first described in bacteria, but it is now widely recognized that many fungal species also form biofilms (Ganguly and Mitchell, 2011; Ramage *et al.*, 2009). Bacterial, fungal, and mixed-species biofilms are found in an enormous variety of natural and manmade environments. Although many biofilms are beneficial or cause no harm to their environments/hosts, there is a subset of biofilms that are detrimental to human health. Indeed, recent estimates by the NIH in PA-03-047 indicate that biofilms are responsible for over 80% of all microbial infections (http://grants.nih.gov/grants/guide/pa-files/PA-03-047.html) (Fox and Nobile, 2012). In this chapter, we focus our attention on fungal biofilms formed by species that specifically colonize and cause disease in humans. Fungal biofilms are a particularly serious medical problem, as they can infect nearly every organ of a mammalian host as well as numerous implanted medical devices, and they are inherently resistant to known antifungal agents, making these infections exceedingly difficult to treat (Elliott, 1988; Ganguly and Mitchell, 2011; Habash and Reid, 1999; Jabra-Rizk *et al.*, 2004; Mathé and Van Dijck, 2013; Ramage *et al.*, 2012; Tumbarello *et al.*, 2007).

The most notorious fungal biofilm-former is the yeast, *Candida albicans*, which forms highly structured biofilms composed of multiple cell types (round budding yeast-form cells, elongated hyphal cells, and pseudohyphal cells) encased in a secreted extracellular matrix material (Chandra *et al.*, 2001; Fox and Nobile, 2012; Nobile and Mitchell, 2006; Ramage *et al.*, 2001a, 2009). *Candida* species, in particular, are the fourth most frequent cause of bloodstream infections in hospitals, accounting for 15% of all hospital-acquired sepsis cases, and are the major fungal species isolated from infections of medical devices in humans (Pfaller and Diekema, 2007; Wisplinghoff *et al.*, 2004). Susceptible devices include central venous catheters, urinary catheters, heart valves, pacemakers, joint prostheses, dentures, and contact lenses (Cauda, 2009; Donlan and Costerton, 2002; Kojic and Darouiche, 2004; Sed-diki *et al.*, 2013; Shoham, 2010). Biofilm growth on implanted medical devices, in general, has the potential to seed disseminated bloodstream infections and lead to invasive infections of multiple organs. Over five million central venous catheters are placed each year in the United States (Kojic and Darouiche, 2004). Currently – even with recent improved clinical approaches – biofilm infection occurs in up to 54% of these catheters. With an estimated 100,000 deaths and $6.5 billion in excess expenditure annually in the United States alone, these infections have huge health and economic consequences. Because fungal biofilms are exceptionally resistant to current antifungal drugs, removal of the colonized medical device combined with administration of high doses of antifungal drugs (often cytotoxic to mammalian cells) are generally required to treat the infection (Andes *et al.*, 2012; Cornely *et al.*, 2012; Lepak and Andes, 2011; Mermel *et al.*, 2009; Perlroth *et al.*, 2007). This is a costly and complicated treatment and is sometimes not even possible, as many severely ill patients are unable to tolerate these doses of antifungals, and certain implanted medical devices cannot be safely removed. Indeed, current treatments for biofilm-based infections are simply ineffective at destroying the biofilm reservoir, and, thus, novel therapeutics specifically designed to target the biofilm are urgently needed to treat these prevalent infections.

Recognition of the importance of the biofilm growth state for virulence of *Candida* species has motivated the investigation into the biofilm-forming capabilities of other fungal species, particularly those that are capable of infecting humans (Fridkin and Jarvis, 1996; Perlroth *et al.*, 2007; Singh *et al.*, 2011). Although this research is in its infancy, many fungal species have now been described as forming biofilms both *in vitro* and *in vivo*, and we will discuss examples of these fungal biofilm-formers that represent three phyla: the Ascomycota (to which *Candida* species belong), the Basidiomycota, and the Zygomycota.

Similar to *Candida* species, many non-*albicans* fungal pathogens form biofilms that are also enhanced for drug resistance (Di Bonaventura *et al.*, 2006; Bowyer *et al.*, 2011; Cushion and Collins, 2011; Figueredo *et al.*, 2013; Jabra-Rizk *et al.*, 2004; Martinez and Casadevall, 2006a,b; Pettit *et al.*, 2010; Ramage *et al.*, 2012; Seidler *et al.*, 2008; Zhang *et al.*, 2012). In this chapter, we discuss the currently available antifungal drugs and known mechanisms of biofilm drug resistance for these three fungal phyla. In summary, the antifungal resistance mechanisms of fungal biofilms are multifactorial, involving overlapping mechanisms with those found in planktonic cells, in addition to biofilm-specific mechanisms (Pierce and Lopez-Ribot, 2013; Ramage *et al.*, 2012; Taff *et al.*, 2013). Most research to date on drug resistance of fungal biofilms has focused on *C. albicans*, however it is likely that several of these resistance mechanisms are conserved to both closely and more distantly related fungal species. Moreover, it is also likely that future research will reveal new mechanisms of biofilm drug resistance that are unique to these other fungal pathogens. Biofilm drug resistance is a huge hurdle to overcome in developing new and effective therapeutics for biofilm infections, and we will discuss these challenges, as well as some potential strategies for overcoming them. While most research is currently focused on *Candida* biofilms, this work will need to be expanded in light of the emerging infectious potential of other biofilm-forming fungal species.

Fungal biofilm infections of humans

The incidence of fungal infections has risen gradually over the last 30 years as advances in modern medicine have successfully prolonged the lives of critically ill patients, such as ICU, HIV/AIDS-infected, transplant, cancer, and neonatal ICU patients (Diekema *et al.*, 2012; Gudlaugsson *et al.*, 2003; Pfaller, 1996; Pfaller *et al.*, 1998, 2001, 2012; Picazo *et al.*, 2008; Whimbey *et al.*, 1986). The medical advances prolonging the lives of these patients have coincided with a rise in the use of broad spectrum antibiotics, indwelling catheters, surgical immunosuppression, and chemotherapy, all factors strongly supporting the overgrowth of fungi (Calderone, 2002; Odds, 1988; Ramage *et al.*, 2009). As more and more fungi have emerged as significant pathogens, the existence of these fungi in recalcitrant biofilms, acting as reservoirs to infection, has complicated the abilities of healthcare workers to effectively treat these fungal infections.

Most known fungal mammalian pathogens are from the phyla Ascomycota, Zygomycota, and Basidiomycota (Table 4.1) (Cuomo and Birren, 2010; Fridkin and Jarvis, 1996; Perlroth *et al.*, 2007). The Ascomycota phylum include species from the genera *Acremonium, Aspergillus, Blastomyces, Blastoschizomyces, Candida, Cladosporium, Coccidioides, Fusarium, Histoplasma, Paracoccidioides, Pneumocystis, Saccharomyces* and *Scedosporium*; the Basidiomycota phylum include the genera *Cryptococcus, Malassezia,* and *Trichosporon*; and the Zygomycota phylum include the genera *Apophysomyces, Lichtheimia, Rhizomucor,* and *Rhizopus*. For some of these genera, biofilm formation is clearly implicated in pathogenesis,

Table 4.1 Biofilm formation of representative fungi that infect mammals

Species	Biofilm?	Phylum
Acremonium implicatum	Yes	Ascomycota
Aspergillus fumigatus	Yes	Ascomycota
Blastomyces dermatitidis	Unknown	Ascomycota
Blastoschizomyces capitatus	Yes	Ascomycota
Candida albicans	Yes	Ascomycota
Candida dubliniensis	Yes	Ascomycota
Candida glabrata	Yes	Ascomycota
Candida krusei	Yes	Ascomycota
Candida parapsilosis	Yes	Ascomycota
Candida tropicalis	Yes	Ascomycota
Cladosporium sphaerospermum	Yes	Ascomycota
Coccidioides immitis	Yes	Ascomycota
Fusarium solani	Yes	Ascomycota
Histoplasma capsulatum	Yes	Ascomycota
Paracoccidioides brasiliensis	Unknown	Ascomycota
Pneumocystis jirovecii	Yes	Ascomycota
Saccharomyces cerevisiae	Yes	Ascomycota
Scedosporium prolificans	Unknown	Ascomycota
Cryptococcus gattii	Yes	Basidiomycota
Cryptococcus neoformans	Yes	Basidiomycota
Malassezia pachydermatis	Yes	Basidiomycota
Trichosporon asahii	Yes	Basidiomycota
Apophysomyces elegans	No	Zygomycota
Lichtheimia corymbifera	Yes	Zygomycota
Rhizomucor pusillus	Yes	Zygomycota
Rhizopus oryzae	Yes	Zygomycota

but for others it is unclear as to what extent the infections they cause are biofilm-associated. Notably, many of these fungal biofilm-formers have been tested for drug susceptibility *in vitro*, and cells in biofilms have been found to be enhanced for drug resistance compared with cells grown in planktonic culture. These include members from *Candida, Aspergillus, Pneumocystis, Fusarium, Acremonium, Cladosporium, Cryptococcus, Malassezia,* and *Trichosporon* (Bachmann *et al.*, 2002; Di Bonaventura *et al.*, 2006; Bowyer *et al.*, 2011; Cushion and Collins, 2011; Figueredo *et al.*, 2013; Jabra-Rizk *et al.*, 2004; Martinez and Casadevall, 2006a,b; Pettit *et al.*, 2010; Ramage *et al.*, 2012; Seidler *et al.*, 2008; Zhang *et al.*, 2012). Below, we will describe the infections caused by representative species of each genus and evaluate whether these species are able to form clinically relevant biofilms. This information is summarized in Table 4.1.

Ascomycota

The major fungal pathogen known to form biofilm-mediated infections in mammals is *Candida albicans*, a member of the phylum Ascomycota (Cauda, 2009; Elliott, 1988; Ganguly and Mitchell, 2011; Hawser and Douglas, 1994; Kojic and Darouiche, 2004; Nobile and Mitchell, 2006, 2007). *C. albicans* causes biofilm-associated infections that range from superficial mucosal, cutaneous, and wound infections (vaginal yeast infections, thrush, nappy rash, denture stomatitis, eye infections, combat and burn wound infections) to severe candidaemia (bloodstream infections) and disseminated candidiasis (colonization of internal organs or mucosal surfaces). Candidaemia and disseminated candidiasis are serious medical conditions with estimated mortality rates ranging from 30% to 50% (Beck-Sagué and Jarvis, 1993; Kibbler *et al.*, 2003; Pfaller *et al.*, 1998). Recent work has additionally revealed that many other *Candida* species are also capable of forming clinically relevant biofilms. The most closely related species to *C. albicans*: *Candida dubliniensis, Candida parapsilosis, Candida tropicalis,* and *Candida krusei*, have been similarly implicated in biofilm-associated infections (Fernandes and Dias, 2013; Hasan *et al.*, 2009; Ramage *et al.*, 2001b, 2006; Sullivan *et al.*, 1995). *Candida glabrata*, a more distantly related species to *C. albicans*, has been reported to form thin biofilms *in vitro*, and robust biofilms on biotic and abiotic surfaces associated with the human host (Culakova *et al.*, 2013; Iraqui *et al.*, 2005; Kaur *et al.*, 2005; Kucharíková *et al.*, 2011; Ramage *et al.*, 2006; Silva *et al.*, 2011). The *Candida* species in general are thought to form biofilms on the mucosal epithelium of multiple body sites, including the oral and nasal cavities, gastrointestinal tract, and the urogenital tract (Bouza and Muñoz, 2008; Dongari-Bagtzoglou *et al.*, 2009; Ganguly and Mitchell, 2011; Harriott *et al.*, 2010; Rosenbach *et al.*, 2010; Samaranayake *et al.*, 2009; Sardi *et al.*, 2013). They also form biofilms on implanted medical devices, which are frequently associated with systemic infections (Cauda, 2009; Habash and Reid, 1999; Kojic and Darouiche, 2004; Ramage *et al.*, 2006; Shoham, 2010).

Other members of the phylum Ascomycota that, in seldom instances, have been found to cause biofilm-related infections include *Saccharomyces cerevisiae* and *Blastoschizomyces capitatus*. *S. cerevisiae*, or baker's yeast, rarely causes infections in humans and is generally not considered a pathogen. In a few case studies, however, *S. cerevisiae* has been implicated in catheter-associated infections in mixed-species biofilms of ICU patients, and is able to form a thin biofilm consisting of round budding yeast-form cells and pseudohyphal cells *in vitro*, implying that it may also exist in biofilms within the human host (Chitasombat *et al.*, 2012; Enache-Angoulvant and Hennequin, 2005; Muñoz *et al.*, 2005; Reynolds and Fink, 2001). *B. capitatus* is another rare cause of infection in humans, although more cases have been emerging recently in highly immunocompromised ICU patients. *B. capitatus* can cause catheter-associated bloodstream infections and has the capacity to form a biofilm both *in vitro* and *in vivo* (Birrenbach *et al.*, 2012; D'Antonio *et al.*, 2004).

Aspergillus is a major fungal pathogen that has been described to form biofilms both *in vitro* and *in vivo* (Mowat *et al.*, 2007, 2009; Ramage *et al.*, 2011). *Aspergillus fumigatus* is a common invasive colonizer of the respiratory tract in individuals with cystic fibrosis or other conditions that compromise the immune system (McCormick *et al.*, 2010; Seidler *et al.*, 2008). During infection, airborne conidia (spores) enter the host via the airway and undergo a morphological switch to the filamentous growth form upon contact with host tissues. *A. fumigatus* then continues to proliferate along the lining of the respiratory tract, forming a dense network of hyphal cells, resembling a biofilm or biofilm-like structure. Aspergillomas, or dense hyphal balls, may also form as *A. fumigatus* continues to proliferate in the respiratory tract; these structures share many architectural characteristics with known biofilm architectures (Ramage *et al.*, 2011).

Many other Ascomycota members primarily infect the respiratory tract of mammals. For example, *Pneumocystis jirovecii* causes pneumonia in mammals that is frequently fatal. Morphological examination of lung tissue infected by *P. jirovecii* reveals a biofilm-like growth structure, consisting of round yeast-form clusters, which can be recapitulated *in vitro* under certain laboratory conditions (Cushion and Collins, 2011; Cushion *et al.*, 2009). *Scedosporium prolificans*, a recently emerging human fungal pathogen of the respiratory system, may also form biofilms, as the presence of 'fungus balls', resembling biofilm structures, have been reported in some clinical cases (Cortez *et al.*, 2008). *Histoplasma capsulatum* is another opportunistic pathogen that infects the respiratory system of mammals, causing histoplasmosis. The immune state of the host is the most deterministic factor for the severity of infection by *H. capsulatum*, where infections range from a mild, localized infection, in mildly immunosuppressed individuals, to a lethal, disseminated infection, affecting multiple organs, in severely immunocompromised individuals (Kauffman, 2007; McKinsey and McKinsey, 2011). *H. capsulatum* typically exists in a filamentous form in the environment outside of the host, but is usually found in yeast-form within the host. There is at least one report of *H. capsulatum* forming biofilms consisting of dense clusters of yeast-form cells *in vitro*, suggesting that *H. capsulatum* may exist in architecturally similar biofilms *in vivo* (Pitangui *et al.*, 2012). Along with *H. capsulatum*, *Coccidioides immitis*, *Paracoccidioides brasiliensis*, and *Blastomyces dermatitidis* are all sources of mycoses in immunocompromised individuals and have also caused endemic outbreaks in specific geographical regions (Fridkin and Jarvis, 1996; Marques, 2012). All commonly infect the respiratory system, but are also able to cause invasive infections. Although rare, *C. immitis* has been reported to form biofilms on medical devices (Davis *et al.*, 2002). Whether *B. dermatitidis* and *P. brasiliensis* form biofilms *in vitro* or *in vivo* is currently unknown.

Other ascomycete species, including *Fusarium solani*, *Acremonium implicatum* and *Cladosporium sphaerospermum*, most commonly cause keratitis in mammals, a serious infection of the cornea of the eye, as well as other less common infections. Keratitis is generally thought of as a biofilm infection, and has been associated with the use of contact lenses (Chang *et al.*, 2006; Khor *et al.*, 2006; Rao *et al.*, 2007; Saw *et al.*, 2007). In fact, a *Fusarium* keratitis outbreak occurred not too long ago in the United States as a result of a contact lens multipurpose solution lacking sufficient antifungal efficacy against *F. solani* (Levy *et al.*, 2006). Nonetheless, recent work *in vitro* has shown that *F. solani*, *A. implicatum*, and *C. sphaerospermum* can all form biofilms under laboratory conditions (Zhang *et al.*, 2012).

Basidiomycota

The most prominent mammalian fungal pathogens of the phylum Basidiomycota are *Cryptococcus neoformans* and *Cryptococcus gattii*, which are able to cause meningoencephalitis, a serious inflammation of the brain, spinal cord, and meninges, as well as severe pulmonary infections, in both immunocompromised and immunocompetent individuals (Gullo *et al.*, 2013; Jarvis and Harrison, 2007, 2008). Both *C. neoformans* and *C. gattii* are known to form biofilms *in vitro* and on medical devices *in vivo* (Alvarez *et al.*, 2008; Braun *et al.*, 1994; Walsh *et al.*, 1986). Another fungal pathogen of mammals is *Malassezia pachydermatis*, which is the cause of frequent nosocomial epidemics in neonatal ICUs (Guillot and Bond, 1999). *M. pachydermatis* has been reported to form biofilms *in vitro* and *in vivo* on intravenous catheters (Cannizzo *et al.*, 2007; Curvale-fauchet *et al.*, 2004; Figueredo *et al.*, 2013). The emerging fungal pathogen, *Trichosporon asahii* is another member of the Basidiomycota that is the cause of disseminated trichosporonosis infections, often in organ transplant patients (Netsvyetayeva *et al.*, 2009; Ramage *et al.*, 2009). Trichosporonosis infections are frequently associated with implanted medical devices, probably colonized by a biofilm (Di Bonaventura *et al.*, 2006).

Zygomycota

For decades, mammalian fungal biofilm formers were largely studied in the Ascomycota and Basidiomycota phyla, but recently, species from the phylum Zygomycota have been reported to form biofilms as well. Four members of the Zygomycota: *Rhizopus oryzae*, *Lichtheimia corymbifera*, *Rhizomucor pusillus*, and *Apophysomyces elegans* were recently evaluated for their abilities to form biofilms *in vitro*, and all, except for *A. elegans*, successfully formed biofilms under the conditions tested (Singh *et al.*, 2011). These four fungi belong to the order Mucorales, and are known to cause infections similar to that of *Aspergillus*, ranging from superficial cutaneous infections to more pervasive disseminated and invasive infections (Rammaert *et al.*, 2012; Skiada and Petrikkos, 2013).

Current standard of care for fungal biofilm infections

The standard of care for treating fungal biofilm infections on medical devices varies depending on the age and health status of the patient, the infection site, and the causative fungal species, but generally, the consensus on care involves administration of antifungal drugs and ultimately removal of the presumed infected medical device (Andes *et al.*, 2012; Cornely *et al.*, 2012; Lepak and Andes, 2011). The majority of fungal infections (device- and non-device- associated) involve *Candida* species, and as such, current general guidelines for care tend to recommend treatments most effective against candidaemia and candidiasis (Arendrup *et al.*, 2011; Mikolajewska *et al.*, 2012).

Four major classes of antifungal drugs are used for treatment of fungal infections: azoles, polyenes, nucleoside analogues, and echinocandins (Chen *et al.*, 2011; Cowen, 2008; Jabra-Rizk *et al.*, 2004; Mikolajewska *et al.*, 2012; Ramage *et al.*, 2012; White *et al.*, 1998). Azoles, such as fluconazole, are fungistatic and block ergosterol synthesis by targeting the demethylase enzyme Erg11, leading to an accumulation of toxic sterol pathway intermediates. Polyenes, such as amphotericin B, are the oldest class of antifungals, are fungicidal, and work by intercalating into membranes containing ergosterol, creating rigid pores that destroy the proton gradient, and result in leakage of the cytoplasm and other cell contents. Nucleoside analogues (or antimetabolites), such as 5-fluorocytosine (5-FC), mimic nucleosides during nucleic acid synthesis. 5-FC specifically acts as a pyrimidine analogue, and disrupts fungal RNA, DNA, and protein synthesis, ultimately leading to cell cycle arrest. 5-FC is not intrinsically antifungal itself, but becomes an antifungal when it is converted into 5-fluorouracil (5-FU) inside fungal cells by a fungal cytosine deaminase. Echinocandins, such as caspofungin, are the newest class of fungicidal drugs and work by targeting the synthesis of 1,3-β-D-glucan, an important component of the fungal cell wall.

In adult patients with normal neutrophil bloodstream levels (non-neutropenic adults), most fungal infections are treated with an azole-based antifungal, such as voriconazole or fluconazole (Chalmers and Bal, 2011; Drew *et al.*, 2013; Glöckner and Cornely, 2013; Pappas *et al.*, 2009). Diagnosis of the causative fungal species is usually determined by blood culture, and treatment is typically continued for two weeks. Use of echinocandins, rather than azoles, is recommended if the patient has had prior azole exposure, or if the infection is noticeably severe. Medical devices are especially prone to fungal biofilm infections. If a device-mediated infection is suspected, general treatment typically includes administration of antifungal drugs as well as removal and replacement of the device (Mermel *et al.*, 2009; Shoham, 2010). Removal of the potentially infected medical device is always recommended for non-neutropenic adult patients who can tolerate the removal procedure (Pappas *et al.*, 2009). In the case of catheters, antifungal lock therapy using amphotericin B, echinocandins, or ethanol may be tried under some circumstances, particularly if the device is surgically implanted and difficult to remove or replace (Pieroni *et al.*, 2013; Walraven and Lee, 2013). Echinocandins, such as caspofungin are the first line of treatment for neutropenic patients or those infected with

C. glabrata, which is generally azole resistant (Chahoud *et al.*, 2013; Pappas *et al.*, 2009). Fungal infections in neonates are usually treated with amphotericin B or fluconazole, but the use of echinocandins in neonatal ICUs is increasing (Hope *et al.*, 2012; Pappas *et al.*, 2009). Although somewhat controversial, and not currently a standard of care in hospital settings, there is evidence to suggest that prophylactic use of fluconazole may be beneficial for transplant recipients, ICU patients, preterm neonates, and other high-risk patient populations (Blumberg *et al.*, 2001; Ho *et al.*, 2005; Lipsett, 2004; Manzoni *et al.*, 2006; Rex and Sobel, 2001; Ullmann *et al.*, 2012).

The presently available antifungal drugs have been effective in reducing morbidity and mortality in a range of patients; however, fungal biofilm infections remain a serious issue. Currently, device-associated, biofilm-mediated candidaemia has an estimated mortality rate of 19–50% and up to 15% of ICU patients are afflicted with this type of infection (Eloy *et al.*, 2006; Gudlaugsson *et al.*, 2003; Hajjeh *et al.*, 2004; Morgan *et al.*, 2005; Pittet and Wenzel, 1995; Wey *et al.*, 1988). New drug and vaccine development are essential for the effective treatment of biofilm-associated infections, especially given the emerging drug resistance characteristics observed in clinically relevant fungi (Cowen, 2008; Pierce and Lopez-Ribot, 2013; Vermeulen *et al.*, 2013). It is particularly important that therapeutics specific to biofilm infections be developed, in light of the prevalence of this type of infection, the increasing number of fungal species recognized to form biofilms, and the inherent drug resistance of biofilms (Ramage *et al.*, 2009, 2012; Tumbarello *et al.*, 2007).

Mechanisms of fungal biofilm drug resistance

There are several general mechanisms of biofilm drug resistance that confer (or are thought to confer) resistance to multiple classes of antifungals in fungal biofilms. These mechanisms include: (1) the presence of an extracellular matrix, (2) constitutive up-regulation of efflux pumps, (3) increased cell density, (4) changes in metabolism, (5) the presence of persister cells, and (6) cellular signalling and stress responses (Akins, 2005; Mathé and Van Dijck, 2013; Ramage *et al.*, 2012; Taff *et al.*, 2013). Below we describe and evaluate these mechanisms in light of current research in the fungal biofilm field. The drug resistance mechanisms most supported for fungal biofilms are summarized in Fig. 4.1.

Extracellular matrix

Cells within a biofilm are embedded in a matrix composed of extracellular material, often forming very elaborate structures primarily composed of carbohydrates, proteins, and nucleic acids (Al-Fattani and Douglas, 2004, 2006; Lal *et al.*, 2010; Martins *et al.*, 2010, 2012). Extracellular DNA is a key component of the biofilm matrix that provides both structural integrity and maintenance to the matrix, and accumulates in a time-dependent manner (Martins *et al.*, 2010). Moreover, treatment of mature biofilms with DNAse has been reported to enhance the activity of caspofungin and amphotericin B, in both *C. albicans* and *A. fumigatus* (Martins *et al.*, 2012; Rajendran *et al.*, 2013). Although it is typically thought that biofilms prevent drugs from penetrating through their complex structure, in *Candida* biofilms, fluconazole is able to permeate quite rapidly and, at least in this case, penetration does not account for biofilm resistance to fluconazole (Al-Fattani and Douglas, 2004). Instead, the constituents of the matrix itself contribute to biofilm drug resistance, such that increased matrix production enhances drug resistance of *C. albicans* biofilms (Al-Fattani and Douglas, 2006). For example, glucans comprise a major portion of the fungal cell wall and are thought to be a significant constituent of the biofilm matrix. Additional research has found significantly higher levels of β-1,3-glucans in the cell wall and biofilm matrix of *C. albicans* biofilm versus planktonic cells, and that the cell walls of *C. albicans* biofilm cells bind four- to five-fold more fluconazole than planktonic cell walls (Nett *et al.*, 2007). Furthermore, disruption of β-1,3-glucans or β-1,3-glucanase treatment have been shown to result in increased susceptibility of biofilms to fluconazole. Conversely, the simple addition of exogenous β-1,3-glucans has been found to increase resistance to fluconazole in planktonic cells (Mitchell *et al.*, 2013). It is likely that biofilms can also sequester amphotericin B since it

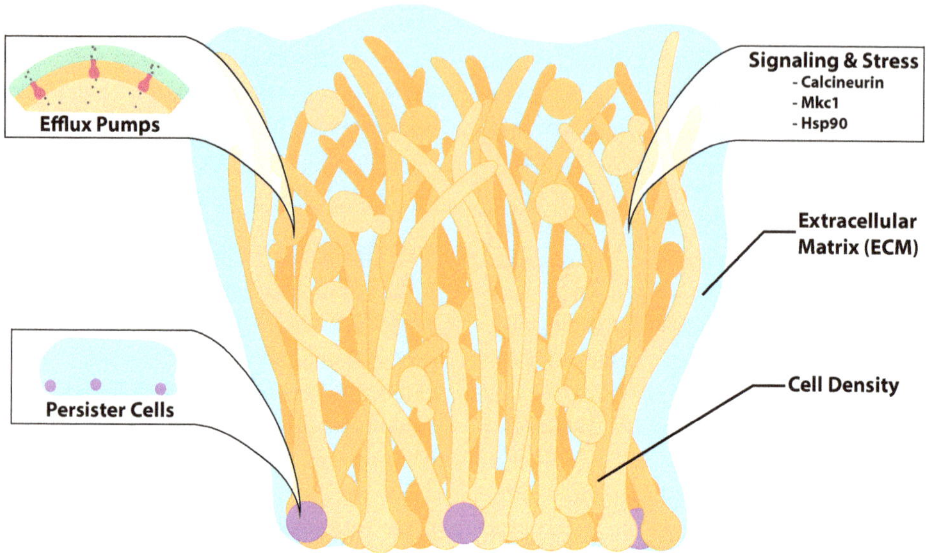

Figure 4.1 Drug resistance mechanisms for fungal biofilms.

has recently been shown that β-1,3-glucans can also bind specifically to this polyene drug (Vediyappan *et al.*, 2010). Compromised expression of the glucan synthase gene *FKS1* has also been shown to enhance the efficacy of amphotericin B, anidulafungin, and flucytosine (Nett *et al.*, 2010). Together these data suggest a clear role for glucans in biofilm drug resistance, possibly by sequestering compounds and/or by structurally protecting cells.

Efflux pumps

In *C. albicans* there are two main classes of efflux pumps that are contributors to antifungal drug resistance: the ATP binding cassette (ABC) transporter superfamily containing *CDR1* and *CDR2*, and the major facilitator (MF) class containing *MDR1* (Akins, 2005; Anderson, 2005; Cowen, 2008). While treatment with antifungals, such as fluconazole, can result in the up-regulation of these transporters in planktonic cells, cells in biofilms up-regulate their transporters within six hours of surface contact both *in vitro* and *in vivo*, even in the absence of drug (Andes *et al.*, 2004; Mukherjee *et al.*, 2003; Nett *et al.*, 2009; Ramage, 2002a). In biofilms, surface adherence alone appears sufficient to increase expression of the genes encoding the efflux pumps (Mateus *et al.*, 2004). Efflux pumps are also up-regulated in mature biofilms, indicating that they continue to mediate drug resistance throughout biofilm development (Nobile *et al.*, 2012; Ramage, 2002; Yeater *et al.*, 2007).

Cell density

When testing *C. albicans* for drug susceptibility using the microtitre method, there is an optimal inoculum size range that is used as a clinical standard since it is well established that inoculum size can affect susceptibility outcomes (Nguyen and Yu, 1999; National Committee for Clinical Laboratory Standards, 1997). In fact, if the cell concentration is increased in microtitre drug resistance assays, resistance to the azoles, fluconazole and ketoconazole; the echinocandin, caspofungin; and the polyene, amphotericin B, is increased up to twenty-fold (Perumal *et al.*, 2007). In contrast, when biofilms are dissociated and assayed at a lower density in the same microtitre drug resistance assays, they exhibit drug susceptibilities at the level of planktonic cells assayed at the same cellular density. A similar phenomenon was observed with *A. fumigatus*, *Aspergillus flavus*, *Aspergillus niger*, and *Aspergillus terreus* in that increasing inoculum size resulted in increased minimum inhibitory concentrations of two azoles, voriconazole and itraconazole, by up to 6-fold (Lass-Flörl *et al.*, 2003).

In the dense biofilm environment, fungal cells are also capable of communicating with each other via quorum sensing through various signalling molecules in a manner that is dependent on cell density (Hornby *et al.*, 2001). Treatment of *C. albicans* with one such molecule, farnesol, has been reported to induce global gene expression changes, including the activation of drug resistance genes through the two-component regulatory system of Chk1 (Cao *et al.*, 2005; Enjalbert and Whiteway, 2005; Kruppa *et al.*, 2004). However, dissociated biofilms composed of a *chk1* deletion mutant strain that is unable to respond to farnesol, display wild-type fluconazole susceptibility, suggesting that quorum sensing signalling through farnesol is not a key contributor to fluconazole resistance in *C. albicans* biofilms (Perumal *et al.*, 2007). Additionally, the echinocandins are the most effective therapeutic class against fungal biofilms, thus cell density and quorum sensing likely play little to no role in biofilm echinocandin resistance (Bachmann *et al.*, 2002). It remains possible that quorum sensing through farnesol may play a role in resistance to other classes of antifungals or that quorum sensing through additional molecules and pathways may be important for biofilm drug resistance.

Metabolism

It is has been hypothesized that the reduced growth rate and metabolism of cells within biofilms, as a result of limited nutrient availability, contribute to their increased drug resistance. Altered metabolism and growth in bacterial biofilms is linked to drug resistance, but it is not clear whether this is an important factor for drug resistance in fungal biofilms (Evans *et al.*, 1991; Gilbert *et al.*, 1990; Martínez and Rojo, 2011). In fungi, planktonic cells are more resistant to amphotericin B at lower growth rates, and biofilms are equivalently resistant over a range of growth rates, suggesting that growth rate plays only a minor role, if any, in fungal biofilm drug resistance (Baillie and Douglas, 1998a). Additional research has shown that neither glucose nor iron limitation affect fungal biofilm resistance to amphotericin B, however iron limitation did increase the susceptibility of dispersed daughter cells from biofilms to amphotericin B (Baillie and Douglas, 1998b). Chandra *et al.* (2001) found that resistance to amphotericin B, fluconazole, nystatin, and chlorhexidine increases as *C. albicans* biofilms mature over time, and report that there is a corresponding increase in metabolic activity over biofilm maturation, however, cell number is not controlled for in these experiments, so it is unclear whether this is a true representation of metabolic activity. There is evidence that the Krebs cycle is down-regulated over time in developing *C. albicans* biofilms, but it is not known if this shift in metabolism is related to drug resistance (Yeater *et al.*, 2007; Zhu *et al.*, 2012). Additionally, work in bacteria has indicated that there are many subpopulations within biofilms that exist in different metabolic states and that these subpopulations have differing drug susceptibilities (Rani *et al.*, 2007). While this has not been fully examined in fungi, fungal biofilms do contain a subpopulation of metabolically dormant, drug-resistant persister cells (discussed in the following section). Thus, existing evidence suggests that cell growth may not be involved in fungal biofilm drug resistance; however the role of altered metabolism in fungal resistance is yet to be fully elucidated.

Persister cells

Within the biofilm cell population there exist a subset of yeast cells, called persister cells, that form stochastically, are phenotypically dormant, and are highly tolerant to antifungal drugs (LaFleur *et al.*, 2006). In bacterial biofilms, persister cells can comprise as much as 1% of the biofilm population and are thought to be antibiotic resistant due to their dormant state as antibiotics generally exert their affect by inhibiting an active target (Keren *et al.*, 2004; Lewis, 2010). In *C. albicans* biofilms, persister cells were first discovered as a small population of yeast cells within biofilms that were highly drug resistant in a manner that was independent of drug pumps and the composition of the cell membrane (Khot *et al.*, 2006; LaFleur *et al.*, 2006). When biofilms were treated with amphotericin B or the antiseptic, chlorhexidine, most cells were killed at a relatively low drug concentration; however, a subpopulation of highly tolerant sessile cells remained, resulting in a biphasic killing pattern (LaFleur *et al.*, 2006). These highly tolerant cells are thought to be a phenotypic variant of the wild type strain, as subculturing persister cells resulted in a biofilm with new subpopulations. Furthermore, *C. albicans* persister cells are exclusively recovered from biofilms and not from planktonic populations, regardless of their growth phase, and require attachment to a substrate to initiate the dormant phenotype. In clinical isolates of *C. albicans*, persister cells are found at a significantly higher frequency from patients with long-term *Candida* infection versus patients with short-term carriage (Lafleur *et al.*, 2010). Persister cells have also been described in biofilms treated with amphotericin B from various isolates of *C. krusei* and *C. parapsilosis* (Al-Dhaheri and Douglas, 2008). Taken together, persister cells are very clinically relevant in that they serve as a reservoir of naturally occurring highly tolerant cells within biofilms, are likely to be refractory to antifungal therapy in the clinical setting, and thus may contribute to treatment failure.

Cellular signalling and stress responses

Up-regulated stress responses in fungal cells can result in increased drug tolerance to the point of being classified as *bone fide* drug resistance, and is thus another possible mechanism of resistance in biofilms. Over the span of a fungal cell's life, it may encounter a broad range of environments, and must, therefore, be able to respond to potential stresses, for example, the host immune response in a bloodstream infection. Surface contact of *C. albicans* cells results in the activation and accumulation of Mkc1, the terminal mitogen-activated protein (MAP) kinase of the protein kinase C (PKC) pathway that is normally activated in response to cell wall stress (Kumamoto, 2005; Navarro-García *et al.*, 1995, 1998). In fact, an *mkc1* deletion mutant strain forms abnormally structured biofilms with reduced filamentation (Kumamoto, 2005). Furthermore, sessile *mkc1* deletion mutant cells are 100 times more susceptible to the azole fluconazole. Subsequent work in planktonic conditions has shown that Mkc1 is involved in resistance to the azoles and echinocandins, and requires the chaperone Hsp90 for protein stability (LaFayette *et al.*, 2010). Hsp90 also mediates resistance to the azoles and echinocandins through calcineurin, a protein phosphatase that is crucially involved in cellular stress responses and highly conserved across all eukaryotes (Cowen and Lindquist, 2005; Cowen *et al.*, 2009; Singh *et al.*, 2009). Pharmacological inhibition of calcineurin or Hsp90 is highly synergistic with fluconazole, not only against *C. albicans* planktonic cells, but also against *C. albicans* biofilms (Cowen *et al.*, 2009; Uppuluri *et al.*, 2008). Inhibition of Hsp90 is also synergistic with fluconazole and echinocandins against *A. fumigatus* biofilms (Robbins *et al.*, 2011). Hsp90 is additionally important for normal biofilm architecture and matrix glucan levels, and genetic disruption of *HSP90* function reduces dispersal of cells from biofilms, which is thought to be important for device-associated candidiasis (Robbins *et al.*, 2011; Uppuluri *et al.*, 2010). An interesting point to note, however, is that although both calcineurin and Mkc1 require Hsp90 for stability in planktonic conditions, this does not appear to be the case in biofilm conditions (Robbins *et al.*, 2011), suggesting separate cellular circuitries for these distinct morphological states. Given the complexity of the biofilm regulatory network itself (Nobile *et al.*, 2012), it is reasonable to propose that the mechanisms governing antifungal drug resistance in biofilms are equally as complex.

Challenges and strategies to developing therapeutics for fungal biofilm infections

No biofilm-specific drugs exist today for either bacteria or fungi, making treatment of biofilm-based infections particularly problematic. The resistance of fungal biofilms to standard antifungal drugs is multifactorial; not only do biofilms provide physical protection from the host immune defences and antifungal drugs (e.g. through the production of the extracellular matrix), cells in biofilms become intrinsically resistant to drugs due to their constitutive up-regulation of drug efflux pumps and their altered metabolic states (e.g. through the development of a subpopulation of highly drug tolerant and metabolically inactive persister cells) (Bonhomme and d'Enfert, 2013; Jabra-Rizk *et al.*, 2004; LaFleur *et al.*, 2006; Mathé and Van Dijck, 2013; Tobudic *et al.*, 2012). It is these same multifactorial biofilm-specific properties that make developing effective therapeutics for fungal biofilm infections extraordinarily challenging.

A better understanding of the underlying molecular mechanisms behind each of these unique biofilm properties will be the key to developing new antifungals that specifically target the biofilm state. With this information, it should be feasible for researchers to identify promising protein targets of biofilm-specific properties, and subsequently to identify or design therapeutics that influence these targets. For example, identification of the molecular mechanisms behind biofilm dispersal could be promising for developing a compound that specifically triggers biofilm cell dispersal. Such a compound would resolve the physical barrier challenges leading to the tolerance of biofilms to antimicrobial compounds, and could be used in combination with known antifungals that are effective against planktonic cells, thereby allowing complete clearance of the biofilm infection. As another example, a better understanding of the molecular triggers causing metabolic dormancy of subpopulations of cells, such as the formation of persister cells, during biofilm formation, would allow researchers to develop compounds that reverse the cell physiology of these cells, preventing the development of antimicrobial-resistant cells during biofilm formation. Ultimately, it seems likely that a future combinatorial treatment will take advantage of both our developing mechanistic knowledge of the biofilm state, in addition to our current knowledge of the planktonic state, to effectively clear biofilm-based infections and reduce relapse infection rates.

In addition to detailed single pathway approaches to studying biofilm formation, genome-wide systems biology approaches will also guide the discovery of important protein targets for biofilm formation. For example, in *C. albicans*, the transcriptional network that orchestrates the development of *C. albicans* biofilms was recently identified using a systems biology approach (Nobile *et al.*, 2012). This network consists of six master transcription regulators (Bcr1, Tec1, Efg1, Ndt80, Rob1 and Brg1) and over 1000 target genes, whose expression is controlled by these regulators. Deletion of each of these regulators identified six key target genes: *ALS1*, *TPO4*, *ORF19.4000*, *EHT1*, *HWP1* and *CAN2* which, when overexpressed, were able to rescue biofilm defects in a *bcr1*Δ/Δ or *tec1*Δ/Δ mutant (Fox and Nobile, 2012). Als1, Hwp1, Can2, and Tpo4 are all cell-surface proteins that may be involved in adhesion (Als1 and Hwp1) (Coleman *et al.*, 2010; Dwivedi *et al.*, 2011; Ene and Bennett, 2009; Nobile *et al.*, 2006a,b; Sundstrom, 1999) or in small molecule transport (Can2 and Tpo4) (Malinska *et al.*, 2004; Regenberg and Kielland-Brandt, 2001). Based on homology to *S. cerevisiae* Eht1, *C. albicans* Eht1 is predicted to be involved in fatty acid synthesis in lipid vesicles, and may be involved in polarized growth and morphogenesis. Orf19.4000 is a predicted transcription regulator that may control additional genes important for biofilm formation. Deletion of *ALS1*, *HWP1*, and *CAN2* causes notable biofilm defects, and these proteins in particular may be promising drug targets.

By screening a library of transcription factor mutants using an *in vitro* flow cell assay, another study identified 30 transcription regulators important for adherence to a silicone substrate (Finkel *et al.*, 2012). Of the 30 adherence regulators identified, four (Bcr1, Ace2, Snf5, and Arg81) were also required for biofilm formation under standard *in vitro* biofilm conditions, and it is possible that others may be required for biofilm formation under specific environmental conditions. Several intriguing sets of genes are controlled by these regulators: the hyphal- and virulence-associated (HYVIR) genes, the cell surface targets of adherence regulators (CSTAR) genes, the regulators of Ace2 and polarized morphogenesis (RAM) genes, which are the major targets of Ace2 and also are regulated by Snf5 and Bcr1, and the zinc uptake genes and other known targets of the transcription factor Zap1 (ZAPT) genes, which are regulated by both Zap1 and Arg81 (Finkel *et al.*, 2012). In addition to its contributions to adherence, Zap1 is also an important negative regulator of extracellular matrix production that controls the expression of several glucoamylases, alcohol dehydrogenases, and other proteins that govern production of β-1,3-glucan and other matrix constituents (Nobile *et al.*, 2009). In general, there has been extensive work in the field on identifying adherence proteins contributing to the organization of the biofilm structure (Dwivedi *et al.*, 2011; Finkel *et al.*, 2012; Iraqui *et al.*, 2005; Monniot *et al.*, 2013; Nobile *et al.*, 2006a; Zhao *et al.*, 2006). These and other studies are the stepping stones to identifying critical drug targets for the development of biofilm-specific therapeutics.

Other work has used large scale proteomics techniques to identify proteins specific to biofilms (Martínez-Gomariz *et al.*, 2009; Thomas *et al.*, 2006). Hundreds of proteins are differentially expressed between biofilms and planktonic cells, both secreted and at the cell surface. In particular, many of these proteins are metabolic enzymes that may represent the different metabolic states in biofilm and planktonic cultures. Using this information as a starting point to prioritize proteins to study in more detail, it will be possible to determine which protein targets are required for biofilm formation in *C. albicans* and, of these new targets, which are predicted to be potentially 'druggable' based on structure-based drug design methods. In principle, this approach could also be taken for all known biofilm-formers to identify potentially relevant drug targets in these species.

Other than the 'targeted' approaches to drug design described above, non-biased approaches to identify compounds with efficacy against fungal biofilms are also currently being pursued. These approaches include screening existing drug compound libraries looking for compounds that disrupt biofilms, and screening novel natural product libraries to identify candidates that are capable of disrupting the biofilm state. Several labs have started working on this approach, and it remains to be seen whether any of the candidates will be developed into useful therapeutics (Chavez-Dozal *et al.*, 2014; LaFleur *et al.*, 2011; Srinivasan *et al.*, 2013; Zeidler *et al.*, 2013). The screening of existing drug compound libraries, if successful, could allow for the identification of already-existing drugs that could be repurposed for use against fungal biofilms. The benefits of such compounds are that they are already approved as drugs, and, as such, the regulatory hurdles associated with their use will be smaller than a new drug. By screening novel natural product libraries or by designing new drugs using the targeted approaches, if successful, could lay the groundwork for the development of a new class of biofilm-specific drugs that could change our current clinical practices for treating infections.

Acknowledgements

This study is supported by National Institutes of Health (NIH) grants K99AI100896 and R01AI083311. SDSB is supported by a Canadian Institutes of Health Research (CIHR) Banting Postdoctoral Fellowship.

References

Akins, R.A. (2005). An update on antifungal targets and mechanisms of resistance in *Candida albicans*. Med. Mycol. 43, 285–318.

Al-Dhaheri, R.S., and Douglas, L.J. (2008). Absence of amphotericin B-tolerant persister cells in biofilms of some Candida species. Antimicrob. Agents Chemother. 52, 1884–1887.

Al-Fattani, M.A., and Douglas, L.J. (2004). Penetration of Candida biofilms by antifungal agents. Antimicrob. Agents Chemother. 48, 3291–3297.

Al-Fattani, M.A., and Douglas, L.J. (2006). Biofilm matrix of *Candida albicans* and *Candida tropicalis*: chemical composition and role in drug resistance. J. Med. Microbiol. 55, 999–1008.

Alvarez, M., Saylor, C., and Casadevall, A. (2008). Antibody action after phagocytosis promotes *Cryptococcus neoformans* and *Cryptococcus gattii* macrophage exocytosis with biofilm-like microcolony formation. Cell. Microbiol. 10, 1622–1633.

Anderson, J.B. (2005). Evolution of antifungal-drug resistance: mechanisms and pathogen fitness. Nat. Rev. Microbiol. 3, 547–556.

Andes, D., Nett, J., Oschel, P., Albrecht, R., Marchillo, K., and Pitula, A. (2004). Development and characterization of an *in vivo* central venous catheter *Candida albicans* biofilm model. Infect. Immun. 72, 6023–6031.

Andes, D.R., Safdar, N., Baddley, J.W., Playford, G., Reboli, A.C., Rex, J.H., Sobel, J.D., Pappas, P.G., and Kullberg, B.J. (2012). Impact of treatment strategy on outcomes in patients with candidemia and other forms of invasive candidiasis: a patient-level quantitative review of randomized trials. Clin. Infect. Dis. 54, 1110–1122.

Arendrup, M.C., Sulim, S., Holm, A., Nielsen, L., Nielsen, S.D., Knudsen, J.D., Drenck, N.E., Christensen, J.J., and Johansen, H.K. (2011). Diagnostic issues, clinical characteristics, and outcomes for patients with fungemia. J. Clin. Microbiol. 49, 3300–3308.

Bachmann, S.P., VandeWalle, K., Ramage, G., Patterson, T.F., Wickes, B.L., Graybill, J.R., and López-Ribot, J.L. (2002). *In vitro* activity of caspofungin against *Candida albicans* biofilms. Antimicrob. Agents Chemother. 46, 3591–3596.

Baillie, G.S., and Douglas, L.J. (1998a). Effect of growth rate on resistance of *Candida albicans* biofilms to antifungal agents. Antimicrob. Agents Chemother. 42, 1900–1905.

Baillie, G.S., and Douglas, L.J. (1998b). Iron-limited biofilms of *Candida albicans* and their susceptibility to amphotericin B. Antimicrob. Agents Chemother. *42*, 2146–2149.

Beck-Sagué, C., and Jarvis, W.R. (1993). Secular trends in the epidemiology of nosocomial fungal infections in the United States, 1980–1990. National Nosocomial Infections Surveillance System. J. Infect. Dis. *167*, 1247–1251.

Birrenbach, T., Bertschy, S., Aebersold, F., Mueller, N.J., Achermann, Y., Muehlethaler, K., and Zimmerli, S. (2012). Emergence of *Blastoschizomyces capitatus* yeast infections, Central Europe. *18*, 98–101.

Blumberg, H.M., Jarvis, W.R., Soucie, J.M., Edwards, J.E., Patterson, J.E., Pfaller, M.A., Rangel-Frausto, M.S., Rinaldi, M.G., Saiman, L., Wiblin, R.T., *et al.* (2001). Risk factors for candidal bloodstream infections in surgical intensive care unit patients: the NEMIS prospective multicenter study. The National Epidemiology of Mycosis Survey. Clin. Infect. Dis. *33*, 177–186.

Bonhomme, J., and d'Enfert, C. (2013). *Candida albicans* biofilms: building a heterogeneous, drug-tolerant environment. Curr. Opin. Microbiol. *16*, 398–403.

Bouza, E., and Muñoz, P. (2008). Epidemiology of candidemia in intensive care units. Int. J. Antimicrob. Agents *32* (Suppl. 2), S87–91.

Bowyer, P., Moore, C.B., Rautemaa, R., Denning, D.W., and Richardson, M.D. (2011). Azole antifungal resistance today: focus on *Aspergillus*. Curr. Infect. Dis. Rep. *13*, 485–491.

Braun, D.K., Janssen, D.A., Marcus, J.R., and Kauffman, C.A. (1994). Cryptococcal infection of a prosthetic dialysis fistula. Am. J. Kidney Dis. *24*, 864–867.

Calderone, R. (2002). Candida and Candidiasis (ASM Press, Washington DC, USA).

Cannizzo, F.T., Eraso, E., Ezkurra, P.A., Villar-Vidal, M., Bollo, E., Castellá, G., Cabañes, F.J., Vidotto, V., and Quindós, G. (2007). Biofilm development by clinical isolates of *Malassezia pachydermatis*. Med. Mycol. *45*, 357–361.

Cao, Y.Y., Cao, Y.B., Xu, Z., Ying, K., Li, Y., Xie, Y., Zhu, Z.Y., Chen, W.S., and Jiang, Y.Y. (2005). cDNA microarray analysis of differential gene expression in *Candida albicans* biofilm exposed to farnesol cDNA microarray analysis of differential gene expression in *Candida albicans* biofilm exposed to farnesol. Antimicrob. Agents Chemother. *49*, 584–589.

Cauda, R. (2009). Candidaemia in patients with an inserted medical device. Drugs *69* (Suppl. 1), 33–38.

Chahoud, J., Kanafani, Z.A., and Kanj, S.S. (2013). Management of candidaemia and invasive candidiasis in critically ill patients. Int. J. Antimicrob. Agents *42* (Suppl.), S29–35.

Chalmers, C.M., and Bal, A.M. (2011). Management of fungal infections in the intensive care unit: a survey of UK practice. Br. J. Anaesth. *106*, 827–831.

Chandra, J., Kuhn, D.M., Mukherjee, P.K., Hoyer, L.L., McCormick, T., and Ghannoum, M.A. (2001). Biofilm formation by the fungal pathogen *Candida albicans*: development, architecture, and drug resistance. J. Bacteriol. *183*, 5385–5394.

Chang, D.C., Grant, G.B., O'Donnell, K., Wannemuehler, K.A., Noble-Wang, J., Rao, C.Y., Jacobson, L.M., Crowell, C.S., Sneed, R.S., Lewis, F.M.T., *et al.* (2006). Multistate outbreak of *Fusarium* keratitis associated with use of a contact lens solution. JAMA *296*, 953–963.

Chavez-Dozal, A.A., Jahng, M., Rane, H.S., Asare, K., Kulkarny, V.V., Bernardo, S.M., and Lee, S.A. (2014). *In vitro* analysis of flufenamic acid activity against *Candida albicans* biofilms. Int. J. Antimicrob. Agents *43*, 86–91.

Chen, S.C.-A., Slavin, M.A., and Sorrell, T.C. (2011). Echinocandin antifungal drugs in fungal infections: a comparison. Drugs *71*, 11–41.

Chitasombat, M.N., Kofteridis, D.P., Jiang, Y., Tarrand, J., Lewis, R.E., and Kontoyiannis, D.P. (2012). Rare opportunistic (non-Candida, non-Cryptococcus) yeast bloodstream infections in patients with cancer. J. Infect. *64*, 68–75.

Clinical and Laboratory Standards Institute (2008). M27-A3. Reference method for broth dilution antifungal susceptibility testing of yeasts (Clinical and Laboratory Standards Institute, Wayne, PA, USA).

Coleman, D.A., Oh, S.-H., Zhao, X., and Hoyer, L.L. (2010). Heterogeneous distribution of *Candida albicans* cell-surface antigens demonstrated with an Als1-specific monoclonal antibody. Microbiology *156*, 3645–3659.

Cornely, O.A., Bassetti, M., Calandra, T., Garbino, J., Kullberg, B.J., Lortholary, O., Meersseman, W., Akova, M., Arendrup, M.C., Arikan-Akdagli, S., *et al.* (2012). ESCMID* guideline for the diagnosis and management of Candida diseases 2012: non-neutropenic adult patients. Clin. Microbiol. Infect. *18* (Suppl. 7), 19–37.

Cortez, K.J., Roilides, E., Quiroz-Telles, F., Meletiadis, J., Antachopoulos, C., Knudsen, T., Buchanan, W., Milanovich, J., Sutton, D.A., Fothergill, A., *et al.* (2008). Infections caused by *Scedosporium* spp. Clin. Microbiol. Rev. *21*, 157–197.

Costerton, J.W., Stewart, P.S., and Greenberg, E.P. (1999). Bacterial biofilms: a common cause of persistent infections. Science *284*, 1318–1322.

Cowen, L.E. (2008). The evolution of fungal drug resistance: modulating the trajectory from genotype to phenotype. Nat. Rev. Microbiol. *6*, 187–198.

Cowen, L.E., and Lindquist, S. (2005). Hsp90 potentiates the rapid evolution of new traits: drug resistance in diverse fungi. Science *309*, 2185–2189.

Cowen, L.E., Singh, S.D., Köhler, J.R., Collins, C., Zaas, A.K., Schell, W.A., Aziz, H., Mylonakis, E., Perfect, J.R., Whitesell, L., *et al.* (2009). Harnessing Hsp90 function as a powerful, broadly effective therapeutic strategy for fungal infectious disease. Proc. Natl. Acad. Sci. U.S.A. *106*, 2818–2823.

Culakova, H., Dzugasova, V., Perzelova, J., Gbelska, Y., and Subik, J. (2013). Mutation of the CgPDR16 gene attenuates azole tolerance and biofilm production in pathogenic *Candida glabrata*. Yeast *30*, 403–414.

Cuomo, C.A., and Birren, B.W. (2010). The fungal genome initiative and lessons learned from genome sequencing. (Elsevier Inc., Cambridge, MA, USA).

Curvale-fauchet, N., Botterel, F., Legrand, P., Guillot, J., and Bretagne, S. (2004). Frequency of intravascular catheter colonization by *Malassezia* spp. in adult patients. Mycoses *47*, 491–494.

Cushion, M.T., and Collins, M.S. (2011). Susceptibility of Pneumocystis to echinocandins in suspension and biofilm cultures. Antimicrob. Agents Chemother. *55*, 4513–4518.

Cushion, M.T., Collins, M.S., and Linke, M.J. (2009). Biofilm formation by *Pneumocystis* spp. Eukaryot. Cell *8*, 197–206.

D'Antonio, D., Parruti, G., Pontieri, E., Di Bonaventura, G., Manzoli, L., Sferra, R., Vetuschi, A., Piccolomini, R., Romano, F., and Staniscia, T. (2004). Slime production by clinical isolates of *Blastoschizomyces capitatus* from patients with hematological malignancies and catheter-related fungemia. Eur. J. Clin. Microbiol. Infect. Dis. *23*, 787–789.

Davis, L.E., Cook, G., and Costerton, J.W. (2002). Biofilm on ventriculo-peritoneal shunt tubing as a cause of treatment failure in coccidioidal meningitis. Emerg. Infect. Dis. *8*, 376–379.

Di Bonaventura, G., Pompilio, A., Picciani, C., Iezzi, M., D'Antonio, D., and Piccolomini, R. (2006). Biofilm formation by the emerging fungal pathogen *Trichosporon asahii*: development, architecture, and antifungal resistance. Antimicrob. Agents Chemother. *50*, 3269–3276.

Diekema, D., Arbefeville, S., Boyken, L., Kroeger, J., and Pfaller, M. (2012). The changing epidemiology of healthcare-associated candidemia over three decades. Diagn. Microbiol. Infect. Dis. *73*, 45–48.

Dongari-Bagtzoglou, A., Kashleva, H., Dwivedi, P., Diaz, P., and Vasilakos, J. (2009). Characterization of mucosal *Candida albicans* biofilms. PLoS One *4*, e7967.

Donlan, R.M. (2001). Biofilm formation: a clinically relevant microbiological process. Clin. Infect. Dis. *33*, 1387–1392.

Donlan, R.M., and Costerton, J.W. (2002). Biofilms : survival mechanisms of clinically relevant microorganisms. *15*, 167–193.

Drew, R.H., Townsend, M.L., Pound, M.W., Johnson, S.W., and Perfect, J.R. (2013). Recent advances in the treatment of life-threatening, invasive fungal infections. Expert Opin. Pharmacother. *14*, 2361–2374.

Dwivedi, P., Thompson, A., Xie, Z., Kashleva, H., Ganguly, S., Mitchell, A.P., and Dongari-Bagtzoglou, A. (2011). Role of Bcr1-activated genes Hwp1 and Hyr1 in *Candida albicans* oral mucosal biofilms and neutrophil evasion. PLoS One *6*, e16218.

Elliott, T.S. (1988). Intravascular-device infections. J. Med. Microbiol. *27*, 161–167.

Eloy, O., Marque, S., Mourvilliers, B., Pina, P., Allouch, P.Y., Pangon, B., Bedos, J.P., and Ghnassia, J.C. (2006). Contribution of the Pittet's index, antigen assay, IgM, and total antibodies in the diagnosis of invasive candidiasis in intensive care unit. J. Mycol. Médicale/J. Med. Mycol. *16*, 113–118.

Enache-Angoulvant, A., and Hennequin, C. (2005). Invasive *Saccharomyces* infection: a comprehensive review. Clin. Infect. Dis. *41*, 1559–1568.

Ene, I.V., and Bennett, R.J. (2009). Hwp1 and related adhesins contribute to both mating and biofilm formation in *Candida albicans*. Eukaryot. Cell *8*, 1909–1913.

Enjalbert, B., and Whiteway, M. (2005). Release from quorum-sensing molecules triggers hyphal formation during *Candida albicans* resumption of growth. *4*, 1203–1210.

Evans, D.J., Allison, D.G., Brown, M.R., and Gilbert, P. (1991). Susceptibility of *Pseudomonas aeruginosa* and *Escherichia coli* biofilms towards ciprofloxacin: effect of specific growth rate. J. Antimicrob. Chemother. *27*, 177–184.

Fernandes, A., and Dias, M. (2013). The microbiological profiles of infected prosthetic implants with an emphasis on the organisms which form biofilms. J. Clin. Diagn. Res. *7*, 219–223.

Figueredo, L.A., Cafarchia, C., and Otranto, D. (2013). Antifungal susceptibility of *Malassezia pachydermatis* biofilm. Med. Mycol. *51*, 863–867.

Finkel, J.S., Xu, W., Huang, D., Hill, E.M., Desai, J.V., Woolford, C.A., Nett, J.E., Taff, H., Norice, C.T., Andes, D.R., *et al.* (2012). Portrait of *Candida albicans* adherence regulators. PLoS Pathog. *8*, e1002525.

Fox, E.P., and Nobile, C.J. (2012). A sticky situation: untangling the transcriptional network controlling biofilm development in *Candida albicans*. Transcription *3*, 1–8.

Fridkin, S.K., and Jarvis, W.R. (1996). Epidemiology of nosocomial fungal infections. Clin. Microbiol. Rev. *9*, 499–511.

Ganguly, S., and Mitchell, A.P. (2011). Mucosal biofilms of *Candida albicans*. Curr. Opin. Microbiol. *14*, 380–385.

Gilbert, P., Collier, P.J., and Brown, M.R. (1990). Influence of growth rate on susceptibility to antimicrobial agents: biofilms, cell cycle, dormancy, and stringent response. Antimicrob. Agents Chemother. *34*, 1865–1868.

Glöckner, A., and Cornely, O.A. (2013). Practical considerations on current guidelines for the management of non-neutropenic adult patients with candidaemia. Mycoses *56*, 11–20.

Gudlaugsson, O., Gillespie, S., Lee, K., Vande Berg, J., Hu, J., Messer, S., Herwaldt, L., Pfaller, M., and Diekema, D. (2003). Attributable mortality of nosocomial candidemia, revisited. Clin. Infect. Dis. *37*, 1172–1177.

Guillot, J., and Bond, R. (1999). Malassezia pachydermatis: a review. Med. Mycol. *37*, 295–306.

Gullo, F.P., Rossi, S.A., Sardi, J.D.C.O., Teodoro, V.L.I., Mendes-Giannini, M.J.S., and Fusco-Almeida, A.M. (2013). Cryptococcosis: epidemiology, fungal resistance, and new alternatives for treatment. Eur. J. Clin. Microbiol. Infect. Dis. *32*, 1377–1391.

Habash, M., and Reid, G. (1999). Microbial biofilms: their development and significance for medical device-related infections. J. Clin. Pharmacol. *39*, 887–898.

Hajjeh, R.A., Sofair, A.N., Harrison, L.H., Lyon, G.M., Arthington-Skaggs, B.A., Mirza, S.A., Phelan, M., Morgan, J., Lee-Yang, W., Ciblak, M.A., *et al.* (2004). Incidence of bloodstream infections due to *Candida* species and *in vitro* susceptibilities of isolates collected from 1998 to 2000 in a population-based active surveillance program. J. Clin. Microbiol. *42*, 1519–1527.

Harriott, M.M., Lilly, E.A., Rodriguez, T.E., Fidel, P.L., and Noverr, M.C. (2010). *Candida albicans* forms biofilms on the vaginal mucosa. Microbiology *156*, 3635–3644.

Hasan, F., Xess, I., Wang, X., Jain, N., and Fries, B.C. (2009). Biofilm formation in clinical *Candida* isolates and its association with virulence. Microbes Infect. *11*, 753–761.

Hawser, S.P., and Douglas, L.J. (1994). Biofilm formation by *Candida* species on the surface of catheter materials *in vitro*. Infect. Immun. *62*, 915–921.

Ho, K.M., Lipman, J., Dobb, G.J., and Webb, S.A.R. (2005). The use of prophylactic fluconazole in immunocompetent high-risk surgical patients: a meta-analysis. Crit. Care *9*, R710–7.

Hope, W.W., Castagnola, E., Groll, A.H., Roilides, E., Akova, M., Arendrup, M.C., Arikan-Akdagli, S., Bassetti, M., Bille, J., Cornely, O.A., *et al.* (2012). ESCMID* guideline for the diagnosis and management of *Candida* diseases 2012: prevention and management of invasive infections in neonates and children caused by Candida spp. Clin. Microbiol. Infect. *18* (Suppl. 7), 38–52.

Hornby, J.M., Jensen, E.C., Lisec, A.D., Tasto, J., Jahnke, B., Shoemaker, R., Nickerson, K.W., Tasto, J.J., and Dussault, P. (2001). Quorum sensing in the dimorphic fungus *Candida albicans* is mediated by farnesol. *67*, 2982–2992.

Iraqui, I., Garcia-Sanchez, S., Aubert, S., Dromer, F., Ghigo, J.-M., D'Enfert, C., and Janbon, G. (2005). The Yak1p kinase controls expression of adhesins and biofilm formation in *Candida glabrata* in a Sir4p-dependent pathway. Mol. Microbiol. *55*, 1259–1271.

Jabra-Rizk, M.A., Falkler, W.A., and Meiller, T.F. (2004). Fungal biofilms and drug resistance. Emerg. Infect. Dis. *10*, 14–19.

Jarvis, J.N., and Harrison, T.S. (2007). HIV-associated cryptococcal meningitis. AIDS *21*, 2119–2129.

Jarvis, J.N., and Harrison, T.S. (2008). Pulmonary cryptococcosis. Semin. Respir. Crit. Care Med. *29*, 141–150.

Kauffman, C.A. (2007). Histoplasmosis: a clinical and laboratory update. Clin. Microbiol. Rev. *20*, 115–132.

Kaur, R., Domergue, R., Zupancic, M.L., and Cormack, B.P. (2005). A yeast by any other name: *Candida glabrata* and its interaction with the host. Curr. Opin. Microbiol. *8*, 378–384.

Keren, I., Shah, D., Spoering, A., Kaldalu, N., and Lewis, K. (2004). Specialized persister cells and the mechanism of multidrug tolerance in *Escherichia coli*. J. Bacteriol. *186*, 8172–8180.

Khor, W.-B., Aung, T., Saw, S.-M., Wong, T.-Y., Tambyah, P.A., Tan, A.-L., Beuerman, R., Lim, L., Chan, W.-K., Heng, W.-J., *et al.* (2006). An outbreak of *Fusarium* keratitis associated with contact lens wear in Singapore. JAMA *295*, 2867–2873.

Khot, P.D., Suci, P.A., Miller, R.L., Nelson, R.D., and Tyler, B.J. (2006). A small subpopulation of blastospores in *Candida albicans* biofilms exhibit resistance to amphotericin B associated with differential regulation of ergosterol and beta-1,6-glucan pathway genes. Antimicrob. Agents Chemother. *50*, 3708–3716.

Kibbler, C.C., Seaton, S., Barnes, R.A., Gransden, W.R., Holliman, R.E., Johnson, E.M., Perry, J.D., Sullivan, D.J., and Wilson, J.A. (2003). Management and outcome of bloodstream infections due to *Candida* species in England and Wales. J. Hosp. Infect. *54*, 18–24.

Kojic, E.M., and Darouiche, R.O. (2004). Candida infections of medical devices. Clin. Microbiol. Rev. *17*, 255–267.

Kruppa, M., Krom, B.P., Chauhan, N., Bambach, A.V., Cihlar, R.L., Richard, A., and Calderone, R.A. (2004). The two-component signal transduction protein Chk1p regulates quorum sensing in *Candida albicans*. *3*, 1062–1065.

Kucharíková, S., Tournu, H., Lagrou, K., Van Dijck, P., and Bujdáková, H. (2011). Detailed comparison of *Candida albicans* and *Candida glabrata* biofilms under different conditions and their susceptibility to caspofungin and anidulafungin. J. Med. Microbiol. *60*, 1261–1269.

Kumamoto, C.A. (2005). A contact-activated kinase signals *Candida albicans* invasive growth and biofilm development. Proc. Natl. Acad. Sci. U.S.A. *102*, 5576–5581.

LaFayette, S.L., Collins, C., Zaas, A.K., Schell, W.A., Betancourt-Quiroz, M., Gunatilaka, A.A.L., Perfect, J.R., and Cowen, L.E. (2010). PKC signaling regulates drug resistance of the fungal pathogen *Candida albicans* via circuitry comprised of Mkc1, calcineurin, and Hsp90. PLoS Pathog. *6*, e1001069.

LaFleur, M.D., Kumamoto, C.A., and Lewis, K. (2006). *Candida albicans* biofilms produce antifungal-tolerant persister cells. Antimicrob. Agents Chemother. *50*, 3839–3846.

LaFleur, M.D., Qi, Q., and Lewis, K. (2010). Patients with long-term oral carriage harbor high-persister mutants of *Candida albicans*. Antimicrob. Agents Chemother. *54*, 39–44.

LaFleur, M.D., Lucumi, E., Napper, A.D., Diamond, S.L., and Lewis, K. (2011). Novel high-throughput screen against *Candida albicans* identifies antifungal potentiators and agents effective against biofilms. J. Antimicrob. Chemother. *66*, 820–826.

Lal, P., Sharma, D., Pruthi, P., and Pruthi, V. (2010). Exopolysaccharide analysis of biofilm-forming *Candida albicans*. J. Appl. Microbiol. *109*, 128–136.

Lass-Flörl, C., Speth, C., Kofler, G., Dierch, M.P., Gunsilius, E., and Würzner, R. (2003). Effect of increasing inoculum sizes of *Aspergillus hyphae* on MICs and MFCs of antifungal agents by broth microdilution method. Int. J. Antimicrob. Agents *21*, 229–233.

Lepak, A., and Andes, D. (2011). Fungal sepsis: optimizing antifungal therapy in the critical care setting. Crit. Care Clin. *27*, 123–147.

Levy, B., Heiler, D., and Norton, S. (2006). Report on testing from an investigation of *Fusarium* keratitis in contact lens wearers. Eye Contact Lens *32*, 256–261.

Lewis, K. (2010). Persister cells. Annu. Rev. Microbiol. *64*, 357–372.

Lipsett, P.A. (2004). Clinical trials of antifungal prophylaxis among patients in surgical intensive care units: concepts and considerations. Clin. Infect. Dis. *39* (Suppl. 4), S193–199.

López, D., Vlamakis, H., and Kolter, R. (2010). Biofilms. Cold Spring Harb. Perspect. Biol. *2*, a000398.

McCormick, A., Loeffler, J., and Ebel, F. (2010). Aspergillus fumigatus: contours of an opportunistic human pathogen. Cell. Microbiol. *12*, 1535–1543.

McKinsey, D.S., and McKinsey, J.P. (2011). Pulmonary histoplasmosis. Semin. Respir. Crit. Care Med. *32*, 735–744.

Malinska, K., Malinsky, J., Opekarova, M., and Tanner, W. (2004). Distribution of Can1p into stable domains reflects lateral protein segregation within the plasma membrane of living *S. cerevisiae* cells. J. Cell Sci. *117*, 6031–6041.

Manzoni, P., Arisio, R., Mostert, M., Leonessa, M., Farina, D., Latino, M.A., and Gomirato, G. (2006). Prophylactic fluconazole is effective in preventing fungal colonization and fungal systemic infections in preterm neonates: a single-center, 6-year, retrospective cohort study. Pediatrics *117*, e22–32.

Marques, S.A. (2012). Paracoccidioidomycosis. Clin. Dermatol. *30*, 610–615.

Martinez, L.R., and Casadevall, A. (2006a). *Cryptococcus neoformans* cells in biofilms are less susceptible than planktonic cells to antimicrobial molecules produced by the innate immune system. Infect. Immun. *74*, 6118–6123.

Martinez, L.R., and Casadevall, A. (2006b). Susceptibility of *Cryptococcus neoformans* biofilms to antifungal agents *in vitro*. Antimicrob. Agents Chemother. *50*, 1021–1033.

Martínez, J.L., and Rojo, F. (2011). Metabolic regulation of antibiotic resistance. FEMS Microbiol. Rev. *35*, 768–789.

Martínez-Gomariz, M., Perumal, P., Mekala, S., Nombela, C., Chaffin, W.L., and Gil, C. (2009). Proteomic analysis of cytoplasmic and surface proteins from yeast cells, hyphae, and biofilms of *Candida albicans*. Proteomics *9*, 2230–2252.

Martins, M., Uppuluri, P., Thomas, D.P., Cleary, I.A., Henriques, M., Lopez-Ribot, J.L., and Oliveira, R. (2010). Presence of extracellular DNA in the *Candida albicans* biofilm matrix and its contribution to biofilms. Mycopathologia *169*, 323–331.

Martins, M., Henriques, M., Lopez-Ribot, J.L., and Oliveira, R. (2012). Addition of DNase improves the *in vitro* activity of antifungal drugs against Candida albicans biofilms. Mycoses *55*, 80–85.

Mateus, C., Crow, S.A., and Ahearn, D.G. (2004). Adherence of *Candida albicans* to silicone induces immediate enhanced tolerance to fluconazole. Antimicrob. Agents Chemother. *48*, 3358–3366.

Mathé, L., and Van Dijck, P. (2013). Recent insights into *Candida albicans* biofilm resistance mechanisms. Curr. Genet. *59*, 251–264.

Mermel, L.A., Allon, M., Bouza, E., Craven, D.E., Flynn, P., O'Grady, N.P., Raad, I.I., Rijnders, B.J.A., Sherertz, R.J., and Warren, D.K. (2009). Clinical practice guidelines for the diagnosis and management of intravascular catheter-related infection: 2009 update by the Infectious Diseases Society of America. Clin. Infect. Dis. *49*, 1–45.

Mikolajewska, A., Schwartz, S., and Ruhnke, M. (2012). Antifungal treatment strategies in patients with haematological diseases or cancer: from prophylaxis to empirical, pre-emptive and targeted therapy. Mycoses *55*, 2–16.

Mitchell, K.F., Taff, H.T., Cuevas, M.A., Reinicke, E.L., Sanchez, H., and Andes, D.R. (2013). Role of matrix β-i,3-glucan in antifungal resistance of non-albicans Candida biofilms. Antimicrob. Agents Chemother. *57*, 1918–1920.

Monniot, C., Boisramé, A., Da Costa, G., Chauvel, M., Sautour, M., Bougnoux, M.-E., Bellon-Fontaine, M.-N., Dalle, F., D'Enfert, C., and Richard, M.L. (2013). Rbt1 protein domains analysis in *Candida albicans* brings insights into hyphal surface modifications and Rbt1 potential role during adhesion and biofilm formation. PLoS One *8*, e82395.

Morgan, J., Meltzer, M.I., Plikaytis, B.D., Sofair, A.N., Huie-White, S., Wilcox, S., Harrison, L.H., Seaberg, E.C., Hajjeh, R.A., and Teutsch, S.M. (2005). Excess mortality, hospital stay, and cost due to candidemia: a case–control study using data from population-based candidemia surveillance. Infect. Control Hosp. Epidemiol. *26*, 540–547.

Mowat, E., Butcher, J., Lang, S., Williams, C., and Ramage, G. (2007). Development of a simple model for studying the effects of antifungal agents on multicellular communities of *Aspergillus fumigatus*. J. Med. Microbiol. *56*, 1205–1212.

Mowat, E., Williams, C., Jones, B., McChlery, S., and Ramage, G. (2009). The characteristics of *Aspergillus fumigatus* mycetoma development: is this a biofilm? Med. Mycol. *47* (Suppl. 1), S120–126.

Mukherjee, P.K., Chandra, J., Kuhn, D.M., and Ghannoum, M.A. (2003). Mechanism of fluconazole resistance in *Candida albicans* biofilms : phase-specific role of efflux pumps and membrane sterols. Infect. Immun. *71*, 4333–4340.

Muñoz, P., Bouza, E., Cuenca-Estrella, M., Eiros, J.M., Pérez, M.J., Sánchez-Somolinos, M., Rincón, C., Hortal, J., and Peláez, T. (2005). Saccharomyces cerevisiae fungemia: an emerging infectious disease. Clin. Infect. Dis. *40*, 1625–1634.

Navarro-García, F., Sánchez, M., Pla, J., and Nombela, C. (1995). Functional characterization of the MKC1 gene of *Candida albicans*, which encodes a mitogen-activated protein kinase homolog related to cell integrity. Mol. Cell. Biol. *15*, 2197–2206.

Navarro-García, F., Alonso-Monge, R., Rico, H., Pla, J., Sentandreu, R., and Nombela, C. (1998). A role for the MAP kinase gene MKC1 in cell wall construction and morphological transitions in *Candida albicans*. Microbiology *144* (Pt 2), 411–424.

Netsvyetayeva, I., Swoboda-Kopeć, E., Paczek, L., Fiedor, P., Sikora, M., Jaworska-Zaremba, M., Blachnio, S., and Luczak, M. (2009). *Trichosporon asahii* as a prospective pathogen in solid organ transplant recipients. Mycoses *52*, 263–265.

Nett, J., Lincoln, L., Marchillo, K., Massey, R., Holoyda, K., Hoff, B., VanHandel, M., and Andes, D. (2007). Putative role of beta-i,3-glucans in *Candida albicans* biofilm resistance. Antimicrob. Agents Chemother. *51*, 510–520.

Nett, J.E., Lepak, A.J., Marchillo, K., and Andes, D.R. (2009). Time course global gene expression analysis of an *in vivo Candida* biofilm. J. Infect. Dis. *200*, 307–313.

Nett, J.E., Crawford, K., Marchillo, K., and Andes, D.R. (2010). Role of Fks1p and matrix glucan in *Candida albicans* biofilm resistance to an echinocandin, pyrimidine, and polyene. Antimicrob. Agents Chemother. *54*, 3505–3508.

Nguyen, M.H., and Yu, C.Y. (1999). Influence of incubation time, inoculum size, and glucose concentrations on spectrophotometric endpoint determinations for amphotericin B, fluconazole, and itraconazole. J. Clin. Microbiol. *37*, 141–145.

Nobile, C.J., and Mitchell, A.P. (2006). Genetics and genomics of *Candida albicans* biofilm formation. Cell. Microbiol. *8*, 1382–1391.

Nobile, C.J., and Mitchell, A.P. (2007). Microbial biofilms: e pluribus unum. Curr. Biol. *17*, R349–353.

Nobile, C.J., Andes, D.R., Nett, J.E., Smith, F.J., Yue, F., Phan, Q.-T., Edwards, J.E., Filler, S.G., and Mitchell, A.P. (2006a). Critical role of Bcr1-dependent adhesins in *C. albicans* biofilm formation *in vitro* and *in vivo.* PLoS Pathog. *2*, e63.

Nobile, C.J., Nett, J.E., Andes, D.R., and Mitchell, A.P. (2006b). Function of *Candida albicans* adhesin Hwp1 in biofilm formation. Eukaryot. Cell *5*, 1604–1610.

Nobile, C.J., Nett, J.E., Hernday, A.D., Homann, O.R., Deneault, J.-S., Nantel, A., Andes, D.R., Johnson, A.D., and Mitchell, A.P. (2009). Biofilm matrix regulation by *Candida albicans* Zap1. PLoS Biol. *7*, e1000133.

Nobile, C.J., Fox, E.P., Nett, J.E., Sorrells, T.R., Mitrovich, Q.M., Hernday, A.D., Tuch, B.B., Andes, D.R., and Johnson, A.D. (2012). A recently evolved transcriptional network controls biofilm development in *Candida albicans*. Cell *148*, 126–138.

Odds, F.C. (1988). Candida and Candidosis (Baillière Tindall, London, UK).

Pappas, P.G., Kauffman, C.A., Andes, D., Benjamin, D.K., Calandra, T.F., Edwards, J.E., Filler, S.G., Fisher, J.F., Kullberg, B.-J., Ostrosky-Zeichner, L., *et al.* (2009). Clinical practice guidelines for the management of candidiasis: 2009 update by the Infectious Diseases Society of America. Clin. Infect. Dis. *48*, 503–535.

Perlroth, J., Choi, B., and Spellberg, B. (2007). Nosocomial fungal infections: epidemiology, diagnosis, and treatment. Med. Mycol. *45*, 321–346.

Perumal, P., Mekala, S., and Chaffin, W.L. (2007). Role for cell density in antifungal drug resistance in *Candida albicans* biofilms. Antimicrob. Agents Chemother. *51*, 2454–2463.

Pettit, R.K., Repp, K.K., and Hazen, K.C. (2010). Temperature affects the susceptibility of *Cryptococcus neoformans* biofilms to antifungal agents. Med. Mycol. *48*, 421–426.

Pfaller, M.A. (1996). Nosocomial candidiasis : emerging species, reservoirs, and modes of transmission. Clin. Infect. Dis. *22*, 89–94.

Pfaller, M.A., and Diekema, D.J. (2007). Epidemiology of invasive candidiasis: a persistent public health problem. Clin. Microbiol. Rev. *20*, 133–163.

Pfaller, M., Neofytos, D., Diekema, D., Azie, N., Meier-Kriesche, H.-U., Quan, S.-P., and Horn, D. (2012). Epidemiology and outcomes of candidemia in 3648 patients: data from the Prospective Antifungal Therapy (PATH Alliance®) registry, 2004–2008. Diagn. Microbiol. Infect. Dis. *74*, 323–331.

Pfaller, M.A., Jones, R.N., Messer, S.A., Edmond, M.B., and Wenzel, R.P. (1998). National surveillance of nosocomial blood stream infection due to *Candida albicans*: frequency of occurrence and antifungal susceptibility in the SCOPE Program. Diagn. Microbiol. Infect. Dis. *31*, 327–332.

Pfaller, M.A., Diekema, D.J., Jones, R.N., Sader, H.S., Fluit, A.C., and Hollis, R.J. (2001). International surveillance of bloodstream infections due to *Candida* species : frequency of occurrence and *in vitro* susceptibilities to fluconazole, ravuconazole, and voriconazole of isolates collected from 1997 through 1999 in the SENTRY antimicrobial surveillance program. J. Clin. Microbiol. *39*, 3254–3259.

Picazo, J.J., González-Romo, F., and Candel, F.J. (2008). Candidemia in the critically ill patient. Int. J. Antimicrob. Agents *32* (Suppl. 2), S83–85.

Pierce, C.G., and Lopez-Ribot, J.L. (2013). Candidiasis drug discovery and development: new approaches targeting virulence for discovering and identifying new drugs. Expert Opin. Drug Discov. *8*, 1117–1126.

Pieroni, K.P., Nespor, C., Poole, R.L., Kerner, J.A., and Berquist, W.E. (2013). Echinocandin and ethanol lock therapy treatment of fungal catheter infections. Pediatr. Infect. Dis. J. *32*, 289–291.

Pitangui, N.S., Sardi, J.C.O., Silva, J.F., Benaducci, T., Moraes da Silva, R.A., Rodríguez-Arellanes, G., Taylor, M.L., Mendes-Giannini, M.J.S., and Fusco-Almeida, A.M. (2012). Adhesion of *Histoplasma capsulatum* to pneumocytes and biofilm formation on an abiotic surface. Biofouling *28*, 711–718.

Pittet, D., and Wenzel, R.P. (1995). Nosocomial bloodstream infections. Secular trends in rates, mortality, and contribution to total hospital deaths. Arch. Intern. Med. *155*, 1177–1184.

Rajendran, R., Williams, C., Lappin, D.F., Millington, O., Martins, M., and Ramage, G. (2013). Extracellular DNA release acts as an antifungal resistance mechanism in mature *Aspergillus fumigatus* biofilms. Eukaryot. Cell *12*, 420–429.

Ramage, G. (2002). Investigation of multidrug efflux pumps in relation to fluconazole resistance in *Candida albicans* biofilms. J. Antimicrob. Chemother. *49*, 973–980.

Ramage, G., Vandewalle, K., Wickes, B.L., and López-Ribot, J.L. (2001a). Characteristics of biofilm formation by *Candida albicans*. Rev. Iberoam. Micol. *18*, 163–170.

Ramage, G., Vande Walle, K., Wickes, B.L., López-Ribot, J.L., and Lo, L. (2001b). Biofilm formation by *Candida dubliniensis*. J. Clin. Microbiol. *39*, 3234–3240.

Ramage, G., Martínez, J.P., and López-Ribot, J.L. (2006). *Candida* biofilms on implanted biomaterials: a clinically significant problem. FEMS Yeast Res. *6*, 979–986.

Ramage, G., Mowat, E., Jones, B., Williams, C., and Lopez-Ribot, J. (2009). Our current understanding of fungal biofilms. Crit. Rev. Microbiol. *35*, 340–355.

Ramage, G., Rajendran, R., Gutierrez-Correa, M., Jones, B., and Williams, C. (2011). *Aspergillus* biofilms: clinical and industrial significance. FEMS Microbiol. Lett. *324*, 89–97.

Ramage, G., Rajendran, R., Sherry, L., and Williams, C. (2012). Fungal biofilm resistance. Int. J. Microbiol. *2012*, 528521.

Rammaert, B., Lanternier, F., Zahar, J.-R., Dannaoui, E., Bougnoux, M.-E., Lecuit, M., and Lortholary, O. (2012). Healthcare-associated mucormycosis. Clin. Infect. Dis. *54* (Suppl. 1), S44–54.

Rani, S.A., Pitts, B., Beyenal, H., Veluchamy, R.A., Lewandowski, Z., Davison, W.M., Buckingham-Meyer, K., and Stewart, P.S. (2007). Spatial patterns of DNA replication, protein synthesis, and oxygen concentration within bacterial biofilms reveal diverse physiological states. J. Bacteriol. *189*, 4223–4233.

Rao, S.K., Lam, P.T.H., Li, E.Y.M., Yuen, H.K.L., and Lam, D.S.C. (2007). A case series of contact lens-associated *Fusarium* keratitis in Hong Kong. Cornea *26*, 1205–1209.

Regenberg, B., and Kielland-Brandt, M.C. (2001). Amino acid residues important for substrate specificity of the amino acid permeases Can1p and Gnp1p in *Saccharomyces cerevisiae*. Yeast *18*, 1429–1440.

Rex, J.H., and Sobel, J.D. (2001). Prophylactic antifungal therapy in the intensive care unit. Clin. Infect. Dis. *32*, 1191–1200.

Reynolds, T.B., and Fink, G.R. (2001). Bakers' yeast, a model for fungal biofilm formation. Science *291*, 878–881.

Robbins, N., Uppuluri, P., Nett, J., Rajendran, R., Ramage, G., Lopez-Ribot, J.L., Andes, D., and Cowen, L.E. (2011). Hsp90 governs dispersion and drug resistance of fungal biofilms. PLoS Pathog. *7*, e1002257.

Rosenbach, A., Dignard, D., Pierce, J.V., Whiteway, M., and Kumamoto, C.A. (2010). Adaptations of *Candida albicans* for growth in the mammalian intestinal tract. Eukaryot. Cell *9*, 1075–1086.

Samaranayake, L.P., Leung, W.K., and Jin, L. (2009). Oral mucosal fungal infections. *49*, 39–59.

Sardi, J.C.O., Scorzoni, L., Bernardi, T., Fusco-Almeida, A.M., and Mendes Giannini, M.J.S. (2013). Candida species: current epidemiology, pathogenicity, biofilm formation, natural antifungal products and new therapeutic options. J. Med. Microbiol. *62*, 10–24.

Saw, S.-M., Ooi, P.-L., Tan, D.T.H., Khor, W.-B., Fong, C.-W., Lim, J., Cajucom-Uy, H.Y., Heng, D., Chew, S.-K., Aung, T., *et al.* (2007). Risk factors for contact lens-related fusarium keratitis: a case–control study in Singapore. Arch. Ophthalmol. *125*, 611–617.

Seddiki, S.M.L., Boucherit-Otmani, Z., Boucherit, K., Badsi-Amir, S., Taleb, M., and Kunkel, D. (2013). Assessment of the types of catheter infectivity caused by *Candida* species and their biofilm formation. First study in an intensive care unit in Algeria. Int. J. Gen. Med. *6*, 1–7.

Seidler, M.J., Salvenmoser, S., and Müller, F.-M.C. (2008). *Aspergillus fumigatus* forms biofilms with reduced antifungal drug susceptibility on bronchial epithelial cells. Antimicrob. Agents Chemother. *52*, 4130–4136.

Shoham, S. (2010). Invasive candidiasis in patients with implants. Curr. Fungal Infect. Rep. *5*, 12–17.

Silva, S., Negri, M., Henriques, M., Oliveira, R., Williams, D.W., and Azeredo, J. (2011). Adherence and biofilm formation of non-Candida albicans Candida species. Trends Microbiol. *19*, 241–247.

Singh, R., Shivaprakash, M.R., and Chakrabarti, A. (2011). Biofilm formation by zygomycetes: quantification, structure and matrix composition. Microbiology *157*, 2611–2618.

Singh, S.D., Robbins, N., Zaas, A.K., Schell, W.A., Perfect, J.R., and Cowen, L.E. (2009). Hsp90 governs echinocandin resistance in the pathogenic yeast *Candida albicans* via calcineurin. PLoS Pathog. *5*, e1000532.

Skiada, A., and Petrikkos, G. (2013). Cutaneous mucormycosis. Skinmed *11*, 155–159; quiz 159–160.

Srinivasan, A., Leung, K.P., Lopez-Ribot, J.L., and Ramasubramanian, A.K. (2013). High-throughput nano-biofilm microarray for antifungal drug discovery. MBio *4*, 1–11.

Sullivan, D.J., Westerneng, T.J., Haynes, K.A., Bennett, D.E., and Coleman, D.C. (1995). *Candida dubliniensis* sp. nov.: phenotypic and molecular characterization of a novel species associated with oral candidosis in HIV-infected individuals. Microbiology *141* (Pt 7), 1507–1521.

Sundstrom, P. (1999). Adhesins in *Candida albicans*. Curr. Opin. Microbiol. *2*, 353–357.

Taff, H.T., Mitchell, K.F., Edward, J.A., and Andes, D.R. (2013). Mechanisms of *Candida* biofilm drug resistance. Future Microbiol. *8*, 1325–1337.

Thomas, D.P., Bachmann, S.P., and Lopez-Ribot, J.L. (2006). Proteomics for the analysis of the *Candida albicans* biofilm lifestyle. Proteomics *6*, 5795–5804.

Tobudic, S., Kratzer, C., Lassnigg, A., and Presterl, E. (2012). Antifungal susceptibility of *Candida albicans* in biofilms. Mycoses *55*, 199–204.

Tumbarello, M., Posteraro, B., Trecarichi, E.M., Fiori, B., Rossi, M., Porta, R., de Gaetano Donati, K., La Sorda, M., Spanu, T., Fadda, G., *et al.* (2007). Biofilm production by *Candida* species and inadequate antifungal therapy as predictors of mortality for patients with candidemia. J. Clin. Microbiol. *45*, 1843–1850.

Ullmann, A.J., Akova, M., Herbrecht, R., Viscoli, C., Arendrup, M.C., Arikan-Akdagli, S., Bassetti, M., Bille, J., Calandra, T., Castagnola, E., *et al.* (2012). ESCMID* guideline for the diagnosis and management of Candida diseases 2012: adults with haematological malignancies and after haematopoietic stem cell transplantation (HCT). Clin. Microbiol. Infect. *18* (Suppl. 7), 53–67.

Uppuluri, P., Nett, J., Heitman, J., and Andes, D. (2008). Synergistic effect of calcineurin inhibitors and fluconazole against *Candida albicans* biofilms. Antimicrob. Agents Chemother. *52*, 1127–1132.

Uppuluri, P., Chaturvedi, A.K., Srinivasan, A., Banerjee, M., Ramasubramaniam, A.K., Köhler, J.R., Kadosh, D., and Lopez-Ribot, J.L. (2010). Dispersion as an important step in the *Candida albicans* biofilm developmental cycle. PLoS Pathog. *6*, e1000828.

Vediyappan, G., Rossignol, T., and d'Enfert, C. (2010). Interaction of *Candida albicans* biofilms with antifungals: transcriptional response and binding of antifungals to beta-glucans. Antimicrob. Agents Chemother. *54*, 2096–2111.

Vermeulen, E., Lagrou, K., and Verweij, P.E. (2013). Azole resistance in *Aspergillus fumigatus*: a growing public health concern. Curr. Opin. Infect. Dis. *26*, 493–500.

Walraven, C.J., and Lee, S.A. (2013). Antifungal lock therapy. Antimicrob. Agents Chemother. *57*, 1–8.

Walsh, T.J., Schlegel, R., Moody, M.M., Costerton, J.W., and Salcman, M. (1986). Ventriculoatrial shunt infection due to *Cryptococcus neoformans*: an ultrastructural and quantitative microbiological study. Neurosurgery *18*, 373–375.

Wey, S.B., Mori, M., Pfaller, M.A., Woolson, R.F., and Wenzel, R.P. (1988). Hospital-acquired candidemia. The attributable mortality and excess length of stay. Arch. Intern. Med. *148*, 2642–2645.

Whimbey, E., Gold, J.W., Polsky, B., Dryjanski, J., Hawkins, C., Blevins, A., Brannon, P., Kiehn, T.E., Brown, A.E., and Armstrong, D. (1986). Bacteremia and fungemia in patients with the acquired immunodeficiency syndrome. Ann. Intern. Med. *104*, 511–514.

White, T.C., Marr, K.A., and Bowden, R.A. (1998). Clinical, cellular, and molecular factors that contribute to antifungal drug resistance. Clin. Microbiol. Rev. *11*, 382–402.

Wisplinghoff, H., Bischoff, T., Tallent, S.M., Seifert, H., Wenzel, R.P., and Edmond, M.B. (2004). Nosocomial bloodstream infections in US hospitals: analysis of 24,179 cases from a prospective nationwide surveillance study. Clin. Infect. Dis. *39*, 309–317.

Wolcott, R., Costerton, J.W., Raoult, D., and Cutler, S.J. (2012). The polymicrobial nature of biofilm infection. Clin. Microbiol. Infect. *19*, 1–6.

Yeater, K.M., Chandra, J., Cheng, G., Mukherjee, P.K., Zhao, X., Rodriguez-Zas, S.L., Kwast, K.E., Ghannoum, M.A., and Hoyer, L.L. (2007). Temporal analysis of *Candida albicans* gene expression during biofilm development. Microbiology *153*, 2373–2385.

Zeidler, U., Bougnoux, M.-E., Lupan, A., Helynck, O., Doyen, A., Garcia, Z., Sertour, N., Clavaud, C., Munier-Lehmann, H., Saveanu, C., *et al.* (2013). Synergy of the antibiotic colistin with echinocandin antifungals in Candida species. J. Antimicrob. Chemother. *68*, 1285–1296.

Zhang, X., Sun, X., Wang, Z., Zhang, Y., and Hou, W. (2012). Keratitis-associated fungi form biofilms with reduced antifungal drug susceptibility. Invest. Ophthalmol. Vis. Sci. *53*, 7774–7778.

Zhao, X., Daniels, K.J., Oh, S., Green, C.B., Yeater, K.M., Soll, D.R., and Hoyer, L.L. (2006). *Candida albicans* Als3p is required for wild-type biofilm formation on silicone elastomer surfaces. Microbiology *152*, 2287–2299.

Zhu, Z., Wang, H., Shang, Q., Jiang, Y., Cao, Y., and Chai, Y. (2012). Time course analysis of *Candida albicans* metabolites during biofilm development. J. Proteome Res. *12*, 2375–2385.

Drug Combinations as a Strategy to Potentiate Existing Antifungal Agents

5

Dominique Sanglard and Leah Cowen

Abstract

Antifungal treatments for combating fungal infections are usually administered as monotherapies. With the few antifungal agents available and the inevitable development of resistance in fungal pathogens, combination therapy may be a future alternative to augment the efficacy of existing agents by synergistic effects. This review gives an overview of attempts to identify effective *in vitro* or *in vivo* combination of known antifungals with each other and with other bioactive molecules. The search for synergistic drug combinations currently involves systematic screening of compounds libraries. The high number of possible combinations with their intensive experimental demand has stimulated *in silico* approaches predictive of drug synergisms. Here we summarize the achievements of these approaches that use mainly chemo-genomic methods in fungal model systems.

Introduction

While superficial fungal infections are common and are usually not life threatening, invasive fungal infections (IFI) have a lower incidence but they are associated with high mortality in at-risk patients (Brown *et al.*, 2012). When considering nosocomial bloodstream infections, the frequency of IFI has been increasing over the last two decades, which is also paralleled by changes in the pattern of the aetiological species (Pfaller and Diekema, 2010). Although *Candida albicans* and *Aspergillus fumigatus* remain predominant pathogens, other species of these two genera, as well as members of the zygomycetes, have gained increasing importance (Maschmeyer, 2006). Several factors are contributing to these observations including the administration of broad-spectrum antibiotics, corticosteroids and cytotoxic agents, intravenous catheters, invasive medical procedures, human immunodeficiency virus infection, haematological malignancy, solid organ or bone marrow transplantation and prolonged neutropenia (Puig-Asensio *et al.*, 2014; Ruhnke, 2014).

Antifungal agents can be used to combat fungal diseases, however, despite an increase in their use over the past two decades, the clinical outcomes remain unsatisfactory (Ruhnke, 2014). Several reasons are behind this conclusion. First, early and directed antifungal therapy cannot be implemented due to delays in fungal disease diagnosis. Antifungal drugs have also restricted routes of administration, spectrum of activity, and bioavailability (Denning and Hope, 2010). Second, only a few antifungal classes are available and the efficacy of these drugs is challenged by toxicity and negative interactions with other prescribed agents. Third, extensive use of antifungal agents for therapeutic or prophylactic purposes tends to favour the emergence of drug resistance (Bowyer *et al.*, 2011; Miceli and Lee, 2011; Rodloff *et al.*, 2011).

Several approaches could be undertaken in order to overcome the numerous problems arising from the management of fungal diseases. Besides the discovery of novel effective agents, one realistic alternative option would be to enhance the activity of existing agents. This strategy could be achieved by combining existing antifungal agents with other bioactive substances (combination therapy) with known activity profiles. In this chapter, the concept of drug combinations will be presented by reviewing not only *in vitro* and *in vivo* approaches, but also by exploring the potential of genome-wide screening technologies.

In vitro and in vivo combinations with known antifungal agents

Two major goals are behind the strategy of antifungal combinations: first, to increase the spectrum of their activity and second to minimize the development of resistance. One of the key issues with antifungal combinations is how to characterize the quality of drug combination *in vitro* or *in vivo*. In fact, when two antifungal drugs are combined, the drugs can interact to enhance the efficacy of the combination (synergy) or to diminish the efficacy (antagonism). The interaction between two antifungals is often quantified by selecting a particular endpoint (inhibition of fungal growth by MIC measurements) and then by measuring the effect of various combinations of the two drugs reaching this endpoint.

The fractional inhibitory concentration (FIC) index (FICI) has long been the most commonly used way to characterize the activity of drug combinations in the laboratory by the use of the chequerboard method (Johnson *et al.*, 2004). FICI values are obtained by the following equation:

$$\mathrm{FICI}=\mathrm{FIC_A}+\mathrm{FIC_B}=\frac{\mathrm{MIC_A^{comb}}}{\mathrm{MIC_A^{alone}}}+\frac{\mathrm{MIC_B^{comb}}}{\mathrm{MIC_B^{alone}}}$$

where

$$\mathrm{MIC_A^{alone}}$$

and

$$\mathrm{MIC_B^{alone}}$$

are the MICs of compound A and B when tested alone and

$$\mathrm{MIC_A^{comb}}$$

and

$$MIC_B^{comb}$$

are the concentrations of compound A with B. As recommended by Odds (2003), synergy between antifungals can be assigned when FICI ≤ 0.5, while it is antagonistic when FICI >4.0. Between FICI values >0.5 and <4, several intermediate states have been proposed (such as additivity and indifference), however, due to the complexity of interpretations (Te Dorsthorst *et al.*, 2002), it is recommended to adopt the term of 'no interactions' for $0.5 < \mathrm{FICI} < 4.0$. Fig. 5.1 illustrates with chequerboard approaches the interpretation of combinations between 2 drugs. Synergistic associations can be identified by systematic FICI calculations deduced from the chequerboard results. Other alternative methods to predict synergistic interactions between 2 drugs exist and rely on deviations from additivity (Yeh *et al.*, 2009). Defining additivity is therefore crucial for classification of drug interactions and two main methods are used for this purpose. The Bliss independence as the first method assumes that the relative effect of a drug is not dependent of the presence of the other drug. As an example, if drugs A and B individually cause growth inhibition of 50% each, then Bliss independence predicts that combined drugs A and B decrease growth by 75% (Keith *et al.*, 2005). Positive or negative deviations from this prediction describe synergistic and antagonistic interactions, respectively. Fig. 5.2 illustrates the representation of synergy using the Bliss independence model with data of Fig. 5.1. Most potent synergies (value above 80% difference between experimental results and predicted values) correspond to the synergistic combinations delimited in Fig. 5.1.

The Loewe additivity as the second method defines a drug as non-interacting with itself. If drugs A and B are in fact the same or similar drugs, then we expect their combined effect at equal concentrations to be identical to the effect of one of the drugs in double the dose. Loewe additivity is the expected response for compounds that act similarly on the same molecular target. In a two-dimensional grid, a drug combined with itself will yield linear contours (isobols), curves such that each point along the curve corresponds to

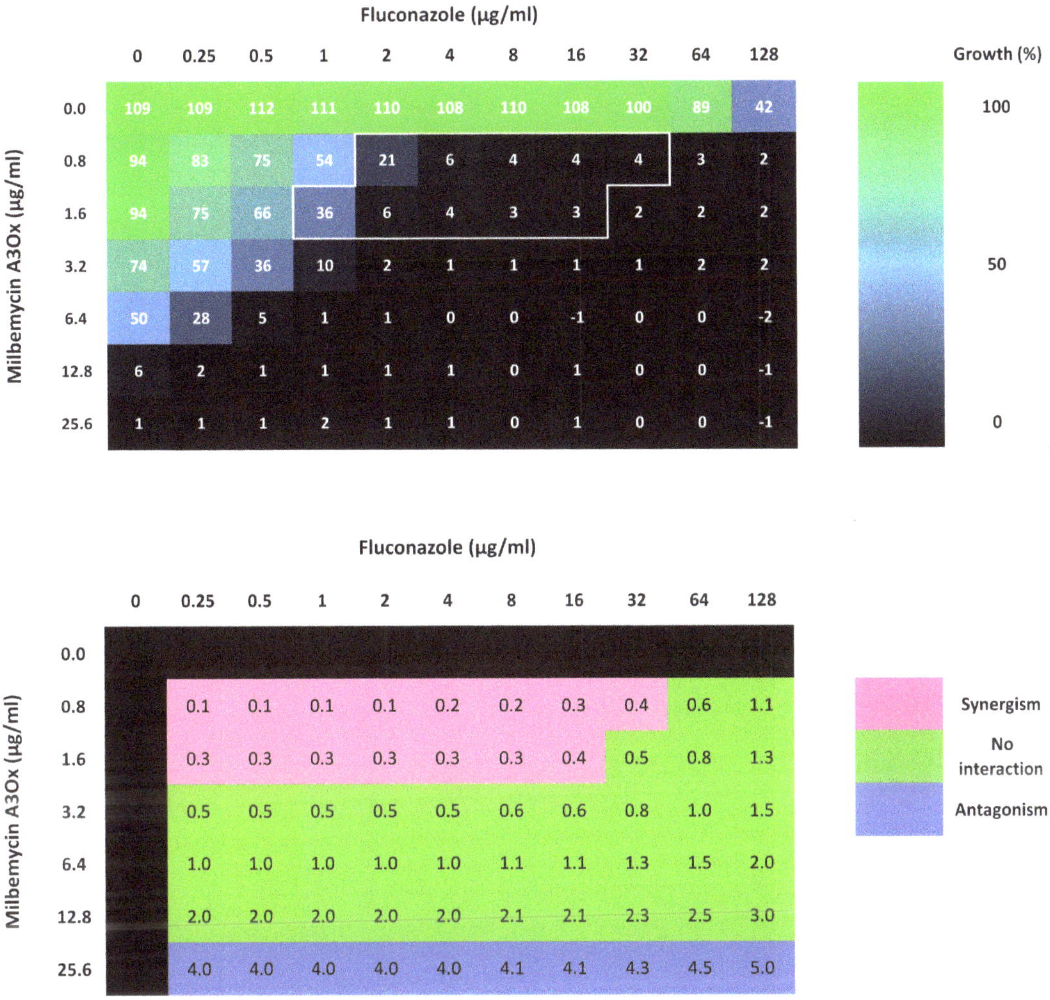

Figure 5.1 Chequerboard assay and interpretation of results. Data were generated from a published study (Silva *et al.*, 2013). Chequerboard data are represented by a heatmap showing growth (as percentage relative to untreated sample) of *C. albicans* in the presence of FLC and/or milbemycin A3 oxim (A3Ox). FLC and A3Ox MICs are 128 and 12.8 µg/ml, respectively. Lower panel shows results from FICI calculations. Synergistic combinations were reported in the upper panel (delimited by a white zone). Synergistic effects (FICI <0.5) are detected when growth is reduced by at least 50% in the presence of both drugs.

a different drug concentration combination that yields the same growth phenotype. However, the contours will be convex when two drugs are synergistic, and concave when they are antagonistic (Keith *et al.*, 2005). Fig. 5.3 shows an example of an isobologram taken from data presented in Fig. 5.1. The shape of the curve produced by FLC/A3Ox combination is convex, thus consistent with the observed synergism detected by the other methods.

To monitor the quality of drug combinations, additional approaches also use chequerboard results. In this case, a three-dimensional surface diagram can be plotted from dose-matrix experiments. From the shape of the surface response, a few assumptions can be made whether or not drug targets are likely part of parallel or same pathways (Lehar *et al.*, 2008).

One important limitation with antifungal combinations is with predicting the outcome of *in vivo* combinations based on *in vitro* measurements. Adding to antifungal combinations the host environment, types of animal model and pharmacokinetics/pharmacodynamic parameters

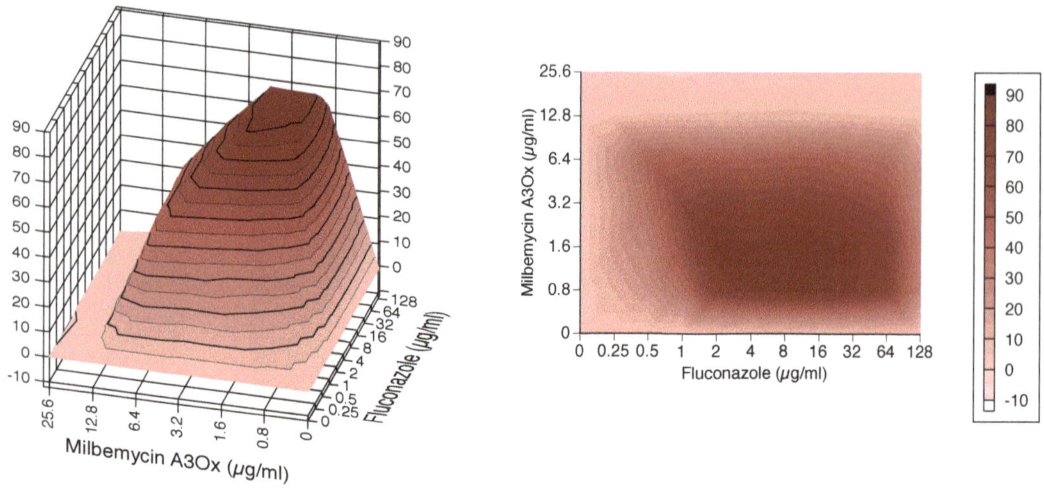

Figure 5.2 Three-dimensional and contour plots of synergy using the Bliss independence theory. The values were obtained using Mac Synergy II (Prichard *et al.*, 1991) for each combination and were depicted on the z-axis to construct a 3-D surface plot. Peaks above and below the zero plane indicate synergistic and antagonistic combinations, respectively, while the zero plane indicates the absence of interaction. The contour plot shows drug concentrations with highest synergistic effect (0.8–1.6 μg/ml for A3Ox; 2–16 μg/ml for FLC).

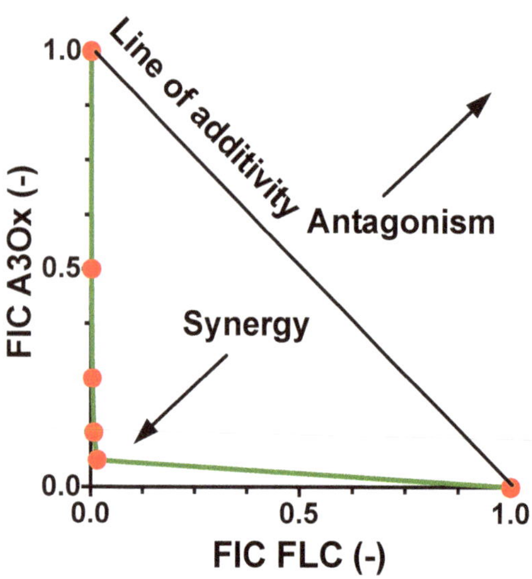

Figure 5.3 Isobologram of FLC and A3Ox combinations. Data were represented according to Prichard *et al.* (1991), and were generated from results of Fig. 5.1. Fractional inhibitory concentration (FIC) was normalized (from 0 to 1) from MIC values of Fig. 5.1. Concave and convex curves indicate synergism and antagonism, respectively (indicated by arrows).

make combination therapy a very challenging experimental field. Clinical trials dealing with combination therapy can be financially prohibitive given that they require several control patients groups. Combination therapy is considered as compassionate for the treatment of recalcitrant infections in patients (Vazquez, 2008).

There are a high number of published reports on *in vitro* antifungal combinations (Ostrosky-Zeichner, 2008). Even if outcomes of the results depend on the tested fungal species, general trends on synergistic and antagonistic combinations can be highlighted. In the following section, combinations of known antifungal agents with each other will be discussed for their *in vitro* and *in vivo* properties as well as their implementation in clinical settings.

Combination of azoles with amphotericin B

The use of azoles and amphotericin B (AmB) in single therapy is common and their combination was debated in several studies. One of the issues about azole/AmB combination is that both agents target sterol metabolites but at different levels. Azoles inhibit ergosterol biosynthesis while AmB binds this sterol with high affinity (Sanglard, 2002). Most *in vitro* studies have demonstrated mixed results of antagonism or indifference, whereas variable results for survival and tissue burden have been demonstrated in animal models of fungal infection (Cuenca-Estrella, 2004; Johnson and Perfect, 2010). One point to consider is the efficacy of sequential combinations of AmB and azoles, a situation that may occur frequently in treated patients. It is believed that administration of azoles will result in ergosterol depletion in the targeted fungal pathogen, which will consequently reduce the efficacy of subsequent AmB treatments, suggesting an antagonistic interaction. This hypothesis has been verified in animal models (Louie *et al.*, 2001). However, evidence of antagonism has generally not been observed in the clinical setting as a consequence of rapid pharmacokinetic clearance of azoles (Ernst *et al.*, 1998). It is generally considered that AmB/azole combination therapy is not superior to monotherapy after analysis of clinical trials (Cuenca-Estrella, 2004), however case reports on successful combined therapy are not infrequent, especially in the case of mould infections (Chaux, 2010).

Combination of 5-FC with amphotericin B

By combining 5-fluorocytosine (5-FC) with AmB, one can expect positive associations since both drugs target different cellular pathways. In addition, 5-FC monotherapy is associated with rapid development of resistance, thus favouring its use in combined therapy (Vandeputte *et al.*, 2011). However, *in vitro* studies performed in *Candida* or *Aspergillus* spp. mostly revealed absence of interactions. The outcome of 5-FC/AmB combinations is more favourable for *Cryptococcus neoformans*, where most studies report synergism (Cuenca-Estrella, 2004). This observation is reflected by animal models and also by clinical practice, where this combination remains the gold standard treatment for cryptococcal meningitis (Perfect *et al.*, 2010).

Combination of candins with amphotericin B

The class of candin antifungals including caspofungin, micafungin and anidulafungin were introduced into clinical practice starting from early 2000. Attempts to combine candins with AmB in *Candida* spp. did not generally result in positive interactions *in vitro* (Tobudic *et al.*, 2010). In contrast, in *Aspergillus* spp. exposed to candin/AmB combinations, synergy is identified in most isolates (Arikan *et al.*, 2002). It is therefore not surprising that most candin/AmB *in vivo* experiments involve experimental aspergillosis. In these experimental models, the combination was found superior to monotherapies (Cuenca-Estrella, 2004). Clinical experience of candin/AmB combinations for treating invasive aspergillosis resulted also in better outcomes than monotherapies (Rojas *et al.*, 2012).

Combination of azoles with 5-FC

Azoles and 5-FC combinations show no interactions *in vitro* in *C. albicans*. In *C. glabrata*, fluconazole (FLC)/5-FC combinations even showed antagonism. This effect was mediated by *CgCDR1* due to induction of this gene by 5-FC (Steier *et al.*, 2013). Azoles and 5-FC combinations are synergistic in *C. neoformans*, as in the case of 5-FC/AmB combinations (Schwarz *et al.*, 2003). The azole/5-FC combination has been validated in animal models. A recent clinical trial established efficacy of oral FLC plus 5-FC compared with FLC monotherapy for therapy of HIV-associated cryptococcal meningitis (Nussbaum *et al.*, 2010).

Combination of candins with 5-FC

The combination of 5-FC with newer agents such as candins has been evaluated in fungal pathogens but in a few reports. Even if candins have limited efficacy against *C. neoformans* due to the presence of cell wall polymers not susceptible to the action of these agents, a report described that 5-FC/candin combination in this species resulted in synergy in most of the investigated isolates (Schwarz *et al.*, 2003). In another study, the 5-FC/candin combination tested *in vitro* in *A. fumigatus* and *A.*

terreus showed synergistic effect in about 60% of the isolates, the remaining without negative interactions (Dannaoui *et al.*, 2004). The clinical use of the 5-FC/candin combination remains limited to case reports, however with successful outcomes, suggesting that this combination may be further exploited in the future (Johnson and Perfect, 2010).

Combination of azoles with candins

Both azoles and candins are widely used in the treatment of fungal diseases, however, their combination is not resulting in similar trends. When combined with FLC, candins have no interactions *in vitro* against *Candida* and *Cryptococcus* spp. (Roling *et al.*, 2002). Synergy had been described for combinations of caspofungin and itraconazole and posaconazole against *A. fumigatus* isolates, however, combinations of caspofungin with voriconazole showed no interactions (Manavathu *et al.*, 2003). Interestingly, voriconazole and caspofungin were synergistic against itraconazole-resistant strains of *A. fumigatus* (Cuenca-Estrella *et al.*, 2005). The same conclusion was made for voriconazole–anidulafungin combinations against azole-resistant *A. fumigatus* isolates (Seyedmousavi *et al.*, 2013). Animal models of infections have further prospected the combination of caspofungin with voriconazole, even if no *in vitro* interactions were measured. In one study, the combination offered slight additional improvements in efficacy as compared to monotherapy (MacCallum *et al.*, 2005). In clinical settings, it was reported that the outcomes of salvage therapy for aspergillosis were superior with the combination of voriconazole and caspofungin as compared to voriconazole alone in patients with invasive aspergillosis (Marr *et al.*, 2004), which, however, was not reproduced in a randomized clinical trial (Marr *et al.*, 2012). This apparent discrepancy illustrates difficulties to translate *in vitro*-generated data into useful clinical guidelines.

Combination of terbinafine with other antifungal agents

Terbinafine, one of the most active allylamine derivatives, acts by preventing fungal ergosterol biosynthesis via specific and selective inhibition of fungal squalene epoxidase (Sanglard and Odds, 2002). Terbinafine has primarily fungicidal activity against a broad spectrum of fungi, including dermatophytes, filamentous fungi, and dimorphic fungi, while against *Candida* spp. it is only fungistatic. Terbinafine/FLC or terbinafine/AmB *in vitro* combinations have revealed in most cases synergistic effect and no antagonism (Barchiesi *et al.*, 1997, 1998). Synergistic interactions are reported in *Aspergillus* spp. (*A. fumigatus*, *A. niger*, *A. terreus*) (Mosquera *et al.*, 2002). In the clinical setting, a potential role of terbinafine combination exists in the treatment of refractory invasive infections due to uncommon pathogens with multidrug resistance pattern. For example, some reports documented *in vitro* synergy for terbinafine with other agents in *Scedosporium prolificans* despite resistance to all available antifungals, leading to successful combination therapy (Li *et al.*, 2008).

In vitro and in vivo combinations of non-antifungals with known antifungal agents

Available antifungals can be systematically combined with each other with the aim of detecting potential synergisms, with the previous section providing an overview of the current achievements. However, only a few positive interactions were found beneficial in patient treatments as compared to single therapy. To increase the number of potential interactions that could be implemented in clinical situations, one approach consists of interrogating active compounds that are used in medicine for their interactions with known antifungal agents. Here we will summarize the current state of this initiative.

Antibacterials and antifungal agents

Tetracycline and its derivatives, which are antibiotics inhibiting protein synthesis, were found to potentiate the activity of FLC in *C. albicans*. For example, minocycline is synergistic with FLC against *C. albicans* isolates resistant to FLC (Liu *et al.*, 2014). Synergism between doxycycline (DOX), another tetracycline derivative, and FLC has also been reported in azole-resistant isolates (Gao *et al.*, 2013). This conclusion was also reported by Fiori and van Dijck (2012) when high DOX concentrations were used (>200 µg/ml). This study also revealed that DOX reduced tolerance of *C. albicans* to FLC (the so-called trailing effect). This effect was dependent on iron availability in the culture medium, which could be modulated by the iron-chelating ability of DOX. Interestingly, DOX had a fungicidal effect at supra-MIC FLC concentrations, while only fungistatic as single agents. This readout of beneficial drug combination is different from traditional FIC indexes calculations defining synergism. In this example, it is microbial survival that served as the key parameter of drug combination efficacy and not simply measurement of growth by cell densities, which does not reflect the degree of viability within a culture.

Ciclopirox is a hydroxypyridone derivative with broad-spectrum activity both in bacteria and fungi (Gupta and Plott, 2004). This derivative has been used in combination with an azole (ketoconazole) in *A. fumigatus* and synergism was identified. It was found as in the case of DOX that ciclopirox has a strong iron chelating activity which could contribute to enhance azole activity (Zarember *et al.*, 2009). Consistent with this hypothesis is that ciclopirox induces several genes involved in iron transport and siderophore transport in *C. albicans* (Lee *et al.*, 2005; Niewerth *et al.*, 2003).

Topoisomerase inhibitors of the quinolone family have been reported to act in combination with azoles with different efficacies, ranging from indifference to synergy. In general, however, ciprofloxacin enhanced the activity of antifungal agents more than moxifloxacin and levofloxacin against both *C. albicans* and *A. fumigatus* (Stergiopoulou *et al.*, 2009).

Triclosan is a chlorinated aromatic compound with antimicrobial activity and has been used in personal care products. Triclosan has intrinsic antifungal activity at high concentrations (4–16 µg/ml) against several *Candida* spp. (Higgins *et al.*, 2012). Triclosan was used in combination with FLC *in vitro* in *C. albicans* in a recent study (Yu *et al.*, 2011). The authors found synergistic interaction between the agents. In addition, the combination resulted in a fungicidal effect in time-kill experiments (Yu *et al.*, 2011). However, these data contradict another study in which the combination was found to be antagonistic in *C. albicans* (Higgins *et al.*, 2012). In this study, the authors found that the antagonism was dependent on morphogenesis and more specifically on hyphal formation. The authors concluded the antagonism could be the consequence of the membranotropic activity of triclosan and the unique composition of hyphal membranes (Higgins *et al.*, 2012).

Recently, Zeidler *et al.* (2013) undertook the screening of agents acting synergistically with candins. They identified colistin, an antibiotic of the polymyxin family of antibiotics that targets membranes in Gram-negative bacteria, which was synergistic with aminocandin, a drug of the candin family, against several *Candida* spp. The combination of colistin and aminocandin led to faster and increased permeabilization of *C. albicans* cells than either colistin or aminocandin alone. The synergistic colistin–aminocandin combination could also be verified in animal models (Zeidler *et al.*, 2013).

Drugs interfering with calcium-dependent pathways and antifungal agents

Calcium is an important messenger in eukaryotes and its homeostasis is strictly regulated. Interfering with calcium homeostasis in eukaryotes can have negative consequences for cell survival in stress conditions. One of the important cellular elements of calcium-dependent cellular systems is the calmodulin-dependent calcineurin system, which requires calcium as co-factor and possesses phosphatase activity when bound to a regulatory subunit. Immunosuppressive drugs (cyclosporin A (CyA), FK506) inhibit calcineurin function. It was discovered that CyA and/or FK506 could potentiate the activity of several antifungal agents including azoles and candins. When FIC indexes are used to categorize the combination of immunosuppressive drugs with azoles or candins, no interactions could be predicted (Marchetti *et al.*, 2000a). When blocking calcineurin activity, however, immunosuppressive drugs convert fungistatic agents such as FLC into fungicidal drugs or increase the fungicidal effect of others such as candins. Generally speaking, it is considered that calcineurin is critical for fungal cells to withstand stress due to the presence of growth inhibitors or exposure to unfavourable environmental conditions (Steinbach *et al.*, 2007). Contrasting with *in vitro* drug combinations, experimental combination treatments performed in animal models demonstrated a high efficiency in curing established infections. For example, a CyA/FLC combined treatment of endocarditis by *C. albicans* in rats resulted in undetectable fungal burdens 3–7 days after onset of therapy, while therapy with single agents was unsuccessful (Marchetti *et al.*, 2000b). Immunosuppressant–azole combinations were also shown to be synergistic in other important fungal pathogens such as *A. fumigatus* and *C. neoformans* and also less common but intrinsically azole-resistant pathogen such as Zygomycetes (Dannaoui *et al.*, 2009; del Poeta *et al.*, 2000; Steinbach *et al.*, 2004, 2011). One of the problem implementing calcineurin inhibitors with azoles in clinical practice is their immunosuppressive effect. Non-immunosuppressive variants of calcineurin inhibitors were described, however, they are still not commercially available (del Poeta *et al.*, 2000).

Amiodarone (AMD) is a membrane-active compound that interferes with calcium homeostasis and causes cell death after short exposure at high drug concentrations in several fungal species including *C. albicans* (Gamarra *et al.*, 2010; Muend and Rao, 2008). Hyperpolarization of membranes by AMD causes calcium influx and release from internal stores, including the endoplasmic reticulum, mitochondria and vacuole (Maresova *et al.*, 2009). Some studies revealed that, although AMD has weak antifungal activity, it exhibited synergistic effects with FLC *in vitro* against FLC-resistant *C. albicans* or *C. tropicalis* isolates. Interestingly, the drug combinations induced rapid cells death in both fungal species (Gamarra *et al.*, 2010; Maresova *et al.*, 2009). AMD, together with other membrane-active compounds, was also shown to exhibit synergism with azoles in *A. fumigatus*. Interestingly, the most potent synergistic effect was identified in itraconazole-resistant isolates, thus raising further interest in potential applications of such combinations (Afeltra *et al.*, 2004).

HSP90 inhibitors in combination with antifungal agents

The molecular chaperone Hsp90 regulates crucial cellular responses to drug-induced stress by stabilizing a diverse set of client proteins, including calcineurin (Cowen, 2008, 2009, 2013). In the context of azoles, impairment of Hsp90 function genetically or pharmacologically with molecules that are well tolerated in humans reduces tolerance of *Candida* species (Cowen *et al.*, 2009). Hsp90 inhibitors that have been tested to date in combination with antifungals include the natural products radicicol and geldanamycin, as well as structural analogues of geldanamycin that are in clinical development for cancer, such as 17-AAG and 17-DMAG (Cowen and Lindquist, 2005; Cowen *et al.*, 2009). Inhibition of Hsp90 can also impair the evolution of resistance to azoles, and abrogate resistance acquired by loss of function of the ergosterol biosynthetic enzyme Erg3 or resistance that evolved in a human host through multiple mechanisms (Cowen *et al.*, 2006; Cowen and Lindquist, 2005). Beyond its impact on azole resistance of planktonic cells, Hsp90 also influences azole resistance of *Candida* biofilms (Robbins *et al.*, 2011). Under planktonic conditions, Hsp90 regulates drug resistance by stabilizing calcineurin and the terminal mitogen activated protein kinase of the Pkc1 cell wall integrity pathway, Mkc1, thereby enabling cellular responses to the membrane stress exerted by azoles (LaFayette *et al.*, 2010; Singh *et al.*, 2009). Strikingly, although calcineurin and Mkc1 influence azole resistance of *C. albicans* biofilms, Hsp90 does not regulate their stability in this cellular context (Robbins *et al.*, 2011). Depletion of Hsp90 reduces biofilm matrix glucan levels, providing a plausible mechanism by which it may influence biofilm azole resistance phenotypes. Inhibition of Hsp90 function with 17-AAG enhances the efficacy of azoles against resistant *Candida* in the invertebrate model host *Galleria mellonella*, as well as in the rat central venous catheter model of biofilm infection, where the infection and drug delivery are localized (Cowen *et al.*, 2009; Robbins *et al.*, 2011). In contrast, toxicity has been observed with 17-AAG in a murine model of disseminated *Candida* infection (Cowen *et al.*, 2009), motivating efforts to identify fungal selective Hsp90 inhibitors or compounds that target components of the Hsp90 chaperone network that can more easily be discriminated between pathogen and host. Lysine deacetylases have recently been discovered to regulate Hsp90 function in *S. cerevisiae*, *C. albicans*, and *A. fumigatus*, providing an attractive alternative strategy to enhance the efficacy of azoles by targeting more divergent cellular regulators (Lamoth *et al.*, 2014b; Robbins *et al.*, 2012).

Hsp90-dependent cellular stress responses are crucial not only for azole resistance, but also for echinocandin resistance. Targeting Hsp90 genetically or pharmacologically reduces basal echinocandin tolerance and resistance of *C. albicans*, *C. glabrata*, and *A. fumigatus* (Cowen and Lindquist, 2005; Cowen *et al.*, 2009; Lamoth *et al.*, 2014a; Singh *et al.*, 2009; Singh-Babak *et al.*, 2012). Inhibition of Hsp90 with geldanamycin reduces echinocandin resistance that was acquired by mutation in the drug target and resistance that evolved in the human host (Singh *et al.*, 2009; Singh-Babak *et al.*, 2012). Geldanamycin also enhances echinocandin efficacy against *A. fumigatus* in the *G. mellonella* infection model (Cowen *et al.*, 2009), and genetic reduction of Hsp90 levels enhances echinocandins efficacy against *C. albicans* in a mouse model of disseminated infection (Singh *et al.*, 2009). As with the azoles, Hsp90 regulates resistance to echinocandins by stabilizing the stress response regulators calcineurin and Mkc1 (LaFayette *et al.*, 2010; Singh *et al.*, 2009).

Recent work has extended the scope of Hsp90s impact on antifungal drug resistance beyond azoles and echinocandins to polyenes as well. AmB-resistant clinical *Candida* isolates remain relatively rare, and this was found to be a consequence of fitness trade-offs, such that the mutations conferring resistance to AmB create diverse cellular stress that demands high levels of Hsp90 for survival, even in the absence of antifungal drug (Vincent *et al.*, 2013). Mutations in *ERG2*, *ERG3*, *ERG6*, and *ERG11* were associated with AmB resistance, as well as hypersensitivity to Hsp90 inhibitors, oxidative stress, febrile temperatures, and killing by neutrophils, illustrating evolutionary constraints that may minimize the evolution of drug resistance.

Efflux inhibitors in combination with antifungal agents

One of the major mechanisms by which fungal pathogen escape the toxicity of antifungal agents and other growth inhibitors is by efflux activity. Drug efflux in fungal pathogens is carried out by two basic transport systems including ATP binding cassette (ABC) transporters and major facilitators (Sanglard *et al.*, 2009). Targeting these efflux systems by specific inhibitors may contribute to enhance antifungal activity and/or decrease the development of resistance. Several studies have investigated the possibility to inhibit efflux-mediated drug transport using specific types of inhibitors. Most of the studies used a combination of inhibitors with FLC and measured the decrease of drug resistance as compared to the effect of single drugs (Hayama *et al.*, 2012; Holmes *et al.*, 2012; Maurya *et al.*, 2013; Monk *et al.*, 2004; Silva *et al.*, 2013). One interesting study exploited the Prestwick chemical library containing about 1280 marketed drugs and identified compounds effective in inhibiting drug efflux via the ABC transporters Cdr1 and Cdr2 (Table 5.1). Flupentixol, prochlorperazine, droperidol, and ethoxyquin that are known inhibitors or inducers of human ABC transporters and four other drugs (ebselen, econazole, sulconazole, and disulfiram) were exhibiting known antifungal activity. The remaining compound, clorgyline, is a monoamine oxidase A selective inhibitor, which

Table 5.1 Hits from the Prestwick chemical library in *S. cerevisiae* strains expressing either the *C. albicans CDR1* or *CDR2* genes as published by Holmes *et al.* (2012)

	Efflux inhibition (%)	
Compound	Cdr1	Cdr2
Clorgyline hydrochloride	98.4	93.8
Ebselen	39.8	100
Flupentixol dihydrochloride	48.9	0
Prochlorperazine dimaleate	46.4	0
Econazole nitrate	0	91.1
Droperidol	0	86.7
Ethoxyquin	0	67
Sulconazole nitrate	0	57.5
Disulfiram	0	54.2

is no longer in use as an antidepressant (Holmes *et al.*, 2012). Clorgyline was shown as an effective efflux inhibitor of both Cdr1 and Cdr2 and was synergistic with FLC against azole-resistant isolates, as deduced from FICI values (Holmes *et al.*, 2012).

While other inhibitors included specific synthetic peptides (Maurya *et al.*, 2013; Niimi *et al.*, 2012), one interesting class of inhibitors included a class of macrocyclic lactones, the milbemycins. Some milbemycin derivatives (milbemycin oxims) are widely used as anti-parasitic agents in animal and have known pharmacology (Silva *et al.*, 2013). A recent study demonstrated their use as drug efflux inhibitors not only in *C. albicans* but also in *C. glabrata*. The drugs were synergistic *in vitro* with FLC in azole-resistant isolates as deduced from FICI values. Of interest in this study is that the synergism was also tested in animal models of infection with *C. albicans* and *C. glabrata*. Combination of FLC with milbemycin oxims could reverse azole resistance with high efficacy, thus suggesting that these type of combination may have some therapeutic utility (Silva *et al.*, 2013).

Miscellaneous substances in combination with antifungal agents

- CTBT (7-chlorotetrazolo[5,1-c]benzo[1,2,4] triazine) is a compound causing intracellular superoxide production and oxidative stress in fungal cells (Batova *et al.*, 2010). CTBT combined with FLC results in synergism in both azole-susceptible and azole-resistant *C. albicans* isolates (Cernicka *et al.*, 2007). Interestingly, the combination of CTBT with itraconazole completely inhibited the growth of filamentous fungi such as *A. fumigatus* and *A. niger* (Culakova *et al.*, 2012).
- Ibuprofen is a non-steroidal anti-inflammatory drug (NSAID) that is commonly used for its antipyretic, analgesic and anti-inflammatory effect. Ibuprofen was found to act synergistically with azole-resistant *C. albicans* isolates (Arai *et al.*, 2005; Ricardo *et al.*, 2009) (see Chapter 7). However, for unknown reasons high ibuprofen concentrations (>500 μg/ml) tend to reverse synergism into an antagonistic behaviour in the presence of FLC (Arai *et al.*, 2005).
- Cationic peptides are part of a vast category of natural defence products from amphibian, animal or human origin and have general antimicrobial activities (Hancock and Scott, 2000). These short peptides can exhibit antifungal activities at different concentrations. Several studies have addressed the use of these antimicrobial peptides in combination with existing antifungal agents. MacCallum *et al.* (MacCallum *et al.*, 2013) investigated the human peptide hMUC7–12 and hLF(1–11) in addition to the amphibian peptide DsS3(1–16) in combination with caspofungin *in vitro* and *in vivo*. All peptides showed *in vitro* synergism with caspofungin. Combination *in vivo* using *Galleria mellonella* infected with *C. albicans* or using the murine infection model revealed enhanced efficacy of the drugs (MacCallum *et al.*, 2013). Similar observations were made with peptides of other origins (dermaseptin S3(1–16), ranalexin magainin) combined with candins in *C. glabrata* and *C. albicans*. In this case, however, no benefit of combined therapy in animal models could be documented (Harris and Coote, 2010).
- Statins are used to reduce blood cholesterol levels. They are competitive inhibitors of 3-hydroxy-3-methylglutaryl-coenzyme A reductase, which catalyses a rate-limiting step in the acetate–mevalonate pathway of the sterol biosynthesis (James K Liao, 2005). In one study, lovastatin, simvastatin, fluvastatin, atorvastatin, rosuvastatin and pravastatin were combined with various azoles in different fungal pathogens (Nyilasi *et al.*, 2010). Lovastatin and rosuvastatin exhibited the strongest synergism with FLC in *C. albicans*. While itraconazole and atorvastatin acted synergistically against *A. fumigatus*, FLC was synergistic with lovastatin and simvastatin against *A. fumigatus*, even though this species is intrinsically resistant to FLC (Nyilasi *et al.*, 2010).
- Chelators for specific metals (iron, calcium) were found to potentiate the activity of antifungal agents. In one study, bathophenanthrolene disulfonic acid (BPS), which is a known iron chelator, was found to increase

susceptibility of *C. albicans* and several other Candida species not only to FLC, but also to nystatin (a compound similar to AmB) (Prasad *et al.*, 2006). Ethylene glycol tetraacetic acid (EGTA), a potent calcium chelator was shown to enhance the activity of azoles in *S. cerevisiae*. (Edlind *et al.*, 2002). Deferasirox is the first orally available iron chelator approved by the FDA, with an indication for the treatment of transfusion-related iron overload (Ibrahim *et al.*, 2007). When combined with liposomal Amb, Deferasirox resulted in enhanced clearance of fungal pathogens such as *A. fumigatus* or *Rhizopus oryzae*, the latest being refractory to monotherapy with AmB (Ibrahim *et al.*, 2010, 2011).

Systematic drug combination screenings

One approach to discover synergistic combinations with known antifungal agents is to systematically screen for compounds deposited in specific libraries. These libraries contain large panel of FDA-approved drugs, pharmacologically active compounds or natural products. Of course, some of the above-mentioned synergisms were recapitulated with compound library screenings, however, several novel interactions have been revealed.

In one study, a microbial natural product library was screened with the tester *species C. parapsilosis* for hits that synergize the effect of a low dosage of ketoconazole (KTC) that alone shows little detectable fungicidal activity. The screening of 20,000 microbial extracts was performed with *C. parapsilosis* and yielded substances with activity combined with KTC and including beauvericin, berberine CyA, geldanamycin, lovastatin and radicicol (Zhang *et al.*, 2007). Beauvericin (BEA), originating from *Fusarium proliferatum*, was the most potent component and it synergized KTC activity against diverse fungal pathogens including *Candida* spp. and *Aspergillus* spp. Combinations of BEA and KTC in an immuno-compromised mouse model prolonged survival of the host infected with *C. parapsilosis* and reduced fungal colony counts in animal organs including kidneys, lungs, and brains (Zhang *et al.*, 2007).

In another report, the authors used compounds deposited in the Prestwick library, which consists of 1280 off-patent drugs and other bioactive agents (Spitzer *et al.*, 2011). In their approach, several fungal species were used including *C. neoformans Cryptococcus gattii*, *C. albicans* as well as *Saccharomyces cerevisiae*. To identify compounds that potentiate the effect of FLC but possess minimal antifungal activity on their own, the screening procedure was performed with subinhibitory FLC concentrations and a single library compound concentration (30 μM). A total of 148 positive hits were found, however, only 12 were active in combination with FLC in all tested species (Fig. 5.4). Noteworthy is that no systematic detailed FIC index listing was published and thus the nature of combination could not be established. These compounds were: mefloquine hydrochloride, an antimalarial that is a 4-quinolinemethanol derivative chemical structural analogue of quinine; the antidepressant sertraline (Zolofts); the monoamine oxygenase inhibitor pirlindole, also known to have antidepressant activity; the allylamine antifungal naftifine; the antibiotic pro-drug pivampicillin; the antinausea

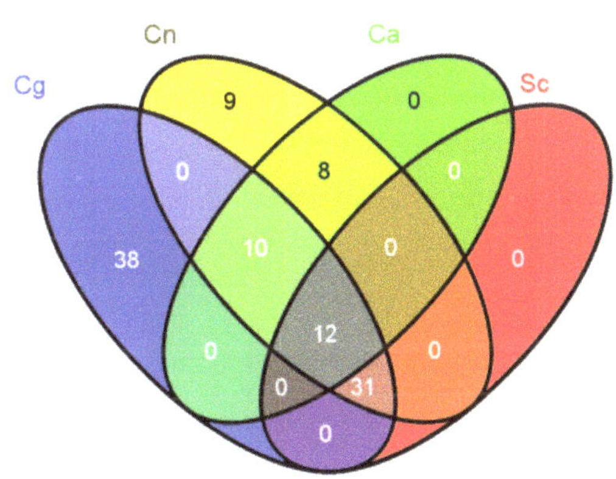

Figure 5.4 Drugs increasing FLC activity in different fungal species. The following drugs were common hits for all species: pirlindole mesylate, sertraline, zuclopenthixol hydrochloride, pivampicillin, naftifine hydrochloride, thiethylperazine malate, mefloquine hydrochloride, perphenazine, trifluoperazine dihydrochloride, perhexiline maleate, clofilium tosylate, fluphenazine dihydrochloride. Venn diagram was obtained by using the online program Venny (http://bioinfogp.cnb.csic.es/tools/venny/index.html). Cg, *C. gattii*; Cn, *C. neoformans*; Ca, *C. albicans*; Sc, *S. cerevisiae*.

drug thiethylperazine (Torecans); perhexiline acting by inhibiting mitochondrial carnitine palmitoyltransferase-1; clofilium tosylate, a K^+ channel blocker; zuclopenthixol, trifluoperazine dihydrochloride perphenazine and fluphenazine, all antipsychotic drugs belonging to a large family of phenothiazines that have antipsychotic and other central nervous system (CNS) activities.

The same 1280 off-patent drugs were tested for intrinsic antifungal activity and combination with FLC in *C. neoformans*. The antihypertensive suloctodil, the antibiotic polymyxin E, the antidepressant setraline and the anti-psychotic fluspirilen were synergistic with FLC based on FICI thresholds (Butts *et al.*, 2013). FLC synergism with either tamoxifen or trifluoperazine was not observed in the study by Spitzer *et al.* (2011), this discrepancy likely resulting from differences in strains and types of susceptibility test employed.

As mentioned above, Zeidler *et al.* (2013) used a compound library containing 38758 chemicals obtained from Prestwick, ChemDiv and the French 'Chimiothèque Nationale'. These libraries were screened with *C. albicans* in the presence of aminocandin. Five hundred and fifty-five compounds were identified whose combination with aminocandin resulted in 100% growth inhibition and were clustered into chemical families. From these chemical families, 156 compounds were selected and 98 showed antifungal activity independently of the addition of aminocandin, while 58 had no significant antifungal activity on their own but acted synergistically with aminocandin. In their study, Zeidler *et al.* (2013) investigated in detail only colistin as a synergizer (see below).

A further study explored the combination of 640 FDA-approved drugs with FLC. The authors identified 10 compounds enhancing the activity of FLC (Kaneko *et al.*, 2013). As in the previous study, no detailed FIC indexes were provided. The 10 compounds were grouped according to their mode of action. These included HMG-CoA reductase inhibitors (lovastatin, mevastatin, simvastatin), antifungal agents (butenafine, terbinafine) and other miscellaneous agents (CyA, 5-FC, artesunat, carvedilol, bortezomib).

Borisy *et al.* (2003) developed a high-throughput screening method for identifying effective combinations of therapeutic compounds. The systematic screening used 120,000 different two-component combinations of reference listed drugs, and used both azole-susceptible and azole-resistant strains. The synergy between drugs was calculated by established approaches. The authors first identified compounds (out of a 500 compounds depository) with moderate to high antifungal activity and found 30 substances, nine of them known as antifungal agents. These compounds were then tested in 435 possible two-component combinations in six doses for each compound. Fig. 5.5 gives an overview of the interacting partners. None of the combinations comprised two known antifungal agents; however, six comprised antifungal and non-antifungal agents,and 16 comprised two non-antifungal agents. Tamoxifen (an anticancer agent) and pentolinium (a ganglionic blocking agent which acts as a nicotinic acetylcholine receptor antagonist) were agents with the strongest synergistic interactions with other agents. Another screen used FLC in combination with the same compound library. Among 20 synergistic interactions, the combination of FLC with phenazopyridine, a urinary tract analgesic, was successful in a FLC-resistant *C. albicans* isolate. The combination of phenazopyridine and FLC was fungicidal. The molecular basis for this combined effect is still unknown (Borisy *et al.*, 2003).

A recent study explored drug combinations not simply with planktonic cells, a system used in most drug combination studies, but with cells forming biofilms. The rationale behind this choice was that biofilms exhibit resistance to antifungal agents (here miconazole) and any drug combined with an azole that could decrease drug resistance in biofilms would be considered as a potentiator of miconazole activity (LaFleur *et al.*, 2011, 2013). The authors screened large compound libraries (containing 63286 substances) and identified 244 hits. According to the data provided, 15% of the hits possessed intrinsic antifungal activity, while others potentiated miconazole. The authors investigated in more details a specific compound, 2-adamantanamine, known as amantadine-like antiviral and inhibitor of the M2 proton channel of the influenza A virus that prevents viral uncoating and replication (Lafleur

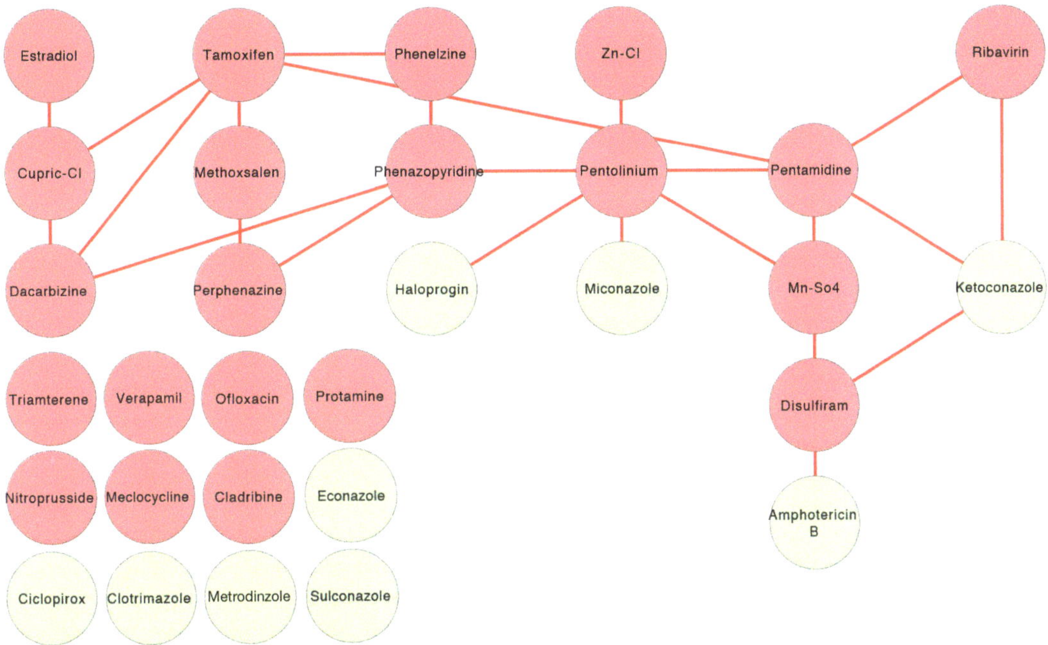

Figure 5.5 Synergistic interaction network of drugs in *C. albicans*. Data were obtained from Borisy *et al.* (2003). Cytoscape 2.8 was used to show synergistic interactions. Substances not interacting with others are indicated at bottom.

et al., 2013). The miconazole-2-adamantanamine combination was effective in reducing biofilm viability. Interestingly, the combination of azoles with 2-adamantanamine was fungicidal in planktonic-grown cells, while single agents were only fungistatic, thus resembling to the DOX-mediated effect as mentioned above. The authors investigated the azole-2-adamantanamine combination in an animal model of infection (guinea pig model of cutaneous candidiasis) and could observe that the combination was 3-fold more effective than single therapy (Lafleur *et al.*, 2013).

Given the extensive literature documenting the synergistic combination of different agents, Chen *et al.* (2014) established a database listing a compilation of these interactions (the Antifungal Synergistic Drug Combination Database, ASDCD). It is possible in this database to inspect 210 antifungal synergistic drug combinations and 1225 drug–target interactions, involving 105 individual drugs from more than 12,000 references (Chen *et al.*, 2014). Fig. 5.6 illustrates interactions that were extracted from the database using known antifungal agents. In such an interaction map, it is possible to highlight that several agents (including CyA, FK506, amiodarone placed at centre of the interaction network) interact synergistically with different antifungal drugs. Interestingly, these agents control calcium homeostasis, which was already mentioned as controlling survival during exposure to antifungal agents.

Chemogenetic approaches in the exploration of drug interaction mechanisms

One challenge in the discovery of synergistic drug combinations is that a vast number of possible combinations of drug pairs need to be tested. Large-scale experimental drug synergy screens have found that synergistic drug pairs are not frequent (1–4% success rate) (Borisy *et al.*, 2003; Zhang *et al.*, 2007). Chemogenetics is a concept that relies on the ability of genetics to predict drug targets using different screening methods. The power of yeast genetics (including for example haploinsufficiency screens, multicopy suppression profiling, and systematic synthetic lethality

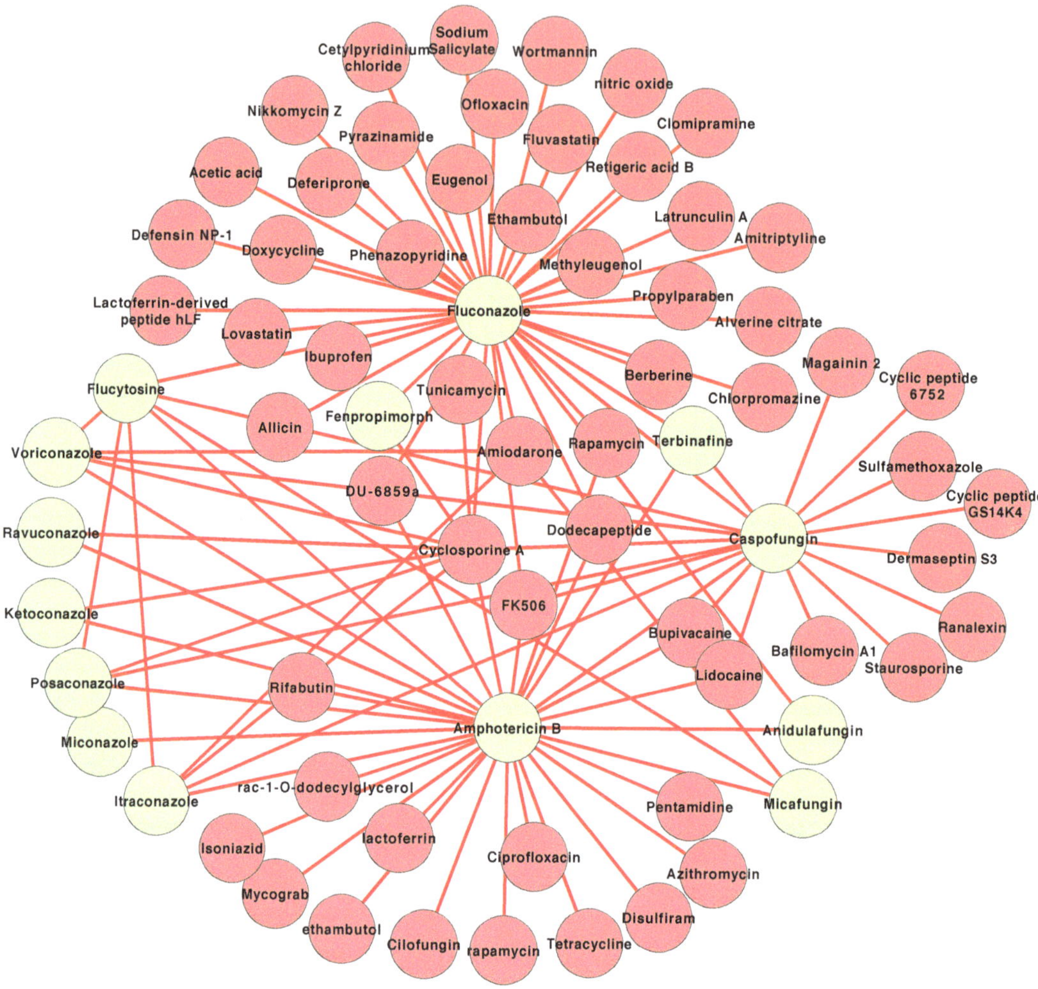

Figure 5.6 Interaction network deduced from data available in the Antifungal Synergistic Drug Combination Database (ASDCD). The network was constructed with published synergistic data of antifungals with other agents as published by Chen *et al.* (2014). Cytoscape 2.8 was used to show interactions.

screens) has been used in chemogenetics for this purpose using large compound libraries. Fig. 5.7 explains the basic principle of chemogenetics and how it can be used to predict synergistic drug interactions. This concept can also be used in fungal pathogens such as *C. albicans* (Lee *et al.*, 2014; Oh *et al.*, 2010). In a proof-of-principle study for the ability of chemogenetics to predict synergies, the authors collected sensitivity-based chemogenomic profiles from the literature and generated a genetic profile in *S. cerevisiae* for the widely used fungistatic drug FLC using the entire gene deletion set and FLC fungicidality as a phenotypic readout (Jansen *et al.*, 2009). The FLC profile was compared with other drug profiles. The expectation was that FLC synergism with other drugs could be predicted from profile similarities. The authors identified *in silico* eight different compounds predicted to be synergistic with FLC including alverine citrate, caspofungin; CyA, FK506, fenpropimorph, latrunculin A, tunicamycin and wortmannin. Out of these predicted synergistic combination, five were confirmed experimentally and included alverine citrate, fenpropimorph, FK506, CyA and wortmannin. These compounds were also synergistic between each other. Interestingly, the synergisms identified in *S. cerevisiae* were also valid in *C. albicans*.

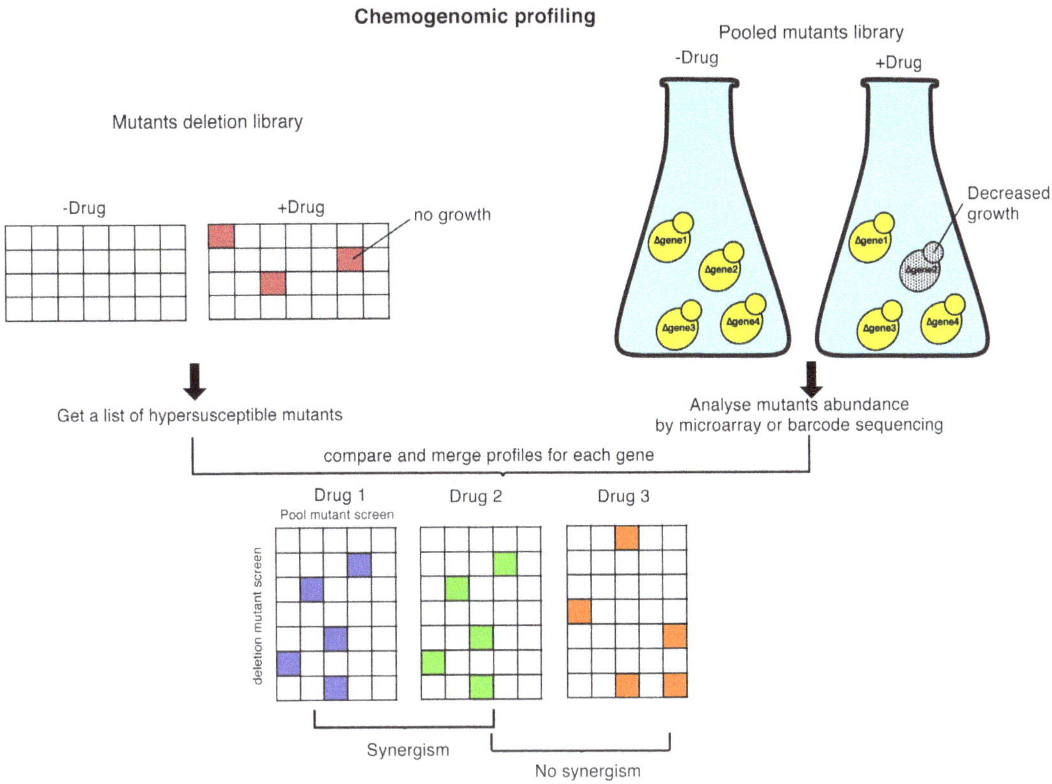

Figure 5.7 The prediction of drug synergies by chemogenetics. Chemogenetic profiling is performed by screening drugs against collections of yeast deletion mutants (left). The collection is arrayed in microplates and the growth of each mutant is recorded in the presence or absence of the drug. With pooled mutants, tagged mutants are grown in a single culture with or without drug and the relative proportion of each tag is quantified. Two drugs with similar chemogenetic profiles deduced from the screening approaches possibly exhibit a synergistic interaction (drug 1 and drug 2).

Chemogenetic approaches can also be used to decipher the mechanisms behind drug interactions. The principle behind this approach is given in Fig. 5.8. As an example, the study of Spitzer *et al.* (2011) first interrogated a group of drugs belonging to different classes synergistic with FLC (see Fig. 5.3); analysis of chemical–genetic interactions suggested two related modes of action for the tested combinations. Trifluoperazine, tamoxifen, clomiphene, sertraline and suloctidil appeared to cause general perturbation of membrane, vesicle trafficking and lipid biosynthesis functions, whereas L-cycloserine preferentially interfered with an early step in sphingolipid biosynthesis (Spitzer *et al.*, 2011). The authors next integrated the chemical–genetic profiles for FLC-synergistic compounds with a global genetic interaction network compiled from the literature. The individual profiles for tamoxifen, trifluoperazine, clomiphene, sertraline and suloctidil showed a significant enrichment of genetic interactions with the FLC profile, whereas L-cycloserine did not, thus being consistent with their different modes of action. To test the predictive power of chemogenetic profiles of the membrane active compounds which were deduced from the *S. cerevisiae* gene deletion collection, Spitzer *et al.* (Spitzer *et al.*, 2011) retrospectively analysed chemical–genetic profiles for 81 psychoactive drugs known to impair yeast growth. Of this set, 7 (fluoxetine, metergoline, clozapine, cyproheptadine, paroxetine, sertraline) were predicted to synergize with FLC based on their effect on deletion strains and overlap with the FLC chemogenetic profile. *In vitro*, fluoxetine, metergoline paroxetine and sertraline were actually synergistic

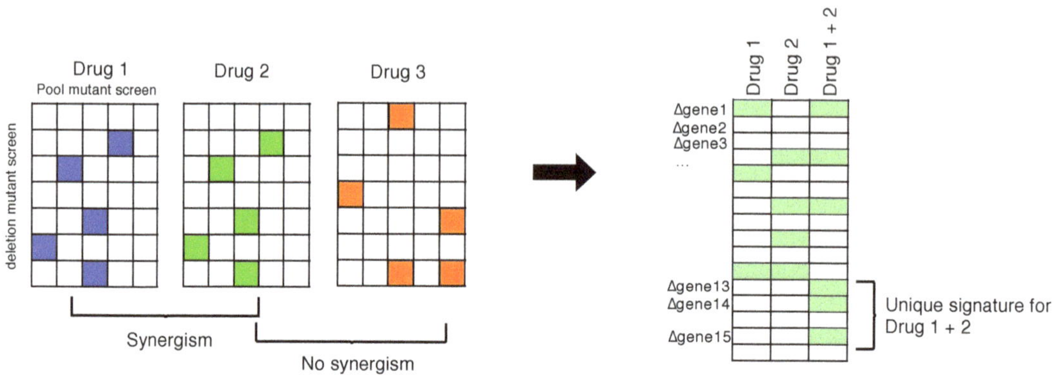

Figure 5.8 Deciphering the mechanism of drug synergism. For a given synergistic drug pair predicted as shown in Fig. 5.7, chemogenetic profiling can be performed on cells treated with each drug and also on cells simultaneously treated with both drugs. This screen can be performed on mutant deletion libraries. A group of mutants unique to both drugs may be obtained (green colour indicates a growth defect phenotype in the screening array). The gene deletion set obtained by the combination profile may provide information on the pathways affected or targeted by the drug combination.

with FLC. Thus, chemical–genetic interaction profiles can help to predict synergistic drug combinations.

A further example in the understanding of synergistic combination can be given by the above-mentioned study of Zeidler *et al.* (2013). The use of the antimicrobial colistin with aminocandin was synergistic and the authors used genome-wide fitness profiling of *S. cerevisiae* deletion mutant libraries with colistin, aminocandin and their combination. The screen yielded genes (37) that formed a network with genetic and physical interactions (Fig. 5.9). Colistin treatment alone yielded genes enriched for 'glycosphingolipid biosynthetic processes' suggesting that colistin affected membrane structures. The aminocandin/colistin combination revealed enrichment in 'fungal-type cell wall chitin biosynthetic processes' and 'cellular cell wall organization or biogenesis'. The authors rationalized from these experiments that the synergistic effect of aminocandin and colistin was due to the weakening of cell wall surfaces by aminocandin, which then enabled facilitated access of colistin to cell membranes and their destabilization. According to this hypothesis, cells treated with aminocandin and colistin were effectively more permeable than those treated by single drugs (Zeidler *et al.*, 2013).

One interesting further development in the use of chemogenetics in synergistic predictions is the estimation of the specificity of the synergism (Cokol *et al.*, 2011). The concept of specific synergism, emerges from two basic mechanistic understandings of synergism. One model for drug synergy is the parallel pathway inhibition model, which suggests that two drugs will be synergistic if they inhibit two proteins in parallel pathways essential for an observed phenotype. To identify protein pairs acting in parallel pathways, the results of systematic genetic interactions can be used, in which two combined genetic perturbations result in a phenotype that is more severe than expected (Tong *et al.*, 2004). Since the action of a drug on a target gene product can yield effects similar to a loss of the gene encoding the drug target, two drugs may be synergistic if they have targets that have a synergistic genetic interaction. The other model that has been proposed to explain the mechanism of synergism is the bioavailability model. This model suggests that two drugs can be synergistic if one drug helps the other drug to reach the target within cells. This can be achieved either by facilitating the entry of the other drug or increasing its accumulation in the cell (Cokol *et al.*, 2011). Drug synergy as a result of the bioavailability model are not specific and may be not favourable since they may increase the bioavailability many other agents in target cells with probable side-effects. The example of such non-specific synergy was given above

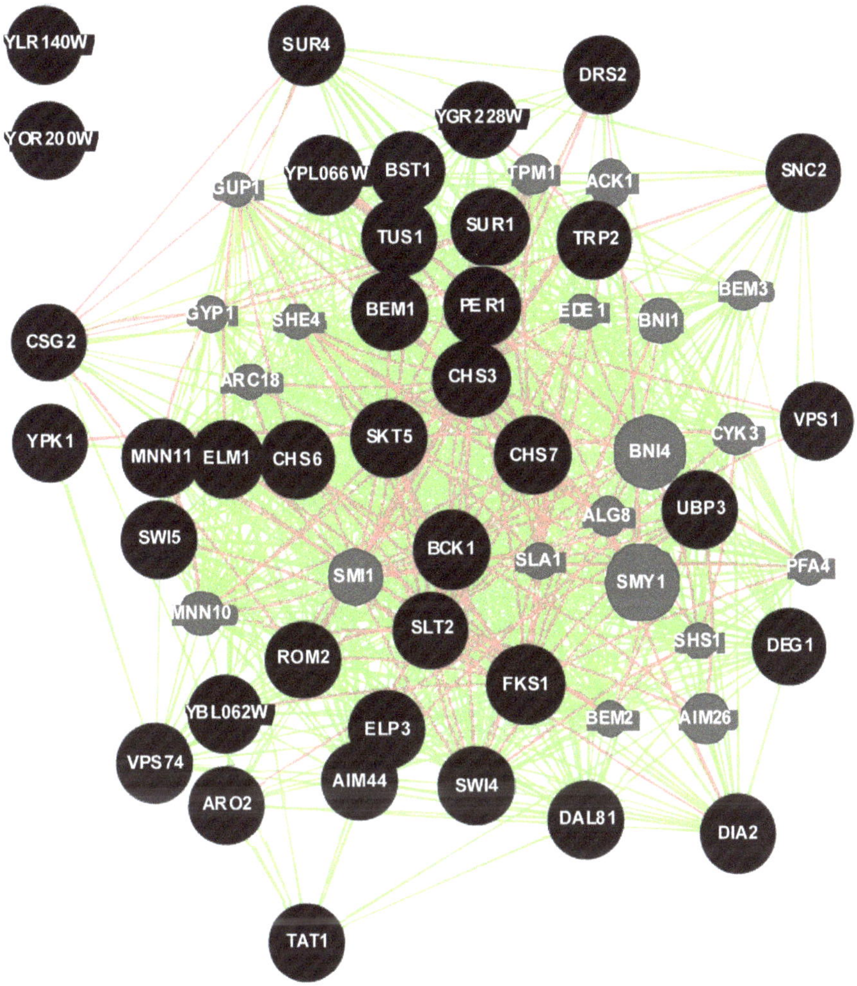

Figure 5.9 Interaction network of genes involved in the synergistic aminocandin/colistin combination. Data were obtained from Zeidler *et al.* (Zeidler *et al.*, 2013) and network obtained by using the online tool GeneMania. Green edges and red edges signify genetic and physical interactions, respectively. Black nodes indicate query genes, grey nodes indicate target genes. Two genes (YLR140W, YOR200W) from the initial genes list published by Zeidler *et al.* (2013) did not identify interactions in the network.

in the synergistic combination of colistin with aminocandin (Zeidler *et al.*, 2013). Using again chemogenetic approaches, it has been possible to predict the synergistic mechanism behind the association of drug pairs (Cokol *et al.*, 2011). The authors examined 200 pairs of antifungal drugs in *S. cerevisiae* (using their chemogenetic profiles) and found 38 antifungal synergies. While 14 cases of drug synergies corresponded to genetic interactions, 92% of the discovered synergies involved six drugs including pentamidine, tacrolimus, dyclonine, fenpropimorph, haloperidol and terbinafine, which involved the bioavailability model. It appeared from this study that the majority of drug synergies result from non-specific synergy attributed to the bioavailability model.

Conclusions

In this chapter, we summarized the current status of drug combinations with the purpose of increasing drug efficacy. While monotherapeutic approaches in clinical settings are still the standard, combined therapies may find future applications. This assumption is based on several factors: first, there are still too few effective antifungal agents used in medicine to successfully treat common and rare fungal infections; second, the threat of antifungal resistance may become more serious in the coming years. From a strategic point of view, combination therapy is a challenge in terms of clinical trial design. It might also not be an attractive option for the drug manufacturers since combined therapy uses several agents, some of them still under protected licence. In any case, despite several challenges for the development of combined therapies, this option cannot simply be ignored and deserves further systematic investigations. We here envisaged the combination of natural and synthetic compounds which each other. However, combination therapy could be extended to other therapeutic options. One of them is the use of host factors such as effectors of the immune system with antifungal agents (see Chapter 11). Combining an improved status of antifungal immunity in infected patients together with the use of antifungal agents is a realistic option, especially given recent advances in understanding of the host response. Increasing the efficacy of antifungal agents can also be seen from another angle. Early and rapid diagnosis of fungal pathogens at the species level, integration in patient management of careful drug monitoring for dosage adjustments and profiling the patient response to antifungal agents with genome-wide analysis could be also targeted for improving the management of antifungal therapies.

Future directions in the area of drug combinations will certainly take advantage of genome-wide studies and high-throughput technologies (see Chapter 9). The power of yeast genetics is a critical parameter in this context. A recent study has revealed the chemogenetic profiling of several thousand compounds (Lee *et al.*, 2014). This existing knowledge will certainly help the discovery of novel specific synergisms with poor interference with human metabolism, an approach that could not simply be apprehended experimentally.

References

Afeltra, J., Vitale, R.G., Mouton, J.W., and Verweij, P.E. (2004). Potent synergistic *in vitro* interaction between nonantimicrobial membrane-active compounds and itraconazole against clinical isolates of Aspergillus fumigatus resistant to itraconazole. Antimicrob. Agents Chemother. *48*, 1335–1343.

Arai, R., Sugita, T., and Nishikawa, A. (2005). Reassessment of the *in vitro* synergistic effect of fluconazole with the non-steroidal anti-inflammatory agent ibuprofen against *Candida albicans*. Mycoses *48*, 38–41.

Arikan, S., Lozano-Chiu, M., Paetznick, V., and Rex, J.H. (2002). In vitro synergy of caspofungin and amphotericin B against *Aspergillus* and *Fusarium* spp. Antimicrob. Agents Chemother. *46*, 245–247.

Barchiesi, F., Falconi Di Francesco, L., and Scalise, G. (1997). *In vitro* activities of terbinafine in combination with fluconazole and itraconazole against isolates of *Candida albicans* with reduced susceptibility to azoles. Antimicrob. Agents Chemother. *41*, 1812–1814.

Barchiesi, F., Di Francesco, L., Compagnucci, P., Arzeni, D., Giacometti, A., and Scalise, G. (1998). *In vitro* interaction of terbinafine with amphotericin B, fluconazole and itraconazole against clinical isolates of *Candida albicans*. J. Antimicrob. Chemother. *41*, 59–65.

Batova, M., Klobucnikova, V., Oblasova, Z., Gregan, J., Zahradnik, P., Hapala, I., Subik, J., and Schüller, C. (2010). Chemogenomic and transcriptome analysis identifies mode of action of the chemosensitizing agent CTBT (7-chlorotetrazolo[5,1-c]benzo[1,2,4] triazine). BMC Genomics *11*, 153.

Borisy, A.A., Elliott, P.J., Hurst, N.W., Lee, M.S., Lehar, J., Price, E.R., Serbedzija, G., Zimmermann, G.R., Foley, M.A., Stockwell, B.R., *et al.* (2003). Systematic discovery of multicomponent therapeutics. Proc. Natl. Acad. Sci. U.S.A. *100*, 7977–7982.

Bowyer, P., Moore, C.B., Rautemaa, R., Denning, D.W., and Richardson, M.D. (2011). Azole antifungal resistance today: focus on *Aspergillus*. Curr. Infect. Dis. Rep. *13*, 485–491.

Brown, G.D., Denning, D.W., Gow, N.A., Levitz, S.M., Netea, M.G., and White, T.C. (2012). Hidden killers: human fungal infections. Sci. Transl. Med. *4*, 165rv113.

Butts, A., DiDone, L., Koselny, K., Baxter, B.K., Chabrier-Rosello, Y., Wellington, M., and Krysan, D.J. (2013). A repurposing approach identifies off-patent drugs with fungicidal cryptococcal activity, a common structural chemotype, and pharmacological properties relevant to the treatment of cryptococcosis. Eukaryot. Cell *12*, 278–287.

Cernicka, J., Kozovska, Z., Hnatova, M., Valachovic, M., Hapala, I., Riedl, Z., Hajós, G., and Subik, J. (2007). Chemosensitisation of drug-resistant and drug-sensitive yeast cells to antifungals. Int. J. Antimicrob. Agents *29*, 170–178.

Chaux, G.E. (2010). Combination antifungal therapy for invasive mold infections involving polyenes. Infect. Dis. Clin. Pract. *18*, 7–15.

Chen, X., Ren, B., Chen, M., Liu, M.-X., Ren, W., Wang, Q.-X., Zhang, L.-X., and Yan, G.-Y. (2014). ASDCD: antifungal synergistic drug combination database. PLoS One *9*, e86499.

Cokol, M., Chua, H.N., Tasan, M., Mutlu, B., Weinstein, Z.B., Suzuki, Y., Nergiz, M.E., Costanzo, M., Baryshnikova, A., Giaever, G., *et al.* (2011). Systematic exploration of synergistic drug pairs. Mol. Syst. Biol. *7*, 544–544.

Cowen, L.E. (2008). The evolution of fungal drug resistance: modulating the trajectory from genotype to phenotype. Nat. Rev. Microbiol. *6*, 187–198.

Cowen, L.E. (2009). Hsp90 orchestrates stress response signaling governing fungal drug resistance. PLoS Pathog. *5*, e1000471.

Cowen, L.E. (2013). The fungal Achilles' heel: targeting Hsp90 to cripple fungal pathogens. Curr. Opin. Microbiol. *16*, 377–384.

Cowen, L.E., and Lindquist, S. (2005). Hsp90 potentiates the rapid evolution of new traits: drug resistance in diverse fungi. Science *309*, 2185–2189.

Cowen, L.E., Carpenter, A.E., Matangkasombut, O., Fink, G.R., and Lindquist, S. (2006). Genetic architecture of Hsp90-dependent drug resistance. Eukaryot. Cell *5*, 2184–2188.

Cowen, L.E., Singh, S.D., Kohler, J.R., Collins, C., Zaas, A.K., Schell, W.A., Aziz, H., Mylonakis, E., Perfect, J.R., Whitesell, L., *et al.* (2009). Harnessing Hsp90 function as a powerful, broadly effective therapeutic strategy for fungal infectious disease. Proc. Natl. Acad. Sci. U.S.A. *106*, 2818–2823.

Cuenca-Estrella, M. (2004). Combinations of antifungal agents in therapy-what value are they? J. Antimicrob. Chemother. *54*, 854–869.

Cuenca-Estrella, M., Gomez-Lopez, A., Garcia-Effron, G., Alcazar-Fuoli, L., Mellado, E., Buitrago, M.J., and Rodriguez-Tudela, J.L. (2005). Combined activity *in vitro* of caspofungin, amphotericin B, and azole agents against itraconazole-resistant clinical isolates of *Aspergillus fumigatus*. Antimicrob. Agents Chemother. *49*, 1232–1235.

Culakova, H., Dzugasova, V., Gbelska, Y., and Subik, J. (2012). CTBT (7-chlorotetrazolo[5,1-c]benzo[1,2,4] triazine) producing ROS affects growth and viability of filamentous fungi. FEMS Microbiol. Lett. *328*, 138–143.

Dannaoui, E., Lortholary, O., and Dromer, F. (2004). *In vitro* evaluation of double and triple combinations of antifungal drugs against *Aspergillus fumigatus* and *Aspergillus terreus*. Antimicrob. Agents Chemother. *48*, 970–978.

Dannaoui, E., Schwarz, P., and Lortholary, O. (2009). *In vitro* interactions between antifungals and immunosuppressive drugs against Zygomycetes. Antimicrob. Agents Chemother. *53*, 3549–3551.

Denning, D.W., and Hope, W.W. (2010). Therapy for fungal diseases: opportunities and priorities. Trends Microbiol. *18*, 195–204.

Edlind, T., Smith, L., Henry, K., Katiyar, S., and Nickels, J. (2002). Antifungal activity in *Saccharomyces cerevisiae* is modulated by calcium signalling. Mol. Microbiol. *46*, 257–268.

Ernst, E.J., Klepser, M.E., and Pfaller, M.A. (1998). In vitro interaction of fluconazole and amphotericin B administered sequentially against Candida albicans: effect of concentration and exposure time. Diagn. Microbiol. Infect. Dis. *32*, 205–210.

Fiori, A., and van Dijck, P. (2012). Potent synergistic effect of doxycycline with fluconazole against *Candida albicans* is mediated by interference with iron homeostasis. Antimicrob. Agents Chemother. *56*, 3785–3796.

Gamarra, S., Rocha, E.M.F., Zhang, Y.-Q., Park, S., Rao, R., and Perlin, D.S. (2010). Mechanism of the synergistic effect of amiodarone and fluconazole in *Candida albicans*. Antimicrob. Agents Chemother. *54*, 1753–1761.

Gao, Y., Zhang, C., Lu, C., Liu, P., Li, Y., Li, H., and Sun, S. (2013). Synergistic effect of doxycycline and fluconazole against *Candida albicans* biofilms and the impact of calcium channel blockers. FEMS Yeast Res. *13*, 453–462.

Gupta, A.K., and Plott, T. (2004). Ciclopirox: a broad-spectrum antifungal with antibacterial and anti-inflammatory properties. Int. J. Dermatol. *43* (Suppl. 1), 3–8.

Hancock, R.E., and Scott, M.G. (2000). The role of antimicrobial peptides in animal defenses. Proc. Natl. Acad. Sci. U.S.A. *97*, 8856–8861.

Harris, M.R., and Coote, P.J. (2010). Combination of caspofungin or anidulafungin with antimicrobial peptides results in potent synergistic killing of *Candida albicans* and *Candida glabrata in vitro*. Int. J. Antimicrob. Agents 35, 347–356.

Hayama, K., Ishibashi, H., Ishijima, S.A., Niimi, K., Tansho, S., Ono, Y., Monk, B.C., Holmes, A.R., Harding, D.R.K., Cannon, R.D., *et al.* (2012). A D-octapeptide drug efflux pump inhibitor acts synergistically with azoles in a murine oral candidiasis infection model. FEMS Microbiol. Lett. *328*, 130–137.

Higgins, J., Pinjon, E., Oltean, H.N., White, T.C., Kelly, S.L., Martel, C.M., Sullivan, D.J., Coleman, D.C., and Moran, G.P. (2012). Triclosan antagonizes fluconazole activity against *Candida albicans*. J. Dent. Res. *91*, 65–70.

Holmes, A.R., Keniya, M.V., Ivnitski-Steele, I., Monk, B.C., Lamping, E., Sklar, L.A., and Cannon, R.D. (2012). The monoamine oxidase A inhibitor clorgyline is a broad-spectrum inhibitor of fungal ABC and MFS transporter efflux pump activities which reverses the azole resistance of *Candida albicans* and *Candida glabrata* clinical isolates. Antimicrob. Agents Chemother. *56*, 1508–1515.

Ibrahim, A.S., Gebermariam, T., Fu, Y., Lin, L., Husseiny, M.I., French, S.W., Schwartz, J., Skory, C.D., Edwards, J.E., and Spellberg, B.J. (2007). The iron chelator deferasirox protects mice from mucormycosis through iron starvation. J. Clin. Invest. *117*, 2649–2657.

Ibrahim, A.S., Gebremariam, T., French, S.W., Edwards, J.E., and Spellberg, B. (2010). The iron chelator deferasirox enhances liposomal amphotericin B efficacy in treating murine invasive pulmonary aspergillosis. J. Antimicrob. Chemother. *65*, 289–292.

Ibrahim, A.S., Gebremariam, T., Luo, G., Fu, Y., French, S.W., Edwards, J.E., and Spellberg, B. (2011). Combination therapy of murine mucormycosis or aspergillosis with iron chelation, polyenes, and echinocandins. Antimicrob. Agents Chemother. *55*, 1768–1770.

James K Liao, U.L. (2005). Pleiotropic effects of statins. Annu. Rev. Pharmacol. Toxicol. *45*, 89–118.

Jansen, G., Lee, A.Y., Epp, E., Fredette, A., Surprenant, J., Harcus, D., Scott, M., Tan, E., Nishimura, T., Whiteway, M., *et al.* (2009). Chemogenomic profiling predicts antifungal synergies. Mol. Syst. Biol. *5*, 338.

Johnson, M.D., and Perfect, J.R. (2010). Use of antifungal combination therapy: agents, order, and timing. Curr. Fungal Infect. Rep. *4*, 87–95.

Johnson, M.D., MacDougall, C., Ostrosky-Zeichner, L., Perfect, J.R., and Rex, J.H. (2004). Combination antifungal therapy. Antimicrob. Agents Chemother. *48*, 693–715.

Kaneko, Y., Fukazawa, H., Ohno, H., and Miyazaki, Y. (2013). Combinatory effect of fluconazole and FDA-approved drugs against *Candida albicans*. J. Infect. Chemother. *19*, 1141–1145.

Keith, C.T., Borisy, A.A., and Stockwell, B.R. (2005). Multicomponent therapeutics for networked systems. Nat. Rev. Drug Discov. *4*, 71–78.

LaFayette, S.L., Collins, C., Zaas, A.K., Schell, W.A., Betancourt-Quiroz, M., Gunatilaka, A.A., Perfect, J.R., and Cowen, L.E. (2010). PKC signaling regulates drug resistance of the fungal pathogen *Candida albicans* via circuitry comprised of Mkc1, calcineurin, and Hsp90. PLoS Pathog. *6*, e1001069.

LaFleur, M.D., Lucumi, E., Napper, A.D., Diamond, S.L., and Lewis, K. (2011). Novel high-throughput screen against *Candida albicans* identifies antifungal potentiators and agents effective against biofilms. J. Antimicrob. Chemother. *66*, 820–826.

Lafleur, M.D., Sun, L., Lister, I., Keating, J., Nantel, A., Long, L., Ghannoum, M., North, J., Lee, R.E., Coleman, K., *et al.* (2013). Potentiation of azole antifungals by 2-adamantanamine. Antimicrob. Agents Chemother. *57*, 3585–3592.

Lamoth, F., Juvvadi, P.R., Gehrke, C., Asfaw, Y.G., and Steinbach, W.J. (2014a). Transcriptional activation of heat shock protein 90 mediated via a proximal promoter region as trigger of caspofungin resistance in *Aspergillus fumigatus*. J. Infect. Dis. *209*, 473–481.

Lamoth, F., Juvvadi, P.R., Soderblom, E.J., Moseley, M.A., Asfaw, Y.G., and Steinbach, W.J. (2014b). Identification of a key lysine residue in heat shock protein 90 required for azole and echinocandin resistance in *Aspergillus fumigatus*. Antimicrob. Agents Chemother. *58*, 1889–1896.

Lee, A.Y., St Onge, R.P., Proctor, M.J., Wallace, I.M., Nile, A.H., Spagnuolo, P.A., Jitkova, Y., Gronda, M., Wu, Y., Kim, M.K., *et al.* (2014). Mapping the cellular response to small molecules using chemogenomic fitness signatures. Science *344*, 208–211.

Lee, R., Liu, T., Barker, K., and Rogers, P. (2005). Genome-wide expression profiling of the response to ciclopirox olamine in *Candida albicans*. J. Antimicrob. Chemother. *55*, 655–662.

Lehar, J., Stockwell, B.R., Giaever, G., and Nislow, C. (2008). Combination chemical genetics. Nat. Chem. Biol. *4*, 674–681.

Li, J.Y.Z., Yong, T.Y., Grove, D.I., and Coates, P.T.H. (2008). Successful control of *Scedosporium prolificans* septic arthritis and probable osteomyelitis without radical surgery in a long-term renal transplant recipient. Transpl. Infect. Dis. *10*, 63–65.

Liu, S., Hou, Y., Chen, X., Gao, Y., Li, H., and Sun, S. (2014). Combination of fluconazole with non-antifungal agents: a promising approach to cope with resistant *Candida albicans* infections and insight into new antifungal agent discovery. Int. J. Antimicrob. Agents *43*, 395–402.

Louie, A., Kaw, P., Banerjee, P., Liu, W., Chen, G., and Miller, M.H. (2001). Impact of the order of initiation of fluconazole and amphotericin B in sequential or combination therapy on killing of *Candida albicans in vitro* and in a rabbit model of endocarditis and pyelonephritis. Antimicrob. Agents Chemother. *45*, 485–494.

MacCallum, D.M., Whyte, J.A., and Odds, F.C. (2005). Efficacy of caspofungin and voriconazole combinations in experimental aspergillosis. Antimicrob. Agents Chemother. *49*, 3697–3701.

MacCallum, D.M., Desbois, A.P., and Coote, P.J. (2013). Enhanced efficacy of synergistic combinations of antimicrobial peptides with caspofungin versus *Candida albicans* in insect and murine models of systemic infection. Eur. J. Clin. Microbiol. Infect. Dis. *32*, 1055–1062.

Manavathu, E.K., Alangaden, G.J., and Chandrasekar, P.H. (2003). Differential activity of triazoles in two-drug combinations with the echinocandin caspofungin against *Aspergillus fumigatus*. J. Antimicrob. Chemother. *51*, 1423–1425.

Marchetti, O., Entenza, J.M., Sanglard, D., Bille, J., Glauser, M.P., and Moreillon, P. (2000a). Fluconazole plus cyclosporine: a fungicidal combination effective against experimental endocarditis due to *Candida albicans*. Antimicrob. Agents Chemother. *44*, 2932–2938.

Marchetti, O., Moreillon, P., Glauser, M.P., Bille, J., and Sanglard, D. (2000b). Potent synergism of the combination of fluconazole and cyclosporine in *Candida albicans*. Antimicrob. Agents Chemother. *44*, 2373–2381.

Maresova, L., Muend, S., Zhang, Y.Q., Sychrova, H., and Rao, R. (2009). Membrane hyperpolarization drives cation influx and fungicidal activity of amiodarone. J. Biol. Chem. *284*, 2795–2802.

Marr, K.A., Boeckh, M., Carter, R.A., Kim, H.W., and Corey, L. (2004). Combination antifungal therapy for invasive aspergillosis. Clin. Infect. Dis. *39*, 797–802.

Marr, K.A., Schlamm, H., Rottinghaus, S.T., and Jagannatha, S. (2012). A randomised, double-blind study of combination antifungal therapy with voriconazole and anidulafungin versus voriconazole monotherapy for primary treatment of invasive aspergillosis. Paper presented at: 22nd European Congress of Clinical Microbiology and Infectious Diseases (ECCMID) (London, UK).

Maschmeyer, G. (2006). The changing epidemiology of invasive fungal infections: new threats. Int. J. Antimicrob. Agents 27 (Suppl. 1), 3–6.

Maurya, I.K., Thota, C.K., Verma, S.D., Sharma, J., Rawal, M.K., Ravikumar, B., Sen, S., Chauhan, N., Lynn, A.M., Chauhan, V.S., *et al.* (2013). Rationally designed transmembrane peptide mimics of the multidrug transporter protein Cdr1 act as antagonists to selectively block drug efflux and chemosensitize azole-resistant clinical isolates of *Candida albicans*. J. Biol. Chem. *288*, 16775–16787.

Miceli, M.H., and Lee, S.A. (2011). Emerging moulds: epidemiological trends and antifungal resistance. Mycoses *54*, 666–678.

Monk, B.C., Niimi, K., Lin, S., Knight, A., Kardos, T.B., Cannon, R.D., Parshot, R., King, A., Lun, D., and Harding, D.R.K. (2004). Surface-active fungicidal D-peptide inhibitors of the plasma membrane proton pump that block azole resistance. Antimicrob. Agents Chemother. *49*, 57–70.

Mosquera, J., Sharp, A., Moore, C.B., Warn, P.A., and Denning, D.W. (2002). *In vitro* interaction of terbinafine with itraconazole, fluconazole, amphotericin B and 5-flucytosine against *Aspergillus* spp. J. Antimicrob. Chemother. *50*, 189–194.

Muend, S., and Rao, R. (2008). Fungicidal activity of amiodarone is tightly coupled to calcium influx. FEMS Yeast Res. *8*, 425–431.

Niewerth, M., Kunze, D., Seibold, M., Schaller, M., Korting, H.C., and Hube, B. (2003). Ciclopirox olamine treatment affects the expression pattern of *Candida albicans* genes encoding virulence factors, iron metabolism proteins, and drug resistance factors. Antimicrob. Agents Chemother. *47*, 1805–1817.

Niimi, K., Harding, D.R., Holmes, A.R., Lamping, E., Niimi, M., Tyndall, J.D., Cannon, R.D., and Monk, B.C. (2012). Specific interactions between the *Candida albicans* ABC transporter Cdr1p ectodomain and a D-octapeptide derivative inhibitor. Mol. Microbiol. *85*, 747–767.

Nussbaum, J.C., Jackson, A., Namarika, D., Phulusa, J., Kenala, J., Kanyemba, C., Jarvis, J.N., Jaffar, S., Hosseinipour, M.C., Kamwendo, D., *et al.* (2010). Combination flucytosine and high-dose fluconazole compared with fluconazole monotherapy for the treatment of cryptococcal meningitis: a randomized trial in Malawi. Clin. Infect. Dis. *50*, 338–344.

Nyilasi, I.I., Kocsubé, S.S., Krizsán, K.K., Galgóczy, L.L., Pesti, M.M., Papp, T.T., and Vágvölgyi, C.C. (2010). *In vitro* synergistic interactions of the effects of various statins and azoles against some clinically important fungi. FEMS Microbiol. Lett. *307*, 175–184.

Odds, F.C. (2003). Synergy, antagonism, and what the chequerboard puts between them. J. Antimicrob. Chemother. *52*, 1–1.

Oh, J., Fung, E., Schlecht, U., Davis, R.W., Giaever, G., St Onge, R.P., Deutschbauer, A., and Nislow, C. (2010). Gene annotation and drug target discovery in *Candida albicans* with a tagged transposon mutant collection. PLoS Pathog. *6*, e1001140.

Ostrosky-Zeichner, L. (2008). Combination antifungal therapy: a critical review of the evidence. Clin. Microbiol. Infect. *14* (Suppl. 4), 65–70.

Perfect, J.R., Dismukes, W.E., Dromer, F., Goldman, D.L., Graybill, J.R., Hamill, R.J., Harrison, T.S., Larsen, R.A., Lortholary, O., Nguyen, M.H., *et al.* (2010). Clinical practice guidelines for the management of cryptococcal disease: 2010 update by the infectious diseases society of america. Clin. Infect. Dis. *50*, 291–322.

Pfaller, M.A., and Diekema, D.J. (2010). Epidemiology of invasive mycoses in North America. Crit. Rev. Microbiol. *36*, 1–53.

del Poeta, M., Cruz, M.C., Cardenas, M.E., Perfect, J.R., and Heitman, J. (2000). Synergistic antifungal activities of bafilomycin A(1), fluconazole, and the pneumocandin MK-0991/caspofungin acetate (L-743,873) with calcineurin inhibitors FK506 and L-685,818 against *Cryptococcus neoformans*. Antimicrob. Agents Chemother. *44*, 739–746.

Prasad, T., Chandra, A., Mukhopadhyay, C.K., and Prasad, R. (2006). Unexpected link between iron and drug resistance of *Candida* spp.: iron depletion enhances membrane fluidity and drug diffusion, leading to drug-susceptible cells. Antimicrob. Agents Chemother. *50*, 3597–3606.

Prichard, M.N., Prichard, L.E., Baguley, W.A., Nassiri, M.R., and Shipman, C. (1991). Three-dimensional analysis of the synergistic cytotoxicity of ganciclovir and zidovudine. Antimicrob. Agents Chemother. *35*, 1060–1065.

Puig-Asensio, M., Padilla, B., Garnacho-Montero, J., Zaragoza, O., Aguado, J.M., Zaragoza, R., Montejo, M., Munoz, P., Ruiz-Camps, I., Cuenca-Estrella, M., *et al.* (2014). Epidemiology and predictive factors for early and late mortality in *Candida* bloodstream infections: a population-based surveillance in Spain. Clin. Microbiol. Infect. *20*, O245–254.

Ricardo, E., Costa-de-Oliveira, S., Dias, A.S., Guerra, J., Rodrigues, A.G., and Pina-Vaz, C. (2009). Ibuprofen reverts antifungal resistance on *Candida albicans* showing overexpression of CDR genes. FEMS Yeast Res. *9*, 618–625.

Robbins, N., Uppuluri, P., Nett, J., Rajendran, R., Ramage, G., Lopez-Ribot, J.L., Andes, D., and Cowen, L.E. (2011). Hsp90 governs dispersion and drug resistance of fungal biofilms. PLoS Pathog. *7*, e1002257.

Robbins, N., Leach, M.D., and Cowen, L.E. (2012). Lysine deacetylases Hda1 and Rpd3 regulate Hsp90 function thereby governing fungal drug resistance. Cell Rep. *2*, 878–888.

Rodloff, A.C., Koch, D., and Schaumann, R. (2011). Epidemiology and antifungal resistance in invasive candidiasis. Eur. J. Med. Res. *16*, 187–195.

Rojas, R., Molina, J.R., Jarque, I., Montes, C., Serrano, J., Sanz, J., Besalduch, J., Carreras, E., Tomas, J.F., Madero, L., *et al.* (2012). Outcome of antifungal combination therapy for invasive mold infections in hematological patients is independent of the chosen combination. Mediterr. J. Hematol. Infect. Dis. *4*, e2012011.

Roling, E.E., Klepser, M.E., Wasson, A., and Lewis, R.E. (2002). Antifungal activities of fluconazole, caspofungin (MK0991), and anidulafungin (LY 303366) alone and in combination against *Candida* spp. and *Crytococcus neoformans* via time-kill methods. Diagn. Microbiol. Infect. Dis. *43*, 13–17.

Ruhnke, M. (2014). Antifungal stewardship in invasive *Candida infections*. Clin. Microbiol. Infect. *20* (Suppl. 6), 11–18.

Sanglard, D. (2002). Resistance of human fungal pathogens to antifungal drugs. Curr. Opin. Microbiol. *5*, 379–385.

Sanglard, D., and Odds, F.C. (2002). Resistance of *Candida* species to antifungal agents: molecular mechanisms and clinical consequences. Lancet Infect. Dis. *2*, 73–85.

Sanglard, D., Coste, A., and Ferrari, S. (2009). Antifungal drug resistance mechanisms in fungal pathogens from the perspective of transcriptional gene regulation. FEMS Yeast Res. *9*, 1029–1050.

Schwarz, P., Dromer, F., Lortholary, O., and Dannaoui, E. (2003). *In vitro* interaction of flucytosine with conventional and new antifungals against *Cryptococcus neoformans* clinical isolates. Antimicrob. Agents Chemother. *47*, 3361–3364.

Seyedmousavi, S., Meletiadis, J., Melchers, W.J.G., Rijs, A.J.M.M., Mouton, J.W., and Verweij, P.E. (2013). *In vitro* interaction of voriconazole and anidulafungin against triazole-resistant *Aspergillus fumigatus*. Antimicrob. Agents Chemother. *57*, 796–803.

Silva, L.V., Sanguinetti, M., Vandeputte, P., Torelli, R., Rochat, B., and Sanglard, D. (2013). Milbemycins: more than efflux inhibitors for fungal pathogens. Antimicrob. Agents Chemother. *57*, 873–886.

Singh, S.D., Robbins, N., Zaas, A.K., Schell, W.A., Perfect, J.R., and Cowen, L.E. (2009). Hsp90 governs echinocandin resistance in the pathogenic yeast *Candida albicans* via calcineurin. PLoS Pathog. *5*, e1000532.

Singh-Babak, S.D., Babak, T., Diezmann, S., Hill, J.A., Xie, J.L., Chen, Y.L., Poutanen, S.M., Rennie, R.P., Heitman, J., and Cowen, L.E. (2012). Global analysis of the evolution and mechanism of echinocandin resistance in *Candida glabrata*. PLoS Pathog. *8*, e1002718.

Spitzer, M., Griffiths, E., Blakely, K.M., Wildenhain, J., Ejim, L., Rossi, L., De Pascale, G., Curak, J., Brown, E., Tyers, M., *et al.* (2011). Cross-species discovery of syncretic drug combinations that potentiate the antifungal fluconazole. Mol. Syst. Biol. *7*, 1–14.

Steier, Z., Vermitsky, J.-P., Toner, G., Gygax, S.E., Edlind, T., and Katiyar, S. (2013). Flucytosine antagonism of azole activity versus *Candida glabrata*: role of transcription factor Pdr1 and multidrug transporter Cdr1. Antimicrob. Agents Chemother. *57*, 5543–5547.

Steinbach, W.J., Singh, N., Miller, J.L., Benjamin, D.K., Schell, W.A., Heitman, J., and Perfect, J.R. (2004). *In vitro* interactions between antifungals and immunosuppressants against *Aspergillus fumigatus* isolates from transplant and nontransplant patients. Antimicrob. Agents Chemother. *48*, 4922–4925.

Steinbach, W.J., Reedy, J.L., Cramer, R.A., Perfect, J.R., and Heitman, J. (2007). Harnessing calcineurin as a novel anti-infective agent against invasive fungal infections. Nat. Rev. Microbiol. *5*, 418–430.

Steinbach, W.J., Juvvadi, P.R., Fortwendel, J.R., and Rogg, L.E. (2011). Newer combination antifungal therapies for invasive aspergillosis. Med. Mycol. *49* (Suppl. 1), S77–81.

Stergiopoulou, T., Meletiadis, J., Sein, T., Papaioannidou, P., Tsiouris, I., Roilides, E., and Walsh, T.J. (2009). Comparative pharmacodynamic interaction analysis between ciprofloxacin, moxifloxacin and levofloxacin and antifungal agents against *Candida albicans* and *Aspergillus fumigatus*. J. Antimicrob. Chemother. *63*, 343–348.

Te Dorsthorst, D.T.A., Verweij, P.E., Meletiadis, J., Bergervoet, M., Punt, N.C., Meis, J.F.G.M., and Mouton, J.W. (2002). *In vitro* interaction of flucytosine combined with amphotericin B or fluconazole against thirty-five yeast isolates determined by both the fractional inhibitory concentration index and the response surface approach. Antimicrob. Agents Chemother. *46*, 2982–2989.

Tobudic, S., Kratzer, C., Lassnigg, A., Graninger, W., and Presterl, E. (2010). *In vitro* activity of antifungal combinations against *Candida albicans* biofilms. J. Antimicrob. Chemother. *65*, 271–274.

Tong, A.H.Y., Lesage, G., Bader, G.D., Ding, H., Xu, H., Xin, X., Young, J., Berriz, G.F., Brost, R.L., Chang, M., *et al.* (2004). Global mapping of the yeast genetic interaction network. Science *303*, 808–813.

Vandeputte, P., Pineau, L., Larcher, G., Noël, T., Brèthes, D., Chabasse, D., and Bouchara, J.-P. (2011). Molecular mechanisms of resistance to 5-fluorocytosine in laboratory mutants of *Candida glabrata*. Mycopathologia *171*, 11–21.

Vazquez, J.A. (2008). Clinical practice: combination antifungal therapy for mold infections: much ado about nothing? Clin. Infect. Dis. *46*, 1889–1901.

Vincent, B.M., Lancaster, A.K., Scherz-Shouval, R., Whitesell, L., and Lindquist, S. (2013). Fitness trade-offs restrict the evolution of resistance to amphotericin B. PLoS Biol. *11*, e1001692.

Yeh, P.J., Hegreness, M.J., Aiden, A.P., and Kishony, R. (2009). Drug interactions and the evolution of antibiotic resistance. Nat. Rev. Microbiol. *7*, 460–466.

Yu, L., Ling, G., Deng, X., Jin, J., Jin, Q., and Guo, N. (2011). In vitro interaction between fluconazole and triclosan against clinical isolates of fluconazole-resistant *Candida albicans* determined by different methods. Antimicrob. Agents Chemother. *55*, 3609–3612.

Zarember, K.A., Cruz, A.R., Huang, C.-Y., and Gallin, J.I. (2009). Antifungal activities of natural and synthetic iron chelators alone and in combination with azole and polyene antibiotics against *Aspergillus fumigatus*. Antimicrob. Agents Chemother. *53*, 2654–2656.

Zeidler, U., Bougnoux, M.E., Lupan, A., Helynck, O., Doyen, A., Garcia, Z., Sertour, N., Clavaud, C., Munier-Lehmann, H., Saveanu, C., *et al.* (2013). Synergy of the antibiotic colistin with echinocandin antifungals in *Candida* species. J. Antimicrob. Chemother. *68*, 1285–1296.

Zhang, L., Yan, K., Zhang, Y., Huang, R., Bian, J., Zheng, C., Sun, H., Chen, Z., Sun, N., An, R., *et al.* (2007). High-throughput synergy screening identifies microbial metabolites as combination agents for the treatment of fungal infections. Proc. Natl. Acad. Sci. U.S.A. *104*, 4606–4611.

Approaches to Detect Alternative Mechanisms of Resistance to Systemic Antifungals

6

Patrick Vandeputte

Abstract

Resistance to antifungals is a major concern in the management of fungal infections, especially while the incidence of pathogens with a poor susceptibility to current treatments is rising. New therapeutic strategies could be developed through the discovery of completely new fungal-specific targets or through the identification of new effectors of resistance to existing antifungals. In addition to basic molecular resistance mechanisms that are well understood, there are also numerous additional effectors able to modulate fungi susceptibility to the four main classes of antifungals. These effectors are unable to drive resistance alone, but they are now believed to be crucial for the establishment and maintenance of drug resistance, as they constitute key modulators allowing the phenotypic expression of resistance acquired by basic mechanisms. Formerly limited, the approaches to detect such alternative resistance mechanisms to antifungals were profoundly renewed with the 'omics' era, allowing the study of whole organism's response. This chapter will focus on the main strategies implemented in the last two decades, with a particular emphasis on high throughput technologies such as whole genome sequencing, transcriptomics, proteomics, and large-scale mutant collections screening.

Introduction

In late 1990s, Joseph-Horne and his co-workers mentioned that

> the technologies required to examine and define azole resistance mechanisms at the molecular level exist, but research appears distinctly lacking in this most important area.
>
> (Joseph-Horne and Hollomon, 1997)

This statement, being equally valid for other classes of antifungals (ATFs), was highly visionary since almost twenty years later the basal molecular mechanisms responsible for ATF resistance became widely investigated and are now largely understood. The majority of scientific research on ATF aimed to understand the molecular mechanisms responsible for resistance to systemic ATFs, those used to treat severe forms of mycoses. Here we will not discuss the mechanisms responsible for resistance to more anecdotal ATFs that often belong to the same classes than systemically acting molecules and thus involve the same biological responses and resistance mechanisms. Systemic ATFs belong to four classes of molecules: (i) azoles, targeting the biosynthesis pathway of ergosterol, one of the most important lipid compound of the fungal membrane, (ii) polyenes, targeting ergosterol itself, (iii) echinocandins, which inhibit the synthesis of a major component of the fungal cell wall and (iv) pyrimidine analogues, which inhibit protein synthesis and DNA replication.

Resistance to systemic ATFs always results from four basic mechanisms: (i) up-regulation of genes encoding efflux pumps that decreases ATF intracellular concentration, (ii) presence of point mutations that decrease affinity of ATFs to their target, (iii) metabolic adaptations that minimize the physiological importance of ATF targets, and (iv) up-regulation of the genes encoding ATF's targets. These basic mechanisms of resistance are described in Chapters 1 to 3.

The real issue in these last 10 years was to understand how they are settled in the fungal cell. In fact, alternative pathways also participate into fungal response to antibiotic stresses and the cellular actors finely tuning fungal susceptibility to ATFs, referred to herein as 'alternative' resistance mechanisms, are almost countless. Moreover, a more precise understanding of the whole pathogen's response to ATFs would be a source for the discovery of new fungal-specific targets (Jack, 1998).

There are two ways of reasoning for the discovery of new therapeutic strategies. The first one is to discover completely new fungal targets. This can be achieved by screening chemical compounds collections for ATF activity or through analysis of the mode of action of candidate ATFs. These two approaches are described in Chapters 7 and 9 of this book. The second rationale is to find new fungal effectors that can be targeted to enhance the effect of existing ATFs and counteract resistance (see Chapter 5). Fungal response to currently available ATFs is indeed more complex than it appeared at the end of the last century and involves multiple pathways, often interconnected with pathways regulating virulence and pathogenicity, stress response and more generally, response to host environment, which constitutes a very promising field of investigation (Homann *et al.*, 2009; Shapiro *et al.*, 2011).

The study of such complex responses is only possible when considering whole organism responses, with methodologies able to give a complete picture of the involved mechanisms (Cowen and Steinbach, 2008; Isaacson, 2002; Monk and Cannon, 2002). This implies access to large scale technologies like transcriptomics, proteomics, genomics, or mutant collections screening (Agarwal *et al.*, 2012). This chapter is an overview of the most recent findings regarding these alternative resistance mechanisms, with a particular emphasis on the strategies employed since the advent of high throughput technologies, allowing the study of whole organism response.

'Omics' approaches

Transcriptomics

Transcriptome analysis methods

One of the most frequent mechanisms of resistance to ATF is the overexpression of effectors genes like those encoding major facilitator (MFS) and ATP-binding cassette (ABC) efflux pumps or ATF's targets. Historically, these mechanisms were investigated by northern blotting, which permitted the analysis of the expression of a restricted number of genes in different conditions or strains. Then in the late 1990s, the technology of real time reverse transcription quantitative PCR was introduced to allow a simultaneous expression analysis of a larger number of genes. At the same period, microarrays technology was developed and the first array spotted with the full genome of an organism was for *Saccharomyces cerevisiae* in 1997. Finally, the advent of next generation sequencing allowed whole transcriptome shotgun sequencing, now referred to as RNAseq. These four transcriptome analysis methods share the same starting point that is RNA extraction (Fig. 6.1).

Northern blot

Northern blot was developed in 1977 (Alwine *et al.*, 1977) and participated in the discovery of the basic molecular mechanisms of resistance to azole drugs (the overexpression of ABC or MFS transporters genes and of *ERG11*). Total RNA is separated on a non-denaturing gel by migration and transferred by capillary or vacuum blotting on a positively charged nylon membrane. A specific probe then detects mRNAs of interest. Radioactive labelling is the most sensitive technique for detection but chemiluminescent or fluorescent probes are now widely used. Northern blotting is a tedious method but has the ability to detect splicing variants. Moreover, northern blotting has a higher sensitivity than microarrays but lower than RT-qPCR.

Real-time quantitative PCR

Gene expression analysis methods were revolutionized in late 1990s with the first quantitative PCR experiments (Becker-André and Hahlbrock,

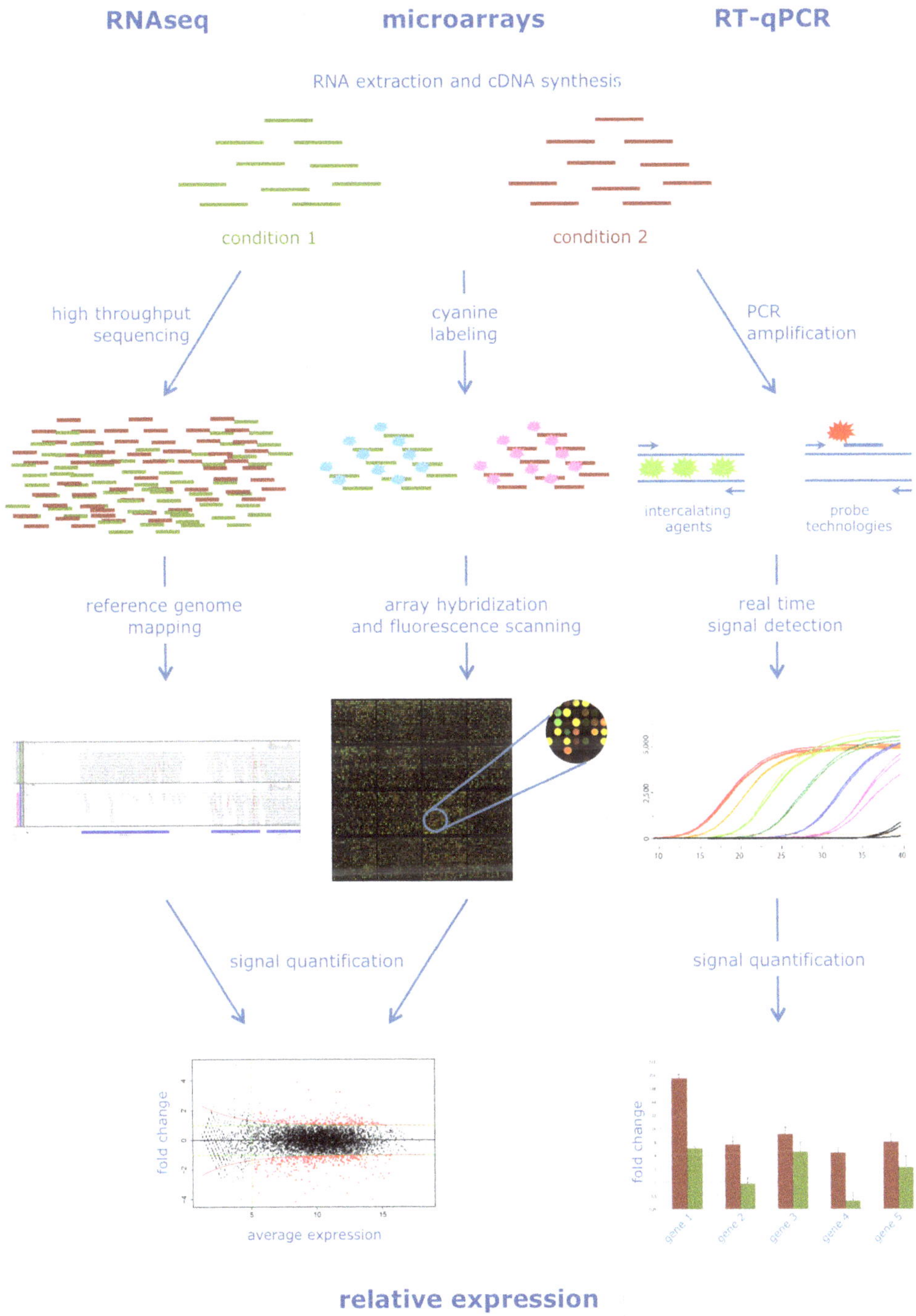

Figure 6.1 Transcriptome analysis methods. Two high throughput technologies are now available for the analysis of gene expression in fungi, microarrays and RNAseq. However, the high reliability of RT-qPCR makes this technique commonly used for the validation of the results obtained by high throughput technologies. First, RNAs extracted from the two conditions that have to be compared are reverse transcribed into cDNAs. For microarrays, these cDNAs are differentially labelled with Cy3 and Cy5 before being hybridized on the array, while for RNAseq, they are sequenced by a next generation sequencing technology. The signals, i.e. fluorescence intensities and sequencing data for microarrays and RNAseq, respectively, are integrated to produce expression ratio for each gene in the two conditions.

1989; Chiang *et al.*, 1996; Heid *et al.*, 1996; Wang *et al.*, 1989), which allowed the study of conditional expression of several genes at the same time. In 1996, equipment limited the number of simultaneous qPCRs to only 32, whereas now the 384-well format is more frequently used. A new technology with higher throughput, based on the multiplexed detection of colour-coded probe pairs and allowing the simultaneous quantification of up to 800 transcripts, was developed in late 2000s (Geiss *et al.*, 2008). Up to now, this technology has never been applied to the discovery of ATF resistance mechanisms. Total RNA is extracted from fungal samples and retrotranscribed into cDNA by reverse transcriptase. In eukaryotic organisms, reverse transcription can be performed by the use of random primers or oligo-dT primers that specifically act on polyadenylated mRNA. Two main technologies are available for the quantification of cDNA of interest: intercalating agents such as SYBR green and hydrolysis probe such as Taqman. RT-qPCR is the transcriptome analysis method with the highest sensitivity and specificity and thereby it is used to validate the results of high throughput experiments (i.e. RNAseq and microarrays).

DNA microarrays

The second revolution in gene expression analysis, concomitant with the increase in the number of genomes fully sequenced, was the availability of genome-wide microarrays (Barker and Rogers, 2005). Probes specific to each ORF of a microorganism are coated onto a glass slide. cDNA produced in two conditions is compared by labelling with cyanines and hybridization to the array. Fluorescence scanning at specific cyanines emission wavelengths (570 nm (green) and 670 nm (red) for Cy3 and Cy5, respectively) quantifies the signal for each individual gene. The signal ratio between the two conditions reflects the expression level of each gene. Microarrays can be of two types, either a single-colour based cDNA labelling in which a single cyanine is used and the ratio of fluorescence intensity between the two conditions (e.g. wild type versus mutant strains) would reflect the expression levels, or a two-colour based labelling that allows the direct observation of the effect of a modifier condition (e.g. identification of the genes differentially regulated between a wild type and a mutant strain upon an ATF treatment).

RNAseq

Finally, the transcriptome analysis method generating the most data is RNAseq, in which two libraries of cDNA are produced in the conditions that have to be compared. The high throughput sequencing of these libraries generates millions of reads that have to be mapped on the organism's reference genome. The determination of the ratio of reads between the two conditions studied allows the identification of the genes differentially expressed. Nowadays, five technologies of high throughput sequencing are available (Liu *et al.*, 2012). They considerably diminished the sequencing cost (from $0.1 to $10 per million bp, depending on the technology, as compared to the $2400 needed by the classical Sanger method by chain termination) and also, by definition, considerably increased the sequencing throughput. As an example, the most processive method allows sequencing of up to 600 billions bp in 1 to 10 days while the last generation of Sanger sequencing devices, available in 384-well format, allow the sequencing of up to 300,000 bp in 24 hours.

One of the first technologies made available to laboratories was the Polony method commercialized by Life Technologies under the platform SOLiD®. This method combines emulsion PCR and ligation and allows the sequencing of 1.2–1.4 billions of reads of up to 100 bp in one to two weeks. It benefits from a high sequencing accuracy (99.9%) and is one of the least expensive methods, but the run time is relatively large.

The massively parallelized pyrosequencing, combining emulsion PCR and light detection, is commercialized by Roche Diagnostics under the name 454®. This high throughput technology is the intermediate between classical Sanger sequencing and the other high throughput methods since it allows the sequencing of approximately 1 million of relatively long reads (700 bp) in around 24 hours but the cost per bp remains high.

The high-throughput technology generating the highest number of reads is the Illumina platform. This technology is based on sequencing by synthesis, where nucleic acids fragments are linked to a flow cell coated with specific adaptors at a precise density. After amplification, DNA clusters are formed, denatured, and the nature of each nucleotide incorporated while the complementary strand is synthesized is determined by image capture. The Illumina technology is the least expensive method when considering the number of data that could be generated, but the run time could be long depending on the instrument and on the specified read length. Moreover, the length of the reads is limited to a maximum of 300 bp.

The IonTorrent® technology commercialized by Life Technologies is based on the classic reaction of sequencing by synthesis. The difference lies in the detection system that is able to electrochemically quantify the incorporation of a nucleotide when added to a DNA template through the detection of the H^+ ion released during the elongation process. The main advantage of this ion semiconductor technology is the cost of the equipment that remains relatively less expensive when compared to the other technologies. Moreover, the run time is relatively short, allowing the sequencing of up to 80 million reads of approximately 400 bp in a couple of hours.

The major disadvantage of all the preceding technologies resides in the relatively short length of the sequences obtained complicating the assembly of reads into contigs, especially for *de novo* genome assembly. Some alternative methods can be used to circumvent that issue, an example is the so-called 'mate-pairs' Illumina libraries, in which only the extremities of larger fragments (up to 4 kbp) are sequenced. However, the most recently released high-throughput technology, commercialized by Pacific Biosciences under the name single-molecule real-time sequencing, has been developed to palliate this issue. Based on sequencing by synthesis and identification of the nucleotide incorporated at each position by fluorescence detection, this technology allows the full sequencing of fragments of up to 30,000 bp in a short time (less than 2 hours). It also allows the detection of base modifications such as methylation. However, the throughput is moderate with a maximum of 800 million bp and a low accuracy (around 90%) which necessitates the combination of another more accurate high-throughput technology.

Paradoxically, RNAseq has rarely been used to detect alternative resistance mechanisms to ATF up to now. This must be due to the fact that it is a recent method, which is not yet fully generalized and standardized in fungi. Thus, most of the advances in the field of discovery of alternative resistance mechanisms to ATF by transcriptomic approaches were made up-to-now by microarrays experiments. More generally, transcriptomic studies are by far the first strategy considering the number of alternative resistance mechanisms to ATF that have been identified.

Contribution of transcriptomics in ATF alternative resistance mechanisms discovery

Azoles

Historically, the basic mechanisms of resistance to azoles, which are the overexpression of efflux pumps and of azole target, were identified by northern blotting or RT-qPCR. However, a large number of alternative effectors of resistance to azole drugs were also identified by large-scale transcriptomics.

Most investigations were conducted in *Candida albicans*, one of the most clinically relevant yeast species. One could cite for example the discovery of the *RTA2* gene and the deciphering of its involvement in azole resistance (Jia *et al.*, 2008). After exposure for around 30 generations to the combination of fluconazole and fluphenazine, a mutant *C. albicans* strain devoid of the three major azole resistance effectors (i.e. *CDR1*, *CDR2* and *MDR1*) developed resistance to this azole drug. Microarray analysis of the genes differentially expressed in this resistant mutant was compared to the parental strain and revealed, among others, the up-regulation of *RTA2*. This finding depicted the probable role of lipid homeostasis in azole resistance since, by analogy with *S. cerevisiae*, *RTA2* would be a sphingolipid translocase. This hypothesis was further confirmed in later studies, in which the authors also evidenced that *RTA2* mediates azole resistance in a calcineurin-dependent manner (Jia *et al.*, 2009, 2012). The genome of *C. albicans* contains four *RTA* genes (*RTA1* to *RTA4*), among which *RTA2* but also *RTA3* are regulated upon exposure to azole ATFs

(Barker, 2004; Coste *et al.*, 2004; Liu *et al.*, 2005), suggesting this gene family is involved in a more general response to lipid homeostasis perturbation. Moreover, the *RTA2*-up-regulating strain also became resistant to calcineurin inhibition. Calcineurin is a calcium-dependent phosphatase with critical roles in intracellular signalling, it is activated by the chaperone Hsp90, and is essential for establishment of ATF resistance in many fungal species including *C. albicans*, *C. glabrata*, *Aspergillus fumigatus*, *Cryptococcus neoformans* and *S. cerevisiae*, mainly through activation of the transcription factors Crz1 and Hph1 (Chen *et al.*, 2011; Cowen, 2009; Cowen and Lindquist, 2005; Cowen *et al.*, 2006; McClellan *et al.*, 2007; Miyazaki *et al.*, 2010; Onyewu *et al.*, 2004; Sanglard *et al.*, 2003; Singh *et al.*, 2009). Moreover, the study of Jia *et al.* (2009) further reinforced the central role of calcineurin in ATF resistance.

Recently, RNAseq was applied to discover the genes involved in azole resistance of *C. albicans* mitochondrial electron transport chain mutants (Sun *et al.*, 2013). The involvement of respiratory metabolism in azole susceptibility was long known, in *C. albicans* as well as in non-*albicans Candida* species such as *C. glabrata petite* mutants (Brun *et al.*, 2003). In these respiratory-deficient strains, the up-regulation of ABC transporters encoding genes, *CDR1* and *CDR2*, was well characterized, but the molecular mechanisms governing this up-regulation were not known. The transcriptome analysis of *C. albicans* mutants for subunits (or regulatory proteins) of the complex I of the mitochondrial electron transport chain evidenced the involvement of peroxisomal energy production in azole susceptibility (Sun *et al.*, 2013). However, this work showed that impairment of aerobic metabolism leads to *CDR1* and *CDR2* down-regulation and subsequent azole hypersusceptibility. Therefore, the molecular mechanisms linking the up-regulation of ABC encoding genes to respiratory status remain to be identified.

Another example is the application of genome-wide microarrays that led to the discovery of the importance of ADP-ribosylation factor (ARF) cycling in azole tolerance in *C. albicans*. In this yeast species, azole drugs are only fungistatic and not fungicidal, a phenomenon called tolerance. Reverse genetics, systematically testing the susceptibility of the complete *S. cerevisiae* mutant collection, identified 22 genes the deletion of which lead to fluconazole fungicidal activity. The deletion of the orthologous genes in *C. albicans* identified 12 of them as potentially involved in fluconazole tolerance. Among these genes, *AGE3*, which is involved in ARF cycling, abrogates tolerance to azoles through perturbation of intracellular vesicles trafficking and ergosterol targeting to membranes, maybe in relation to the calcineurin pathway (Epp *et al.*, 2010).

A more anecdotal strategy used for the identification of azole-responsive genes in *C. albicans* is suppression subtractive hybridization (Diatchenko *et al.*, 1999). This technique was applied to identify genes the expression of which is regulated by the transcription factors Efg1 and Cph1 and revealed that these major regulators of *C. albicans* morphology also control the expression of *ERG3*. Therefore the deletion of *EFG1* leads to an increase in azole susceptibility in *C. albicans* (Lo *et al.*, 2005).

Transcriptomic approach was also the source of the discovery of master transcriptional regulators of the expression of genes falling into the category of basal resistance mechanisms when overexpressed. This is the case of the zinc cluster transcription factor Mrr1p, which is the major regulator of the expression of the MFS transporter Mdr1 and discovered as commonly up-regulated in a series of clinical isolates resistant to azole ATFs by *MDR1* overexpression (Morschhäuser *et al.*, 2007).

In *C. albicans*, transcriptomic studies comparing clinical isolates or laboratory mutants resistant to azoles to their respective parental susceptible relatives ended with the same conclusions as transcriptomic identification of genes differentially regulated upon exposure to azole antifungals. Indeed such transcriptomic experiments often re-identified previously well-characterized azole resistance effectors such as ABC and MFS genes, ergosterol biosynthesis pathway genes, *RTA2*, *HSP12*, and more generally all the genes known to contain a drug-responsive element sequence in their promoter (De Backer *et al.*, 2001; Barker, 2004; Coste *et al.*, 2004; Liu *et al.*, 2005; Rogers and Barker, 2002, 2003). Besides, a clear link between the azole resistance and expression level of oxidative stress effectors has been evidenced on several occasions (Karababa *et al.*, 2004; Rogers and Barker, 2003; Xu *et al.*, 2006). Moreover, some of these transcriptomic approaches allowed us to correlate between iron homeostasis and azoles resistance (Rogers and Barker, 2003; Xu *et al.*, 2006).

Several transcriptomic studies were conducted in the budding yeast *S. cerevisiae* after its complete genome and the first commercially released microarrays were made available. Despite the non-pathogenic properties of this species, the molecular mechanisms underlying its response to azole ATFs could give valuable information when transposed to pathogenic species. Obviously, the results observed in *S. cerevisiae* have to be moderated given transcriptional networks rewiring that occurred in pathogenic species, probably as a consequence of their adaptation to host niches. Thus, the number of transcripts differentially regulated upon exposure of *S. cerevisiae* cells to azole ATFs is most of the time drastically lower than in pathogenic fungal species. In 2000, Bammert and Fostel (2000) conducted complete microarray analysis of the genes differentially expressed upon *S. cerevisiae* exposure to six azoles drugs, as well as to two other sterol-targeting ATFs (i.e. terbinafine and amorolfine). To further complete their observations, they also compared the transcriptomes of three strains lacking each a gene of the ergosterol biosynthesis pathway (Bammert and Fostel, 2000). Ergosterol biosynthesis pathway genes and other sterol metabolism-related genes were commonly regulated between the different conditions tested, as confirmed by DeRisi *et al.* (2000) and later by Agarwal *et al.* (2003). Interestingly, this study is one of the first to establish a link between ergosterol biosynthesis impairment, either by azoles drugs or by gene deletion, and the expression of genes involved in oxidative stress and mitochondrial energy production. As another example, in the study of Barker *et al.* (2003) the transcriptome of an azole-resistant isolate obtained after successive passages on itraconazole-containing plates was compared to this of the parent strain and, along with genes known to be involved in azole resistance, identified some putative new effectors of drug efflux. These microarray experiments identified 116 genes differentially regulated between the azole-resistant and azole-susceptible strains. Genes involved in sterol metabolism, stress response and cell wall maintenance were up-regulated while some other genes involved in energy production and ion homeostasis were significantly down-regulated. Interestingly, two efflux genes previously not matched with azole resistance were identified, namely *YHK8* and *AUS1*. Based on previous observations, it is possible that Aus1 is an ABC transporter responsible for both efflux of azole drugs and regulation of sterol transport, similarly to the well-characterized Cdr1 transporter. In contrast to what is observed in azole-resistant *C. albicans* clinical isolates, several sterol metabolism genes were down-regulated in this *S. cerevisiae* mutant. This observation highlights the differences between pathogenic and non-pathogenic species but also demonstrates that naturally occurring and chemically induced resistances can have drastically different consequences on gene expression patterns.

A. fumigatus is one of the most prevalent species responsible for fungal infections, as it is the principal agent of airborne lung infections. Therein the molecular mechanisms responsible for its resistance to azole ATFs have been dissected at the transcriptomic level by microarrays. The main findings deduced from these experiments have recently been reviewed by Alcazar-Fuoli and Mellado (2012). Without entering into details, the study of da Silva Ferreira *et al.* (2006) is particularly informative as it demonstrated in *A. fumigatus* a relationship between the cyclic AMP/protein kinase A (cAMP/PKA) signalling pathway and azole susceptibility as already established in *S. cerevisiae* (Kontoyiannis and Rupp, 2000) and *C. albicans* (Jain *et al.*, 2003).

The incidence of cryptococcal meningitis in HIV patients is of major importance, with an estimated one billion cases per year, accounting for almost 600,000 deaths at the global level (Park *et al.*, 2009). Therefore, molecular mechanisms underlying resistance to azoles were widely investigated in *C. neoformans* and several transcriptomic studies were performed. After short-term exposure to fluconazole, the responsive genes of *C. neoformans* resembled those identified in other pathogenic fungal species, such as ergosterol biosynthesis genes, cell wall maintenance, and stress response (Florio *et al.*, 2011). Other transcriptomic studies in *C. neoformans* showed more precisely the involvement of alternative resistance mechanisms already identified in other species. For example, a crucial role for iron homeostasis in fluconazole susceptibility was demonstrated by RNAseq experiments comparing the genes differentially expressed between a strain lacking a permease of the major iron uptake system (namely the ferroxidase Cfo1) and its parental strain. Therein authors suggested that iron chelation constitutes an interesting target for the development of a combination treatment for cryptococcosis (Kim *et al.*, 2012). Similarly, transcriptomic study of *C. neoformans* mutants lacking major effectors of the Ras and cAMP signalling pathways confirmed the involvement of these pathways in the regulation of ergosterol biosynthetic process (Maeng *et al.*, 2010). Consequently, the deletion of the genes coding these effectors was accompanied with an increased susceptibility to azole and polyene drugs.

Other antifungal classes

Studies analysing the genes differentially regulated in fungi upon exposure to systemic ATFs other than azoles are less frequent. However some studies gave interesting insights into the multiple effectors of the transcriptional response to non-azole ATFs. Thus, in *S. cerevisiae*, Zhang *et al.* (2002a) analysed by microarray the transcriptional response to polyene exposure and their results confirmed the mode of action of polyenes. As such, amphotericin B or nystatin exposure concurrently induced the expression of genes of the ergosterol biosynthesis pathway, and in relation with the plasma membrane leakage, of genes involved in the transport of small molecules and in stress responses. This has been further confirmed in a later study by Agarwal *et al.* who also evidenced a compensatory activation of cell wall integrity pathway (Agarwal *et al.*, 2003). Likewise, a similar induction of ergosterol biosynthesis and stress response genes have been observed in *C. albicans* (Barker, 2004). Despite what was thought about polyenes that they do not need to enter the fungal cell to exert their action, the study of Zhang *et al.* suggested that an increased expression of efflux pumps might contribute to polyenes resistance, highlighting a possible link between the pleiotropic drug resistance network of *S. cerevisiae* and acquisition of polyenes resistance. Moreover, in opposition to some observations in *Candida* spp. (Carrillo-Muñoz *et al.*, 2006) or in *Aspergillus terreus* (Blum *et al.*, 2013), these microarray experiments ruled out a possible contribution of oxidative stress in polyenes mode of action in *S. cerevisiae* since none of the major oxidative stress-responsive genes were induced. This observation appeared to be restricted to non-pathogenic fungal species since transcriptomic analysis of the consequences of exposure to amphotericin B in *C. albicans* (Liu *et al.*, 2005) led to the conclusion that the expression of oxidative stress-responsive genes is regulated. Moreover, the same observations were made by Gautam *et al.* in *A. fumigatus* (Gautam *et al.*, 2008), using a combination of proteomics and transcriptomics, a clear effect of amphotericin B on oxidative stress-responsive genes was shown. This last study also suggested that amphotericin B mode-of-action is dependent on the intracellular vesicle trafficking, which was never observed before.

Despite the decrease in treatments with 5-FC, its usage in developing countries remains frequent. For this reason, transcriptomic experiments analysing the response to 5-FC are of primordial importance, particularly for species such as *C. neoformans* whose incidence in developing countries remains relatively high. After comparing the transcriptome of *C. neoformans* two-components system mutants through microarray experiments, Song *et al.* evidenced the involvement of the transcription factor Mbs1 in response to 5-FC (Song *et al.*, 2012). This transcription factor is activated by the two-component system protein Tco2 through Hog1 signalling pathway, it controls the expression of genes involved in genotoxic and oxidative stresses as well as ergosterol biosynthesis genes which explains why *TCO2* deletion affected *C. neoformans* azoles and polyenes susceptibility. Transcriptomic experiments were also performed in *Trichophyton rubrum*, the species most commonly responsible for onychomycosis (Zhao *et al.*, 2009). In this dermatophyte, nucleotide metabolism-related genes were among the most differentially expressed following 5-FC exposure. This study confirmed the hypothesis that 5-FC induces a drastic perturbation of fungal cell physiology through impairment of protein synthesis. Indeed, the expression of genes involved in signal transduction, protein chaperoning, ion and amino acid transport, and secondary metabolites biosynthesis in *T. rubrum* was affected by this pyrimidine analogue. In *S. cerevisiae* (Agarwal *et al.*, 2003; Zhang *et al.*, 2002b) and *C. albicans* (Liu *et al.*, 2005), transcriptomic experiments conducted following exposure to 5-FC showed an up-regulation in the expression of genes involved in DNA repair and synthesis, and protein synthesis. Unexpectedly, in *S. cerevisiae*, 5-FC exposure reduced the expression of two genes encoding cell wall degrading enzymes responsible for the cell separation after mitosis. This response constituted an adaptive mechanism by which yeast cells increase their tolerance to the presence of 5-FC (Zhang *et al.*, 2002b).

Echinocandins target the enzyme responsible for the biosynthesis of the major cell wall polysaccharides β-glucans. Predictably, genome-wide analyses of the genes differentially regulated upon echinocandins exposure identified those involved in cell wall maintenance in *C. albicans* (Liu *et al.*, 2005), *S. cerevisiae* (Agarwal *et al.*, 2003), and *Aspergillus niger* (Meyer *et al.*, 2007). In the budding yeast, echinocandins triggered expression of genes involved in heat, oxidative, osmotic, and ethanol stresses as well as small molecules transport. Surprisingly, caspofungin exposure down-regulated the expression of genes of the mating response (Agarwal *et al.*, 2003). In *A. niger*, sublethal concentrations of caspofungin induced changes in growth polarity, cytoskeleton organization and induction of related genes (Meyer *et al.*, 2007). As a consequence, counteracting these adaptive mechanisms would constitute an interesting approach to enhance the efficacy of echinocandins.

Proteomics

Proteome analysis methods

The protein content of a cell at a given time reflects cell adaptations to a stimulus more accurately than that of nucleic acids. In fact, the post-transcriptional modifications of mRNA are often misleading but on the other hand mRNA analysis is less tedious than protein expression study. Historically, western blot was the sole method available to study protein expression; and it permitted the analysis of only one or few proteins simultaneously. Moreover, immunodetection of a specific protein necessitated the availability of a monoclonal antibody directed against this protein or to obtain a tagged, functional protein that would be detected through a specific antibody directed against the small peptide tag added. The sequencing of complete genomes together with the enhancement of bioinformatics tools available to identify protein sequences greatly improved the proteome analysis. The major improvement was the development of two-dimensional electrophoresis (2D) coupled with matrix-assisted LASER desorption ionization time-of-flight mass spectrometry (MALDI-TOF/MS) for protein identification and the parallel enrichment of proteins databases (Fig. 6.2) (Wilkins *et al.*, 1996).

Two-dimensional electrophoresis was developed in 1975 by O'Farrell (1975), but was not applied to ATF resistance research until 2000 since complete proteome data were not available before that date. I invite the reader to refer to a review by Hooshdaran *et al.* for a valuable overview of the experimental procedures needed for the analysis of the proteome of *C. albicans* (Hooshdaran *et al.*, 2005). Briefly, after extraction of the specific subpopulation to be studied, proteins are separated depending on their isoelectric point in a first dimension (isoelectric focusing, IEF) due to the application of an electric field on a strip containing a pH gradient. For the second dimension, proteins are separated according to their molecular weight by classical denaturing polyacrylamide gel electrophoresis (SDS-PAGE). After separation and comparison of their relative quantity in each tested condition, proteins of interest are cut out from the polyacrylamide gel and in-gel digested with trypsin. The resulting peptides are extracted from gel slices, desalted and their m/z ratios determined by MALDI-TOF/MS. The comparison of the obtained peptide mass fingerprints with databases allows protein identification.

An improvement of the classical 2D electrophoresis consists of using cyanine-labelled

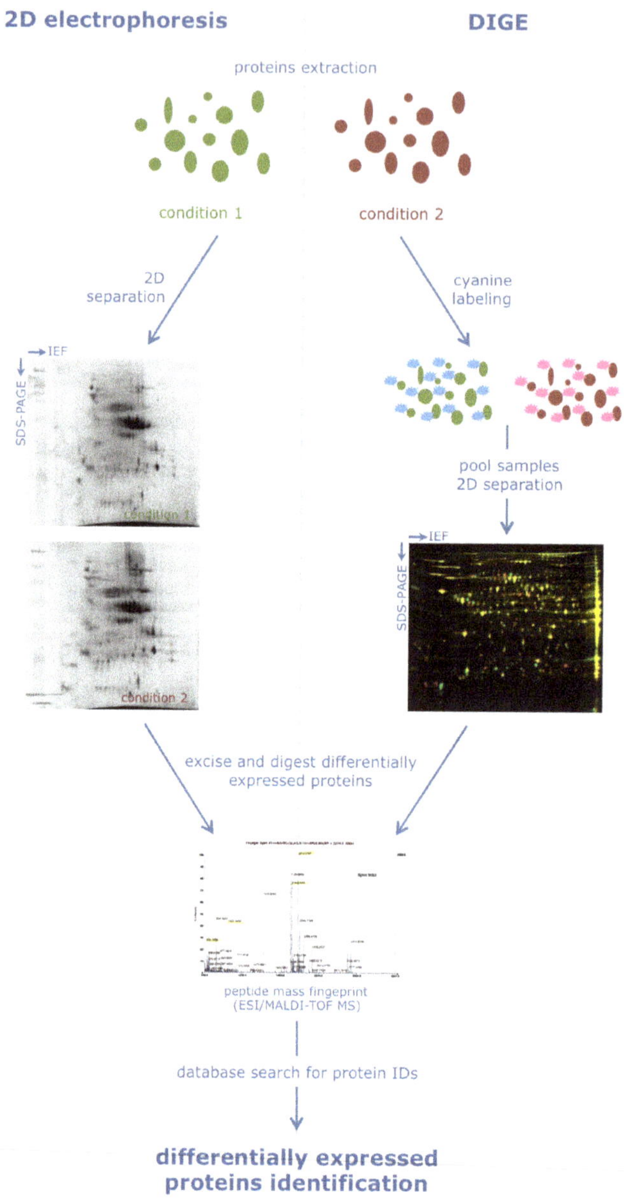

Figure 6.2 Proteome analysis methods. The identification of the proteins differentially synthesized upon antifungal exposure or following antifungal-responsive gene deletion necessitates in a first step to extract the proteins of the desired cell compartment. While soluble proteins could be extracted directly from whole-cell preparations, the extraction of the proteins linked to the different cell constituents needs the fractionation of fungal cell compartments. In classical two-dimensions (2D) electrophoresis, the proteins of the two conditions that have to be compared are separated on individual gels. On the other hand, proteins are differentially labelled with cyanines before being pooled and separated on a unique gel in the case of differential gel electrophoresis (DIGE). This strongly decreases the bias of poor reproducibility between gels. In the first dimension, the isoelectric focusing, the protein extracts are separated according to their isoelectric point by means of electric migration along strips coated with an immobilized pH gradient. In the second dimension, the proteins are separated according to their molecular weight by classical denaturing polyacrylamide gel electrophoresis. After data acquisition (i.e. visible or fluorescence scanning) and quantification of relative protein abundances, proteins differentially expressed are excised and in-gel digested with trypsin. Then peptides are extracted from gel, desalted, and their masses are determined by MALDI-TOF/MS. At the end, the proteins are identified by comparison of peptide mass fingerprints with whole-proteome databases.

proteins. This technology, namely the differential gel electrophoresis (DIGE), was developed in the late 1990s (Unlü *et al.*, 1997). It allows the comparison of up to three protein populations in a single 2D gel (Fig. 6.2). The main advantage of DIGE is that the simultaneous migration of samples to be compared overcomes the main classical 2D electrophoresis limitation that is inter-gel variability. Proteins from the conditions to be compared are differentially labelled with cyanines and then processed through IEF and SDS-PAGE as in classical 2D electrophoresis. After separation, the gel is scanned at the respective emission wavelengths of the cyanines and superimposed images processing allows the identification of differentially expressed proteins to be identified by MALDI-TOF/MS.

Contribution of proteomics to discovery of alternative resistance mechanisms to ATF

As in transcriptomics, efforts in the field of alternative resistance mechanisms discovery by proteomics have focused on the most clinically relevant fungal species, *C. albicans* whose genome was already available at the time. Ene *et al.* (2012) have studied the effects of carbon source on the secretome and the cell wall proteome of *C. albicans* and they observed that when lactate was added as a sole carbon source to *C. albicans* cultures susceptibility to caspofungin and amphotericin B, unlike susceptibility to miconazole, decreased. In an effort to identify the cell wall and secreted proteins differentially expressed responsible for this phenotype, observations similar to those in transcriptomic approaches were made. For example, when grown on lactate, the increase in cell wall integrity-related proteins seemed to be responsible for the decreased susceptibility to caspofungin. Additional data were also provided by the study of Sorgo *et al.* in which the effect of fluconazole on the secretome and cell wall proteome of *C. albicans* was studied. Indeed, this last study showed a clear effect of fluconazole not only on plasma membrane, as suggested by fluconazole mode of action, but also on the cell wall since major cell wall integrity-related proteins were differentially expressed (Sorgo *et al.*, 2011). The influence of carbon source on *C. albicans* ATF susceptibility was furthermore comforted by the study of Yan *et al.* (2007) who suggested that fluconazole induced a metabolic shift. In this study, whole proteome analysis essentially identified proteins involved in energy production. The analysis of proteins differentially expressed after experimental induction of fluconazole resistance suggested that *C. albicans* metabolism is greatly impacted. The same conclusions were made in azole-resistant clinical isolates (Hooshdaran *et al.*, 2004; Kusch *et al.*, 2004) as well as in azole-exposed susceptible isolates (Bruneau *et al.*, 2003; Hoehamer *et al.*, 2010). The technical limitations responsible for the relatively low number of proteins identified compared to the number of genes differentially expressed are the main reason why high discrepancies between transcriptomic and proteomic studies exist. Extraction of insoluble, cell wall, or membrane proteins is particularly fastidious and may explain why proteomics fail to identify the major effectors of azole resistance (i.e. ABC and MFS transporters). However, similarly to transcriptomics, proteomics highlighted the induction of an oxidative stress by azole ATFs (Hoehamer *et al.*, 2010). Besides, proteomics presented evidence for the induction of some ergosterol biosynthesis genes, and particularly *ERG10*, *ERG6*, and *ERG13* (Hoehamer *et al.*, 2009, 2010; Hooshdaran *et al.*, 2004; Yan *et al.*, 2007). Similar results were obtained with *C. glabrata*, the second most prevalent *Candida* species, where proteins of the ergosterol biosynthesis pathway, hydrocarbons metabolism, and oxidative stress were differentially expressed between a strain in which azole resistance was experimentally induced and its parental strain (Rogers *et al.*, 2006). Recently, an extensive, time-course study of protein expression changes driven by azole exposure has been performed in the second most important agent of cryptococcosis, *C. gattii* (Chong *et al.*, 2012). Proteins involved in energy production were identified as responsive to azole in this species. The specific enzymes detected (e.g. subunits of mitochondrial ATP synthase) suggested not only that azole drugs induce a rewiring of fungal metabolism, but also that cellular machinery is mobilized upon azole exposure to maintain a level of ATP production sufficient for an adequate increase in ATP-dependent efflux of the ATF. Analysis of this ubiquitous mechanism in *S. cerevisiae* Pdr1/Pdr3 mutants proteomes revealed, among others, the overexpression of proteins involved in mitochondrial metabolism (Nawrocki *et al.*, 2001).

Proteome changes induced by other classes of ATFs have been less extensively studied. However, the consequences of echinocandin treatment on protein expression have been studied in *C. albicans* (Bruneau *et al.*, 2003; Hoehamer *et al.*, 2010) and the filamentous fungus *A. fumigatus* (Cagas *et al.*, 2011). In fungal pathogens, echinocandins, through inhibition of the synthesis of β-glucans, induce a remodelling of the cell wall together with a redistribution of cell wall glucans and especially an increase in chitin content (Cota *et al.*, 2008; Dichtl *et al.*, 2012; Jiménez-Ortigosa *et al.*, 2012; Lee *et al.*, 2012; Moreno *et al.*, 2010; Vavala *et al.*, 2013; Walker *et al.*, 2013). This cell wall remodelling upon echinocandins exposure, as shown by transcriptomics, has been further confirmed by proteomics. Likewise, amphotericin B exposure in both *C. albicans* (Hoehamer *et al.*, 2010) and *A. fumigatus* (Cagas *et al.*, 2011) induced changes in the expression of proteins involved in the oxidative and osmotic stresses, and ergosterol biosynthesis pathway, as polyene mode of action came to light.

Lipidomics

Azoles and polyenes target directly ergosterol, which is the main sterol constituent of fungal plasma membrane, or target its biosynthesis. Thus, the study of the lipid composition changes induced after an exposure to these sterol inhibitors could give valuable information regarding the adaptive mechanisms settle by fungal cells to cope with the presence of these drugs. Moreover, it is well characterized that the lipid composition of fungal plasma membrane, and particularly the presence of lipid rafts comprising ergosterol and sphingolipids, highly influences the targeting of the most important effectors of azole resistance, the ABC efflux pumps (Kohli *et al.*, 2002; Mukhopadhyay *et al.*, 2002, 2004; Pasrija *et al.*, 2005). Historically, the consequences of the ergosterol depletion were studied on a limited number of lipid molecules, but the improvement of analytical techniques and the increasing knowledge of lipid biochemistry in the main fungal pathogens opened up new perspectives to the study of subtle changes into whole lipid composition of fungal cells exposed to ergosterol biosynthesis inhibitors or in strains resistant to these ATFs. The idea behind this recent alternative approach, namely lipidomics, is the discovery of adaptive mechanisms that allow fungal cells to counteract the mode of action of these widely used ATFs. Ultimately, these adaptive mechanisms would be targeted to enhance the efficacy of ergosterol biosynthesis inhibitors. Lipidomics was developed by Prasad and his coworkers and was essentially applied to understand the lipid changes associated with azole resistance in *C. albicans* (Singh and Prasad, 2011). It is based on the high-throughput analysis of the lipid constituents by electron spray ionization tandem mass spectrometry against standard lipid molecules. In azole-resistant clinical isolates, a dramatic remodelling of lipid composition was observed. Interestingly, the changes in lipid composition differed following the primary resistance mechanism settled by clinical isolates, i.e. MFS or ABC increased expression. This suggested that azole resistance, through overexpression of efflux pumps, can influence the lipid composition of the plasma membrane, which in turn influences the ATF transport and translocation (Shahi and Moye-Rowley, 2009; Singh and Prasad, 2011). Moreover, azole resistance-associated lipid changes pointed out a link between mitochondria and cell wall integrity. The rationale for this observation would be that azole-induced lipid changes of mitochondrial membrane, through a yet unidentified signalling network, triggers multidrug resistance and cell wall modifications (Singh *et al.*, 2012). This constitutes an attractive explanation for the relationship between mitochondria and both azole resistance observed for example in respiratory-deficient mutants and cell wall integrity (Shingu-Vazquez and Traven, 2011).

Interestingly, lipidomics revealed some differences between *in vivo* (i.e. in patients treated with azoles) and *in vitro* acquisition of resistance to fluconazole. Neither changes in cell wall composition nor in mitochondrial activity were observed *in vitro*, suggesting that azole drugs trigger substantially different secondary mechanisms of resistance depending on the environment (Singh *et al.*, 2013). Moreover, lipidomics can possibly lead to the discovery of alternative resistance mechanisms in non-ergosterol biosynthesis inhibitors ATFs; take the example of *C. glabrata* in which the deletion of sphingolipids biosynthesis genes alters susceptibility to echinocandins through a mechanism that remains to be identified (Healey *et al.*, 2012).

Genomics

In the last 10 years, next generation sequencing technologies profoundly changed the study of whole genomes, mostly in terms of time and cost to achieve the sequencing of a complete genome (Sanmiguel, 2011). Despite that, a relatively low number of studies have applied this cutting edge technology to the analysis of molecular mechanisms of resistance to ATFs. Whole genome sequencing, performed after artificial stepwise acquisition of resistance to azole and Hsp90 or calcineurin inhibitor, identified the molecular mechanisms that allow the transition from calcineurin dependent to calcineurin independent azole resistance in *C. albicans* and *S. cerevisiae* (Hill *et al.*, 2013). Four of the 14 lineages that acquired cross-resistance to fluconazole and calcineurin inhibitor selected aneuploidies as resistance mechanisms. Besides, some other lineages with cross-resistance acquired mutations in sphingolipids biosynthesis, a mechanism already highlighted in lipidomics. In *C. glabrata*, genome-wide sequencing identified the compensatory mechanisms that take place in echinocandin-resistant isolates to counteract fitness costs (Singh-Babak *et al.*, 2012). Even though the *petite* phenotype had no effect on echinocandin susceptibility, it was selected as it precluded the synergistic effect between echinocandin and a calcineurin inhibitor seen in *C. albicans* (Singh *et al.*, 2009).

The recent implement of omics tools has allowed the expansion of knowledge on fungal biology and the discovery of major alternative resistance mechanisms to ATFs. Moreover, these large-scale investigation methods increased drastically the knowledge on fungal genomes composition and organization. Together with the enhancement of genetic tools, these genomic data allowed the creation of genome-wide mutant collections that constitute another powerful tool for the study of alternative resistance mechanisms to ATFs.

Mutants collections screening

Mutants production in fungi

Targeted gene deletion

Fungi and bacteria were the first organisms to be genetically modified due to the feasibility of the existing transformation methods. The fungal model organism *S. cerevisiae* was the first fungal microorganism genetically modified in 1978 by spheroplasts transformation (Hinnen *et al.*, 1978). Since then, other transformation procedures have been designed and improved. The spheroplasts transformation remains the method of choice for the transformation of filamentous fungi whereas lithium acetate/polyethylene glycol procedure remains the gold standard for yeasts. Electroporation is also frequently used for the transformation of both yeasts and filamentous fungi (Kawai *et al.*, 2010). Likewise, *Agrobacterium tumefaciens*-mediated transformation is sometime the most efficient transformation method in filamentous fungi, depending on the species (de Groot *et al.*, 1998)

The ability to perform gene replacement in fungi relies on homologous recombination (Fig. 6.3). This capacity of eukaryotic cells to exchange DNA at a specific locus depends on a finely regulated nuclear enzymatic complex, and the efficiency of this process is species dependent. When a couple of hundred of base pair homology are sufficient to mediate a targeted gene replacement in most yeast species, up to one thousand base pairs are necessary to ensure correct integration in some filamentous fungi species. Likewise, filamentous fungi possess a very efficient system of non-specific DNA repair machinery called non-homologous end joining (NHEJ), which drastically decrease targeted gene replacement efficiency. Depending on the species, the deletion of one essential gene of the NHEJ DNA repair machinery (namely *KU70* or *KU80*) is a prerequisite to obtain mutants at a high frequency (Kück and Hoff, 2010). To circumvent this issue, episomal plasmids are used to introduce extra

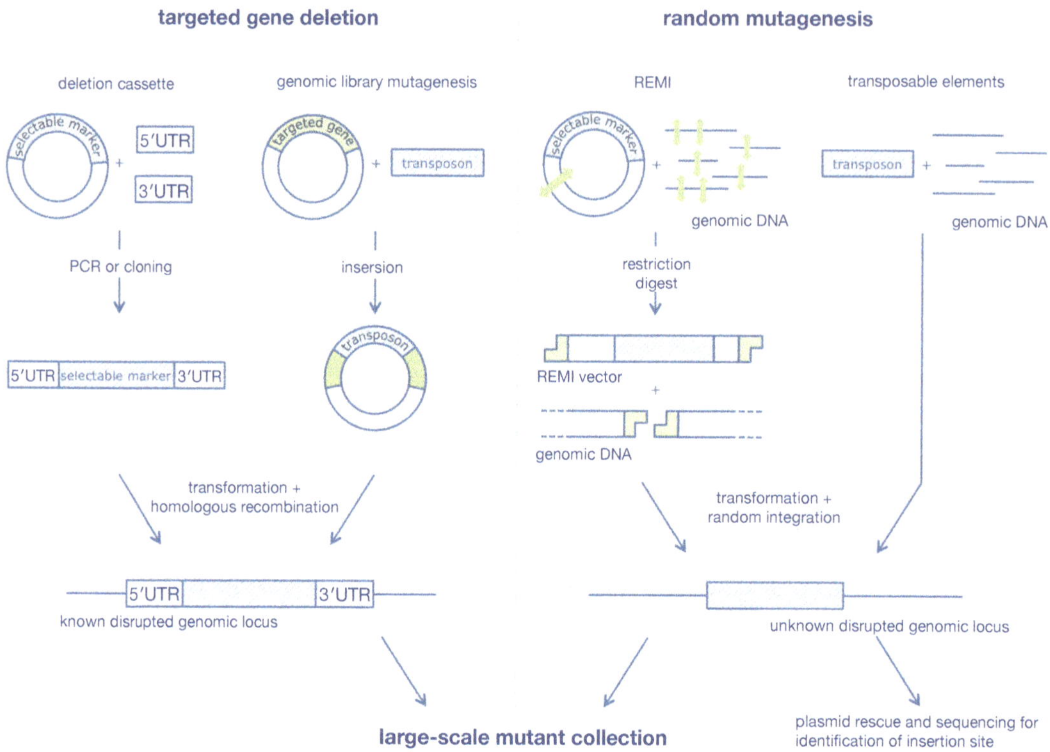

Figure 6.3 Large-scale mutants production in fungi. Mutant collections can be produced either by targeted deletion of all the ORFs identified within a genome or by random mutagenesis. Targeted strategy necessitates the construction of deletion cassettes containing a selectable marker (i.e. a gene conferring resistance to an antibiotic or restoring an auxotrophy) flanked by sequences identical to the regions surrounding the targeted locus (5′ and 3′ UTRs). Once the deletion cassette introduced into the recipient strain, these sequences allow the locus replacement by homologous recombination. Targeted deletion cassettes can also be constructed by transposon mutagenesis of a genomic library, where a naturally transposable element, modified to contain a marker that will be selectable in the recipient strain, is randomly introduced in a plasmid-borne genomic library. On the other side, random mutagenesis can be produced either by restriction enzyme-mediated integration (REMI) or by the use of transposable elements. The REMI strategy uses the fungal non-homologous end joining DNA repair machinery to introduce a selectable marker at non-specific sites. During the transformation of the recipient strain some units of the restriction enzyme used to linearize the REMI vector are added and thus favour its integration through the formation of breaks in genomic DNA. Fungi-adapted transposable elements can also be introduced directly into a recipient strain. If mutant collections produced by random mutagenesis remain valuable in fungal species for which genomic data are not available, this approach has the main disadvantage to necessitate a large number of mutants to ensure a good representativeness of each gene and, once candidate mutants selected, to perform additional identification of the disrupted gene.

copies, or modified copies, of the desired genes. However, the use of episomal plasmids does not allow the replacement of genes within the genome (Christianson *et al.*, 1992).

Integration of exogenous DNA into the genome necessitates the introduction of a selectable marker along with the DNA fragment of interest. This selectable marker belongs to one of two categories, positive selection markers or auxotrophic markers. Positive transformation markers often correspond to open reading frames placed behind a constitutive promoter and coding a protein conferring resistance to an antifungal. Among these positive transformation markers are *NAT1* which encodes the nourseothricin acetyl transferase (Krügel *et al.*, 1993), *SH BLE* which confers resistance to phleomycin (Benko and Zhao, 2011) and *HPH* which confers resistance to hygromycin B (Kaster *et al.*, 1984). The second category of transformation marker is auxotrophic markers that consist of genes restoring a biosynthesis deficiency, more commonly in amino acid biosynthesis pathways. The most important disadvantage of these markers is that they require working on genetically modified strains, in which a biosynthesis pathway have been blocked through the deletion of one essential gene, e.g. *LEU2*, *HIS1*, *ARG4*, or *URA3*. The study of wild-type strains, and particularly of clinical isolates, is not possible with this type of markers.

The dissection of alternative resistance mechanisms to ATF often relies on the phenotypic characterization of mutants for genes that have been experimentally shown to be involved directly or indirectly in ATF response. Therefore, gene targeted deletion is a very powerful tool for the discovery of alternative resistance mechanisms. Moreover, in a limited number of species, some large-scale mutant collections are available. The most important collection of such mutants is available for the fungal model *S. cerevisiae* in which the systematic replacement by a kanamycin resistance marker of all ORFs was undertaken (Giaever *et al.*, 2002). However, the deletion of the entire set of predicted ORFs of a genome is a long-term process. Therefore mutant collections represent only portions of the gene content of a microorganism, or comprise a subclass of genes with specialized function such as the transcription factor mutant collection available in *C. albicans*.

With the advent of microarrays and high throughput sequencing, a new strategy for the study of genome-wide drug–gene interactions, so-called chemogenomics, has emerged (Ericson *et al.*, 2010; Pierce *et al.*, 2007; Smith *et al.*, 2012). This approach relies on the screening of heterozygous (Giaever *et al.*, 2004; Oh *et al.*, 2010; Roemer *et al.*, 2003; Xu *et al.*, 2007), homozygous (Chen *et al.*, 2012; Noble *et al.*, 2010; Parsons *et al.*, 2006), or overexpression (Butcher and Schreiber, 2006; Cabral *et al.*, 2012; Douglas *et al.*, 2012; Ho *et al.*, 2009; Prelich, 2012) mutant collections for decreased fitness in the presence of a chemical compound. The introduction of a unique barcode during the deletion process allows the identification of the candidate mutants through high throughput sequencing (Smith *et al.*, 2012) or specific microarrays (Xu *et al.*, 2009). This large-scale approach has been applied in *S. cerevisiae*, *C. albicans* (Oh *et al.*, 2010; Roemer *et al.*, 2003; Xu *et al.*, 2007), *Schizosaccharomyces pombe* (Kim *et al.*, 2010) and *C. glabrata* (Rai *et al.*, 2012). Some chemogenomic studies analysed the effect of ATF classes with known mode-of-action in order to validate the screening strategy but investigations were not extended to analyse putative alternative resistance mechanisms. Likewise, chemogenomics was mostly used to study the mode-of-action of new compounds with putative ATF activity and this is further detailed in Chapters 7 and 9.

Random mutagenesis

The production of random mutants collection allows the identification of genes involved in a particular process. However, this strategy necessitates a high number of mutants to be screened to assume a correct representativeness of each gene, which could rapidly generate a huge number of mutants and therefore complicate the screening of such collections. Therefore, most of mutant collections produced by random mutagenesis represent only a portion of the genome. The easiest way and therefore the most used strategy to produce random mutations within a genome is the use of exogenous transposable elements (Fig. 6.3), such as the *Escherichia coli* transposons Tn3, Tn5, and Tn7, the *Drosophila melanogaster* mariner transposon, or the impala transposon of *Fusarium oxysporum* (Biery *et al.*, 2000; Castano *et al.*, 2003; Firon *et al.*, 2003; Gueiros-Filho and Beverley, 1997; Kumar and Snyder, 2001; Kumar *et al.*, 2004; Ross-Macdonald *et al.*, 1999; Seifert *et al.*, 1986; Xu *et al.*, 2011).

Lastly, the restriction enzyme mediated integration (REMI) strategy (Fig. 6.3), in which a selectable marker-bearing plasmid integrates the genome by illegitimate recombination, is a very efficient method to produce random mutants collection, especially in filamentous fungi (Brown *et al.*, 1998; Dmytruk *et al.*, 2006; Jianping *et al.*, 2012; Schiestl and Petes, 1991; Schroder *et al.*, 2007).

An interesting alternative strategy for the identification of new effector genes is the so-called 'variomics'. Huang and coworkers recently designed this smart approach in the model yeast

S. cerevisiae (Huang *et al.*, 2013). Briefly, libraries of plasmids carrying allelic variants of all *S. cerevisiae* genes were produced and transformed into the corresponding heterozygous diploid mutant. After haploid conversion, a pool of yeast cells, each containing an allelic variant of the targeted gene, was tested for the desired phenotype. Demonstrating the reliability of the variomics approach, a functional screening of 5847 libraries, each corresponding to one single *S. cerevisiae* gene and containing at least 2×10^5 allelic variants of this gene, identified ergosterol biosynthesis genes to be responsible for amphotericin B resistance. More interestingly, the variomics tool also identified some genes that were never associated to amphotericin B resistance, such as *PMP3*, a gene of unknown function the product of which is be able to bind directly to amphotericin B and prevent yeast membrane permeation.

Large-scale mutant collections screening

Genome-wide mutant collections

The genome of the fission yeast *S. pombe* was released in 2002 (Wood *et al.*, 2002). An almost complete set of deletion strains was first achieved in 2010 (Kim *et al.*, 2010) and screened for the identification of genes involved in clotrimazole and micafungin susceptibility (Fang *et al.*, 2012). This very informative screening highlighted some similarity with other fungal species where the involvement of ergosterol biosynthesis, the mitochondrial metabolism, and the cell wall integrity was brought up again.

In *S. cerevisiae*, four different complete sets of deletion strains are now publicly available. The haploid deletion set screened for altered susceptibility to caspofungin unsurprisingly identified genes of the wall integrity pathway. More interestingly, the deletion of genes involved in response to osmotic stress or in the biosynthesis of chitin also lead to an altered susceptibility to caspofungin (Markovich *et al.*, 2004). The deletion mutant set was also screened for genes responsible for the low susceptibility of fungal biofilms to azole ATFs (see Chapter 4). Some mechanisms identified through this screening, such as the mitochondrial respiratory chain, were not specific to biofilms. On the other hand, genes involved in peroxisome organization were recognized as responsible not only for biofilms formation but also for their low susceptibility to azoles (Vandenbosch *et al.*, 2013). Hence, this method constitutes a promising approach for the development of ATF strategies especially directed against the difficult-to-treat biofilm-associated fungal infections.

Despite the great promise chemogenomic offers for the discovery of alternative resistance mechanisms to systemic ATFs, its use for this purpose was relatively meagre. In *S. cerevisiae*, Lum *et al.* (2004) screened the heterozygous deletion mutant set for haploinsufficiency to two azole drugs and amphotericin B and Giaever *et al.* (2004) tested the same mutant collection with 5-FC, amphotericin B and clotrimazole but none of these two studies extended their investigations to genes unknown to participate in the mode of action of these ATFs. The same can be said for *C. albicans*, where Xu *et al.* (2007) and Smith *et al.* (2011) performed genome-wide chemical screening of azoles, echinocandins, amphotericin B and 5-FC among a large number of compounds, but restricted their analysis on well-characterized effector genes.

One major drawback of the conditional overexpression mutant collection is the relative difficulty in reading out results given that the expected phenotype is an increased growth capacity of a strain overexpressing a relevant gene. In *S. cerevisiae*, an overexpression strain collection was screened for altered susceptibility to caspofungin and successfully correlated the Golgi protein Sbe2 to echinocandin susceptibility (Osherov *et al.*, 2002). In fact, this protein is involved in the transport of cell wall components and the hypersusceptibility of the *SBE2* mutant to caspofungin highlighted the importance of cell wall assembly in the echinocandins' mode of action. Thus, chemogenomics using overexpression mutant collections is complementary to the screening of deletion mutant collections.

Specialized mutant collections

As mentioned earlier, ATF stress mobilizes the fungal cell machinery to produce an adapted response. This response is essential for the fungus to cope with the presence of the drug and involves alternative resistance mechanisms defined above as the controlled expression (induction or repression) of resistance effectors. The fine-tuning of ATF resistance therefore necessitates the integration by transcription factors (TFs) of signals generated following ATF exposure. Thus, the study of TFs and of their transcriptional networks constitutes a promising source for the discovery of alternative resistance mechanisms to ATFs. Moreover, the response pathways involved in these alternative resistance mechanisms constitute the perfect target for the design of new therapeutic strategies. Systematic deletion of the genes encoding signal transducers has been conducted in *C. albicans*. Two collections of TF mutants were created independently. A partial collection, containing 143 deletion strains, was screened for about 40 phenotypes relevant to *C. albicans* pathogenicity, including susceptibility to fluconazole and 5-FC (Homann *et al.*, 2009). The second collection was almost exhaustive and contained 241 homozygous deletion strains for 278 proteins containing a DNA-binding domain identified in *C. albicans*. This collection was selectively screened for altered susceptibility to the four classes of systemic ATFs (Vandeputte *et al.*, 2012). Some of the TFs identified through the screening of these two mutant sets were already characterized as involved in response to ATFs. More interestingly, the two studies identified TFs not previously related to ATF resistance as involved in general stress response or specifically activated by ATFs. This is the case for example of TF Rca1, a b-Zip DNA-binding domain-containing protein, capable of modulating the *C. albicans* susceptibility to fluconazole and caspofungin in a carbon dioxide-dependent manner through its control of the carbonic anhydrase expression and the cAMP/PKA signalling pathway (Vandeputte *et al.*, 2012). This study further reinforces previous observations, linking cAMP and/or respiratory and mitochondrial statuses, to ATF susceptibility in several fungal species including *S. cerevisiae*, *C. albicans*, *C. glabrata* and *C. neoformans* (Brun *et al.*, 2003; Cheng *et al.*, 2007; Dagley *et al.*, 2011; Fu *et al.*, 2012; Geraghty and Kavanagh, 2003; Kontoyiannis, 2000; Peng *et al.*, 2012; Rogers *et al.*, 2004; Sarinová *et al.*, 2007; Shingu-Vazquez and Traven, 2011; Vandeputte *et al.*, 2009; Yang *et al.*, 2012; Zhang and Moye-Rowley, 2001).

The zinc cluster family TFs, containing the fungal-specific Zn2Cys6 DNA binding domain, control the expression of genes playing key roles in virulence-associated traits (Vandeputte *et al.*, 2011) and ATF resistance development (Le Crom *et al.*, 2002; Hikkel *et al.*, 2003). Gain-of-function mutations in three of these Zn2Cys6 TFs (*TAC1*/*PDR1*, *MRR1* and *UPC2*) result in constitutive overexpression of azole responsive-genes and are frequently responsible for azole resistance. To get new insights into the involvement of this type of TFs in *C. albicans* susceptibility to azoles, a collection of strains expressing a hyperactive form of all the identified Zn2Cys6 TFs was created through fusion with the Gal4 activation domain and screened for fluconazole resistance (Schillig and Morschhäuser, 2013). As expected, this screening identified Tac1, Mrr1 and Upc2 as the TFs producing the strongest azole resistance in their hyperactive form. More interestingly, four TFs never previously linked to azole resistance (i.e. Cta4, Stb5, Mrr2 and Znc1) were also identified. Microarray analysis of the genes regulated by these TFs demonstrated a significant overlap between their regulatory networks and those of the TFs known to be involved in azole resistance. Furthermore, this work pointed out that the expression of the ABC efflux pump *CDR1* is under the control of Mrr2 and that Tac1-dependent activation of *CDR1* overexpression is partially dependent on Mrr2 activity. This observation may explain why *TAC1* deletion is not sufficient to mediate hypersusceptibility to azole drugs, as frequently observed (Coste *et al.*, 2008).

Random mutant collections

The *C. glabrata* master regulator of multidrug resistance Pdr1 transcription factor was identified through its sequence homology with the *S. cerevisiae* closest orthologue (Vermitsky *et al.*, 2006) and concomitantly through the creation of an insertion mutants library with a Tn5-derived transposon that was screened for increased susceptibility to azole drugs (Tsai *et al.*, 2006). The screening of this mutant library also identified *ERG1* among the genes the deletion of which increase susceptibility of *C. glabrata* to azoles

(Tsai *et al.*, 2004). The deletion of genes involved in mitochondrial, calcium and protein kinase C signalling was also associated to *C. glabrata* increased susceptibility to azoles through the screening of a second random mutants library, created with a Tn7-based transposon (Borah *et al.*, 2011; Kaur *et al.*, 2004).

By analogy, a set of random mutants in *A. fumigatus*, created by REMI, was also screened for altered susceptibility to azole ATFs (Bowyer *et al.*, 2012). Thus, six genes not previously linked to azoles response were identified, i.e. novel efflux transporters, transcriptional regulator of the expression of genes of the ergosterol biosynthesis pathway and genes involved in mitochondrion metabolism.

In *C. albicans* a set of random mutants containing a relatively low number of strains was constructed on the basis of a Tn7-UAU1 transposon (Davis *et al.*, 2002) and screened for altered susceptibility to fluconazole. The protein kinase CK2 was identified as a mediator of azole response through its control of *CDR1* and *CDR2* expression level in a calcineurin/Crz1-dependent manner (Bruno and Mitchell, 2005).

In *S. cerevisiae*, a mutant collection obtained by random mutagenesis following exposure to ethane methyl sulfonate was screened for altered susceptibility to the two main polyenic systemic ATFs, amphotericin B and nystatin (Hapala *et al.*, 2005). Even if the interrupted gene could not be identified through chemical mutagenesis, two mutants with different profiles of susceptibility to polyenes were obtained. Their further phenotypic characterization suggested that amphotericin B and not nystatin, despite their closely related chemical structures, have different mode of action. Indeed, in these mutants, only amphotericin B mediated a cell wall stress.

Comparison with others species

Genomic libraries screening

Decisive findings in the field of ATF resistance mechanisms are often inferred by analogy with mechanisms discovered in other fungal species and even in non-fungal organisms. Historically, the absence of genomic data in the most important fungal pathogens forces researchers to screen genomic libraries. This necessitates possessing a mutant strain in which a gene of interest, known to be involved in ATF resistance, is deleted. A genomic library, representative of the genome of the organism to be studied, is transformed into this recipient strain in search for phenotypic complementation (Fig. 6.3). Once the clones are obtained, the genes responsible for the phenotype reversion are identified. Following this strategy, Talibi *et al.* (Talibi and Raymond, 1999) searched for *C. albicans* genes able to restore the expression of the ABC protein *PDR5* coding gene in a *PDR1*/*PDR3 S. cerevisiae* mutant hypersusceptible to azole drugs. This genomic library screening identified three transcription factors (namely Fcr1, Fcr2 and Fcr3) but unfortunately, in *C. albicans*, none of these genes share the same function; the deletion of one of these transcription factors (i.e. *FCR1*) had an opposite effect given that it increased resistance to azole. The same method was applied later and failed to identify Pdr1/Pdr3 homologues in *C. albicans* since the genes complementing *S. cerevisiae* azole hypersusceptibility displayed divergent functions in *C. albicans* (Coste *et al.*, 2008). Genomic library screening, however, underlined the involvement of a major transcriptional regulator of *C. albicans*, Ndt80, in azole susceptibility. Using a *S. cerevisiae* strain in which the promoter of a major effector gene of azole resistance, *CDR1*, was placed behind the β-galactosidase reporter gene, Chen *et al.* (2004) identified the transcription factor Ndt80 as an alternative pathway for the regulation of azole resistance genes expression in *C. albicans*. The existence of an alternative pathway mediating azole resistance was further confirmed since it was previously demonstrated that the overexpression of azole resistance genes (i.e. *CDR1*, *CDR2*, *MDR1* and *ERG11*) driven by the transcription factors Tac1, Mrr1 and Upc2 is not dependent on Ndt80 (Sasse *et al.*, 2011; Sellam *et al.*, 2009). Another example is the identification of the transcriptional repressor Rep1, which increased the susceptibility of *S. cerevisiae* to fluconazole and controlled together with Mrr1 the expression of the MFS transporter *MDR1* in *C. albicans* (Chen *et al.*, 2009).

Bioinformatic orthologs identification

The increasing number of fully sequenced genomes has greatly facilitated the identification of ATF resistance effectors. While genomic library screening is a fastidious approach, simple bioinformatics analysis looking for gene homologues known to be involved in ATF resistance in other species often leads to major discoveries. The main transcriptional regulators of azole resistance genes were identified through bioinformatic search for homologous genes in *S. cerevisiae*, among them Upc2 (MacPherson *et al.*, 2005; Silver *et al.*, 2004) in *C. albicans*, and Pdr1 in *C. glabrata* (Vermitsky *et al.*, 2006). Alternative effectors genes were also identified by simple genome search for homologues. As such, the *C. glabrata* Stb5 (Noble *et al.*, 2013) and the *C. albicans* Cap1 (Alarco and Raymond, 1999) and Mcm1 (Riggle and Kumamoto, 2006) transcription factors, based on similarity to their *S. cerevisiae* homologues, were identified as regulators of azole resistance. In *C. glabrata*, the ABC transporter Snq2 (Torelli *et al.*, 2008) and the proton antiporter Qdr2 (Costa *et al.*, 2013a), transcriptionally regulated by Pdr1 and mediating azole resistance, were also identified based on their homology with their *S. cerevisiae* homologues. Another example is the characterization of the Aqr1 antiporter involved both in *S. cerevisiae* and *C. glabrata* susceptibility to acetic acid, flucytosine and azole drugs (Costa *et al.*, 2013b). In *C. albicans*, the involvement of the two-components signal transduction system in azole resistance was discovered by analogy with prokaryotes, where β-lactams resistance is mediated by this two-component system (Chauhan *et al.*, 2007). Thus several *trans*-acting elements, controlling the expression of ATF resistance effectors, were identified through their sequence homology to *S. cerevisiae* orthologues. This is the case of Ncb2 protein, a component of the negative cofactor that represses the transcriptional activity of the RNA polymerase II and controls the expression of *CDR1* in *C. albicans* (Shukla *et al.*, 2011). Likewise, a phosphoinositide-dependent kinase 1 orthologue participates in *C. neoformans* resistance to azoles and polyenes although the exact underlying mechanisms are not yet identified (Chabrier-Roselló *et al.*, 2013). More importantly, by analogy with the mammalian nuclear receptor PXR, it was demonstrated that Pdr1, the central transcriptional regulator of multidrug resistance genes, is able to directly bind azole drugs in the nucleus of both *S. cerevisiae* and *C. glabrata* cells (Thakur *et al.*, 2008), opening a new prospects for the development of antifungal strategies.

Concluding remarks

The omics era along with the implementation of large-scale and genome-wide mutant collections have revolutionized the comprehension of fungal biology. Twenty years ago, the availability of genomic data of the main causative agents of fungal infections was very limited and imposed investigating the molecular mechanisms involved in ATF response from a targeted point of view, relying on previous observations. Nowadays, high throughput strategies allow the study of whole organism's responses to a broad range of environmental conditions or chemical compounds.

Genome-wide analyses are definitely the most powerful tools currently available to discover alternative resistance mechanisms that could be targeted to define new antifungal strategies. However, the comprehension of the fungal response to ATF stress remains complicated since these tools tend to dramatically increase the number of known effectors to a given response. Add to this, potential new antifungal molecules should not only minimize side-effects in humans but also should have a broad-range activity against fungal pathogens and limit the possibilities of acquired resistance given the high plasticity of fungal genomes. In conclusion, the release of new ATFs, despite the remarkable improvement of methods, will necessitate the deciphering of a large part of fungal biology that is not decrypted yet.

Acknowledgements

I want to thank the French non-profit organization 'Vaincre la Mucoviscidose' (research grant RF20130500877) as well as the 'Institut de Parasitologie de l'Ouest' for their financial support.

I also would like to thank Sarah Ghamrawi for proofreading the manuscript and apologize to the colleagues whose valuable work is not cited.

References

Agarwal, A.K., Rogers, P.D., Baerson, S.R., Jacob, M.R., Barker, K.S., Cleary, J.D., Walker, L.A., Nagle, D.G., and Clark, A.M. (2003). Genome-wide expression profiling of the response to polyene, pyrimidine, azole, and echinocandin antifungal agents in *Saccharomyces cerevisiae*. J. Biol. Chem. *278*, 34998–35015.

Agarwal, A.K., Tripathi, S.K., Xu, T., Jacob, M.R., Li, X.-C., and Clark, A.M. (2012). Exploring the molecular basis of antifungal synergies using genome-wide approaches. Front. Microbiol. *3*, 115.

Alarco, A.M., and Raymond, M. (1999). The bZip transcription factor Cap1p is involved in multidrug resistance and oxidative stress response in Candida albicans. J. Bacteriol. *181*, 700–708.

Alcazar-Fuoli, L., and Mellado, E. (2012). Ergosterol biosynthesis in *Aspergillus fumigatus*: its relevance as an antifungal target and role in antifungal drug resistance. Front. Microbiol. *3*, 439.

Alwine, J.C., Kemp, D.J., and Stark, G.R. (1977). Method for detection of specific RNAs in agarose gels by transfer to diazobenzyloxymethyl-paper and hybridization with DNA probes. Proc. Natl. Acad. Sci. U.S.A. *74*, 5350–5354.

Bammert, G.F., and Fostel, J.M. (2000). Genome-wide expression patterns in *Saccharomyces cerevisiae*: comparison of drug treatments and genetic alterations affecting biosynthesis of ergosterol. Antimicrob. Agents Chemother. *44*, 1255–1265.

Barker, K.S. (2004). Genome-wide expression profiling reveals genes associated with amphotericin B and fluconazole resistance in experimentally induced antifungal resistant isolates of *Candida albicans*. J. Antimicrob. Chemother. *54*, 376–385.

Barker, K.S., and Rogers, P.D. (2005). Application of deoxyribonucleic acid microarray analysis to the study of azole antifungal resistance in *Candida albicans*. Methods Mol. Med. *118*, 45–56.

Barker, K.S., Pearson, M.M., and Rogers, P.D. (2003). Identification of genes differentially expressed in association with reduced azole susceptibility in *Saccharomyces cerevisiae*. J. Antimicrob. Chemother. *51*, 1131–1140.

Becker-André, M., and Hahlbrock, K. (1989). Absolute mRNA quantification using the polymerase chain reaction (PCR). A novel approach by a PCR aided transcript titration assay (PATTY). Nucleic Acids Res. *17*, 9437–9446.

Benko, Z., and Zhao, R.Y. (2011). Zeocin for selection of bleMX6 resistance in fission yeast. BioTechniques *51*, 57–60.

Biery, M.C., Stewart, F.J., Stellwagen, A.E., Raleigh, E.A., and Craig, N.L. (2000). A simple *in vitro* Tn7-based transposition system with low target site selectivity for genome and gene analysis. Nucleic Acids Res. *28*, 1067–1077.

Blum, G., Hörtnagl, C., Jukic, E., Erbeznik, T., Pümpel, T., Dietrich, H., Nagl, M., Speth, C., Rambach, G., and Lass-Flörl, C. (2013). New insight into amphotericin B resistance in *Aspergillus terreus*. Antimicrob. Agents Chemother. *57*, 1583–1588.

Borah, S., Shivarathri, R., and Kaur, R. (2011). The Rho1 GTPase-activating protein CgBem2 is required for survival of azole stress in *Candida glabrata*. J. Biol. Chem. *286*, 34311–34324.

Bowyer, P., Mosquera, J., Anderson, M., Birch, M., Bromley, M., and Denning, D.W. (2012). Identification of novel genes conferring altered azole susceptibility in *Aspergillus fumigatus*. FEMS Microbiol. Lett. *332*, 10–19.

Brown, J.S., Aufauvre-Brown, A., and Holden, D.W. (1998). Insertional mutagenesis of *Aspergillus fumigatus*. Mol. Gen. Genet. MGG *259*, 327–335.

Brun, S., Aubry, C., Lima, O., Filmon, R., Bergès, T., Chabasse, D., and Bouchara, J.-P. (2003). Relationships between respiration and susceptibility to azole antifungals in *Candida glabrata*. Antimicrob. Agents Chemother. *47*, 847–853.

Bruneau, J.-M., Maillet, I., Tagat, E., Legrand, R., Supatto, F., Fudali, C., Caer, J.-P.L., Labas, V., Lecaque, D., and Hodgson, J. (2003). Drug induced proteome changes in *Candida albicans*: comparison of the effect of beta(1,3) glucan synthase inhibitors and two triazoles, fluconazole and itraconazole. Proteomics *3*, 325–336.

Bruno, V.M., and Mitchell, A.P. (2005). Regulation of azole drug susceptibility by *Candida albicans* protein kinase CK2. Mol. Microbiol. *56*, 559–573.

Butcher, R.A., and Schreiber, S.L. (2006). A microarray-based protocol for monitoring the growth of yeast overexpression strains. Nat. Protoc. *1*, 569–576.

Cabral, V., Chauvel, M., Firon, A., Legrand, M., Nesseir, A., Bachellier-Bassi, S., Chaudhari, Y., Munro, C.A., and d' Enfert, C. (2012). Modular gene over-expression strategies for *Candida albicans*. Methods Mol. Biol. *845*, 227–244.

Cagas, S.E., Jain, M.R., Li, H., and Perlin, D.S. (2011). Profiling the *Aspergillus fumigatus* proteome in response to caspofungin. Antimicrob. Agents Chemother. *55*, 146–154.

Carrillo-Muñoz, A.J., Giusiano, G., Ezkurra, P.A., and Quindós, G. (2006). Antifungal agents: mode of action in yeast cells. Rev. Esp. Quimioter. *19*, 130–139.

Castano, I., Kaur, R., Pan, S., Cregg, R., Penas, A.D.L., Guo, N., Biery, M.C., Craig, N.L., and Cormack, B.P. (2003). Tn7-based genome-wide random insertional mutagenesis of *Candida glabrata*. Genome Res. *13*, 905–915.

Chabrier-Roselló, Y., Gerik, K.J., Koselny, K., DiDone, L., Lodge, J.K., and Krysan, D.J. (2013). *Cryptococcus neoformans* phosphoinositide-dependent kinase 1 (PDK1) ortholog is required for stress tolerance and survival in murine phagocytes. Eukaryot. Cell *12*, 12–22.

Chauhan, N., Kruppa, M., and Calderone, R. (2007). The Ssk1p response regulator and Chk1p histidine kinase mutants of *Candida albicans* are hypersensitive to fluconazole and voriconazole. Antimicrob. Agents Chemother. *51*, 3747–3751.

Chen, B.-R., Hale, D.C., Ciolek, P.J., and Runge, K.W. (2012). Generation and analysis of a barcode-tagged insertion mutant library in the fission yeast *Schizosaccharomyces pombe*. BMC Genomics *13*, 161.

Chen, C.-G., Yang, Y.-L., Shih, H.-I., Su, C.-L., and Lo, H.-J. (2004). CaNdt80 is involved in drug resistance in *Candida albicans* by regulating CDR1. Antimicrob. Agents Chemother. *48*, 4505–4512.

Chen, C.-G., Yang, Y.-L., Tseng, K.-Y., Shih, H.-I., Liou, C.-H., Lin, C.-C., and Lo, H.-J. (2009). Rep1p negatively regulating MDR1 efflux pump involved in drug resistance in *Candida albicans*. Fungal Genet. Biol. *46*, 714–720.

Chen, Y.-L., Brand, A., Morrison, E.L., Silao, F.G.S., Bigol, U.G., Malbas, F.F., Jr, Nett, J.E., Andes, D.R., Solis, N.V., Filler, S.G., *et al.* (2011). Calcineurin controls drug tolerance, hyphal growth, and virulence in *Candida dubliniensis*. Eukaryot. Cell *10*, 803–819.

Cheng, S., Clancy, C.J., Nguyen, K.T., Clapp, W., and Nguyen, M.H. (2007). A *Candida albicans* petite mutant strain with uncoupled oxidative phosphorylation overexpresses MDR1 and has diminished susceptibility to fluconazole and voriconazole. Antimicrob. Agents Chemother. *51*, 1855–1858.

Chiang, P.W., Song, W.J., Wu, K.Y., Korenberg, J.R., Fogel, E.J., Van Keuren, M.L., Lashkari, D., and Kurnit, D.M. (1996). Use of a fluorescent-PCR reaction to detect genomic sequence copy number and transcriptional abundance. Genome Res. *6*, 1013–1026.

Chong, H.S., Campbell, L., Padula, M.P., Hill, C., Harry, E., Li, S.S., Wilkins, M.R., Herbert, B., and Carter, D. (2012). Time-course proteome analysis reveals the dynamic response of *Cryptococcus gattii* cells to fluconazole. PloS One *7*, e42835.

Christianson, T.W., Sikorski, R.S., Dante, M., Shero, J.H., and Hieter, P. (1992). Multifunctional yeast high-copy-number shuttle vectors. Gene *110*, 119–122.

Costa, C., Pires, C., Cabrito, T.R., Renaudin, A., Ohno, M., Chibana, H., Sá-Correia, I., and Teixeira, M.C. (2013a). *Candida glabrata* drug:H+ antiporter CgQdr2 confers imidazole drug resistance, being activated by transcription factor CgPdr1. Antimicrob. Agents Chemother. *57*, 3159–3167.

Costa, C., Henriques, A., Pires, C., Nunes, J., Ohno, M., Chibana, H., Sá-Correia, I., and Teixeira, M.C. (2013b). The dual role of *Candida glabrata* drug:H+ antiporter CgAqr1 (ORF CAGL0J09944g) in antifungal drug and acetic acid resistance. Front. Microbiol. *4*, 170.

Coste, A.T., Karababa, M., Ischer, F., Bille, J., and Sanglard, D. (2004). TAC1, transcriptional activator of CDR genes, is a new transcription factor involved in the regulation of *Candida albicans* ABC transporters CDR1 and CDR2. Eukaryot. Cell *3*, 1639–1652.

Coste, A.T., Ramsdale, M., Ischer, F., and Sanglard, D. (2008). Divergent functions of three *Candida albicans* zinc-cluster transcription factors (CTA4, ASG1 and CTF1) complementing pleiotropic drug resistance in Saccharomyces cerevisiae. Microbiology *154*, 1491–1501.

Cota, J.M., Grabinski, J.L., Talbert, R.L., Burgess, D.S., Rogers, P.D., Edlind, T.D., and Wiederhold, N.P. (2008). Increases in SLT2 expression and chitin content are associated with incomplete killing of *Candida glabrata* by caspofungin. Antimicrob. Agents Chemother. *52*, 1144–1146.

Cowen, L.E. (2009). Hsp90 orchestrates stress response signaling governing fungal drug resistance. PLoS Pathog. *5*, e1000471.

Cowen, L.E., and Lindquist, S. (2005). Hsp90 potentiates the rapid evolution of new traits: drug resistance in diverse fungi. Science *309*, 2185–2189.

Cowen, L.E., and Steinbach, W.J. (2008). Stress, drugs, and evolution: the role of cellular signaling in fungal drug resistance. Eukaryot. Cell *7*, 747–764.

Cowen, L.E., Carpenter, A.E., Matangkasombut, O., Fink, G.R., and Lindquist, S. (2006). Genetic architecture of Hsp90-dependent drug resistance. Eukaryot. Cell *5*, 2184–2188.

Dagley, M.J., Gentle, I.E., Beilharz, T.H., Pettolino, F.A., Djordjevic, J.T., Lo, T.L., Uwamahoro, N., Rupasinghe, T., Tull, D.L., McConville, M., *et al.* (2011). Cell wall integrity is linked to mitochondria and phospholipid homeostasis in *Candida albicans* through the activity of the post-transcriptional regulator Ccr4-Pop2. Mol. Microbiol. *79*, 968–989.

Davis, D.A., Bruno, V.M., Loza, L., Filler, S.G., and Mitchell, A.P. (2002). *Candida albicans* Mds3p, a conserved regulator of pH responses and virulence identified through insertional mutagenesis. Genetics *162*, 1573–1581.

De Backer, M.D., Ilyina, T., Ma, X.J., Vandoninck, S., Luyten, W.H., and Vanden Bossche, H. (2001). Genomic profiling of the response of Candida albicans to itraconazole treatment using a DNA microarray. Antimicrob. Agents Chemother. *45*, 1660–1670.

DeRisi, J., van den Hazel, B., Marc, P., Balzi, E., Brown, P., Jacq, C., and Goffeau, A. (2000). Genome microarray analysis of transcriptional activation in multidrug resistance yeast mutants. FEBS Lett. *470*, 156–160.

Diatchenko, L., Lukyanov, S., Lau, Y.-F.C., and Siebert, P.D. (1999). Suppression subtractive hybridization: a versatile method for identifying differentially expressed genes. Methods Enzymol. *303*, 349–380.

Dichtl, K., Helmschrott, C., Dirr, F., and Wagener, J. (2012). Deciphering cell wall integrity signalling in *Aspergillus fumigatus*: identification and functional characterization of cell wall stress sensors and relevant Rho GTPases. Mol. Microbiol. *83*, 506–519.

Dmytruk, K.V., Voronovsky, A.Y., and Sibirny, A.A. (2006). Insertion mutagenesis of the yeast *Candida famata* (*Debaryomyces hansenii*) by random integration of linear DNA fragments. Curr. Genet. *50*, 183–191.

Douglas, A.C., Smith, A.M., Sharifpoor, S., Yan, Z., Durbic, T., Heisler, L.E., Lee, A.Y., Ryan, O., Göttert, H., Surendra, A., *et al.* (2012). Functional analysis with a barcoder yeast gene overexpression system. G3 (Bethesda) *2*, 1279–1289.

Ene, I.V., Heilmann, C.J., Sorgo, A.G., Walker, L.A., de Koster, C.G., Munro, C.A., Klis, F.M., and Brown, A.J.P. (2012). Carbon source-induced reprogramming of the cell wall proteome and secretome modulates the adherence and drug resistance of the fungal pathogen *Candida albicans*. Proteomics *12*, 3164–3179.

Epp, E., Vanier, G., Harcus, D., Lee, A.Y., Jansen, G., Hallett, M., Sheppard, D.C., Thomas, D.Y., Munro, C.A., Mullick, A., *et al.* (2010). Reverse genetics in *Candida albicans* predicts ARF cycling is essential for drug resistance and virulence. PLoS Pathog. *6*, e1000753.

Ericson, E., Hoon, S., St Onge, R.P., Giaever, G., and Nislow, C. (2010). Exploring gene function and drug action using chemogenomic dosage assays. Methods Enzymol. *470*, 233–255.

Fang, Y., Hu, L., Zhou, X., Jaiseng, W., Zhang, B., Takami, T., and Kuno, T. (2012). A genomewide screen in *Schizosaccharomyces pombe* for genes affecting the sensitivity of antifungal drugs that target ergosterol biosynthesis. Antimicrob. Agents Chemother. *56*, 1949–1959.

Firon, A., Villalba, F., Beffa, R., and D'Enfert, C. (2003). Identification of essential genes in the human fungal pathogen *Aspergillus fumigatus* by transposon mutagenesis. Eukaryot. Cell 2, 247–255.

Florio, A.R., Ferrari, S., De Carolis, E., Torelli, R., Fadda, G., Sanguinetti, M., Sanglard, D., and Posteraro, B. (2011). Genome-wide expression profiling of the response to short-term exposure to fluconazole in *Cryptococcus neoformans* serotype A. BMC Microbiol. *11*, 97.

Fu, M.-S., De Sordi, L., and Mühlschlegel, F.A. (2012). Functional characterization of the small heat shock protein Hsp12p from *Candida albicans*. PloS One 7, e42894.

Gautam, P., Shankar, J., Madan, T., Sirdeshmukh, R., Sundaram, C.S., Gade, W.N., Basir, S.F., and Sarma, P.U. (2008). Proteomic and transcriptomic analysis of *Aspergillus fumigatus* on exposure to amphotericin B. Antimicrob. Agents Chemother. *52*, 4220–4227.

Geiss, G.K., Bumgarner, R.E., Birditt, B., Dahl, T., Dowidar, N., Dunaway, D.L., Fell, H.P., Ferree, S., George, R.D., Grogan, T., *et al.* (2008). Direct multiplexed measurement of gene expression with color-coded probe pairs. Nat. Biotechnol. *26*, 317–325.

Geraghty, P., and Kavanagh, K. (2003). Disruption of mitochondrial function in *Candida albicans* leads to reduced cellular ergosterol levels and elevated growth in the presence of amphotericin B. Arch. Microbiol. *179*, 295–300.

Giaever, G., Chu, A.M., Ni, L., Connelly, C., Riles, L., Véronneau, S., Dow, S., Lucau-Danila, A., Anderson, K., André, B., *et al.* (2002). Functional profiling of the *Saccharomyces cerevisiae* genome. Nature *418*, 387–391.

Giaever, G., Flaherty, P., Kumm, J., Proctor, M., Nislow, C., Jaramillo, D.F., Chu, A.M., Jordan, M.I., Arkin, A.P., and Davis, R.W. (2004). Chemogenomic profiling: identifying the functional interactions of small molecules in yeast. Proc. Natl. Acad. Sci. U.S.A. *101*, 793–798.

de Groot, M.J., Bundock, P., Hooykaas, P.J., and Beijersbergen, A.G. (1998). Agrobacterium tumefaciens-mediated transformation of filamentous fungi. Nat. Biotechnol. *16*, 839–842.

Gueiros-Filho, F.J., and Beverley, S.M. (1997). Transkingdom transposition of the *Drosophila* element mariner within the protozoan *Leishmania*. Science *276*, 1716–1719.

Hapala, I., Klobucníková, V., Mazánová, K., and Kohút, P. (2005). Two mutants selectively resistant to polyenes reveal distinct mechanisms of antifungal activity by nystatin and amphotericin B. Biochem. Soc. Trans. *33*, 1206–1209.

Healey, K.R., Katiyar, S.K., Raj, S., and Edlind, T.D. (2012). CRS-MIS in *Candida glabrata*: sphingolipids modulate echinocandin–Fks interaction. Mol. Microbiol. *86*, 303–313.

Heid, C.A., Stevens, J., Livak, K.J., and Williams, P.M. (1996). Real time quantitative PCR. Genome Res. *6*, 986–994.

Hikkel, I., Lucau-Danila, A., Delaveau, T., Marc, P., Devaux, F., and Jacq, C. (2003). A general strategy to uncover transcription factor properties identifies a new regulator of drug resistance in yeast. J. Biol. Chem. *278*, 11427–11432.

Hill, J.A., Ammar, R., Torti, D., Nislow, C., and Cowen, L.E. (2013). Genetic and genomic architecture of the evolution of resistance to antifungal drug combinations. PLoS Genet. *9*, e1003390.

Hinnen, A., Hicks, J.B., and Fink, G.R. (1978). Transformation of yeast. Proc. Natl. Acad. Sci. U.S.A. *75*, 1929–1933.

Ho, C.H., Magtanong, L., Barker, S.L., Gresham, D., Nishimura, S., Natarajan, P., Koh, J.L.Y., Porter, J., Gray, C.A., Andersen, R.J., *et al.* (2009). A molecular barcoded yeast ORF library enables mode-of-action analysis of bioactive compounds. Nat. Biotechnol. *27*, 369–377.

Hoehamer, C.F., Cummings, E.D., Hilliard, G.M., Morschhäuser, J., and David Rogers, P. (2009). Upc2p-associated differential protein expression in *Candida albicans*. Proteomics 9, 4726–4730.

Hoehamer, C.F., Cummings, E.D., Hilliard, G.M., and Rogers, P.D. (2010). Changes in the proteome of *Candida albicans* in response to azole, polyene, and echinocandin antifungal agents. Antimicrob. Agents Chemother. *54*, 1655–1664.

Homann, O.R., Dea, J., Noble, S.M., and Johnson, A.D. (2009). A phenotypic profile of the *Candida albicans* regulatory network. PLoS Genet. *5*, e1000783.

Hooshdaran, M.Z., Barker, K.S., Hilliard, G.M., Kusch, H., Morschhäuser, J., and Rogers, P.D. (2004). Proteomic analysis of azole resistance in *Candida albicans* clinical isolates. Antimicrob. Agents Chemother. *48*, 2733–2735.

Hooshdaran, M.Z., Hilliard, G.M., and Rogers, P.D. (2005). Application of proteomic analysis to the study of azole antifungal resistance in Candida albicans. Methods Mol. Med. *118*, 57–70.

Huang, Z., Chen, K., Zhang, J., Li, Y., Wang, H., Cui, D., Tang, J., Liu, Y., Shi, X., Li, W., *et al.* (2013). A functional variomics tool for discovering drug-resistance genes and drug targets. Cell Rep. *3*, 577–585.

Isaacson, R.E. (2002). Genomics and the prospects for the discovery of new targets for antibacterial and antifungal agents. Curr. Pharm. Des. *8*, 1091–1098.

Jack, D.B. (1998). The search for new antifungal drugs and strategies continues. Drug News Perspect. *11*, 306–309.

Jain, P., Akula, I., and Edlind, T. (2003). Cyclic AMP signaling pathway modulates susceptibility of *Candida* species and *Saccharomyces cerevisiae* to antifungal azoles and other sterol biosynthesis inhibitors. Antimicrob. Agents Chemother. *47*, 3195–3201.

Jia, X.-M., Ma, Z.-P., Jia, Y., Gao, P.-H., Zhang, J.-D., Wang, Y., Xu, Y.-G., Wang, L., Cao, Y.-Y., Cao, Y.-B., *et al.* (2008). RTA2, a novel gene involved in azole resistance in *Candida albicans*. Biochem. Biophys. Res. Commun. *373*, 631–636.

Jia, X.M., Wang, Y., Jia, Y., Gao, P.H., Xu, Y.G., Wang, L., Cao, Y.Y., Cao, Y.B., Zhang, L.X., and Jiang, Y.Y. (2009). RTA2 is involved in calcineurin-mediated azole resistance and sphingoid long-chain base release in *Candida albicans*. Cell. Mol. Life Sci. *66*, 122–134.

Jia, Y., Tang, R.-J., Wang, L., Zhang, X., Wang, Y., Jia, X.-M., and Jiang, Y.-Y. (2012). Calcium-activated-calcineurin reduces the *in vitro* and *in vivo* sensitivity of fluconazole to *Candida albicans* via Rta2p. PloS One 7, e48369.

Jianping, Z., Guifang, D., Kai, Z., Yongjun, Z., Yongliang, L., and Liuqing, Y. (2012). Screening and identification of insertion mutants from *Bipolaris eleusines* by mutagenesis based on restriction enzyme-mediated integration. FEMS Microbiol. Lett. *330*, 90–97.

Jiménez-Ortigosa, C., Aimanianda, V., Muszkieta, L., Mouyna, I., Alsteens, D., Pire, S., Beau, R., Krappmann, S., Beauvais, A., Dufrêne, Y.F., *et al.* (2012). Chitin synthases with a myosin motor-like domain control the resistance of *Aspergillus fumigatus* to echinocandins. Antimicrob. Agents Chemother. *56*, 6121–6131.

Joseph-Horne, T., and Hollomon, D.W. (1997). Molecular mechanisms of azole resistance in fungi. FEMS Microbiol. Lett. *149*, 141–149.

Karababa, M., Coste, A.T., Rognon, B., Bille, J., and Sanglard, D. (2004). Comparison of gene expression profiles of *Candida albicans* azole-resistant clinical isolates and laboratory strains exposed to drugs inducing multidrug transporters. Antimicrob. Agents Chemother. *48*, 3064–3079.

Kaster, K.R., Burgett, S.G., and Ingolia, T.D. (1984). Hygromycin B resistance as dominant selectable marker in yeast. Curr. Genet. *8*, 353–358.

Kaur, R., Castaño, I., and Cormack, B.P. (2004). Functional genomic analysis of fluconazole susceptibility in the pathogenic yeast *Candida glabrata*: roles of calcium signaling and mitochondria. Antimicrob. Agents Chemother. *48*, 1600–1613.

Kawai, S., Hashimoto, W., and Murata, K. (2010). Transformation of *Saccharomyces cerevisiae* and other fungi: methods and possible underlying mechanism. Bioeng. Bugs *1*, 395–403.

Kim, D.-U., Hayles, J., Kim, D., Wood, V., Park, H.-O., Won, M., Yoo, H.-S., Duhig, T., Nam, M., Palmer, G., *et al.* (2010). Analysis of a genome-wide set of gene deletions in the fission yeast *Schizosaccharomyces pombe*. Nat. Biotechnol. *28*, 617–623.

Kim, J., Cho, Y.-J., Do, E., Choi, J., Hu, G., Cadieux, B., Chun, J., Lee, Y., Kronstad, J.W., and Jung, W.H. (2012). A defect in iron uptake enhances the susceptibility of *Cryptococcus neoformans* to azole antifungal drugs. Fungal Genet. Biol. *49*, 955–966.

Kohli, A., Smriti, Mukhopadhyay, K., Rattan, A., and Prasad, R. (2002). *In vitro* low-level resistance to azoles in *Candida albicans* is associated with changes in membrane lipid fluidity and asymmetry. Antimicrob. Agents Chemother. *46*, 1046–1052.

Kontoyiannis, D.P. (2000). Modulation of fluconazole sensitivity by the interaction of mitochondria and erg3p in *Saccharomyces cerevisiae*. J. Antimicrob. Chemother. *46*, 191–197.

Kontoyiannis, D.P., and Rupp, S. (2000). Cyclic AMP and fluconazole resistance in *Saccharomyces cerevisiae*. Antimicrob. Agents Chemother. *44*, 1743–1744.

Krügel, H., Fiedler, G., Smith, C., and Baumberg, S. (1993). Sequence and transcriptional analysis of the nourseothricin acetyltransferase-encoding gene nat1 from *Streptomyces noursei*. Gene *127*, 127–131.

Kück, U., and Hoff, B. (2010). New tools for the genetic manipulation of filamentous fungi. Appl. Microbiol. Biotechnol. *86*, 51–62.

Kumar, A., and Snyder, M. (2001). Genome-wide transposon mutagenesis in yeast. Curr. Protoc. Mol. Biol. *55* (1:13.3), 13.3.1–13.3.15.

Kumar, A., Seringhaus, M., Biery, M.C., Sarnovsky, R.J., Umansky, L., Piccirillo, S., Heidtman, M., Cheung, K.-H., Dobry, C.J., Gerstein, M.B., *et al.* (2004). Large-scale mutagenesis of the yeast genome using a Tn7-derived multipurpose transposon. Genome Res. *14*, 1975–1986.

Kusch, H., Biswas, K., Schwanfelder, S., Engelmann, S., Rogers, P.D., Hecker, M., and Morschhäuser, J. (2004). A proteomic approach to understanding the development of multidrug-resistant *Candida albicans* strains. Mol. Genet. Genomics *271*, 554–565.

Le Crom, S., Devaux, F., Marc, P., Zhang, X., Moye-Rowley, W.S., and Jacq, C. (2002). New insights into the pleiotropic drug resistance network from genome-wide characterization of the YRR1 transcription factor regulation system. Mol. Cell. Biol. *22*, 2642–2649.

Lee, K.K., Maccallum, D.M., Jacobsen, M.D., Walker, L.A., Odds, F.C., Gow, N.A.R., and Munro, C.A. (2012). Elevated cell wall chitin in *Candida albicans* confers echinocandin resistance *in vivo*. Antimicrob. Agents Chemother. *56*, 208–217.

Liu, L., Li, Y., Li, S., Hu, N., He, Y., Pong, R., Lin, D., Lu, L., and Law, M. (2012). Comparison of next-generation sequencing systems. J. Biomed. Biotechnol. *2012*, 251364.

Liu, T.T., Lee, R.E.B., Barker, K.S., Lee, R.E., Wei, L., Homayouni, R., and Rogers, P.D. (2005). Genome-wide expression profiling of the response to azole, polyene, echinocandin, and pyrimidine antifungal agents in *Candida albicans*. Antimicrob. Agents Chemother. *49*, 2226–2236.

Lo, H.-J., Wang, J.-S., Lin, C.-Y., Chen, C.-G., Hsiao, T.-Y., Hsu, C.-T., Su, C.-L., Fann, M.-J., Ching, Y.-T., and Yang, Y.-L. (2005). Efg1 involved in drug resistance by regulating the expression of ERG3 in *Candida albicans*. Antimicrob. Agents Chemother. *49*, 1213–1215.

Lum, P.Y., Armour, C.D., Stepaniants, S.B., Cavet, G., Wolf, M.K., Butler, J.S., Hinshaw, J.C., Garnier, P., Prestwich,

G.D., Leonardson, A., *et al.* (2004). Discovering modes of action for therapeutic compounds using a genome-wide screen of yeast heterozygotes. Cell *116*, 121–137.

McClellan, A.J., Xia, Y., Deutschbauer, A.M., Davis, R.W., Gerstein, M., and Frydman, J. (2007). Diverse cellular functions of the Hsp90 molecular chaperone uncovered using systems approaches. Cell *131*, 121–135.

MacPherson, S., Akache, B., Weber, S., De Deken, X., Raymond, M., and Turcotte, B. (2005). *Candida albicans* zinc cluster protein Upc2p confers resistance to antifungal drugs and is an activator of ergosterol biosynthetic genes. Antimicrob. Agents Chemother. *49*, 1745–1752.

Maeng, S., Ko, Y.-J., Kim, G.-B., Jung, K.-W., Floyd, A., Heitman, J., and Bahn, Y.-S. (2010). Comparative transcriptome analysis reveals novel roles of the Ras and cyclic AMP signaling pathways in environmental stress response and antifungal drug sensitivity in *Cryptococcus neoformans*. Eukaryot. Cell *9*, 360–378.

Markovich, S., Yekutiel, A., Shalit, I., Shadkchan, Y., and Osherov, N. (2004). Genomic approach to identification of mutations affecting caspofungin susceptibility in *Saccharomyces cerevisiae*. Antimicrob. Agents Chemother. *48*, 3871–3876.

Meyer, V., Damveld, R.A., Arentshorst, M., Stahl, U., van den Hondel, C.A.M.J.J., and Ram, A.F.J. (2007). Survival in the presence of antifungals: genome-wide expression profiling of *Aspergillus niger* in response to sublethal concentrations of caspofungin and fenpropimorph. J. Biol. Chem. *282*, 32935–32948.

Miyazaki, T., Yamauchi, S., Inamine, T., Nagayoshi, Y., Saijo, T., Izumikawa, K., Seki, M., Kakeya, H., Yamamoto, Y., Yanagihara, K., *et al.* (2010). Roles of calcineurin and Crz1 in antifungal susceptibility and virulence of *Candida glabrata*. Antimicrob. Agents Chemother. *54*, 1639–1643.

Monk, B.C., and Cannon, R.D. (2002). Genomic pathways to antifungal discovery. Curr. Drug Targets Infect. Disord. *2*, 309–329.

Moreno, I., Martinez-Esparza, M., Laforet, L.C., Sentandreu, R., Ernst, J.F., and Valentin, E. (2010). Dosage-dependent roles of the Cwt1 transcription factor for cell wall architecture, morphogenesis, drug sensitivity and virulence in *Candida albicans*. Yeast *27*, 77–87.

Morschhäuser, J., Barker, K.S., Liu, T.T., BlaB-Warmuth, J., Homayouni, R., and Rogers, P.D. (2007). The transcription factor Mrr1p controls expression of the MDR1 efflux pump and mediates multidrug resistance in *Candida albicans*. PLoS Pathog. *3*, e164.

Mukhopadhyay, K., Kohli, A., and Prasad, R. (2002). Drug susceptibilities of yeast cells are affected by membrane lipid composition. Antimicrob. Agents Chemother. *46*, 3695–3705.

Mukhopadhyay, K., Prasad, T., Saini, P., Pucadyil, T.J., Chattopadhyay, A., and Prasad, R. (2004). Membrane sphingolipid–ergosterol interactions are important determinants of multidrug resistance in *Candida albicans*. Antimicrob. Agents Chemother. *48*, 1778–1787.

Nawrocki, A., Fey, S.J., Goffeau, A., Roepstorff, P., and Larsen, P.M. (2001). The effects of transcription regulating genes PDR1, pdr1-3 and PDR3 in pleiotropic drug resistance. Proteomics *1*, 1022–1032.

Noble, J.A., Tsai, H.-F., Suffis, S.D., Su, Q., Myers, T.G., and Bennett, J.E. (2013). STB5 is a negative regulator of azole resistance in *Candida glabrata*. Antimicrob. Agents Chemother. *57*, 959–967.

Noble, S.M., French, S., Kohn, L.A., Chen, V., and Johnson, A.D. (2010). Systematic screens of a *Candida albicans* homozygous deletion library decouple morphogenetic switching and pathogenicity. Nat. Genet. *42*, 590–598.

O'Farrell, P.H. (1975). High resolution two-dimensional electrophoresis of proteins. J. Biol. Chem. *250*, 4007–4021.

Oh, J., Fung, E., Schlecht, U., Davis, R.W., Giaever, G., St Onge, R.P., Deutschbauer, A., and Nislow, C. (2010). Gene annotation and drug target discovery in *Candida albicans* with a tagged transposon mutant collection. PLoS Pathog. *6*, e1001140.

Onyewu, C., Wormley, F.L., Jr, Perfect, J.R., and Heitman, J. (2004). The calcineurin target, Crz1, functions in azole tolerance but is not required for virulence of *Candida albicans*. Infect. Immun. *72*, 7330–7333.

Osherov, N., May, G.S., Albert, N.D., and Kontoyiannis, D.P. (2002). Overexpression of Sbe2p, a Golgi protein, results in resistance to caspofungin in *Saccharomyces cerevisiae*. Antimicrob. Agents Chemother. *46*, 2462–2469.

Park, B.J., Wannemuehler, K.A., Marston, B.J., Govender, N., Pappas, P.G., and Chiller, T.M. (2009). Estimation of the current global burden of cryptococcal meningitis among persons living with HIV/AIDS. AIDS *23*, 525–530.

Parsons, A.B., Lopez, A., Givoni, I.E., Williams, D.E., Gray, C.A., Porter, J., Chua, G., Sopko, R., Brost, R.L., Ho, C.-H., *et al.* (2006). Exploring the mode-of-action of bioactive compounds by chemical-genetic profiling in yeast. Cell *126*, 611–625.

Pasrija, R., Prasad, T., and Prasad, R. (2005). Membrane raft lipid constituents affect drug susceptibilities of *Candida albicans*. Biochem. Soc. Trans. 33, 1219–1223.

Peng, Y., Dong, D., Jiang, C., Yu, B., Wang, X., and Ji, Y. (2012). Relationship between respiration deficiency and azole resistance in clinical *Candida glabrata*. FEMS Yeast Res. *12*, 719–727.

Pierce, S.E., Davis, R.W., Nislow, C., and Giaever, G. (2007). Genome-wide analysis of barcoded *Saccharomyces cerevisiae* gene-deletion mutants in pooled cultures. Nat. Protoc. *2*, 2958–2974.

Prelich, G. (2012). Gene overexpression: uses, mechanisms, and interpretation. Genetics *190*, 841–854.

Rai, M.N., Balusu, S., Gorityala, N., Dandu, L., and Kaur, R. (2012). Functional genomic analysis of *Candida glabrata*–macrophage interaction: role of chromatin remodeling in virulence. PLoS Pathog. *8*, e1002863.

Riggle, P.J., and Kumamoto, C.A. (2006). Transcriptional regulation of MDR1, encoding a drug efflux determinant, in fluconazole-resistant *Candida albicans* strains through an Mcm1p binding site. Eukaryot. Cell *5*, 1957–1968.

Roemer, T., Jiang, B., Davison, J., Ketela, T., Veillette, K., Breton, A., Tandia, F., Linteau, A., Sillaots, S., Marta, C., *et al.* (2003). Large-scale essential gene identification in *Candida albicans* and applications to antifungal drug discovery. Mol. Microbiol. *50*, 167–181.

Rogers, P.D., and Barker, K.S. (2002). Evaluation of differential gene expression in fluconazole-susceptible and -resistant isolates of *Candida albicans* by cDNA microarray analysis. Antimicrob. Agents Chemother. *46*, 3412–3417.

Rogers, P.D., and Barker, K.S. (2003). Genome-wide expression profile analysis reveals coordinately regulated genes associated with stepwise acquisition of azole resistance in *Candida albicans* clinical isolates. Antimicrob. Agents Chemother. *47*, 1220–1227.

Rogers, K.M., Pierson, C.A., Culbertson, N.T., Mo, C., Sturm, A.M., Eckstein, J., Barbuch, R., Lees, N.D., and Bard, M. (2004). Disruption of the *Candida albicans* CYB5 gene results in increased azole sensitivity. Antimicrob. Agents Chemother. *48*, 3425–3435.

Rogers, P.D., Vermitsky, J.-P., Edlind, T.D., and Hilliard, G.M. (2006). Proteomic analysis of experimentally induced azole resistance in *Candida glabrata*. J. Antimicrob. Chemother. *58*, 434–438.

Ross-Macdonald, P., Coelho, P.S., Roemer, T., Agarwal, S., Kumar, A., Jansen, R., Cheung, K.H., Sheehan, A., Symoniatis, D., Umansky, L., *et al.* (1999). Large-scale analysis of the yeast genome by transposon tagging and gene disruption. Nature *402*, 413–418.

Sanglard, D., Ischer, F., Marchetti, O., Entenza, J., and Bille, J. (2003). Calcineurin A of *Candida albicans*: involvement in antifungal tolerance, cell morphogenesis and virulence. Mol. Microbiol. *48*, 959–976.

Sanmiguel, P. (2011). Next-generation sequencing and potential applications in fungal genomics. Methods Mol. Biol. *722*, 51–60.

Sarinová, M., Tichá, E., Obernauerová, M., and Gbelská, Y. (2007). Impact of mitochondrial function on yeast susceptibility to antifungal compounds. Folia Microbiol. (Praha) *52*, 223–229.

Sasse, C., Schillig, R., Dierolf, F., Weyler, M., Schneider, S., Mogavero, S., Rogers, P.D., and Morschhäuser, J. (2011). The transcription factor Ndt80 does not contribute to Mrr1-, Tac1-, and Upc2-mediated fluconazole resistance in *Candida albicans*. PloS One *6*, e25623.

Schiestl, R.H., and Petes, T.D. (1991). Integration of DNA fragments by illegitimate recombination in *Saccharomyces cerevisiae*. Proc. Natl. Acad. Sci. U.S.A. *88*, 7585–7589.

Schillig, R., and Morschhäuser, J. (2013). Analysis of a fungus-specific transcription factor family, the *Candida albicans* zinc cluster proteins, by artificial activation. Mol. Microbiol. *89*, 1003–1017.

Schroder, L.A., Glick, B.S., and Dunn, W.A. (2007). Identification of pexophagy genes by restriction enzyme-mediated integration. Methods Mol. Biol. *389*, 203–218.

Seifert, H.S., Chen, E.Y., So, M., and Heffron, F. (1986). Shuttle mutagenesis: a method of transposon mutagenesis for *Saccharomyces cerevisiae*. Proc. Natl. Acad. Sci. U.S.A. *83*, 735–739.

Sellam, A., Tebbji, F., and Nantel, A. (2009). Role of Ndt80p in sterol metabolism regulation and azole resistance in *Candida albicans*. Eukaryot. Cell *8*, 1174–1183.

Shahi, P., and Moye-Rowley, W.S. (2009). Coordinate control of lipid composition and drug transport activities is required for normal multidrug resistance in fungi. Biochim. Biophys. Acta *1794*, 852–859.

Shapiro, R.S., Robbins, N., and Cowen, L.E. (2011). Regulatory circuitry governing fungal development, drug resistance, and disease. Microbiol. Mol. Biol. Rev. *75*, 213–267.

Shingu-Vazquez, M., and Traven, A. (2011). Mitochondria and fungal pathogenesis: drug tolerance, virulence, and potential for antifungal therapy. Eukaryot. Cell *10*, 1376–1383.

Shukla, S., Yadav, V., Mukhopadhyay, G., and Prasad, R. (2011). Ncb2 is involved in activated transcription of CDR1 in azole-resistant clinical isolates of *Candida albicans*. Eukaryot. Cell *10*, 1357–1366.

da Silva Ferreira, M.E., Malavazi, I., Savoldi, M., Brakhage, A.A., Goldman, M.H.S., Kim, H.S., Nierman, W.C., and Goldman, G.H. (2006). Transcriptome analysis of *Aspergillus fumigatus* exposed to voriconazole. Curr. Genet. *50*, 32–44.

Silver, P.M., Oliver, B.G., and White, T.C. (2004). Role of *Candida albicans* transcription factor Upc2p in drug resistance and sterol metabolism. Eukaryot. Cell *3*, 1391–1397.

Singh, A., and Prasad, R. (2011). Comparative lipidomics of azole sensitive and resistant clinical isolates of *Candida albicans* reveals unexpected diversity in molecular lipid imprints. PloS One *6*, e19266.

Singh, A., Yadav, V., and Prasad, R. (2012). Comparative lipidomics in clinical isolates of *Candida albicans* reveal crosstalk between mitochondria, cell wall integrity and azole resistance. PloS One *7*, e39812.

Singh, A., Mahto, K.K., and Prasad, R. (2013). Lipidomics and *in vitro* azole resistance in *Candida albicans*. OMICS *17*, 84–93.

Singh, S.D., Robbins, N., Zaas, A.K., Schell, W.A., Perfect, J.R., and Cowen, L.E. (2009). Hsp90 governs echinocandin resistance in the pathogenic yeast *Candida albicans* via calcineurin. PLoS Pathog. *5*, e1000532.

Singh-Babak, S.D., Babak, T., Diezmann, S., Hill, J.A., Xie, J.L., Chen, Y.-L., Poutanen, S.M., Rennie, R.P., Heitman, J., and Cowen, L.E. (2012). Global analysis of the evolution and mechanism of echinocandin resistance in *Candida glabrata*. PLoS Pathog. *8*, e1002718.

Smith, A.M., Durbic, T., Oh, J., Urbanus, M., Proctor, M., Heisler, L.E., Giaever, G., and Nislow, C. (2011). Competitive genomic screens of barcoded yeast libraries. J. Vis. Exp. *Aug 11*(54), 2864.

Smith, A.M., Durbic, T., Kittanakom, S., Giaever, G., and Nislow, C. (2012). Barcode sequencing for understanding drug–gene interactions. Methods Mol. Biol. *910*, 55–69.

Song, M.-H., Lee, J.-W., Kim, M.S., Yoon, J.-K., White, T.C., Floyd, A., Heitman, J., Strain, A.K., Nielsen, J.N., Nielsen, K., *et al.* (2012). A flucytosine-responsive Mbp1/Swi4-like protein, Mbs1, plays pleiotropic roles in antifungal drug resistance, stress response, and virulence of *Cryptococcus neoformans*. Eukaryot. Cell *11*, 53–67.

Sorgo, A.G., Heilmann, C.J., Dekker, H.L., Bekker, M., Brul, S., de Koster, C.G., de Koning, L.J., and Klis, F.M. (2011). Effects of fluconazole on the secretome, the wall proteome, and wall integrity of the clinical fungus *Candida albicans*. Eukaryot. Cell *10*, 1071–1081.

Sun, N., Fonzi, W., Chen, H., She, X., Zhang, L., Zhang, L., and Calderone, R. (2013). Azole susceptibility and transcriptome profiling in Candida albicans mitochondrial electron transport chain complex I mutants. Antimicrob. Agents Chemother. *57*, 532–542.

Talibi, D., and Raymond, M. (1999). Isolation of a putative *Candida albicans* transcriptional regulator involved in pleiotropic drug resistance by functional complementation of a pdr1 pdr3 mutation in Saccharomyces cerevisiae. J. Bacteriol. *181*, 231–240.

Thakur, J.K., Arthanari, H., Yang, F., Pan, S.-J., Fan, X., Breger, J., Frueh, D.P., Gulshan, K., Li, D.K., Mylonakis, E., *et al.* (2008). A nuclear receptor-like pathway regulating multidrug resistance in fungi. Nature *452*, 604–609.

Torelli, R., Posteraro, B., Ferrari, S., La Sorda, M., Fadda, G., Sanglard, D., and Sanguinetti, M. (2008). The ATP-binding cassette transporter-encoding gene CgSNQ2 is contributing to the CgPDR1-dependent azole resistance of *Candida glabrata*. Mol. Microbiol. *68*, 186–201.

Tsai, H.-F., Bard, M., Izumikawa, K., Krol, A.A., Sturm, A.M., Culbertson, N.T., Pierson, C.A., and Bennett, J.E. (2004). *Candida glabrata* erg1 mutant with increased sensitivity to azoles and to low oxygen tension. Antimicrob. Agents Chemother. *48*, 2483–2489.

Tsai, H.-F., Krol, A.A., Sarti, K.E., and Bennett, J.E. (2006). *Candida glabrata* PDR1, a transcriptional regulator of a pleiotropic drug resistance network, mediates azole resistance in clinical isolates and petite mutants. Antimicrob. Agents Chemother. *50*, 1384–1392.

Unlü, M., Morgan, M.E., and Minden, J.S. (1997). Difference gel electrophoresis: a single gel method for detecting changes in protein extracts. Electrophoresis *18*, 2071–2077.

Vandenbosch, D., De Canck, E., Dhondt, I., Rigole, P., Nelis, H.J., and Coenye, T. (2013). Genomewide screening for genes involved in biofilm formation and miconazole susceptibility in *Saccharomyces cerevisiae*. FEMS Yeast Res. *13*, 720–730.

Vandeputte, P., Tronchin, G., Rocher, F., Renier, G., Bergès, T., Chabasse, D., and Bouchara, J.-P. (2009). Hypersusceptibility to azole antifungals in a clinical isolate of *Candida glabrata* with reduced aerobic growth. Antimicrob. Agents Chemother. *53*, 3034–3041.

Vandeputte, P., Ischer, F., Sanglard, D., and Coste, A.T. (2011). *In vivo* systematic analysis of *Candida albicans* Zn2-Cys6 transcription factors mutants for mice organ colonization. PloS One *6*, e26962.

Vandeputte, P., Pradervand, S., Ischer, F., Coste, A.T., Ferrari, S., Harshman, K., and Sanglard, D. (2012). Identification and functional characterization of Rca1, a transcription factor involved in both antifungal susceptibility and host response in *Candida albicans*. Eukaryot. Cell *11*, 916–931.

Vavala, E., Mignogna, G., Spano, F., Stringaro, A., Colone, M., Sanguinetti, M., Maras, B., and Angiolella, L. (2013). The cell wall protein Rhd3/Pga29 is over-expressed in *Candida albicans* upon micafungin treatment. J. Chemother. *25*, 332–340.

Vermitsky, J.-P., Earhart, K.D., Smith, W.L., Homayouni, R., Edlind, T.D., and Rogers, P.D. (2006). Pdr1 regulates multidrug resistance in *Candida glabrata*: gene disruption and genome-wide expression studies. Mol. Microbiol. *61*, 704–722.

Walker, L.A., Gow, N.A.R., and Munro, C.A. (2013). Elevated chitin content reduces the susceptibility of *Candida* species to caspofungin. Antimicrob. Agents Chemother. *57*, 146–154.

Wang, A.M., Doyle, M.V., and Mark, D.F. (1989). Quantitation of mRNA by the polymerase chain reaction. Proc. Natl. Acad. Sci. U.S.A. *86*, 9717–9721.

Wilkins, M.R., Pasquali, C., Appel, R.D., Ou, K., Golaz, O., Sanchez, J.C., Yan, J.X., Gooley, A.A., Hughes, G., Humphery-Smith, I., *et al.* (1996). From proteins to proteomes: large scale protein identification by two-dimensional electrophoresis and amino acid analysis. Biotechnol. (NY) *14*, 61–65.

Wood, V., Gwilliam, R., Rajandream, M.-A., Lyne, M., Lyne, R., Stewart, A., Sgouros, J., Peat, N., Hayles, J., Baker, S., *et al.* (2002). The genome sequence of *Schizosaccharomyces pombe*. Nature *415*, 871–880.

Xu, D., Jiang, B., Ketela, T., Lemieux, S., Veillette, K., Martel, N., Davison, J., Sillaots, S., Trosok, S., Bachewich, C., *et al.* (2007). Genome-wide fitness test and mechanism-of-action studies of inhibitory compounds in *Candida albicans*. PLoS Pathog. *3*, e92.

Xu, Q., Schlabach, M.R., Hannon, G.J., and Elledge, S.J. (2009). Design of 240,000 orthogonal 25mer DNA barcode probes. Proc. Natl. Acad. Sci. U.S.A. *106*, 2289–2294.

Xu, T., Bharucha, N., and Kumar, A. (2011). Genome-wide transposon mutagenesis in *Saccharomyces cerevisiae* and *Candida albicans*. Methods Mol. Biol. *765*, 207–224.

Xu, Z., Zhang, L.-X., Zhang, J.-D., Cao, Y.-B., Yu, Y.-Y., Wang, D.-J., Ying, K., Chen, W.-S., and Jiang, Y.-Y. (2006). cDNA microarray analysis of differential gene expression and regulation in clinically drug-resistant isolates of *Candida albicans* from bone marrow transplanted patients. Int. J. Med. Microbiol. *296*, 421–434.

Yan, L., Zhang, J.-D., Cao, Y.-B., Gao, P.-H., and Jiang, Y.-Y. (2007). Proteomic analysis reveals a metabolism shift in a laboratory fluconazole-resistant *Candida albicans* strain. J. Proteome Res. *6*, 2248–2256.

Yang, D.-H., Maeng, S., Strain, A.K., Floyd, A., Nielsen, K., Heitman, J., and Bahn, Y.-S. (2012). Pleiotropic roles of the Msi1-like protein Msl1 in *Cryptococcus neoformans*. Eukaryot. Cell *11*, 1482–1495.

Zhang, X., and Moye-Rowley, W.S. (2001). *Saccharomyces cerevisiae* multidrug resistance gene expression inversely correlates with the status of the F(0) component of the mitochondrial ATPase. J. Biol. Chem. *276*, 47844–47852.

Zhang, L., Zhang, Y., Zhou, Y., An, S., Zhou, Y., and Cheng, J. (2002a). Response of gene expression in *Saccharomyces cerevisiae* to amphotericin B and nystatin measured by microarrays. J. Antimicrob. Chemother. *49*, 905–915.

Zhang, L., Zhang, Y., Zhou, Y., Zhao, Y., Zhou, Y., and Cheng, J. (2002b). Expression profiling of the response of *Saccharomyces cerevisiae* to 5-fluorocytosine using a DNA microarray. Int. J. Antimicrob. Agents *20*, 444–450.

Zhao, R., Bin, W., Diao, Y., Yang, J., Liu, T., Peng, J., and Jin, Q. (2009). Global transcriptional profiles of *Trichophyton rubrum* in response to Flucytosine. Sci. China C. Life Sci. *52*, 1173–1185.

7 New Antifungal Discovery from Existing Chemical Compound Collections

Olihile M. Sebolai and Adepemi O. Ogundeji

Abstract

Immunosuppressive conditions, either as a result of HIV infection or unintended consequences of using anti-rejection drugs in organ transplant patients, have led to an increase in host vulnerability to fungal diseases. These conditions create optimal environments for opportunistic infections to occur, and thus it is not coincidental that previously non-pathogenic fungal species have become new mycotic agents. The emergence of these new mycotic agents in clinical settings is also in parallel to a rise in drug resistance, which is an inherent trait in natural selection. Therefore, it is a challenge to keep up with these agents (some of which may initiate a mode of infection that is not targeted by current antifungal drugs) given the time for development and approval period required to have a new drug on the market. And to compound this, the usage of conventional drugs such as azoles and polyenes is often times characterized by clinical failure, particularly in persons who are immunocompromised.

Thus a solution to the above could be the repurposing of existing chemical compounds (already FDA-approved) as 'new' antimicrobial agents, whose previous mechanisms of action were thought to be negative. To date, reports have emerged documenting the antifungal activity of non-traditional antimicrobial drugs, or where they enhance the activity of other traditional antimicrobial drugs. This text will therefore explore literature concerning the application of non-traditional antimicrobial drugs such as the prototypical anti-inflammatory drug acetylsalicylic acid, and other compounds in controlling human fungal diseases.

Introduction

Fungi are essential for sustaining human life, i.e. in their application in the preparation of food or as sources of pharmacological agents (Alba-Lois and Segal-Kischinevzky, 2010; Cauvain and Young, 2007; Johnson and Echavarri-Erasun, 2011; Shiby and Mishra, 2013). Paradoxically, this group of microorganisms also acts as agents that mediate infectious processes that may end life in susceptible hosts. Over the years, yeasts, and in particular, species belonging to *Cryptococcus* and *Candida* genera have emerged as important disease-causing fungi (Calderone, 2002; Cooper, 2011; Heitman *et al.*, 2011). As a result, these microbes have been extensively studied, often with an attempt to understanding their biology in the context of controlling infections (Ashbee and Bignell, 2010; Brown and Netea, 2007; Kurtzman *et al.*, 2011). Such intimate insight into their cellular pathology and physiology may reveal strategies for identifying potential drug targets.

Fungi can cause fatal diseases in individuals with apparently well-functioning immune systems, a case in point being *Cryptococcus gattii* (Bartlett *et al.*, 2008; Kidd *et al.*, 2004). Simultaneously, some fungal species normally exist in harmony with their hosts, i.e. as commensals in healthy individuals, where they can grow without causing any disease symptoms (Soll, 2002). However, given the right conditions, i.e. in hosts exhibiting impaired cell immunity, species such as *Cryptococcus neoformans* and *Candida albicans* can become 'opportunistic' and establish disseminated infections (Calderone and Clancy, 2012; Casadevall and Perfect, 1998). Therefore, it is not surprising that the rise in the number of people with immunosuppressive conditions, primarily due to HIV, has given infectious diseases such as cryptococcosis and candidiasis a 'new' lease on life. To illustrate this point, it is estimated that well over 1 million cases of cryptococcosis (considered to be an AIDS-defining illness) are reported annually (Park *et al.*, 2009). In addition, 10% of persons living in Asia, Africa and Latin America who are infected with HIV have oral candidiasis (Armstrong-James *et al.*, 2014).

While there has been an interest in fungal patho-biology for more than a century, there have been many more recent studies that sought to demonstrate the usefulness of a number of chemical compounds in controlling fungal growth. The latter is primarily informed by a number of limitations concerning the clinical usage of conventional drugs such as polyenes and azoles (which act on eukaryotic-specific drug targets and not fungi-specific drug targets) in managing fungal infections (Ghannoum and Rice, 1999). These limitations include (1) a narrow therapeutic index, (2) side-effects, (3) clinical failure, more so, in immunocompromised individuals, and (4) development of drug resistance – due to either the widespread usage of drugs or poor drug-compliance by patients (Armstrong-James *et al.*, 2014; Ghannoum and Rice, 1999). Hence, the constant need to find other compounds that may possess antifungal qualities. An interesting development in this field has been the rise in the number of texts exploring non-traditional antimicrobial agents. Importantly, some of these studies aligned their protocols to standard *in vitro* susceptibility reference methods (Clinical Laboratory Standards Institute or European Committee on Antimicrobial Susceptibility Testing). And as such, results obtained from these studies have given the scientific community a sense of the efficacy of the tested compounds, which is important in improving patient outcomes.

Further to the discussion, some of the compounds that have been tested include anti-inflammatory drugs (Alem and Douglas, 2004), which are relatively cheap and are widely distributed across the world. It is unarguable that management of fungal infectious diseases is strongly dependent on capital resources available to a specific region (Perfect *et al.*, 2010). In a region like sub-Saharan Africa, with its uniquely complex geo-political and socio-economic challenges, it has proven difficult for governments to provide the necessary life-saving drugs, which are often expensive. Corollary, incidences of disseminated cryptococcal diseases, for an example, has remained a major cause of morbidity and mortality compared to developed countries (Park *et al.*, 2009). It is therefore pleasing to note that multi-national companies, through their drug donation programmes, have made medicines such as fluconazole readily available (Collett and Parrish, 2007; Wertheimer *et al.*, 2004). This, in conjunction with the expanded antiretroviral programme in countries like South Africa, has greatly assisted in managing fungal infections among HIV-infected populations and general public. This gesture has unfortunately also lead to unintended consequences, where such drugs can be used widely as curative agents leading to development of drug resistance. To overcome these challenges, it is time that existing chemical compounds (especially those that are cheap and widely distributed) with demonstrable antimicrobial activity and therapeutic benefit be considered, particularly in these parts of the world.

Therefore, in this text, we explore the current literature and make a case for non-traditional antimicrobial drugs, such as acetylsalicylic acid, as possible alternative antifungal drugs. The reader is also referred to excellent texts detailing the current treatment guidelines for the management of cryptococcosis (Perfect *et al.*, 2010) and candidiasis (Pappas *et al.*, 2009) as the discussion in the following subsections is mainly dedicated to anti-inflammatory compounds like acetylsalicylic acid. We also give the necessary attention to other compounds such as other non-steroidal anti-inflammatory drugs, natural products (plant extracts and lipids) and others in controlling human fungal diseases. Finally, in sum, we also explore a few high throughput assays were among others, small molecules, have been screened for their potential antifungal qualities.

A new career for acetylsalicylic acid as an antifungal

Definition, structure and function

Anti-inflammatory drugs, and in particular non-steroidal anti-inflammatory drugs, are a group of chemical compounds that share similar mechanisms of action (Busti *et al.*, 2005; Rao and Knaus, 2008). The prototypical anti-inflammatory drug, acetylsalicylic acid, was discovered by chance when plant extracts were successfully employed to, among others, alleviate pain (Rainsford, 2007; Rao and Knaus, 2008). Since this discovery, advances in chemistry have allowed for mass production of this compound from as early as the late 1800s. The actual mode of action of this drug remained elusive until the 1970s when John Vane demonstrated that acetylsalicylic acid inhibited prostaglandin production (Vane, 1971). The discovery of two targeted enzymes, cyclooxygenase (COX) 1 and 2, whose products regulate important physiological and pathological processes, and the establishment that their selective inhibition could yield a specific desired outcome (i.e. control of inflammation) has led to the development of many other novel anti-inflammatory drugs, including indomethacin and ibuprofen as a few examples (Fig. 7.1).

While anti-inflammatory drugs have successfully been used to treat conditions such as rheumatic diseases, like any other drug they have shortcomings. For instance, one major shortcoming is gastrointestinal toxicity (Rao and Knaus, 2008). As a result, there are efforts to design more effective anti-inflammatory drugs with little to no side-effects. More encouraging is that further investigations in these compounds have implicated new mechanisms of action, which have since shed light upon additional beneficial applications (i.e. antimicrobial activity).

The discovery of antifungal activity of acetylsalicylic acid

Just over 20 years ago, the Kock group initiated bioprospecting studies aimed at testing for the presence of prostaglandins in yeasts (Kock *et al.*, 1991). This led to a chance discovery of a mitochondrial-produced, acetylsalicylic acid-sensitive compound named 3-hydroxy eicosatetraenoic acid, which is characterized by the hydroxyl group on the beta carbon from the carboxylic end (Van Dyk *et al.*, 1991, 1993). Interestingly, this molecule is structurally similar to arachidonic acid (Fig. 7.2), which is a precursor for prostaglandins in mammalian cells. In elegant biochemical studies, Ciccoli *et al.* demonstrated that this 3-hydroxy molecule could mimic arachidonic acid and serve as a substrate to mammalian COX enzymes leading to production of prostaglandins (Ciccoli *et al.*, 2005).

In yeast biology, this 3-hydroxy molecule is important for development, and in particular this molecule has been documented to be intimately involved in yeast sexual phases (Kock *et al.*, 2003, 2007). The effects of acetylsalicylic acid were at the time speculatively attributed to diminishing mitochondrial function, which was previously said to be important during the sexual phase

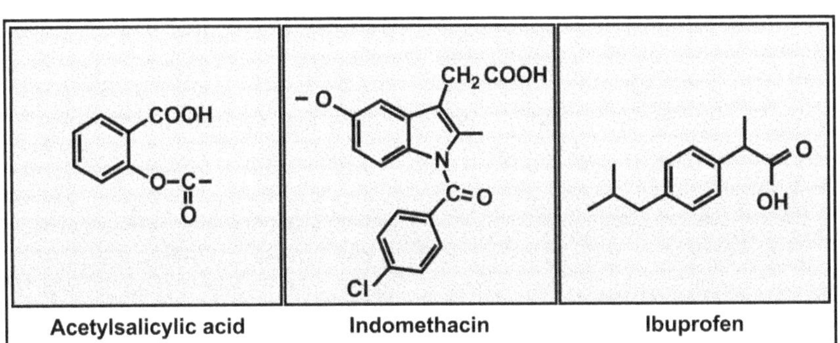

Figure 7.1 The schematic representation of some chemical compounds representing anti-inflammatory drugs.

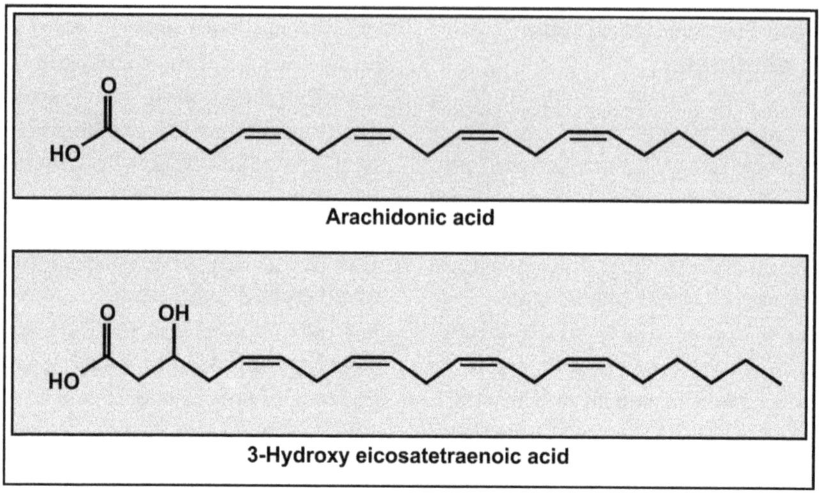

Figure 7.2 Chemical structures of arachidonic acid and 3-hydroxy eicosatetraenoic acid (with a hydroxyl group on the beta carbon). Although 3-hydroxy eicosatetraenoic acid is derived from fungal cells, this acetylsalicylic acid-sensitive compound serves as a substrate to mammalian cyclooxygenase enzyme leading to production of prostaglandins (Ciccoli *et al.*, 2005).

of yeasts (Codon *et al.*, 1995; Marmiroli *et al.*, 1983).

This assertion was validated with additional biochemical studies, demonstrating the effects of acetylsalicylic acid on mitochondria. In 1999, Glasgow and co-workers showed that the active component of acetylsalicylic acid, salicylate, inhibits mitochondrial beta-oxidation (Glasgow *et al.*, 1999), which is the proposed biosynthetic pathway for 3-hydroxy acids (Sebolai *et al.*, 2012). In particular, this is ascribed to acetylsalicylic acid metabolites having structural similarities to the acyl-portions of the substrate and product of the 3-hydroxyacyl-CoA dehydrogenase activity of the beta-oxidation pathway. In addition, acetylsalicylic acid was shown to inhibit mitochondrial activity by uncoupling mitochondrial oxidative phosphorylation and/or inhibiting electron transport (Norman *et al.*, 2004; Somasundaram *et al.*, 1997). A recent study also showed that acetylsalicylic acid induced microbial apoptosis via accumulation of reactive oxygen species in mitochondria and cytoplasm as well as oxidation of mitochondrial NAD(P)H (Farrugia *et al.*, 2013). Thus it is not surprising that the yeast sexual cycle (which is dependent on active mitochondria) is blocked by acetylsalicylic acid. This insight into the biology of yeast 3-hydroxy molecules and the effects of acetylsalicylic acid led to a hypothesis that yeast with a mitochondrion-dependent strict aerobic metabolism will be more sensitive to acetylsalicylic acid than those that can also produce energy through an alternative anaerobic glycolytic fermentative pathway in which mitochondria are not involved (Kock *et al.*, 2007). The hypothesis set a stage for screening microbes that would fit into the model.

Cryptococcus species are non-fermentative (Kwon-Chung, 2011), requiring functional mitochondria to sustain growth and other energy-driven cellular processes (Litter, 2004). Mitochondria are reported to produce approximately 95% of the cell's ATP (Nadanaciva *et al.*, 2007). The high dependency of *Cryptococcus* on mitochondrial-derived ATP for survival highlights these organelles as susceptible targets for potential pharmacological intervention. To this end, a study assessing the *in vitro* susceptibility of a number of *Cryptococcus* species towards acetylsalicylic acid revealed a dose-dependent response – with growth inhibition being noted at concentrations as low as 1 mmol/ml (Sebolai *et al.*, 2008a).

The recommended concentrations of acetylsalicylic acid in the blood for the treatment of a number of physiological conditions in humans ranges from 20 to 30 mg/dl, which corresponds

Table 7.1 Acetylsalicylic acid concentration range generally accepted as optimum for safe and effective therapy in the blood (Levy, 1976)

Physiological condition/ indication	Therapeutic dosage
Arthritis	20 mg/dl to 30 mg/dl
Rheumatic fever	20 mg/dl to 30 mg/dl
Antifungal activity against *Cryptococcus* species	18 mg/dl (approximately 1 mmol/ml)

to approximately 1 to 1.5 mmol/ml (Table 7.1) (Levy, 1976). The growth of some of the *Cryptococcus* species tested was slightly inhibited at this concentration. A clear inhibition of growth of all *Cryptococcus* species tested was seen at 3 mmol/ml only, which would lead to deleterious effect in humans (Levy, 1976). However, this highlights the potential practical implementation of using acetylsalicylic acid in the treatment of cryptococcal infections, where acetylsalicylic acid would be used at physiological concentration in combined therapy with conventional antifungal to enhance their efficacy.

In addition to the above, formation of cryptococcal capsules is influenced by acetylsalicylic acid treatment (Sebolai *et al.*, 2008b). This ability of salicylates to reduce capsule formation is a phenotypic consequence that has also been reported elsewhere in the literature (Domenico *et al.*, 1989). These authors reported that the salicylate derivative sodium salicylate reduced encapsulation in the bacterium *Klebsiella pneumoniae*. This capsule chelating property of sodium salicylate could be partially reversed by addition of calcium and magnesium (Domenico *et al.*, 1989).

In order to test the wider effects of acetylsalicylic acid, another study assessing acetylsalicylic acid susceptibility among fungi was conducted (Leeuw *et al.*, 2007). In brief, the tested species were categorized based on their ability to ferment. Thus some species from *Eremothecium, Ascoidea, Cryptococcus, Lipomyces, Rhodotorula, Dipodascus, Dipodascopsis* constituted the fermentation negative grouping and fermentation positive grouping was made up of some species from *Candida, Pichia, Kluyveromyces, Saccharomyces, Schizosaccahromyces and Zygosaccharomyces.* The results of this study revealed that indeed organisms with a strict-mitochondrion-dependent aerobic metabolism (*Eremothecium, Ascoidea, Cryptococcus, Lipomyces, Rhodotorula, Dipodascus, Dipodascopsis*) were more susceptible to acetylsalicylic acid (MIC = 1mmol/ml) compared to those that can ferment. All species with fermentation capability were shown to be resistant to acetylsalicylic acid.

The effects of acetylsalicylic acid are not limited to planktonic cells but also extend to fungal biofilms. These structured microbial communities that are linked by a matrix of exopolymeric material are of clinical significance as some community members can evade host immuno-processing and display drug resistance properties (Ramage *et al.*, 2012). Alem and Douglas (2004) reported on the ability of acetylsalicylic acid to inhibit *Candida* biofilms at concentrations as low as 1 mmol/ml. These authors further implicated acetylsalicylic acid as having the ability to exert a dual function by inhibiting prostaglandin production, which are molecules that are important for promoting development into the pathogenic hyphal stage (Alem and Douglas, 2004). A synergistic and an indifferent outcome were reported when acetylsalicylic acid was applied in combined therapy with amphotericin B in an *in vitro* assay against *Candida* planktonic cells and biofilms (Zhou *et al.*, 2012). The latter is of clinically significance as these drugs, by acting on multiple target sites, may counter the development of drug resistance. Another derivative of salicylate, sodium salicylate, was shown to act in synergy with fluconazole (fractional inhibitory index, < 0.5) against *Candida albicans* planktonic cells (Scott *et al.*, 1995). The findings by Scott *et al.* (1995) are encouraging as azoles are usually the first-line of drugs administered and also the ones against which resistance is more often encountered.

Possible role of anti-inflammatory antifungal drugs in controlling pathogen-induced immune reconstitution inflammatory syndrome

Another important ancillary benefit of acetylsalicylic acid, in addition to antimicrobial activity, could be the control of microbial-induced inflammatory conditions. It is well established that certain pathologies can lead to an out-of-control production of pro-inflammatory cytokines, which may overcome susceptible hosts (Boulware *et al.*, 2010; Haddow *et al.*, 2010). The latter can be observed in the case of cryptococcal-immune reconstitution inflammatory syndrome, where, for example, the host immune system recognizes pathogen antigens that were never cleared, in part, due to the fungistatic nature of drugs such as fluconazole.

Is it possible, based on the discussion above, to consider administering anti-inflammatory drugs in combined therapy with antiretroviral drugs, to restrain the out-of control inflammatory response while at the same time clearing fungal cells? On the face of this, the argument seems logical. However, the human response is always dynamic and may produce an undesired expression to such a set up. Nonetheless, these newly discovered functionalities of anti-inflammatory drugs provide a premise for new studies in the quest to improve patient outcomes.

Other non-traditional antimicrobial agents

Other non-steroidal anti-inflammatory drugs

The clinical application of non-steroidal anti-inflammatory drugs is mainly limited by their undesired side-effects, namely peptic ulcers. Nonetheless, most of the anti-inflammatory drugs currently on the market were developed with the sole aim of finding a new drug that would have fewer side-effects than acetylsalicylic acid (Rainsford, 2007; Rao and Knaus, 2008). Similarly to acetylsalicylic acid, novel mechanisms of action have since been ascribed to these drugs, and thus they have found 'new' careers.

Among the first reports detailing the antifungal activity of ibuprofen was a 1993 study by Sanyal and co-workers (Sanyal *et al.*, 1993). In their study, these authors note that ibuprofen could be employed as an antimicrobial agent against dermatophytes, which are a group of fungi that colonize the skin due to their dependence on keratin (Weitzman and Summerbell, 1995). In this report, ibuprofen was shown to significantly inhibit dermatophyte growth at MIC ranging from 5 to 40 μg/ml. Based on the above, it is perhaps not surprising that these molecules (acetylsalicylic acid and ibuprofen) have been included as constituents of topical creams, where they may have an ancillary benefit of controlling skin microbiota.

In 2000, Pina Vaz and co-authors offered the first possible mode of action for ibuprofen in yeasts (Pina Vaz *et al.*, 2000). These authors were able to show that ibuprofen achieved cell death via cell lysis. To be specific, in addition to using imaging technology, they demonstrated that ibuprofen targeted the cytoplasmic membrane causing cell-architectural structure damage that was associated with subsequent intracellular leakage. This study further demonstrated that the effects of ibuprofen are dose-dependent, where at a high concentration (10 mg/ml) *Candida* cells were killed, whereas at a low concentration, 5 mg/ml, the drug was fungistatic. Ibuprofen has also been shown to act in synergy with fluconazole when tested in combined therapy against *Candida albicans* (fractional inhibitory index, <0.5) (Scott *et al.*, 1995). The latter is of clinically significance as these drugs, by acting on multiple target sides, may counter the development of drug resistance.

Other reports have shown that non-steroidal anti-inflammatory drugs possess antifungal activity. The list extends to, among others, diclofenac, indomethacin, piroxicam, celecoxib, etodolac, meloxicam, nimesulide (Abdelmegeed and Shaaban, 2013; Alem and Douglas, 2004; Ncango *et al.*, 2011).

Natural products

The usage of natural products to treat ailments is an old-age practice, which dates back to ancient civilisations (Biggs, 1995; Nunn, 1997). Since time immemorial, humans and other primates have repeated looked to nature to find answers, to be specific, to discover natural products (either plant extracts or lipids, i.e. fatty acids and essential oils) that may possess pharmacological qualities (Baker, 1996; Biggs, 1995). Over time, after exhaustive experimentation (as no prior knowledge existed about the product's curative properties), information regarding the performance of these products (whether they worked or not) was integrated into the local's indigenous knowledge. Interestingly, even today, a significant proportion of the population in developing countries, which is not covered by public and private health care systems, still appropriate this knowledge in order to attend to their primary health care needs (Farnsworth, 1988; WHO, 1998). This may also in part be attributed to the cost of conventional medicines, which are oftentimes astronomical.

Plant extracts

By definition, a plant extract can be regarded as a natural product derivative whose chemical constituents have not been fully defined (Negri *et al.*, 2014). Some of these chemical constituents (phytomolecules) may be secreted as a defence mechanism to counter predation, including predation pressure exerted by microbes (Fraenkel, 1959; Heil and Bostock, 2002). Therefore, these molecules are rather secondary metabolites, which seem to have no apparent function in the primary metabolism of the plant (Fraenkel, 1959).

In many reported *in vitro* susceptibility bioassays, researchers often refer to the antimicrobial activity of crude extracts, and deem it necessary to not elucidate the full chemical composition of the active compound(s). However, it is quite crucial to pinpoint the active compound(s) that contributes to the plant extract's antimicrobial activity for purposes of mass production. But more importantly, to also demonstrate the compound's therapeutic effect in an *in vivo* study. Nonetheless, based on their chemistry, phytomolecules have been categorized into a number of classes (Aqil *et al.*, 2010; Meena *et al.*, 2013). Some of these include (Fig. 7.3):

1. Glycosides: small organic molecules in which a functional group is attached to a sugar moiety via a glycosidic bond. In plants, these molecules require an additional activation step, i.e. hydrolysis, before they are utilized. An example is cyanoglycoside from cherry fruits.
2. Alkaloids: chemical compounds characterized by a nitrogen ring and may carry other elements such as oxygen and sulfur. These compounds are primarily toxic to many organisms. Examples include stimulant cocaine and caffeine.
3. Polyphenols: organic compounds characterized by repeating phenolic rings, which seem to attribute a particular function, i.e. therapeutic, among others. Examples include isoflavones and tannins.
4. Terpenes: aromatic organic compounds, which are major biosynthetic building blocks (isoprene unit) found in almost all living organisms. Examples include steroids and vitamin A.

From literature, it is clear that plant extracts exert a demonstrable pharmacological effect in humans (Ficker *et al.*, 2003; Gur *et al.*, 2006). And as pointed out by Ficker *et al.* the traditional uses of plants, i.e. as food sources, implies their extracts would have a low toxicity profile in humans. To illustrate this point, extracts arising from edible plants such as *Zingiber officinale* (ginger), *Curcumae longae* (tumeric) and *Juglans cinerea* (butternut) have been examined and shown to cause growth inhibition (Ficker *et al.*, 2003). More to the point, the above-mentioned extracts were also shown to inhibit growth of fungal strains that were highly resistant to amphotericin B and ketoconazole. Atai and co-workers reported that ethanolic extracts of ginger exert an inhibitory effect on *Candida albicans* at concentrations as low as 2 mg/ml (Atai *et al.*, 2009).

An interesting extract is the polyphenolic compound curcumin, which originates from rhizomes of the plant *Curcuma longa*. This compound has been shown to inhibit growth of *Cryptococcus neoformans* and some clinical isolates of *Candida* (Martins *et al.*, 2008). Additionally, the same study reported on the ability of this compound to limit the adherence of *Candida* cells to human buccal epithelial cells (Martins *et al.*, 2008). This observation is important, as the adhesion of microbes to host mucosal surfaces is a prerequisite for colonization and infection (Lyon and de Resende, 2006; Martins *et al.*, 2008).

An interesting application of this compound was in photodynamic therapy against clinical isolates of *Candida albicans*, *Candida tropicalis* and *Candida glabrata* (Dovigo *et al.*, 2011). Photodynamic therapy is mainly used in cancer treatment,

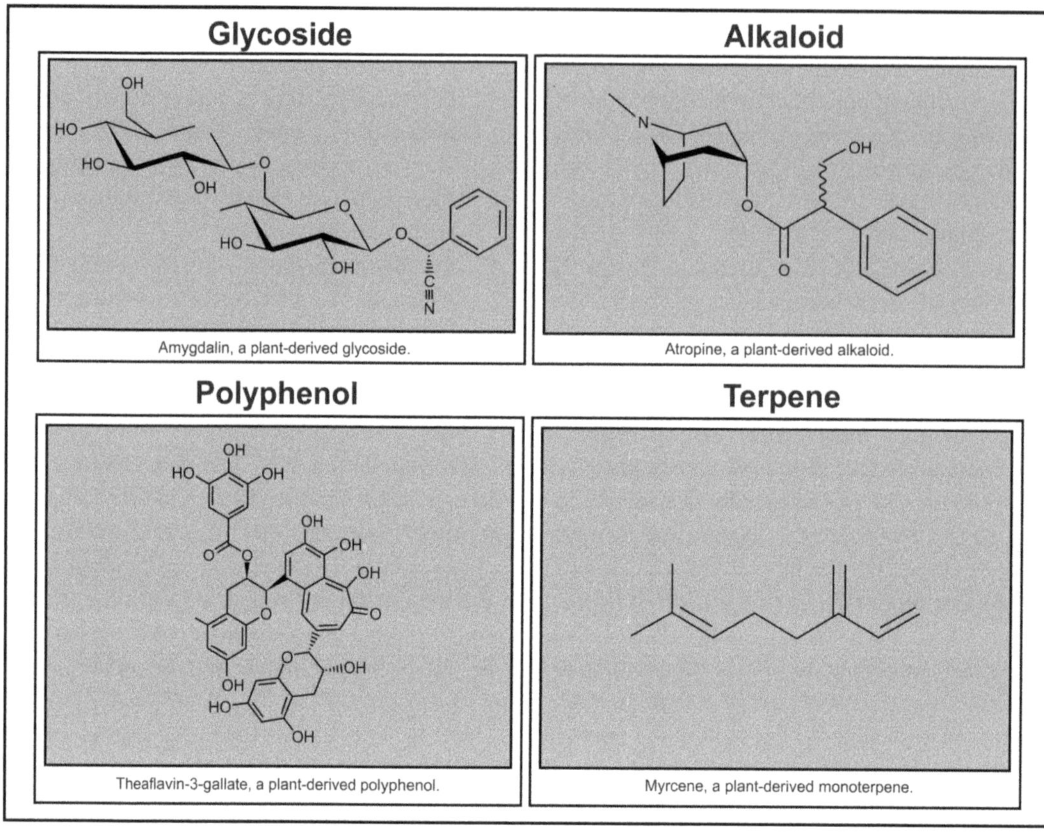

Figure 7.3 Some examples of plant-based extracts that possess biological activity.

and its since been extended to microbes. Here, a non-toxic compound like curcumin (regarded as a photosensitizer) is activated by light of a particular wavelength leading to the production of reactive oxygen species (ROS). The resultant oxidative burst can then target cellular components (mitochondria, cell membranes, etc.) leading to cell death (Gonzales and Maisch, 2012). In their study, Dovigo *et al.* (2011) reported optical properties of curcumin as a function of illumination fluence. In brief, their data showed that when compared with the control group, a statistically significant reduction in *Candida albicans* viability was observed after photodynamic therapy ($P < 0.05$), for both planktonic cells and biofilms.

Lipids: fatty acids and essential oils

Fatty acids and essential oils are lipid-based molecules that naturally occur in nature. As lipids, they are defined primarily on the basis of their solubility in solvents. Fatty acids exist as constituents of triglycerides or in a 'free-form' where they are not attached to other molecules (Kock and Botha, 1998). These molecules are generally regarded as inert structural components of membrane or as sources of fuel were their catabolism results in energy generation. Some of these molecules are regarded as essential fatty acids, and they play an important role in human physiology (Sanders and Emery, 2003). Thus these molecules are often taken in as food supplements.

In addition to the above, fatty acids have been shown to exert antimicrobial activities against microbes. Recently, Thibane *et al.* (2012) reported on the antimicrobial effects of stearidonic acid, eicosapentaenoic acid and docosapentaenioc acid: these fatty acids reduced the metabolic activity and biofilm biomass production by *Candida albicans* and *Candida dubliniensis*. In another study, fatty acids (in particular stearidonic acid) acted in synergy with amphotericin B against *Candida albicans* and *Candida dubliniensis* (Thibane, 2012). In the latter study, Thibane (2012) point out that fractional inhibitory concentration index values of 0.25 and 0.38 were observed, which respectively correlated to 99% and 96% growth inhibition.

Another class of fatty acids, i.e. 3-hydroxy fatty acids have also been documented to act as antimicrobial agents (Sjogren *et al.*, 2003). These authors show that lactic acid bacterial 3-hydroxy fatty acids (3-OH 10:0, 3-OH 11:0, 3-OH 12:0 and 3-OH 14:0) exhibit antifungal activity with MIC ranging from 10 μg/ml to 100 μg/ml against species belonging to *Aspergillus*, *Penicillium*, *Kluyveromyces*, *Pichia* and *Rhodotorula*. These 3-hydroxy molecules may be 'antimicrobial agents', which are deliberately secreted to appropriate an environmental advantage against yeasts and moulds.

Concerning essential oils, these molecules are defined as volatile aromatic organic compounds (made up of hydrocarbon molecules) that are considered to carry the 'essence or scent' of a plant (Burt, 2004). Like fatty acids, their lipophilic nature allows them to alter the fluidity of membrane and integrity leading to cell leakage. This may be a possible mode of action for killing microbes. In their study, Pinto and co-workers showed that essential oils from *Thymus pulegioides* (thymol and carvacrol; as determined by GC and GC-MS) displayed broad-spectrum antifungal activity against clinical *Candida*, *Aspergillus* and *dermatophyte* species (Pinto *et al.*, 2006). These authors attributed the killing effect on altered membrane function. Thus authors caution that considering the negative effects of these oils on fungi, which are of eukaryotic origin like humans, studies into toxicity and improved formulations, among others, should be undertaken in order to avert negative outcome when prescribed.

Other studies have also demonstrated the antifungal effect of essential oils against non-traditional pathogenic fungi, i.e. those that colonize the built environment (Verma *et al.*, 2011) and spoilage fungi of bakery products (Guynot *et al.*, 2005). Here, oils extracted from among others, included eucalyptus, cinnamon, rosemary, thyme and clove, were tested.

Other compounds

Anti-malarial drug: chloroquine

Chloroquine is primarily used as a prophylactic drug for treating malaria (Krafts, 2012). The drug's mode of action involves binding to haem proteins and forming a complex that is toxic to the cell and ultimately targets membrane function (Islahudin *et al.*, 2013). The chemosensitising properties of chloroquine have also been explored in other studies. Ma *et al.* (2006) reported that cryptococcal cells could take up residency inside macrophages and successfully replicate without evoking an immune response. More to this point, Webber and co-workers documented that treatment of parasitized-macrophages with chloroquine leads to accumulation of this drug inside macrophages forming iron-complexes that kill cryptococcal cells and *Histoplasma capsulatum* (Webber *et al.*, 2000). It should now be possible to combat such cells by chloroquine treatment. This sensitizing capability has also been seen when treating the usually drug-resistant *Candida* biofilms (Shinde *et al.*, 2013). By adding chloroquine in combined therapy with drugs such as azoles, biofilms became susceptible to these conventional antifungals.

A recent paper by Huang *et al.* (2012) demonstrated via using a functional genomics approach that chloroquine achieves fungal growth inhibition via blocking thiamine transportation. This vitamin plays a crucial role in glucose metabolism, where cells can derive energy to support cellular processes (Lonsdale, 2006).

Anti-psychotic drugs

The drug chlorpromazine is a dopamine antagonist, and works by acting on a number of receptors in the brain (David, 2002). This drug is mainly prescribed for the treatment of schizophrenia. However, a recent study has shown that chlorpromazine also possesses antifungal activity (Vitale *et al.*, 2007). This drug, together with another anti-psychotic agent, trifluoperazine, was shown to inhibit fungal, bacterial and protozoan species. The inhibitory effects of these drugs were noted at concentrations ranging from 16 μg/ml to 64 μg/ml. These authors did not elucidate on the possible mode of action employed by these drugs.

It is, nonetheless, conceivable that microbial death was achieved via destabilization of the membrane integrity leading to increased membrane permeability (Maruoka *et al.*, 2007).

In a recent study, Zhai *et al.* (2012) highlighted that the ineffectiveness of current antifungals may in part be due to poor central nervous system penetration, particularly in the case of disseminated cryptococcal infections. To circumvent this limitation, these authors sought to test the antimicrobial activity of sertraline (Zoloft®, Pfizer), which is an antidepressant drug, against *Cryptococcus neoformans*. Here, *in vitro* analysis revealed sertraline to have a fungicidal effect as well as act in synergy with fluconazole. From the above examples, it is clear that the localization of infection is an important consideration for drug delivery, and thus, a drug that is able to cross the blood–brain barrier may be a viable treatment option in case of a disseminated cryptococcal infection.

In an interesting study, Karababa *et al.* demonstrated that improved penetration or delivery of azoles could be achieved by addition of antipsychotic drugs like fluphenazine (Karababa *et al.*, 2004). This drug was shown, in elegant molecular studies, to up-regulate the multidrug transporter genes in azole-resistant *Candida albicans* strains – leading to enhanced efflux of azoles.

The picture emerging above makes for a convincing case concerning the practical application of antipsychotic drugs in the treatment of fungal infectious diseases.

Metal complexes

A metal complex is a chemical species that is formed when a ligand, i.e. functional group (molecule) is attached to a central metal to form a coordination complex (Fig. 7.4) (Greenwood and Earnshaw, 1997). Here, the ligand determines the capability of the metal to undergo a chemical reaction. Metal complexes have been studied quite extensively over the years, particularly in light of these metals and their ligands pronouncing strong biological activities (upon chelation) against cancer cells (Chohan *et al.*, 2006; Hossain *et al.*, 2009). An encouraging observation about complexes is that, they are more potent and less toxic compared with their parent compound (Swamy *et al.*, 2012).

Recently, the chelating quality of complexes has also been documented to cause microbial cell death in some studies. This is reflected in a paper by Hossain *et al.* (2009) where via using a disc diffusion method, they showed that metal complexes of copper(II), cobalt(III), rhodium(III) and platinum(IV) ions with dibasic acids displayed antifungal activity against two fungal species, *Botryodiplodia theobromae* and *Colletotrichum gloespariodes*. In another study, amino acid-derived compounds and their cobalt(II), copper(II), nickel(II), and zinc(II) metal complexes were screened for their *in vitro* antifungal activity against and shown to inhibit growth of *Trichophyton longifusus, Candida albicans, Aspergillus flavus, Microsporum canis, Fusarium solani* and *Candida glabrata* (Chohan et al., 2006).

The platinum complex cisplatin has also been shown to display antifungal activity against *Candida* species (Watanabe *et al.*, 2000). These complexes are reported to effect cell death by damaging the cell membrane. Conventionally, cisplatin is regarded as an anti-cancer drug and is therefore used in cancer chemotherapy (Watanabe *et al.*, 2000). The application of this drug as

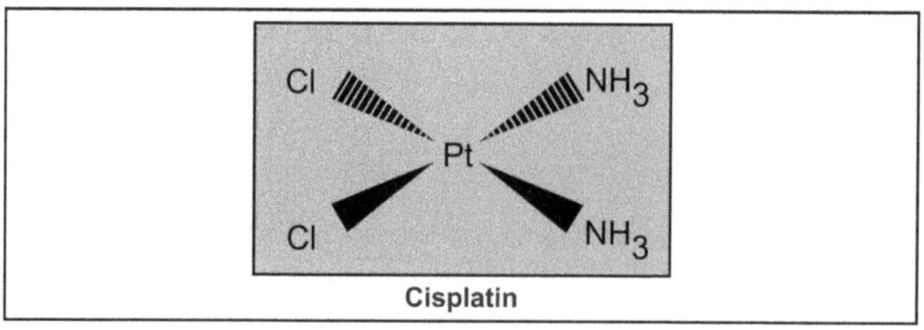

Figure 7.4 An example of a metal complex. Here, cisplatin is constituted by a metal (platinum) that is linked to four ligands, i.e. ($(NH_3)_2$ and $(Cl)_2$) via dipolar bonds to form a mirror-image.

an antifungal agent may be further limited due to inducing immunosuppression.

Molecules identified by high-throughput screening

High-throughput screening is a rapid and systematic process (primarily applied in drug discovery) were large numbers of compounds are mapped for their activity and their targets identified (Liu *et al.*, 2004). By design, the screening assays are carried out in a microtitre plate, where a logical entity like yeast cells, are added and allowed to interact with other components (drug) over time. In some cases, these targeted entities may be enzymes that catalyse key metabolic pathways. Subsequently, the resultant response (or non-response) is then measured using an automated analysis machine that has robotic capability to, among others, mix and add reagents. At the heart of it, HTS combines statistics and experiments and, thus, a researcher should be able to sort and identify important biochemical information (hit) from the background noise (Liu *et al.*, 2004; Mayr and Bojanic, 2009).

In recent years there has been a number of studies looking at identifying novel antifungal compounds through large screening of chemical libraries. Towards this end, we highlight a few of these. In their study, Krysan and Didone designed a luminescent assay that detected small molecules that disrupt yeast cell integrity (Krysan and Didone, 2008). The screened libraries [held at the Institute of Chemistry and Cell Biology-Longwood (ICCB-L)] included known compounds (Biomol, Prestwick I), commercial compounds (ChemDiv5) as well as biological extracts (NCCDG1 and ICBG6), which target among others, second messenger modulators, gene regulation agents and kinase inhibitors. Using *Saccharomyces cerevisiae* and *Candida albicans* as model yeasts; they screened over 4500 molecules and mapped the release of the intracellular enzyme adenylate kinase into the culture medium as a reporter for cell lysis. In another study, Breger and co-workers screened over 1200 compounds from ICCB-L that prolonged the survival of *Candida albicans*-infected nematodes and inhibited *in vivo* filamentation (Breger *et al.*, 2007). Here, they used nematodes to model and explain the pathogenesis of *Candida* in mammalian cells. Based on their screening results, caffeic acid phenethyl ester and enoxacin were shown to possess antifungal activity in a murine model. By exploring if nutrient-deprived host environment may assist in clearing disseminated cryptococcal infections, Rabjohns *et al.* (2013) showed in a HTS assay that a potential chemical scaffold, i.e. 10058-F4 displayed a cidal effect at low micromolar range. A Library of Pharmacologically Active Compounds (LOPAC – constituted by active compounds that target lipid biosynthesis, phosphorylation, ion channels, etc.) was screened and measured residual metabolic activity using alamarBlue (profluorescent dye) (Rabjohns *et al.*, 2013). Further to the concept of nutrient deprivation, Dehdashti and co-authors assessed the viability of *Cryptococcus neoformans*, under these conditions, using a miniaturized ATP-content based assay involving over 1200 compounds also obtained from LOPAC (Dehdashti *et al.*, 2013). In the end, they were able to identify four active compounds viz. niclosamide, malonoben, 6-bromoindirubin-3′-oxime, and 5-[(-4ethylphenyl)methylene]-2-thioxo-4-thiazolidinone, and further validated their data when they reproduced the cidal effect against clinical isolates. LaFleur and co-workers were able to identify and validate antifungal potentiators that enhanced activity of clotrimazole or acted on fungal biofilms by inhibiting their metabolic activity; after screening over 120000 small molecule compounds from the NIH Molecular Libraries Small Molecule Repository (LaFleur *et al.*, 2011). The library has sets of known molecules and natural products. Some of these molecules (i.e. 1,3-benzothiazole scaffold) were, in this study, shown to target biofilm formation. In an interesting study, Chaturvedi *et al.* (2011) studied the cidal effect of antifungal compounds and biocides against *Geomyces destructans. Geomyces* causes a deadly disease (white nose syndrome) in bats. After screening a SpectrumPlus compound library (made up of known compounds (including imidazoles) and natural products, they concluded that *Geomyces destructans* was susceptible to some drugs at a concentration gradient similar to that of human pathogenic fungi.

Conclusion and future perspectives

The rise in the emergence of new mycotic agents and drug resistance drives the need to find new and/or alternative drugs. Today, a picture is emerging where non-traditional antimicrobial agents, which were previously prescribed to treat non-infectious conditions, show antimicrobial properties. It is worth our time to explore these compounds further. Our ability to also conduct high throughput screening at institutions of higher learning (not only at pharmaceutical companies), has added impetus to this cause.

In this text, we presented reports where compounds ranging from anti-inflammatory drugs to anti-psychotic drugs are documented to control fungal growth. However, due to toxicities attributed to some of these drugs, their clinical application may be limited. Nonetheless, there are on-going efforts to modify and minimize the side-effects of some of these drugs. For example, new derivatives of anti-inflammatory drugs are being synthesized, and thus, in the future these new derivatives should also be tested for their antimicrobial properties. It is also important to demonstrate if these 'new' antifungals will work in host cells, and probably in some cases, in situations of advanced disease. Towards this end, the usage of animals in modelling human disease should be considered in order to clearly establish the therapeutic benefits of these chemical compounds.

Another ancillary effect of these drugs that should be explored is the application of anti-inflammatory drugs in the treatment of pathogen-induced immune reconstitution inflammatory syndrome. The duality in function of these molecules as antifungal agents and anti-inflammatory drugs may provide an additional beneficial therapeutic outcome. To this end, studies are required to test the possible clinical outcomes of anti-inflammatory use to fungal disease.

The major limitation in providing appropriate medical treatment is the cost of drugs, which is a pertinent issue in developing countries. Therefore, the therapeutic benefit of anti-inflammatory drugs (aspirin) and natural products (plant extracts), which are relatively cheap and widely distributed, should be explored.

Moreover, the possibility to take some of the above-mentioned non-traditional antimicrobial drugs as prophylactic agents should also be considered. This may be so in the case of natural products; were some are already taken in as supplements.

An important note is that some of these compounds in *in vitro* assays act in synergy with traditional antimicrobial agents, such as azoles. This is one attribute that is encouraging in that it implies these compounds can be added at low concentrations where clinically they could not exert any negative physiological outcomes. Additionally, the added effect of these chemical compounds in combination therapy enhances the efficiency of traditional antimicrobial drugs, which under normal conditions may be fungistatic. More intriguing to the above discussion, is the application of antimalarial drugs that seem to enhance immuno-processing of parasitizing fungal cells in macrophages instead of enhancing the effect of another drug. It is also worth the time to establish if clinical populations taking antimalarial drugs may be protected from fungal disease.

More to this point, the concept of also extending the conventional application of chemosensitizers and photosensitizers from treating cancer cells to combating microbial infections should be explored. Special consideration should be placed in the defined scope of enhancing the host's immune response to microbial infection. To date, reports have emerged that show natural compounds like curcumin, when chemically or optically activated, can kill microbes by releasing reactive oxygen species, which in turn, target cellular components. The latter is in some respect, similar to the cytotoxic 'oxidative burst' effect, which is expressed by immune cells in response to microbial infection. By necessary extension, metals and their complexes, which are also used to treat cancer cells, should be examined for their antimicrobial properties.

Once again we point out that careful consideration should be taken when attempting such approaches – especially in the case of (i) compounds that generally target eukaryotic organisms, and (ii) complexes and sensitizers that may induce immunosuppression, in order to realize the desired therapeutic outcome to the exclusion of adverse effects.

Acknowledgements

The authors wish to thank Professor Alex Idnurm, University of Missouri-Kansas City, for his critical reading of the draft paper and valuable comments.

References

Abdelmegeed, E., and Shaaban, M.I. (2013). Cyclooxygenase inhibitors reduce biofilm formation and yeast-hypha conversion of fluconazole resistant *Candida albicans*. J. Microbiol. *51*, 598–604.

Alba-Lois, L., and Segal-Kischinevzky, C. (2010). Beer and wine makers. Nat. Edu. 3, 17.

Alem, M.A.S., and Douglas, L.J. (2004). Effects of aspirin and other nonsteroidal anti-inflammatory drugs on biofilms and planktonic cells of *Candida albicans*. Antimicrob. Agents Chemother. *48*, 41–47.

Aqil, F., Zahin, M., Ahmad, I., Owais, M., Khan, M.S.A., Bansal, S.S., and Farooq, S. (2010). Antifungal activity of medicinal plant extracts and phytocompounds: a review. In Combating Fungal Infections: Problems and Remedies, I. Ahmad, M. Owais, M. Shahid, F. Aqil, eds. (Springer, Berlin, Germany), pp. 449–484.

Armstrong-James, D., Meintjes, G., and Brown, G.D. (2014). A neglected epidemic: fungal infections in HIV/AIDS. Trends Microbiol. 22, 120–127.

Ashbee, H.R., and Bignell, E.M. (2010). The Yeast Handbook: Pathogenic Yeasts. (Springer-Verlag, Berlin, Germnay).

Atai, Z., Atapour, M., and Mohseni, M. (2009). Inhibitory effect of ginger extract on *Candida albicans*. Am. J. Appl. Sci. *6*, 1067–1069.

Baker, M. (1996). Fur rubbing: use of medicinal plants by Capuchin monkeys. Am. J. Primatol. *38*, 263–270.

Bartlett, K.H., Kidd, S.E., and Kronstad, J.W. (2008). The emergence of *Cryptococcus gattii* in British Columbia and the Pacific Northwest. Curr. Infect. Dis. Rep. *10*, 58–65.

Biggs, R.D. (1995). Medicine, surgery, and public health in ancient Mesopotamia. In Civilizations of the Ancient Near East, J.M. Sasson, ed. (Charles Scribner's Sons, New York, NY), pp. 1911–1924.

Boulware, D.R., Bonham, S.C., Meya, D.B., Weisner, D.L., Park, G.S., Kambugu, A., Janoff, E.N., and Bohjanen, P.R. (2010). Paucity of initial cerebrospinal fluid inflammation in cryptococcal meningitis is associated with subsequent immune reconstitution inflammatory syndrome. J. Infect. Dis. *202*, 962–970.

Breger, J., Fuchs, B.B., Aperis, G., Moy, T.I., Ausubel, F.M., and Mylonakis, E. (2007). Antifungal chemical compounds identified using a *C. elegans* pathogenicity assay. PLoS Pathog. 3, e18.

Brown, G.D., and Netea, M.G. (2007). Immunology of Fungal Infections. (Springer, Dordrecht, the Netherlands).

Burt, S. (2004). Essential oils: their antibacterial properties and potential applications in foods – a review. Int. J. Food Microbiol. *94*, 223–253.

Busti, A.J., Justin, S., Hooper, J.S., Christopher, J., Amaya, C.J., and Kazi, S. (2005). Effects of perioperative antiinflammatory and immunomodulating therapy on surgical wound healing. Pharmacotherapy *25*, 1566–1591.

Calderone, R.A. (2002). *Candida* and Candidiasis. (ASM Press, Washington, DC).

Calderone, R.A., and Clancy, C.J. (2012). *Candida* and Candidiasis, 2nd edn. (ASM Press, Washington, DC).

Casadevall, A., and Perfect, J.R. (1998). *Cryptococcus neoformans*. (ASM Press, Washington, DC).

Cauvain, S.P., and Young, L.S. (2007). Technology of Breadmaking, 2nd edn. (Springer, New York, NY).

Chaturvedi, S., Rajkumar, S.S., Li, X., Hurteau, G.J., Shtutman, M., and Chaturvedi, V. (2011). Antifungal testing and high-throughput screening of compound library against *Geomyces destructans*, the etiologic agent of geomycosis (wns) in bats. PLoS One *6*, e17032.

Chohan, Z.H., Arif, M., Akhtar, M.A., and Supuran, C.T. (2006). Metal-based antibacterial and antifungal agents: synthesis, characterization, and *in vitro* biological evaluation of Co(II), Cu(II), Ni(II), and Zn(II) complexes with amino acid-derived compounds. Bioinorg. Chem. Appl. *2006*, 1–14.

Ciccoli, R., Sahi, S., Singh, S., Prakash, H., Zafiriou, M.-P., Ishdorj, G., Kock, J.L.F., and Nigam, S. (2005). Oxygenation by cyclooxygenase-2 (COX-2) of 3-hydroxyeicosatetraenoic acid (3-HETE), a fungal mimetic of arachidonic acid, produces a cascade of novel bioactive 3-hydroxy-eicosanoids. Biochem. J. *390*, 737–747.

Codon, A.C., Gasent-Ramirez, J.M., and Benitez, T. (1995). Factors which affect the frequency of sporulation and tetrad formation in *Saccharomyces cerevisiae* baker's yeast. Appl. Environ. Microbiol. *61*, 630–638.

Collett, G., and Parrish, A. (2007). Fluconazole donation and outcomes assessment in cryptococcal meningitis. S. Afr. Med. J. 97, 175–176.

Cooper, C.R. (2011). Yeasts pathogenic to humans. In The Yeasts, A Taxonomic Study, 5th edn, C.P. Kurtzman, J.W. Fell, T. Boekhout, eds. (Elsevier B.V., Amsterdam, the Netherlands), pp. 9–20.

David, H. (2002). The Creation of Psychopharmacology. (Harvard University Press, Boston, MA).

Dehdashti, S.J., Abbott, J., Nguyen, D.T., McKew, J.C., Williamson, P.R., and Zheng, W. (2013). A high-throughput screening assay for assessing the viability of *Cryptococcus neoformans* under nutrient starvation conditions. Anal. Bioanal. Chem. *405*, 6823–6829.

Domenico, P., Schwartz, S., and Cunha, B.A. (1989). Reduction of capsular polysaccharide production in *Klebsiella pneumoniae* by sodium salicylate. Infect. Immun. 57, 3778–2782.

Dovigo, L.N., Pavarina, A.C., Carmello, J.C., Machado, A.L., Brunetti, I.L., and Bagnato, V.S. (2011). Susceptibility of clinical isolates of *Candida* to photodynamic effects of curcumin. Laser Surg. Med. 43, 927–934.

Farnsworth, N.R. (1988). Screening plants for new medicines. In Biodiversity, E.O. Wilson, ed. (National Academy Press, Washington, DC), pp. 83–97.

Farrugia, G., Bannister, W.H., Vassallo, N., and Balzan, R. (2013). Aspirin-induced apoptosis of yeast cells is associated with mitochondrial superoxide radical accumulation and NAD(P)H oxidation. FEMS Yeast Res. *13*, 755–768.

Ficker, C.E., Arnason, J.T., Vindas, P.S., Alvarez, L.P., Akpagana, K., Gbeassor, M., De Souza, C., and Smith, M.L. (2003). Inhibition of human pathogenic fungi by ethnobotanically selected plant extracts. Mycoses *46*, 29–37.

Fraenkel, G.S. (1959). The raison d'être of secondary plant substances. Science *129*, 1466–1470.

Ghannoum, M.A., and Rice, L.B. (1999). Antifungal agents: mode of action, mechanisms of resistance, and correlation of these mechanisms with bacterial resistance. Clin. Microbiol. Rev. *12*, 501–517.

Glasgow, J.F.T., Middleton, B., Moore, R., Gray, A., and Hill, J. (1999). The mechanism of inhibition of β-oxidation by aspirin metabolites in skin fibroblasts from Reye's syndrome patients and controls. Biochim. Biophys. Acta *1454*, 115–125.

Gonzales, F.P., and Maisch, T. (2012). Photodynamic inactivation for controlling *Candida albicans*. Fungal Biol. *116*, 1–10.

Greenwood, N.N., and Earnshaw, A. (1997). Chemistry of the Elements, 2nd edn. (Elsevier, Berlin, Germany).

Gur, S., Turgut-Balik, D., and Gur, N. (2006). Antimicrobial activities and some fatty acids of turmeric, ginger root and linseed used in the treatment of infectious diseases. World J. Agric. Sci. *2*, 239–442.

Guynot, M.E., Marín, S., Setó, L., Sanchis, V., and Ramos, A.J. (2005). Screening for antifungal activity of some essential oils against common spoilage fungi of bakery products. Food Sci. Technol. Int. *11*, 25–32.

Haddow, L.J., Colebunders, R., Meintjes, G., Lawn, S.D., Elliott, J.H., Manabe, Y.C., Bohjanen, P.R., Sungjanuparph, S., Easterbrook, P.J., French, M.A., *et al.* (2010). Cryptococcal immune reconstitution inflammatory syndrome in HIV-1-infected individuals: literature review and proposed clinical case definitions. Lancet Infect. Dis. *10*, 791–802.

Heil, M., and Bostock, R.M. (2002). Induced systemic resistance against pathogens in the context of induced plant defences. Ann. Bot. *89*, 503–512.

Heitman, J., Kozel, T.R., Kwon-Chung, J., Perfect, J.R., and Casadevall, A. (2011). *Cryptococcus*: from Human Pathogen to Model Yeast. (ASM Press, Washington, DC).

Hossain, M.B., Islam, M.S., Hassan, M.N., Yousef, M.A., and Salam, M.A. (2009). Antifungal activity of mixed ligand transition metal complexes of some dibasic acid and heterocyclic bases. Daffodil Int. Uni. J. Sci. Technol. *4*, 57–59.

Huang, Z., Srinivasan, S., Zhang, J., Chek, K., Li, Y., Li, W., Quiocho, F.A., and Pan, X. (2012). Discovering thiamine transporters as targets of chloroquine using a novel functional genomics strategy. PLoS Genet. *8*, e1003083.

Islahudin, F., Khozoie, C., Bates, S., Ting, K.-N., Pleass, R.J., and Avery, S.V. (2013). Cell wall perturbation sensitizes fungi to antimalarial drug chloroquine. Antimicrob. Agents Chemother. *57*, 3889–3896.

Johnson, E.A., and Echavarri-Erasun, C. (2011). Yeast biotechnology. In The Yeasts, A Taxonomic Study, 5th edn, C.P. Kurtzman, J.W. Fell, Boekhout T, eds. (Elsevier B.V., Amsterdam, the Netherlands), pp. 22–44.

Karababa, M., Coste, A.T., Rognon, B., Bille, J., and Sanglard, D. (2004). Comparison of gene expression profiles of *Candida albicans* azole-resistant clinical isolates and laboratory strains exposed to drugs inducing multidrug transporters. Antimicrob. Agents Chemother. *48*, 3064–3079.

Kidd, S.E., Hagen, F., Tscharke, R.L., Huynh, M., Bartlett, K.H., Fyfe, M., MacDougall, L., Kwon-Chung, K.J., and Meyer, W. (2004). A rare genotype of *Cryptococcus gattii* caused the cryptococcosis outbreak on Vancouver Island (British Columbia, Canada). Proc. Natl. Acad. Sci. U.S.A. *101*, 17258–17263.

Kock, J., Coetzee, D., Van Dyk, M., Truscott, M., Cloete, P., van Wyk, V., and Augustyn, O.P.H. (1991). Evidence of pharmacologically active prostaglandins in yeasts. S. Afr. J. Sci. *87*, 73–76.

Kock, J.L.F., and Botha, A. (1998). Fatty acids in fungal taxonomy. In Chemical Fungal Taxonomy, J.C. Frisvad, P.D. Bridge, D.K. Arora, eds. (Marcel Dekker, New York, NY), pp. 216–246.

Kock, J.L.F., Strauss, C.J., Pohl, C.H., and Nigam, S. (2003). The distribution of 3-hydroxy oxylipins in fungi. Prostaglandins Other Lipid Mediat. *71*, 85–96.

Kock, J.L.F., Sebolai, O.M., Pohl, C.H., van Wyk, P.W.J., and Lodolo, E.J. (2007). Oxylipin studies expose aspirin as antifungal. FEMS Yeast Res. *7*, 1207–1217.

Krafts, K., Hempelmann, E., and Stania, S. (2012). From methylene blue to chloroquine: a brief review of the development of an antimalarial therapy. Parasitol. Res. *11*, 1–6.

Krysan, D.J., and Didone, L. (2008). A high-throughput screening assay for small molecules that disrupt yeast cell integrity. J. Biomol. Screen *181*, 1073–1081.

Kurtzman, C.P., Fell, J.W., and Boekhout, T. (2011). The Yeasts, A Taxonomic Study, 5th edn. (Elsevier B.V., Amsterdam, the Netherlands).

Kwon-Chung, K.J. (2011). *Filobasidiella* Kwon-Chung (1975). In The Yeasts, A Taxonomic Study, 5th edn, C.P. Kurtzman, J.W. Fell, T. Boekhout, eds. (Elsevier B.V., Amsterdam, the Netherlands), pp. 1443–1456.

LaFleur, M.D., Lucumi, E., Napper, A.D., Diamond, S.L., and Lewis, K. (2011). Novel high-throughput screen against *Candida albicans* identifies antifungal potentiators and agents effective against biofilms. J. Antimicrob. Chemother. *66*, 820–826.

Leeuw, N.J., Swart, C.W., Ncango, D.M., Pohl, C.H., Sebolai, O.M., Strauss, C.J., Botes, P.J., van Wyk, P.W.J., Nigam, S., and Kock, J.L.F. (2007). Acetylsalicylic acid as antifungal in *Eremothecium* and other yeasts. Antonie van Leeuwenhoek *91*, 393–405.

Levy, G. (1976). Pharmacokinetics of aspirin in man. J. Investig. Dermatol. *67*, 667–668.

Litter, J. (2004). Analysis of the mitochondrial genome of the human pathogenic yeast *Cryptococcus neoformans*. Acta Biol. Szeged. *48*, 61.

Liu, B., Li, S., and Hu, J. (2004). Technological advances in high-throughput screening. Am. J. Pharmacogenom. *4*, 263–276.

Lonsdale, D. (2006). A review of the biochemistry, metabolism and clinical benefits of thiamin(e) and its derivatives. Evid. Based Complement Alternat. Med. 3, 49–59.

Lyon, J.P., and de Resende, M.A. (2006). Correlation between adhesion, enzyme production, and susceptibility to fluconazole in *Candida albicans* obtained from denture wearers. Oral Surg. Oral Med. Oral Pathol. Oral Radiol. Endod. *102*, 632–638.

Ma, H., Croudace, J.E., Lammas, D.A., and May, R.C. (2006). Expulsion of live pathogenic yeast by macrophages. Curr. Biol. *16*, 2156–2160.

Marmiroli, N., Ferri, M., and Puglisi, P.P. (1983). Involvement of mitochondrial protein synthesis in sporulation: effects of erythromycin on macromolecular synthesis, meiosis, and ascospore formation in *Saccharomyces cerevisiae*. J. Bacteriol. *154*, 118–129.

Martins, C.V.B., da Silva, D.L., Neres, A.T.M., Magalhaes, T.F.F., Watanabe, G.A., Modolo, L.V., Sabino, A.A., de Fatima, A., and de Resende, M.A. (2008). Curcumin as a promising antifungal of clinical interest. J. Antimicrob. Chem. *63*, 337–339.

Maruoka, N., Murata, T., Omata, N., Takashima, Y., Tanii, H., Yonekara, Y., Fubibayashi, Y., and Wada, Y. (2007). Effects of chlorpromazine on plasma membrane permeability and fluidity in the rat brain: a dynamic position autoradiography and fluorescence polarization study. Prog. Neuropsychiatry Biol. Psychiatry *31*, 178–186.

Mayr, L.M., and Bojanic, D. (2009). Novel trends in high-throughput screening. Curr. Opin. Pharmacol. *9*, 580–588.

Meena A.K., Khan, S., Rao, M.M., Reedy, R.K., and Padhi, M.M. (2013). Antifungal metabolites from medicinal plants used in Ayurvedic system of medicine in India. In Antifungal Metabolites from Plants, M. Razzaghi-Abyaneh, M. Rai, eds. (Springer-Verlag, Berlin, Germany).

Nadanaciva, S., Bernal, A., Aggeler, R., Capaldi, R., and Will, Y. (2007). Target identification of drug induced mitochondrial toxicity using immunocapture based OXPHOS activity assays. Toxicol. In Vitro *21*, 902–911.

Ncango, D.M., Pohl, C.H., van Wyk, P.W.J., and Kock, J.L.F. (2011). Non-steroidal anti-inflammatory drugs (NSAIDs) inhibit growth of yeast pathogens. Afr. J. Microbiol. Res. *5*, 1123–1125.

Negri, M., Salci, T.P., Shinobu-Mesquita, C.S., Capoci, I.R.G., Svidzinski, T.I.E., and Kioshima, E.S. (2014). Review: early state research on antifungal natural products. Molecules *19*, 2925–2956.

Norman, C., Howell, K.A., Millar, H., Whelan, J.M., and Day, D.A. (2004). Salicylic acid is an uncoupler and inhibitor of mitochondrial electron transport. Plant Physiol. *134*, 492–501.

Nunn, J.F. (1996). Ancient Egyptian Medicine. (University of Oklahoma Press, Norman, OK).

Pappas, P.G., Kauffman, C.A., Andes, D., Benjamin, D.K., Calandra, T.F., Edwards, J.E., Filler, S.G., Fisher, J.F., Kullberg, B.-J., Ostrosky-Zeichner, L., *et al.* (2009). Clinical practice guidelines for the management of candidiasis: 2009 update by the Infectious Diseases Society of America. Clin. Infect. Dis. *48*, 503–535.

Park, B.J., Wannemuehler, K.A., Marston, B.J., Govender, N., Pappas, P.G., and Chiller, T.M. (2009). Estimation of the current global burden of cryptococcal meningitis among persons living with HIV/AIDS. AIDS *23*, 525–530.

Perfect, J.R., Dismukes, W.E., Dromer, F., Goldman, D.L., Graybill, J.R., Hamill, R.J., Harrison, T.S., Larsen, R.A., Lortholary, O., Nguyen, M., *et al.* (2010). Clinical practice guidelines for the management of cryptococcal disease: 2010 update by the Infectious Diseases Society of America. Clin. Infect. Dis. *50*, 291–322.

Pina Vaz, C., Sansonetty, F., Rodrigues, A.G., Martinez-De-Oliviera, J., Fonseca, A.F., and Mardh, P.A. (2000). Antifungal activity of ibuprofen alone and in combination with fluconazole against *Candida* species. J. Med. Microbiol. *49*, 831–840.

Pinto, E., Pina-Vaz, P., Salgueiro, L., Goncalves, M.J., Costa-de-Oliveira, S., Cavaleiro, C., Palmeira, A., Rodrigues, A., and Martinez-de-Oliveira, J. (2006). Antifungal activity of the essential oil of *Thymus pulegioides* on *Candida, Aspergillus* and dermatophyte species. J. Med. Microbiol. *55*, 1367–1373.

Rabjohns, J.L., Park, Y.D., Dehdashti, J., Sun, W., Henderson, C., Zelazny, A., Metallo, S.J., Zheng, W., and Williamson, P.R. (2014). A high-throughput screening assay for fungicidal compounds against *Cryptococcus neoformans*. J. Biomol. Screen *19*, 270–277.

Rainsford, K.D. (2007). Anti-inflammatory drugs in the 21st century. Subcell Biochem. *42*, 3–27.

Ramage, G., Rajandran, R., Sherry, L., and Williams, C. (2012). Fungal biofilm resistance. Int. J. Microbiol. *2012*, 528521.

Rao, P.N.P., and Knaus, E.E. (2008). Evolution of nonsteroidal anti-inflammatory drugs (NSAIDs): cyclooxygenase (COX) inhibition and beyond. J. Pharm. Pharmaceut. Sci. *11*, 81s-110s.

Sanders, T., and Emery, P. (2003). The Molecular Basis of Human Nutrition. (Taylor Frances, London, UK).

Sanyal, A.K., Roy, D., Chowdhury, B., and Banaerjee, A.B. (1993). Ibuprofen, a unique anti-inflammatory compound with antifungal activity against dermatophytes. Lett. Appl. Microbiol. *17*, 109–111.

Scott, E.M., Tariq, V.N., and McCrory, R.M. (1995). Demonstration of synergy with fluconazole and either ibuprofen, sodium salicylate, or propylparaben against *Candida albicans in vitro*. Antimicrob. Agents Chemother. *39*, 2610–2614.

Sebolai, O.M., Pohl, C.H., Botes, P.J., van Wyk, P.W.J., Mzizi, R., Swart, C.W., and Kock, J.L.F. (2008a). Distribution of 3-hydroxy oxylipins and acetylsalicylic acid sensitivity in *Cryptococcus* species. Can. J. Microbiol. *54*, 111–118.

Sebolai, O.M., Pohl, C.H., Botes, P.J., van Wyk, P.W.J., Botha, A., and Kock, J.L.F. (2008b). The influence of acetylsalicylic acid on oxylipin migration in *Cryptococcus neoformans* var. *neoformans* UOFS Y-1378. Can. J. Microbiol. *54*, 91–96.

Sebolai, O.M., Pohl, C.H., Kock, J.L.F., Chatuverdi, V., and Del Poeta, M. (2012). The presence of 3-hydroxy oxylipins in pathogenic microbes. Prostaglandins Other Lipid Mediat. *97*, 17–12.

Shiby, V.K., and Mishra, H.N. (2013). Fermented milks and milk products as functional foods – a review. Crit. Rev. Food Sci. Nutr. *53*, 482–496.

Shinde, R.B., Raut, J.S., Chauhan, N.M., and Karuppayil, S.M. (2013). Chloroquine sensitizes biofilms of *Candida albicans* to antifungal azoles. Braz. J. Infect. Dis. *17*, 395–400.

Sjogren, J., Magnusson, J., Broberg, A., Schnurer, J., and Kenne, l. (2003). Antifungal 3-hydroxy fatty acids from *Lactobacillus plantarum* MiLAB 14. Appl. Environ. Microbiol. *69*, 7554–7557.

Soll, D.R. (2002). *Candida* commensalism and virulence: the evolution of phenotypic plasticity. Acta Trop. *81*, 101–110.

Somasundaram, S., Rafi, S., Hayllar, J., Sigthorsson, G., Jacob, M., Price, A.B., Macpherson, A., Mahmod, T., Scott, D., Wrigglesworth, J.M., *et al.* (1997). Mitochondrial damage: a possible mechanism of the "topical" phase of NSAID induced injury to the rat intestine. Gut *41*, 344–353.

Swamy, D.K., Pachling, S.P., and Bhagat, T.M. (2012). Synthesis, characterization, antibacterial and antifungal studies of metal complexes with benzothiazolyl hydrazine. Rayasan J. Chem. *5*, 208–213.

Thibane, V.S. (2012). The effects of fatty acids on biofilm formation, oxidative stress and antifungal susceptibility of *Candida albicans* and *Candida dubliniensis*. Magister Scientiae Dissertation (University of the Free State, Bloemfontein, South Africa).

Thibane, V.S., Ells, R., Hugo, A., Albertyn, J., Janse van Rensburg, W., van Wyk, P.W.J., Kock, J.L.F., and Pohl, C.H. (2012). Polyunsaturated fatty acids cause apoptosis in *C. albicans* and *C. dubliniensis* biofilms. Biochim. Biophys. Acta *1820*, 1463–1468.

Van Dyk, M.S., Kock, J.L.F., Botha, A., Coetzee, D.J., Botes, P.J., Augustyn, O.P.H., and Nigam, S. (1993). 3-Hydroxy-5,8,11,14 (all *cis*)-eicosatetraenoic acid (3-HETE) – a new aspirin sensitive arachidonic acid metabolite from yeasts. In Eicosanoid and Other Bioactive Lipids in Cancer, Inflammation and Radiation Injury, S. Nigam, K.V. Honn, L.J. Marnett, T.L. Walden, eds. (Kluwer Academic Publishers, Boston, MA) pp. 67–70.

Van Dyk, M.S., Kock, J.L.F., Coetzee, D.J., Augustyn, O.P.H., and Nigam, S. (1991). Isolation of a novel aspirin sensitive arachidonic acid metabolite 3-hydroxy-5,8,11,14-eicosatetraenoic acid (3-HETE) from the yeast *Dipodascopsis uninucleata* UOFS-Y128. FEBS Lett. *283*, 195–198.

Vane, J.R. (1971). Inhibition of prostaglandin synthesis as a mechanism of action for aspirin-like drugs. Nat. New Biol. *231*, 232–235.

Verma, R.K., Chaurasia, L., and Kumar, M. (2011). Antifungal activity of essential oils against selected building fungi. Indian J. Nat. Prod. Resour. *2*, 448–451.

Vitale, R.G., Alfeltra, J., Meis, F.J., and Verweij, P.E. (2007). Activity and post antifungal effect of chlorpromazine and trifluopherazine against *Aspergillus, Scedosporium* and zygomycetes. Mycoses *50*, 270–276.

Watanabe, T., Takano, M., Ogasawara, A., Mikami, T., Kobayashi, T., Watabe, M., and Matsumoto, T. (2000). Anti-*Candida* activity of a new platinum derivative. Antimicrob. Agents Chemother. *44*, 2853–2854.

Webber, S.M., Levitz, S.M., and Harrison, T.S. (2000). Chloroquine and the fungal phagosome. Curr. Opi. Microbiol. *3*, 349–353.

Weitzman, I., and Summerbell, R.C. (2005). The dermatophytes. Clin. Microbiol. Rev. *8*, 240–259.

Wertheimer, A.I., Santella, T.M., and Lauver, H.J. (2004). Successful public/private donation programs: a review of the diflucan partnership program in South Africa. J. Int. Assoc. Physicians AIDS Care *3*, 74–79.

World Health Organization (1998). The World Health Report. Life in the 21st Century: A Vision for All. (World Health Organization Geneva, Switzerland).

Zhai, B., Wu, C., Wang, L., Sachs, M.S., and Lin, X. (2012). The antidepressant sertraline provides a promising therapeutic option for neurotropic cryptococcal infections. Antimicrob. Agents Chemother. *56*, 3758–3766.

Zhou, Y., Wang, G., Li, Y., Liu, Y., Song, Y., Zheng, W., Zhang, N., Hu, X., Yan, S., and Jia, J. (2012). *In vitro* interactions between aspirin and amphotericin B against planktonic cells and biofilm cells of *Candida albicans* and *C. parapsilosis*. Antimicrob. Agents Chemother. *56*, 3250–3260.

Exploring New Insights into Fungal Biology as Novel Antifungal Drug Targets

8

Rebecca A. Hall and Robin C. May

Abstract

Although there are currently a number of good antifungal therapies available for the treatment of invasive fungal disease, the growing use of immune suppressive therapies and the ageing population has led to widespread use of antifungals. This overuse has led to the development of antifungal resistance. Therefore, there is an urgent need for the identification of novel drug targets. This chapter highlights some of the recent discoveries in fungal biology including CO_2 sensing, quorum sensing, cell wall biogenesis and intracellular proliferation, and we discuss the potential of these discoveries in the search for novel antifungal drug targets.

Introduction

A recent study that quantified the global burden of fungal infections on healthcare systems, estimated that 2,075,000 people worldwide per year suffer with some form of life-threatening fungal disease (Brown *et al.*, 2012). Although fungal infections substantially contribute to human mortality and morbidity, fungal research is still an under represented area of science. Most fungi are opportunistic pathogens infecting individuals whose immune system has been weakened by a predisposing condition (i.e. cancer, AIDS, trauma, neutropenia and age). The ageing population, together with the increased use of immune suppressive therapies, has greatly enhanced the number of individuals at risk of invasive fungal disease. Current therapeutic options have adverse side-effects and antifungal resistance is a growing concern. In addition, many fungi affect people in developing counties, where there is insufficient medical expertise for the diagnosis and treatment of fungal infections. Therefore, there is a clear need for the development of rapid, cost-effective, efficient diagnostic techniques, novel antifungals and vaccines against fungi.

The requirement for novel antifungals stems from the widespread use of current antifungals, which has led to the development of antifungal resistance. In agreement with this, *Candida glabrata* that was once thought of as non-pathogenic is now a major contributor to mucosal and systemic infections (Hitchcock *et al.*, 1993; Li *et al.*, 2007; vanden Bossche *et al.*, 1992). Depending on the site of infection, *C. glabrata* is now the second or third causative agent of candidiasis (Pfaller *et al.*, 2007; Tan *et al.*, 2010). This increase in persistence at infection sites is a result of *C. glabrata*'s natural resistance to the commonly prescribed azole compounds (i.e. fluconazole).

One of the technical difficulties in the development of new antifungals is the choice of target. This is because, aside from the fungal cell wall, there are very few proteins or metabolic pathways that are specific to fungi and could thus be targeted without damage to the host. Significant efforts have been put into identifying compounds that inhibit cell wall or plasma membrane biogenesis. The most successful compounds to come out of these studies have been the azoles and echinocandins. Azoles have been successfully used in clinics since the 1980s. Despite the early advances in identifying six azole formulations with good *in vivo* efficacy (miconazole, ketoconazole, fluconazole, itraconazole, posaconazole and voriconazole), there have been no recent additions to this class of antifungals. Several additional azole compounds have been tested in clinical trials, but have not made it through to the development stage. However, in the last two years, eight new patents covering imidazole and triazole derivatives have been registered. Many of the azole compounds disclosed in these patents have high activity against *Candida* spp. (including *C. glabrata*), *Cryptococcus* spp. and *Aspergillus* spp., suggesting that these compounds have high therapeutic potential (Castelli *et al.*, 2014). For the echinocandin class of antifungal agents, three have made it into the clinical setting (micafungin, caspofungin and anidulafungin). Although still relatively new, resistance in the clinics is already being noted (Pfaller *et al.*, 2006a,b). In addition to developing new classes of antifungals, there have also been alterations in drug formulations in order to improve delivery and reduce cytotoxic effects. For example, recently developed lipid formulations of the polyene amphotericin B offer reduced renal toxicity and ultimately may offer the promise of oral delivery over intravenous administration. An oral formulation of amphotericin B would be of great benefit, as its administration in developing countries is hampered due to the lack of trained medical professionals able to deliver intravenous drugs. As a result, less efficient antifungals are often prescribed in such resource-poor settings, leading to ineffective patient care.

In an attempt to identify new antifungal targets, many fundamental studies into the mechanisms of fungal biology and pathogenicity have been performed. Not only have these studies enhanced our understanding of fungal biology; they have offered novel targets with the potential to become antifungal drug targets. In this chapter, we discuss key virulence attributes of fungal pathogens including CO_2 sensing, quorum sensing, cell wall biogenesis, intracellular proliferation and antimicrobial peptides as novel mechanisms to target fungal disease.

Fungal CO_2 sensing

In the host, pathogenic fungi will be exposed to increased partial pressures of carbon dioxide. In the blood, CO_2 concentrations are approximately 150-fold higher than atmospheric conditions (5% vs. 0.033%), and can reach even higher concentrations in deep tissues. Therefore, the ability of fungi to adapt and respond to these conditions is essential to their virulence. Fungal CO_2 sensing was initiated by pioneering studies in *C. albicans*, where elevated CO_2 concentrations were found to promote the yeast to hyphal transition of this fungus (Klengel *et al.*, 2005). CO_2 sensing was initially brought to the forefront of fungal research as a rapid diagnostic tool for the identification of *Candida* species. Although other *Candida* species are filamentous, CO_2 only induces hyphal formation in *C. albicans* (Sheth *et al.*, 2005). Therefore, this rapid phenotypic test (<18 hours) provides an efficient non-labour intensive method for quick identification of *C. albicans* from mixed samples. Since then, CO_2 has been shown to regulate a variety of functions involved in fungal virulence including white-opaque switching in *C. albicans* (Huang *et al.*, 2009), capsule synthesis and mating in *Cryptococcus neoformans* (Bahn *et al.*, 2005; Zaragoza *et al.*, 2003), allergenicity of *Aspergillus fumigatus* (Lang-Yona *et al.*, 2013), and aflatoxin synthesis in *Aspergillus flavus* (Giorni *et al.*, 2008).

Fungal CO_2 sensing: a role for soluble adenylyl cyclase

Extensive investigation into CO_2 sensing pathways in *C. neoformans* and *C. albicans* has elucidated two key proteins essential for the adaptation to changes in environmental CO_2 (Fig. 8.1), the soluble adenylyl cyclase (sAC) and carbonic anhydrases (CA) (Mogensen *et al.*, 2006; Klengel *et al.*, 2005). Adenylyl cyclases (AC) are ubiquitous enzymes, which synthesize the secondary messenger cAMP. This secondary messenger then activates the cAMP dependent protein kinase A (PKA), which then goes on to activate/inhibit kinases functioning in multiple signalling cascades. In contrast to mammalian adenylyl cyclases that contain transmembrane domains linking the AC to the inner leaflet of the membrane, the fungal AC lacks any transmembrane domains and is thought to be soluble, although the specific localization of the sAC still remains to be determined. Like ACs the fungal sAC is activated by small GTPases. In *C. albicans*,

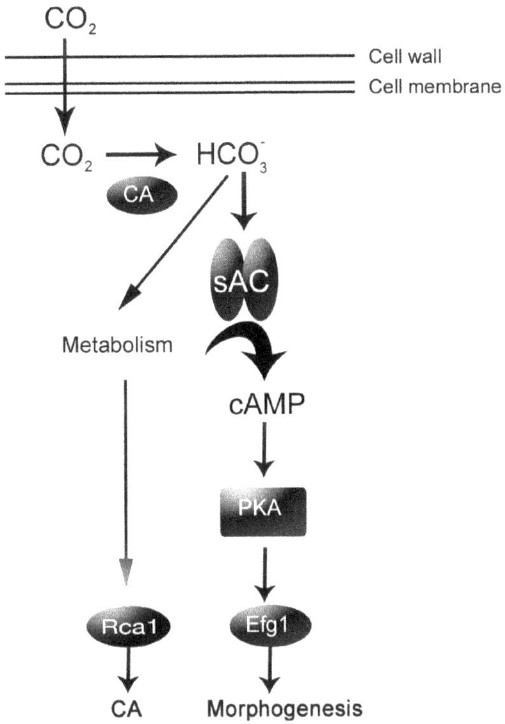

Figure 8.1 Fungal CO_2 sensing. In *Candida albicans* at high concentrations CO_2 diffuses into the cell and is converted into its impermeable form, bicarbonate, by the actions of carbonic anhydrase (CA). The bicarbonate then directly activates the soluble adenylyl cyclase (sAC) to increase cAMP production and induce morphogenesis via the activation of the transcription factor Efg1. Alternatively bicarbonate can signal through metabolism to activate the Rca1 transcription factor to regulate the expression of carbonic anhydrase (CA) and other CO_2 regulated genes in a cAMP independent manner.

environmental signals including nutrient limitation and serum induce the activation of the GTPase, Ras1, which then subsequently activates sAC. In *C. albicans,* sAC is a central regulator of virulence, regulating the switch between yeast and hyphal growth, and deletion of sAC in *C. albicans* severely attenuates pathogenicity (Rocha *et al.*, 2001). However, it appears as though these adaptor proteins are not required for CO_2 sensing (Zou *et al.*, 2010). Deletion of Ras1 in *C. albicans* inhibits hyphal formation in response to nutrient deprivation and the presence of serum. On the other hand, CO_2 was still able to drive morphogenesis in the *ras1Δ* mutant (Klengel *et al.*, 2005), suggesting that the ability to sense elevated CO_2 was not dependent on Ras1 (Fig. 8.1). Biochemical characterization of recombinant fungal sAC confirmed that HCO_3^- directly activated the sAC, stimulating cAMP production in the absence of any other adaptor proteins (Klengel *et al.*, 2005). Indeed, direct HCO_3^- stimulation of sAC has also been shown for bacterial and mammalian sACs (Chen *et al.*, 2000; Masuda and Ono, 2005; Rojas *et al.*, 1992), suggesting evolutionary conservation of this trait.

Adenylyl cyclase inhibitors as novel antifungals

The secondary messenger cAMP regulates multiple cell process and is therefore important for cell function. Indeed, fungal sACs have the potential to be therapeutic targets. In *C. albicans,* deletion of the sAC results in reduced growth rates, inability to switch between yeast and hyphal growth forms, increased sensitivity to antifungal agents, and attenuated virulence in a murine model of disseminated disease (Jain *et al.*, 2003; Rocha *et al.*, 2001). Similar effects are observed in *C. neoformans,* where deletion of sAC results in defective capsule biosynthesis and avirulence (Alspaugh *et al.*, 2002). However, inhibition of mammalian cAMP synthesis have been linked to many diseases. Humans express nine AC isoforms, including a sAC, which is important for sperm motility. Therefore, if fungal sAC is to be pursued as an antifungal target, compounds must have selectivity for the fungal sAC over the mammalian transmembrane AC isoforms and the mammalian sAC. The development of such enzyme specificity should be possible as the fungal sAC is predicted to be structurally unique from the mammalian ACs.

Adenylyl cyclase inhibitors

The most common class of AC inhibitors is the P-site inhibitors that are substrate analogues, which compete for the enzyme's active site with ATP (Tesmer *et al.*, 2000). P-site inhibitors have little enzyme specificity and also inhibit a variety of nucleotide binding proteins *in vivo* (Dessauer *et al.*, 1999). In an attempt to improve these inhibitors the adenine ring was modified by the addition of metal binding groups (Levy *et al.*, 2003). However, these inhibitors still suffer the same non-specificity and off target binding properties as the parental inhibitors. One commercially available AC inhibitor is KH7. KH7 selective targets sAC when used at low concentrations (<10 μM), but non-specifically inhibits transmembrane AC isoforms when used at high concentration (>50 μM). Owing to its selective ability to inhibit sAC, KH7 has been used to induce intracellular cAMP concentrations in sperm and may have a potential role in male contraceptives (Sinclair *et al.*, 2000).

Development of adenylyl cyclase inhibitors as a novel antifungal

The development of potent, specific AC inhibitors has been a challenging field. Many compounds that have good *in vitro* activity are non-selective AC inhibitors and have low *in vivo* efficacy. Recently the crystal structure of the mammalian sAC has been used to design inhibitors that exploit the catechol oestrogen-binding site. This approach identified a series of novel AC inhibitors, two of which were specific for the inhibition of sAC, and can be used as a scaffold for the development of more sAC specific inhibitors (Schlicker *et al.*, 2008). Surprisingly, the sAC from *Pseudomonas aeruginosa* was insensitive to the sAC specific inhibitor, KH7. However, screening of a compound library based on KH7 with alternative functional groups attached identified both activators and inhibitors of this sAC. Interestingly, compounds that inhibited sAC activity contained functional groups similar to the catechol oestrogens, suggesting that they may inhibit enzyme activity via a similar mechanism (Topal *et al.*, 2012).

Studies on fungal sAC are scarce. Recently Chang *et al.* have shown that the petacyclic triterpenen acid from *Lobaria kurokawae* inhibited *C. albicans* morphogenesis through the inhibition of cAMP synthesis (Chang *et al.*, 2012). However, this inhibition was an indirect effect caused by the increased production of farnesol, a quorum-sensing molecule that directly inhibits sAC. In conclusion, if sAC inhibitors are to be used as potential antifungal agents there is a clear need to invest in gaining a deeper understanding of this pathway.

Fungal CO_2 sensing: a role for carbonic anhydrase

The second key mediator in fungal CO_2 sensing is the carbonic anhydrase (CA). Carbonic anhydrases (CAs) catalyse the formation of bicarbonate (HCO_3^-) from the hydration of CO_2. CAs are ubiquitous in all organisms and are the most active enzyme identified to date. So far, five classes of CAs have been identified, including α-, β-, γ-, δ-, ζ-CAs, and to date, sixteen mammalian isoforms have been described. All classes of CA contain a catalytic zinc ion in their active site, which is normally coordinated by three neighbouring histidine or cysteine residues. The Fungal CAs predominantly belong to the β-class of CAs, although *Aspergillus oryzae* has both α- and β-class CAs. The genome of *C. neoformans* predicts two CA isoforms, although expression analysis suggests that only one of the isoforms is expressed under standard laboratory conditions (Mogensen *et al.*, 2006). However, the genomes of most yeast, including *C. albicans*, *Candida glabrata* and *Saccharomyces cerevisiae*, only encode a single isoform (Cottier *et al.*, 2013; Götz *et al.*, 1999; Klengel *et al.*, 2005).

The activity of CAs in fungi is essential for providing bicarbonate, which is required for carboxylation reactions in fatty acid and amino acid biosynthesis (Aguilera *et al.*, 2005; Bahn *et al.*, 2005). Deletion of CA activity results in a CO_2 growth dependent phenotype, where strains will only grow in air with elevated CO_2 partial pressures (Aguilera *et al.*, 2005; Mogensen *et al.*, 2006; Götz *et al.*, 1999; Klengel *et al.*, 2005). However, this phenotype is not specific to fungi, with many organisms experiencing CO_2 dependency under restrictive CA activity (Kusian *et al.*, 2002; Merlin *et al.*, 2003; Mitsuhashi *et al.*, 2004). The restoration of growth at high CO_2 concentrations is thought to be due to the increased diffusion of CO_2 into the cell. At high concentrations, CO_2 is spontaneously hydrated to its impermeable form, bicarbonate, thereby meeting the metabolic demand of the fungus. In *C. albicans* and *C. neoformans*, bicarbonate also directly activates the sAC (see above) inducing morphogenesis and capsule biosynthesis; two key virulence determinates of these fungi, respectively. Indeed, growing CA deficient strains of *C. albicans* in modest CO_2 concentrations (<1%) is sufficient to restore growth, but not to induce morphogenesis (Hall; unpublished), suggesting that HCO_3^- can only be used as a signalling molecule once metabolic demands are met.

Transcriptional analysis of CA expression in *C. albicans, C. glabrata Aspergillus nidulans, A. fumigatus, S. cerevisiae* and *Sordaria macrospora*, confirmed that the expression of fungal CAs is directly regulated by CO_2 partial pressures (Amoroso *et al.*, 2005; Cottier *et al.*, 2012, 2013; Elleuche and Poggeler, 2009; Han *et al.*, 2010). CA expression is repressed under high CO_2 conditions, conditions in which CA activity is dispensable for growth. However, CA expression is rapidly restored under atmospheric conditions (0.033% CO_2). More intriguingly, the expression of the *C. albicans* CA isoform was induced upon phagocytosis, despite the presence of 5% CO_2 in the atmosphere (Cottier *et al.*, 2012; Lorenz *et al.*, 2004), although currently the reason for this expression is unclear. Initial transcript analysis of the *C. neoformans* CA, suggested that the environment did not regulate the expression of this CA (Bahn *et al.*, 2005). On the other hand, analysis of CA protein levels confirmed that the expression of this CA is indeed regulated by CO_2 concentrations (Hall, unpublished). This discrepancy is likely a result of sample processing times, as mRNA levels of the fungal CAs are induced within five minutes of exposure to air.

Despite the fact that fungal CO_2 sensing is regulated via the cAMP-PKA pathway (see above), CA expression in response to environmental CO_2 is cAMP independent (Cottier *et al.*, 2012, 2013). Therefore, fungi have evolved two separate signalling cascades to regulate its response to one important host condition. Recent studies to determine the second fungal CO_2 sensing pathway, identified a novel CO_2 dependent transcription factor, Rca1 (Regulator of Carbonic Anhydrase; Fig. 8.1). Cottier *et al.* show that Rca1 binds to the promoter region of 85 open reading frames in the *C. albicans* genome, 54% of which encode proteins with unidentified functions (Cottier *et al.*, 2012). Although the exact signalling cascade that leads to the activation of Rca1 remains to be elucidated, Rca1 regulates genes involved in morphogenesis, metabolism and cell wall biogenesis (Cottier *et al.*, 2012; Vandeputte *et al.*, 2012), suggesting that Rca1 is an important regulator of fungal virulence factors. An independent study, also seeking the transcriptional regulator of CO_2 sensing in *C. albicans*, identified Flo8 as being essential for CO_2 induced phenotypes including white-opaque switching and hyphal development (Du *et al.*, 2012). The Flo8 signalling cascade was again independent of the cAMP-PKA pathway, suggesting that *C. albicans* employs several important signalling cascades to respond and adapt to CO_2. Whether Flo8 and Rca1 function in the same signalling pathway remains to be determined, but it is clear that the regulation of CO_2 sensing may not be as straightforward as first thought. Deletion of *RCA1* results in increased resistance to some current antifungal agents. Although the mechanism by which Rca1 achieves this is unknown, increased resistance may be an indirect effect resulting from differential regulation of genes involved in cell wall biogenesis (Vandeputte *et al.*, 2012). Therefore, the impact of inhibiting specific components of the CO_2 sensing pathway should be considered when targeting it for antifungal development.

Carbonic anhydrase inhibitors as antifungals

The discovery that deletion or inhibition of CAs in fungal species has a dramatic affect on the proliferation of many fungal species has made the fungal CAs interesting candidates for the development of novel antifungals. However, in the host, fungi are exposed to elevated CO_2 partial pressures, which rescue fungal proliferation. Interestingly though, *C. albicans* CA mutants display attenuated virulence in epithelial infection models (Klengel *et al.*, 2005), suggesting that fungal CA inhibitors have the potential to be powerful antifungals for the treatment of skin and nail infections. Dermatophytes are a collection of fungi that reside in dead keratinized tissues including the skin, hair and nails. The dermatophyte population is comprised of over 40 different species from the *Microsporum*, *Trichophyton* and *Epidermophyton* genera. These fungi do not normally penetrate the skin, and are therefore continuously exposed to environments with low CO_2 partial pressures. Infection is normally the result of an inflammatory response to metabolic by products secreted by the fungus. Although not life threatening infections, 25% of the world's population suffers from dermatophyte infection at a cost of $400 million per year (Havlickova *et al.*, 2008; Smith *et al.*, 1998). In addition to the dermatophytes, *Malassezia* species including *M. globosa, M. furfur* and *M. restricta,* which form part of the skin's natural flora, have been associated with dandruff and atopic dermatitis (Baker, 2006). Again during infection *M. furfur* is continuously exposed to atmospheric concentrations of CO_2, and therefore has the potential to be treated by antifungal therapies based on CA inhibitors.

Carbonic anhydrase inhibitors

Acetazolamide methazolamide, ethoxzolamide, dorzolamide and topiramate are commercially available carbonic anhydrase inhibitors. These inhibitors are used to treat a wide range of diseases/disorders including epilepsy, glaucoma and altitude sickness. In glaucoma, intraocular pressure increases due to the increased production of aqueous humour in response to elevated levels of HCO_3^-. Treatment with CA inhibitors reduces the CA activity, reducing the production of HCO_3^-, and thus reducing the intraocular pressure. More recent research has focused on the design of CA inhibitors for the treatment of cancer, obesity, and fungal and bacterial infections (Capasso and Supuran, 2013; Monti *et al.*, 2013; Scozzafava *et al.*, 2013). However, the broad activity of these CA inhibitors against the multiple mammalian isoforms, coupled with their lack of tissue specificity, result in numerous side-effects including vomiting, dizziness, kidney stones and fatigue. One method to reduce off target effects is to develop different formulations of the inhibitor. For example, ointment formulations of acetazolamide have been developed for the treatment of glaucoma (Antonaroli *et al.*, 1992). This allows the drug to be applied directly to the eye, rather than oral administration, targeting the inhibitor to the isoform expressed in the eye and reducing unwanted side-effects.

Development of carbonic anhydrase inhibitors as a novel antifungal

One hurdle when using a ubiquitous enzyme for the development of novel antifungals is being able to selectively inhibit the fungal protein over the mammalian protein. In the case of CAs, the majority of fungal CAs are from the β-class, which is not present in mammals, allowing the potential to develop selective CA inhibitors. One major difference between the mammalian CA isoforms and the fungal β-CA isoforms is the coordination of the catalytic zinc ion. In mammalian CAs, the zinc is coordinated via three histidine residues, while in fungi the zinc ion is coordinated by an aspartate, histidine and two conserved cysteine residues. A vast amount of effort has been put into the development of novel CA inhibitors that selectively inhibit β-class CAs over α-class CAs. These studies have been largely successful at identifying compounds with selective activity against recombinant protein, with sulfonamides, sulfamates, boronic acids and carboxylates being the most potent inhibitors of these fungal CAs (Carta *et al.*, 2011; Davis *et al.*, 2011; Güzel *et al.*, 2010; Hector and Schaller, 1992; Innocenti *et al.*, 2008). Carboxylates (e.g. acetate, maleate, citrate and formate) are weak inhibitors of the *C. albicans* and *C. neoformans* CAs. However, malate and formate are extremely weak inhibitors of human CA II (K_I 1000–24,000 μM) (Innocenti *et al.*, 2009a), suggesting that these metabolic intermediates from the Krebs cycle may be able to selectively target fungal CAs over mammalian CA isoforms.

Homology modelling of the *C. albicans* CA using the published *C. neoformans* CA structure has provided many insights into structure activity relationships, and may be beneficial for the development of fungal specific CA inhibitors. For example, structure predictions suggest that compounds with less polar zinc binding domains would have greater efficacy against the *C. albicans* CA, while compounds with polar binding groups and bulky side chains would be more potent inhibitors of the *C. neoformans* CA (Innocenti *et al.*, 2009b).

Biochemical characterization and inhibitor studies of the β-CA from *M. globosa* confirmed that this β-CA was resistant to the majority of CA inhibitors, including commercially available CA inhibitors like acetazolamide (Hewitson *et al.*, 2012). In spite of this, 4-aminoethylbenzensulfonamide was found to have a K_I in the nM range (< 100 nM) against recombinant protein. Although this compound had a high MIC against *Malassezia* species (640 μg/ml), 67% of mice showed improved clinical symptoms in a skin infection model (Hewitson *et al.*, 2012). Therefore, this carbonic anhydrase inhibitor has the potential to be developed as an antifungal.

Despite the ongoing efforts into the identification of novel fungal specific CA inhibitors, little attention is focused on drug efficacy. Although many of the sulfamates/sulfonamides have good activity against recombinant protein, these compounds currently lack activity against fungal cells, with many compounds only partially inhibiting fungal growth (Schlicker *et al.*, 2009). This problem is hypothesized to result from the impermeability of the fungal cell wall, preventing the inhibitor from reaching its target, which is cytosolic. Similar problems have been observed with current antifungals (i.e. amphotericin B), and permeability has been improved with the development of lipid formulations (i.e. AmBisome). Therefore, if CA inhibitors are to be pursued as antifungals, future research should be aimed at improving the permeability of these compounds to permit efficient uptake across the fungal cell wall.

Inhibition of fungal cell wall biosynthesis

Fungi are surrounded by a dynamic organelle known as the cell wall. The cell wall plays important functions in maintaining cell shape and the internal turgor pressure, protects the fungus from the external environment and has important implications in innate immune recognition. For polymorphic fungi, such as the human fungal pathogen *Candida albicans*, the cell wall must be sufficiently rigid to resist changes in internal turgor pressure, but adequately plastic to permit surface expansion during growth and morphogenesis. To achieve this, the fungal cell wall is multi-layered (Fig. 8.2), with an internal skeletal layer, to provide the strength and rigidity, and an outer matrix of highly glycosylated cell wall proteins that have important functions in cell adhesion, virulence, cell wall biogenesis and innate immune recognition. Many of the proteins involved in cell wall

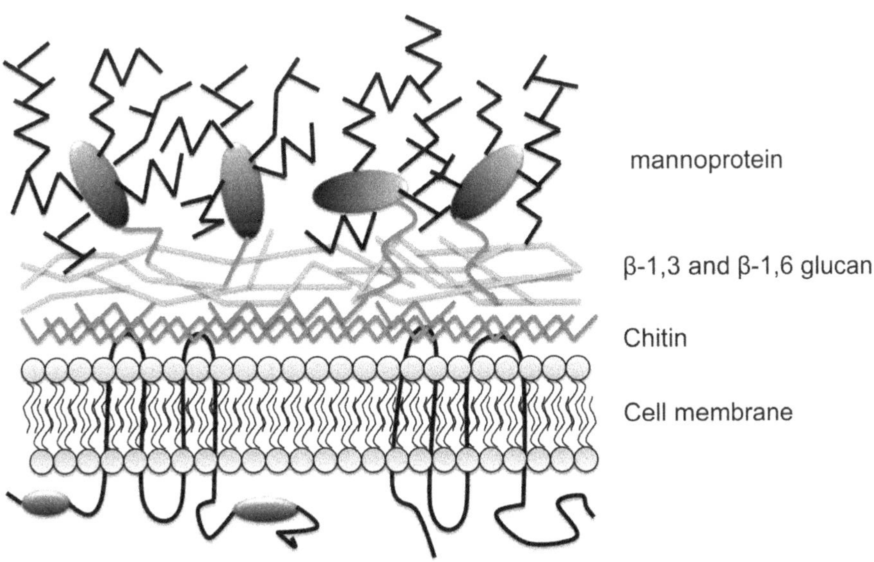

Figure 8.2 The fungal cell wall. Diagrammatic representation of the cell wall from *Candida albicans*. The inner skeletal layer is comprised of chitin and β1,3- and β1,6-glucan, while the outer layer contains heavily glycosylated mannoproteins.

biogenesis are fungal specific and therefore the fungal cell wall is an attractive target for the development of novel antifungal compounds, with a large emphasis of chitin and glucan biosynthesis (reviewed in Munro, 2013).

Targeting chitin biogenesis

The inner skeletal layer is comprised of chitin. In *C. albicans* chitin is only a minor component of the cell wall (1–3% dry weight), while in other fungi such as *Aspergillus*, chitin can account for up to 20% of the total dry weight of the cell wall. Despite the vast differences in chitin composition, chitin is essential to maintain the integrity of the cell wall and hence is vital for fungal viability. This critical dependence on chitin for viability has led to the evolution of functionally redundant chitin biosynthetic enzymes and multiple regulatory pathways, which induce chitin synthesis in response to cell wall stress to increase the integrity of the cell wall (Munro *et al.*, 2007).

Chitin is synthesized by chitin synthases from UDP-*N*-acetylglucosamine. In *C. albicans*, this is achieved by the actions of four chitin synthases, Chs1, Chs2, Chs3 and Chs8 (Au-Young and Robbins, 1990; Bulawa *et al.*, 1995; Gow *et al.*, 1994; Munro *et al.*, 2003). The 1,4-*N*-acetylglucosamine polymers synthesized by the various chitin synthases hydrogen bond to form anti-parallel microfibrils. Chs1 activity is essential for cell viability and generates chitin that is incorporated into the primary septum and lateral cell wall (Munro *et al.*, 2001), while Chs2 activity accounts for the majority of chitin synthase activity detected in membrane preparations (Gow *et al.*, 1994). Chs3 synthesizes the majority of chitin found in *C. albicans* cell wall (Bulawa *et al.*, 1995), and Chs8 is required for the synthesis of longer chitin chains that are incorporated into bud scars and hyphal septa (Munro *et al.*, 2003). Therefore, there are multiple chitin synthases that must be targeted in order to inhibit chitin synthesis.

Due to the essentiality of chitin for fungal viability, compounds that inhibit chitin synthesis have been exploited as potential antifungal agents. As chitin synthesis is a multi-step process involving many enzymatic steps, there are enormous opportunities for the development of potential inhibitors. For example, the substrate (UDP-*N*-acetylglucosamine) for chitin synthases is synthesised from fructose-6-phosphate via the actions of four enzymes, glutamine:Fru-6P acetyltransferase, GlcN-6P acetyltransferase, *N*-acetylphosphoglucosamine mutase and UDP-GlcNAc pyrophosphorylase. Characterization of *N*-acetylphosphoglucosamine mutase (Agm1) from *Aspergillus fumigatus, Saccharomyces cerevisiae* and *C. albicans* confirm that this enzyme is essential for fungal viability (Fang *et al.*, 2013; Horfmann *et al.*, 1994; Nishitani *et al.*, 2006), suggesting that it would make a good non-selective antifungal target. However, the ability to selectively inhibit fungal Agm1 over the mammalian orthologue remains to be explored. Sequence analysis and crystallization of fungal Agm1 identified differences in the amino acid composition around the active sites of fungal and mammalian Agm1, which may be utilized to design fungal specific Agm1 inhibitors (Fang *et al.*, 2013).

Nikkomycins and polyoxins are natural compounds that competitively inhibit chitin synthase activity. In *C. albicans* Nikkomycin Z and polyoxin D inhibit the enzyme activity of Chs1, Chs2 and Chs8 (Kim *et al.*, 2002), and are promising antifungal compounds. Nikkomycin Z has antifungal activity against *C. albicans* in murine infection models, although high doses were required to prolong survival (Kim *et al.*, 2002). There is some evidence to suggest the nikkomycin synergises with the azole family of antifungals (Hector and Schaller, 1992), suggesting that nikkomycin Z could be used as a combinational therapy to treat candidiasis. In agreement with this, nikkomycin Z alone is not effective against *A. fumigatus*, but when used in combination with caspofungin (which is also not effective against *A. fumigatus* alone) killed germinating conidia (Ganesan *et al.*, 2004). More recently, Nikkomycin Z was shown to have *in vivo* activity against respiratory coccidioidomycosis in canines (Shubitz *et al.*, 2013). Therefore, there is potential for the development of Nikkomycin Z as an antifungal agent.

In addition to natural chitin synthase inhibitors, there are some synthetic inhibitors in development. RO-09-3143 is a synthetic compound that has selective activity against Chs1 from *C. albicans*. On whole fungal cells, this synthetic inhibitor only exhibits fungal-static effects (Gaughran *et al.*, 1994; Sudoh *et al.*, 2000), presumably due to residual chitin synthase activity. Currently, the development of structural analogues of UDP-*N*-acetylglucosamine, which have strong fungicidal activity, is being met with varying success (Chang *et al.*, 2006; Gautier-Lefebvre *et al.*, 2005; Magellan *et al.*, 2013). Therefore, the development of chitin synthase inhibitors (or a combinational therapy) that inhibits all isoforms of the enzyme is required.

Other approaches to inhibit chitin biosynthesis include the use of purified chitinases. Chitinases are hydrolytic enzymes, which hydrolyse the glycosidic bonds in the chitin polymers. The GH-46 chitinase from *Bacillus circulans* displays antifungal activity against zygomycete species including *Rhizopus* and *Mucor* (Tomita *et al.*, 2013), suggesting that GH-46 could be a promising antifungal.

Targeting glucan biogenesis

The inner skeletal layer of the *C. albicans* cell wall is composed of β-1,3-glucan (40% dry weight), with β-1,6-linked branches (20% dry weight) that anchor cell wall proteins to the matrix through glycosylphosphatidylinositol (GPI) remnants (Klis *et al.*, 2001). However, the inner core of the cell wall in *A. fumigatus* is composed of branched glucan–chitin complex surrounded by linear α-glucans (Latgé *et al.*, 2005). Glucans are glucose polymers. Glucose polymerization is catalysed by glucan synthase, which use UDP-glucose as a substrate. These enzymes are membrane-bound protein complexes, with the substrate-binding site located on the cytoplasmic side and the extruding polypeptide exiting through the membrane into the cell wall. The glucan synthases complex is comprised of a regulatory GTPase (Rho) and a catalytic (Fks) subunit (Kondoh *et al.*, 1997; Mio *et al.*, 1997). As with the chitin synthases, yeasts express multiple Fks proteins. For example, *S. cerevisiae* possesses three FKS genes (*FKS1*, *FKS2* and *FKS3*). Individually, the FKS genes are dispensable for viability, but deletion of multiple FKS genes is fatal (Mazur and Baginsky, 1996).

To date, three classes of β1,3-glucan synthase inhibitors have been identified including glycolipid papulacandins, acidic terpenoids and the cyclic hexapeptides (including the echinocandins). From these, only the echinocandins are currently used in the clinical setting (caspofungin, micafungin, anidulafungin). The major problem with the echinocandins is the development of antifungal resistance. Echinocandin resistance is a result of dominant point mutations in hot spots of the Fks protein. In *C. albicans*, two Fks1 hot spots have been identified (hot spot 1 amino acids 641–649 and hot spot 2 amino acids 1345–1365). These mutations reduce the catalytic efficiency (V_{max}) of Fks1 (Garcia-Effron *et al.*, 2009). In addition, inhibition of glucan synthase stimulates activation of cell wall salvage pathways, which up-regulate chitin synthesis as a compensatory mechanism. Indeed, activation of chitin synthesis in *C. albicans* is sufficient to induced echinocandin resistance (Drakulovski *et al.*, 2011; Imtiaz *et al.*, 2012; Lee *et al.*, 2012; Walker *et al.*, 2013). Although not used in the clinical setting, preliminary studies suggest that combination therapies of echinocandins and chitin synthase inhibitors may be beneficial for the treatment of strains with a high echinocandin MIC (Walker *et al.*, 2008). In agreement with this, amphotericin B has been shown to have synergistic effects when used in combination with micafungin (Zhou *et al.*, 2013). Therefore, future advances in the treatment of echinocandin-resistant fungal infections may lie in the use of combinational therapy.

Targeting mannan biosynthesis

The outer shield of mannoproteins has many roles in fungal fitness and virulence. Although glycosylation occurs in all eukaryotes, fungal mannan contains complex branched outer chain structures, which are not present on mammalian cells. Therefore, many of the glycotransferases involved in the synthesis of these complex mannan structures are fungal specific and may serve as potential novel therapeutic targets.

One of the major functions of these glycosylated proteins is regulating host–pathogen interactions. The glycan structures protruding from the cell wall act as PAMPs (pathogen-associated molecular patterns) and are recognized by PRRs (pattern recognition receptors) located on myeloid cells of the host's innate immune system. PAMP recognition induces phagocytosis of the fungal particle and activates pro- and anti-inflammatory cytokine responses (reviewed in Netea *et al.*, 2008). The generation of a series of mannosylation mutants has been important to the field for advancing our understanding of these interactions (reviewed extensively in Hall and Gow, 2013). Inhibition of mannan biosynthesis results in increased cell aggregation, decreased cell wall integrity, and alters the interaction with immune effector cells. As a result glycosylation mutants display reduced virulence in murine models of systemic infection. Moreover, some mannosylation mutants show increased sensitivity to clinically relevant antifungals (Bates *et al.*, 2006; Mora-Montes *et al.*, 2007, 2010; Prill *et al.*, 2005), suggesting that inhibiting mannan biosynthesis in combination with antifungal therapy may be a possibility. Currently, there are no commercially available compounds that inhibit mannan biosynthesis. However, studies focusing on biochemical characterization of glycotransferases are on the increase (Díaz-Jiménez *et al.*, 2012; Mora-Montes *et al.*, 2008), and can be utilized to develop specific inhibitors.

Quorum sensing molecules as potential antifungals

Microorganisms communicate using soluble molecules through a cell density dependent process known as quorum sensing. At low cell densities the concentration of the quorum-sensing molecule (QSM) is too low to have an effect, but as the population increases the QSM intensity reaches threshold concentrations and initiates specific effects. Quorum sensing was first identified in *Vibrio fischeri*, a bacterium that resides within the photophore (light organ) of the Hawaiian bobtail squid (Nealson *et al.*, 1970). When grown as a planktonic culture, the concentration of the *V. fischeri* QSM (or autoinducer) in the media is low, due to diffusion. However in the photophore, the cell concentration is high resulting in increased concentrations of the QSM, which lead to transcription of the luciferase gene and bioluminescence. Since then, quorum sensing has been discovered in a number of bacteria including *Pseudomonas aeruginosa*, *Salmonella enterica* and *Burkholderia cenocepacia* (Bassler, 1999; Lewenza *et al.*, 1999; Passador *et al.*, 1993). In these organisms, quorum sensing has been shown to regulate biofilm formation, virulence factors and antibiotic resistance.

The first example of eukaryotic quorum sensing was described in *C. albicans*, where the sesquiterpene farnesol was identified as an inhibitor of hyphal development (Hornby, 2001). Farnesol is produced from farnesyl pyrophosphate, which is an intermediate in the conserved sterol biosynthesis pathway. Manipulation of the farnesyl pyrophosphate pools, through the inhibition of sterol biosynthesis, results in increased farnesol secretion (Hornby *et al.*, 2003). Therefore, at high cell densities *C. albicans* may divert farnesyl pyrophosphate to farnesol biosynthesis by decreasing sterol metabolism. In addition to regulating morphogenesis, farnesol also regulates biofilm formation. In this case, high concentrations of farnesol are thought to weaken the architecture of the biofilm due to a reduction in morphogenesis. As a result, yeast cells are released from the biofilm. As biofilms form on indwelling medial devices like cannulas, the release of yeast cells from the biofilm enables dissemination of *C. albicans* from the cannula directly into the bloodstream of the patient. Therefore, the effects of farnesol on biofilm formation have clinical implications and warrant further investigation.

In addition to farnesol, *C. albicans* produces a second quorum sensing molecule, tyrosol. However, instead of inhibiting hyphal development tyrosol promotes filamentation (Chen *et al.*, 2004). Biofilm studies suggest that tyrosol also aids in biofilm formation by initiating early hyphal development (Alem *et al.*, 2006). Therefore, it would appear that *C. albicans* uses two quorum sensing systems to regulate morphogenesis and biofilm formation, two key virulence traits of the fungus.

Mechanism of fungal quorum sensing

The mechanism by which farnesol exerts its affects on *C. albicans* is largely unknown. Supplementation of exogenous cAMP into the media inhibits the quorum sensing properties of farnesol, suggesting that farnesol targets cAMP synthesis (Davis-Hanna *et al.*, 2008). Biochemical characterization of the catalytic domain of the fungal soluble adenylyl cyclase confirmed that farnesol directly modulates its activity by binding to the active site of the sAC (Hall *et al.*, 2011). As a result cAMP synthesis is inhibited, which is a key secondary messenger in fungal morphogenesis and virulence (Fig. 8.3). This effect was specific for the sAC from *C. albicans*, as sACs from other yeasts including *S. cerevisiae* were non-responsive to farnesol (Hall *et al.*, 2011), despite also being regulated in a similar manner by other environmental stimuli (i.e. CO_2). Other studies have highlighted roles for the two-component histidine kinase signalling cascade, the mitogen activated protein kinase (MAPK) cascade, mitochondria and Tup1 in modulating the response to farnesol (Kebaara *et al.*, 2008; Kruppa *et al.*, 2004; Sato *et al.*, 2004). Therefore, farnesol seems to exert its effects through multiple mechanisms. So far, no receptor, apart from sAC, has been identified for farnesol. This may in part be due to the fact that farnesol is lipophilic and can directly cross the cell membrane and interact with multiple proteins within the cell. A fluorescent analogue of farnesol, which maintains QSM activity, is available (Shchepin *et al.*, 2005), but so far there have been no reports of its use to visualize the localization of farnesol within the fungal cell.

Farnesol is comprised of a 12-carbon chain containing three double bonds, three methyl branches and a hydroxyl group. Due to the double bonds, farnesol exists as multiple stereoisomers, but only the E,E isomer has biological activity in fungal quorum sensing (Shchepin *et al.*, 2003). *C. albicans* is also responsive to QSMs secreted by other microbes. For example, the quorum sensing molecules 3-oxo-C12 homoserine lactone (C12-HSL) secreted by *P. aeruginosa*, *Burkholderia* diffusible signal factor (BDSF) secreted by *B. cenocepacia*, and trans-2-decenoic acid secreted by *Streptococcus mutans* also inhibit *C. albicans* hyphal development and biofilm formation (Boon *et al.*, 2007; Hogan *et al.*, 2004; Vílchez *et al.*, 2010). The biological relevance for evolving a response to so

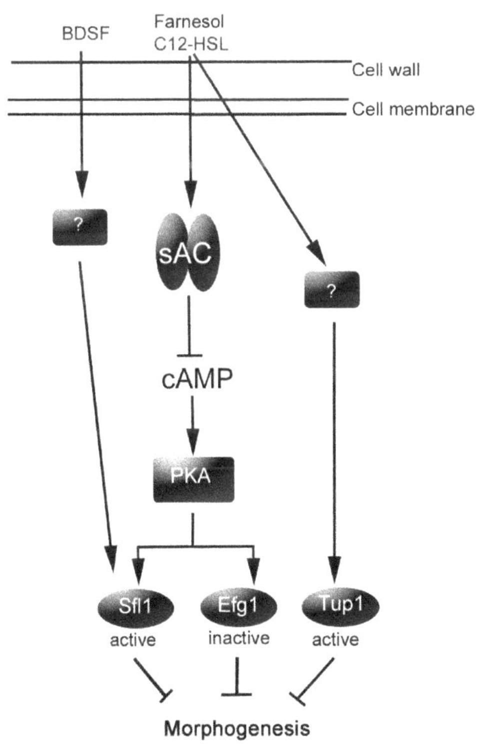

Figure 8.3 Quorum sensing in fungal pathogens. The fungal and bacterial quorum sensing molecules farnesol and C12-HSL directly inhibit the soluble adenylyl cyclase (sAC), reducing cellular cAMP levels and inhibiting morphogenesis. BDSF inhibits morphogenesis in a cAMP dependent manner involving Sfl1, but independently of sAC.

many QSMs is not fully understood. However, *P. aeruginosa* can form biofilms on the surface of *C. albicans* hyphae (Hogan and Kolter, 2002). This interaction is beneficial for the bacterium, but is detrimental to the fungus. Therefore, a possible explanation for QSMs inhibiting hyphal development is that it protects the yeast from being killed by the bacterium. Although structurally very different to farnesol, all these QSMs possess a 12-carbon backbone. Therefore, it was thought that these QS compounds act by occupying the farnesol receptor/binding site. In agreement with this, studies using structural analogues of farnesol suggested that the hydroxyl head group is not required for QSM activity, but the chain length does appear to be important, with a 12-carbon backbone displaying potent QSM activity (Shchepin *et al.*, 2003). Interestingly, a recent report confirmed that C12-HSL directly modulates the activity of sAC along with several other 12-carbon molecules, suggesting that they are in fact interacting with the farnesol binding site. However, dodecanol, a 12-carbon alcohol, that has QSM activity against *C. albicans*, did not directly modulate sAC activity (Hall *et al.*, 2011). Therefore, a 12-carbon backbone is not sufficient for mediating quorum sensing in *C. albicans*. This result is surprising considering that dodecanol has been shown to inhibit fungal morphogenesis in a cAMP dependent manner similar to farnesol and C12-HSL (Davis-Hanna *et al.*, 2008). Further exploration into the mechanism of dodecanol's mode of action confirmed that this QSM mediates its effects through a cAMP dependent pathway involving Sfl1 (Hall *et al.*, 2011). Deletion of *SFL1* confirmed resistance to the morphogenic properties of dodecanol, but not farnesol, suggesting that Sfl1 is not required for the response to farnesol and potentially C12-HSL. On the other hand, BDSF was unable to inhibit hyphal formation of the *sfl1*Δ mutant, suggesting that BDSF exerts its affects in an Sfl1-dependent manner, similar to dodecanol.

Effects of QSMs

In addition to mediating effects on morphogenesis, farnesol exerts other biological processes in *C. albicans*. For example, pre-incubation in culture supernatants increases the resistance of *C. albicans* to reactive oxygen species (ROS) (Westwater *et al.*, 2005). This increased resistance to ROS is a result of inhibition of the cAMP-signalling cascade (Deveau *et al.*, 2010). Other molecules that possess QSM activity against *C. albicans* also increase the fungus' resistance to ROS (Hall *et al.*, 2011). This is important during dissemination from biofilms *in vivo*. Fungal cells within the biofilm will be exposed to high concentrations of farnesol (600 μM), which when released provides enhanced protection from ROS generated by macrophages and neutrophils.

Farnesol has also been shown to induce drug resistance in *C. albicans*. Multi-drug resistance in *C. albicans* is mediated via the ATP binding cassette (ABC) transporters Cdr1, Cdr2 and the facilitator protein Mdr1. These efflux pumps function to exclude drugs from the cell. Farnesol directly influences drug resistance by modulating the efflux potential of Cdr1 and Cdr2 (Sharma and Prasad, 2011). This results in accumulation of the drug inside the fungal cell and reduces the minimal inhibitor concentration (MIC; concentration at which 50% growth inhibition occurs) of the compound. In addition to directly modulating the activity of efflux pumps, farnesol also acts in synergy with antifungals. Addition of farnesol to azole-resistant strains reduced the MIC_{80} (concentration at which 80% growth inhibition occurs) by up to 16-fold (Sharma and Prasad, 2011). This synergistic effect is thought to be due to excessive mitochondrial ROS generation induced by farnesol.

In addition to affecting *C. albicans*, farnesol has been reported to influence the physiology and virulence attributes of other microbes. For example, the addition of exogenous farnesol, or co-culture of *C. albicans* and *P. aeruginosa* results in the repression of *Pseudomonas* quinolone signal and pyocyanin (a key virulence factor of *P. aeruginosa*) production (Cugini *et al.*, 2007). In *Aspergillus niger*, farnesol has been shown to inhibit the formation of conidia (Lorek *et al.*, 2008), while in *Aspergillus nidulans* farnesol induces apoptosis (Semighini *et al.*, 2006). As with *Saccharomyces cerevisiae*, farnesol induced apoptosis was dependent on mitochondrial ROS production (Semighini *et al.*, 2006). In contrast, farnesol does not affect the growth rate of *C.*

albicans at the concentrations that induce apoptosis in other fungi (Hornby, 2001), suggesting that *C. albicans* is resistant to, or has a compensatory mechanism to combat, ROS production.

Fungal QSMs also affect host cells. For example, farnesol has been shown to decrease the antimicrobial actions and induce ROS dependent apoptosis of murine macrophages (Abe *et al.*, 2009). In contrast, farnesol has been shown to up-regulate the expression of Toll-like receptor 2 (TLR2) in oral epithelial models, resulting in increased production of IL-6 and β-defensin-2 (Décanis *et al.*, 2009). These observations, in conjunction with the ability of farnesol to promote stress resistance in *C. albicans*, suggest that the production of QSMs during complex biofilm formation will have severe consequences on host–pathogen interactions. Moreover, QSMs have been shown to affect motility and acrosome formation of sperm (Rennemeier *et al.*, 2009). Considering that *C. albicans* is a frequent colonizer of the genital tract, and that 75 million sexually active women have at least one case of vaginal candidiasis (thrush) per year, QSMs may play an important role in decreased fertility among couples.

Antifungal potential of QSMs

As discussed above, farnesol induces apoptosis in many fungal species and inhibits biofilm formation, suggesting that QSMs have the potential to be utilized as novel antifungal therapies. BDSF has recently been shown to inhibit *C. albicans* biofilm formation on iatrogenic devices (Tian *et al.*, 2013) whilst farnesol induced the detachment of bacterial biofilms (Cerca *et al.*, 2012, 2013). This opens the possibility of coating medical devises with QSMs to prevent cell attachment and biofilm formation within cannulas. Although not technically working as an antimicrobial, coating cannulas and catheters with QSMs would reduce the potential for biofilm formation within these devices. This would reduce the potential for infection in high risk patients, and prevent the need to continually replace medial devices improving patient care and management.

Inoculation of mice with a cocktail of alcohols produced by *C. albicans* (including farnesol and tyrosol) resulted in prolonged survival of mice subsequently infected with *C. albicans* in the disseminated model of candidiasis (Martins *et al.*, 2012). Taking these data together with the reports that farnesol has synergistic killing with commonly used antifungals (e.g. fluconazole), suggests that treatment with exogenous QSMs may provide an alternative method for disease management. However, another report suggests that inhibition of farnesol production reduces fungal burden and enhances survival of mice (Tashiro *et al.*, 2012). Clearly there are many differences between these two approaches. For example, while Martins *et al.* added exogenous farnesol, Tashiro *et al.* used pravastatin to inhibit farnesol production. Pravastatin may also suppress other key virulence factors of *C. albicans*, suggesting that the reduced virulence may not be solely dependent on the depressed farnesol secretion. In agreement with QSMs providing protection to fungal infections, the addition of exogenous farnesol in models of oral candidiasis improved lesion scores and prevented the formation of fungal biomasses on mouse tongues (Hisajima *et al.*, 2008). Determination of the number of colony forming units suggested that improvement was not associated with decreased fungal load, but rather was due to reduced invasion. The most likely explanation for this is that farnesol inhibits the yeast to hyphal transition, maintaining colonization but preventing invasion into the tissue.

Considering that quorum sensing in *C. albicans* regulates many key virulence attributes of the fungus (e.g. morphogenesis and biofilm formation), and that the quorum sensing molecule farnesol has been shown to possess antimicrobial activity against many other fungal and bacterial species, exploration into the potential use of QSMs as novel antimicrobial therapeutics is definitely an important area of future research.

Intracellular proliferation

During infection, phagocytic cells of the host innate immune system clear invading microbes through a process termed phagocytosis. Phagocytosis is a receptor-mediated event in which epitopes on the microbe's surface (pathogen associated molecular pattern; PAMP) are recognized and engaged by phagocytic receptors on the surface of the patrolling macrophage. This epitope–receptor interaction results in receptor clustering on the macrophage surface, inducing the activation of a series of signalling cascades that in turn lead to membrane invagination, and engulfment of the microbe. Once internalized, the microbe is encapsulated in a membranous organelle known as the phagosome. During a series of maturation steps the phagosome fuses with early and late endosomes, and finally lysosomes to form the mature phagolysosome, which is capable of digesting the phagocytosed particle (recently reviewed in Smith and May, 2013).

However, many microbes have evolved the ability to 'hijack' cells of the host's immune system, allowing them to proliferate either within the membrane bound vesicles, or to escape from these vesicles and to replicate in the host cell's cytoplasm. The best examples of this are *Mycobacterium tuberculosis, Streptococcus pyogenes* and *Listeria monocytogenes* (recently reviewed in Smith and May, 2013). Recently, the fungal pathogens *Cryptococcus neoformans, C. gattii* and *Candida glabrata* have been reported to proliferate within host macrophages (Feldmesser *et al.*, 2000; Ma *et al.*, 2009; Seider *et al.*, 2011; Tucker and Casadevall, 2002). The rationale for the evolution of replicating in hosts cells is that it permits the invading pathogen to hide from the immune system and antimicrobial therapy, aids in dissemination and permits reseeding of the infection as cells either lyse the host cells, or escape via a non-lytic mechanism (i.e. vomocytosis; (Ma *et al.*, 2006)). The precise mechanism by which these fungi hijack the macrophage is unknown. *C. glabrata* appears to promote its survival within macrophages by inhibiting phagosome maturation and attenuating reactive oxygen species generation (Seider *et al.*, 2011). On the other hand, intracellular proliferation in *C. gattii* appears to be related to mitochondrial function. Specifically, *C. gattii* mitochondria undergo a change in morphology resulting in tubular networks of mitochondria in the cytoplasm of the fungi (Ma *et al.*, 2009). This phenomenon is not solely attributed to the fungal mitochondrial genotype, as genetic crosses between highly proliferative strains and non-proliferating strains, where the progeny receive their mitochondria solely from the highly proliferating strains, were not always highly proliferating, suggesting that other genetic factors are required (Voelz *et al.*, 2013). The effect the tubular mitochondria have on yeast or phagosome physiology, and how this mediates survival within the macrophage, remains unknown.

Targeting intracellular microbes

Microbes that reside within host cells are shielded from the effects of antimicrobial compounds. Therefore, there is an urgent need to develop antimicrobials that effectively target intracellular pathogens. Compounds that inhibit autophagy, a process whereby cells can degrade damaged organelles releasing fresh nutrients into the cell, have been successful under laboratory conditions for killing intracellular pathogens, by restricting nutrient availability within the host cell (Steele *et al.*, 2013). However, these compounds are not currently used in the clinical setting, due to the unwanted side-effects associated with them. Therefore, a better understanding of the mechanisms bacteria and fungi use to hijack host cells would permit us to exploit these pathways and develop novel antimicrobial compounds, active against intracellular pathogens.

Antimicrobial peptides

Antimicrobial peptides (AMPs) are a conserved component of the host's innate immune system. These peptides are generally between 6 and 100 amino acids long and are divided into subcategories depending on their composition. The majority of antimicrobial peptides are unstructured until they integrate into the membrane of target cells. Once inside the membrane these peptides normally conform to one of four structures, α-helical, β-strands, β-hairpins and extended peptides. These peptides are amphipathic, possessing both polar and non-polar amino acids, to aid insertion into the bipolar lipid membrane. In many cases the mechanism by which these peptides affect microbial viability is unknown. Characterized mechanisms include pore formation within membranes and inhibition of DNA, cell wall and protein synthesis. Antimicrobial peptides are effective against both gram-positive and gram-negative bacteria, fungi and viruses. The major AMPs of the mammalian innate immune system included the cathelicidins, defensins, and the histatins.

LL-37 is an AMP of the cathelicidin family. LL-37 is stored in granulocytes within neutrophils as an inactive peptide. After release, the inactive peptide is cleaved producing the active peptide. At high concentrations (40 μg/ml) LL-37 has fungicidal effects on *C. albicans* (Tsai *et al.*, 2011). Addition of sub-MIC concentrations LL-37 to *C. albicans* significantly reduces fungal attachment to medical devices, the oral epidermis and bladder. This reduction in cell adhesion is thought to be due to the AMP binding the cell wall polysaccharides (Tsai *et al.*, 2011). Fungal cells treated with sub-MIC concentrations of the AMP showed increased flocculation (cell aggregation), suggesting that surface bound LL-37 self-associates sequestering fungal cells.

Histatin-5 (Hst-5) is another AMP with fungicidal activity against *C. albicans*. Owing to the low amphipathic nature of Hst-5, this AMP was thought not to integrate into the membrane but transverse the membrane and target mitochondrial respiration. At low concentrations Hst-5 is taken up by *C. albicans* via receptor mediate endocytosis into vesicles (Mochon and Liu, 2008). At this concentration Hst-5 does not kill the fungal cell, suggesting that vacuolar Hst-5 is non-cytotoxic to *C. albicans*. However, at physiological concentrations Hst-5 rapidly enters the cytoplasm of the cell through a single membrane intrusion (Mochon and Liu, 2008).

Fungal resistance to AMPs

It was once thought that AMPs would be advantageous over current antifungal therapies due to the disruptive method of killing resulting in rapid loss of cell viability. This is important because it could prevent the organism from developing resistance. In conjunction with this, not all AMPs require interaction with proteins or receptors. Therefore, it was thought that it would be less likely that resistance to AMPs would develop due to spontaneous mutation, as is the case for echinocandin resistance (see above), or other traditional antifungals. Despite this, it appears as though fungi have developed methods to combat AMPs. This is likely to be the evolutionary consequence of growth within the mammalian host, which uses AMPs in innate immunity. Amoebae also produce AMPs to kill bacteria (Leippe, 1999). Therefore, it is possible that resistance to AMPs in non-commensal fungi, developed as an evolutionary selective pressure to soil amoebae.

C. glabrata and *S. cerevisiae* are resistant to the actions of Hst-5 (De Smet *et al.*, 2004; Helmerhorst *et al.*, 2005). In *C. glabrata* Hst-5 is localized to vacuolar compartments within the cell even at high concentrations, while Hst-5 was only found on the surface of *S. cerevisiae* (Mochon and Liu, 2008). Therefore, both of these yeasts evade the antifungal activity of Hst-5 by preventing membrane lysis. However, *C. glabrata* is not the only fungus that has developed the ability to withstand AMPs. Recent evidence suggests that the membrane sensor Msb2 from *C. albicans* protects against AMPs (Szafranski-Schneider *et al.*, 2012). Msb2 is a type I transmembrane protein with a single membrane spanning domain separating the large extracellular domain from a smaller cytoplasmic domain. Msb2 functions in membrane integrity and morphogenesis (Csank *et al.*, 1998; Román *et al.*, 2009). The extracellular domain of Mbs2 is cleaved and protects *C. albicans* and bacteria from AMPs including Hst-5 and LL-37 by sequestering the peptides from the environment (Szafranski-Schneider *et al.*, 2012). In addition to Msb2, *C. albicans* secretes aspartyl proteases which inactivate AMPs (Meiller *et al.*, 2009), and can protect itself from AMPs through a Ssd1- and Bcr1-dependent mechanism (Jung *et al.*, 2013).

Antifungal potential of AMPs

Despite reports of antifungal resistance to AMPs, these peptides are still being pursued as novel antifungal compounds. Screening for novel antimicrobial peptides has been greatly enhanced by the development of technologies that permit the synthesis of peptides on cellulose membranes. Peptide synthesis has resulted in the production of vast peptide libraries, which can be screened for antimicrobial activity.

A colorimetric assay based on resazurin has been developed to permit rapid identification of peptides with antimicrobial activity (Yamaguchi *et al.*, 2002). Resazurin is a blue non-fluorescent compound that is converted into a pink fluorescent dye in the presence of metabolic activity. Therefore, for peptides with good antimicrobial properties the media remains blue due to inhibition of fungal metabolic activity, while for peptides that do not function as AMPs the media turns pink. This colour change can easily be observed by eye and quantified by a shift in absorbance from 570 to 600 nm. The screening of peptides for antifungal activity against *A. nidulans* using the assay described above identified Sub5, a 12 amino acid peptide, and Sub3 as having pharmacological potential with MICs in the range of 1–2 μM. Co-incubation of *A. nidulans* with Sub5 resulted in increased branching and hyphal tip splitting, but sub5 did not affect germination (Mania *et al.*, 2010).

Burrows *et al.* developed a series of cationic antimicrobial peptides that mimicked integral sections of membrane proteins, designated Kaxins. These molecules possess no cytotoxic effects against mammalian cells, but displayed antifungal activity against 37 *Candida* isolates (Burrows *et al.*, 2006). This activity was independent of fluconazole resistance. However, *C. glabrata* displayed a degree of resistance to the Kaxins, similar to the histatins (Burrows *et al.*, 2006). Therefore, AMPs may not be suitable for the treatment of *C. glabrata* infections.

One issue with AMPs is their lack of specificity. When targeting fungal infections an ideal treatment would be an AMP whose activity is targeted solely against the fungal pathogen. This type of approach would stop damage to the natural microbial flora and prevent secondary infections. In this regard, Arnusch *et al.* discovered that the addition of lipid moieties including vitamin E and cholesterol to AMPs selectively targeted the AMPs to fungi. These modified AMPs were shown to be functional against *Aspergillus fumigatus*, *C. neoformans* and *C. albicans*. On the other hand, these compounds were ineffective against bacteria and showed no toxicity in mice (Arnusch *et al.*, 2012). Therefore, there is the potential to specifically target AMPs to fungal pathogens.

In silico approaches have also been useful in the identification of novel AMPs. For example, Moradi *et al.* (2013) compared multiple peptide sequences that have been shown to possess antifungal activity to identify common patterns among the AMPs. This approach identified the sequence K-CENLA-DTY (where – can be any amino acid), which was used in the study to identify novel antifungal peptides. Although this peptide showed good activity against protoplasts (fungal cells where the cell wall has been removed), the peptide was fairly inactive against actual fungi (Moradi *et al.*, 2013). This is presumably due to poor penetration of the fungal cell wall. Therefore, although more studies are required to develop novel AMPs with specific activity against fungi, which are capable of penetrating fungal cell walls and polysaccharide capsules, there is a promising future for the use of novel AMPs for the treatment of fungal infections.

Summary

In this chapter we have reviewed the progress of some of the efforts to identify novel antifungal drug targets. One of the limitations in antifungal drug design is the lack of fungal specific targets. For this reason, the fungal membrane and cell wall have received a lot of attention in terms of antifungal drug design. This area has been profitable, identifying some of the most promising antifungals to date including the azoles and the echinocandins. However, their widespread use has led to the development of antifungal resistance, and now alternative therapies are needed. Further exploitation of the fungal cell wall has suggested that chitinases and chitin synthase inhibitors may hold potential for antifungal drug design. These compounds could be used in combination with other cell wall inhibitors to enhance potency. It is clear that as we delve deeper into the architecture of the cell wall more novel targets will be realized.

Initially carbonic anhydrases were thought to be a good antifungal target, but the ability of the fungus to bypass the requirement for CA activity in CO_2-enriched environments, suggests that these enzymes will not be useful in the design of treatments for invasive disease. However, these enzymes may still be beneficial for the treatment of dermatophytes. A more promising target is the fungal soluble adenylyl cyclase. This enzyme plays essential roles in virulence in many fungi and would therefore be an excellent choice of target. One limitation in the use of sAC for antifungal drug development is the ability to synthesis fungal specific inhibitors. Although humans only have one sAC located in sperm, the active sites of sACs and tmACs are fairly homologous. Active research into sAC inhibitors appears to be in its infancy compared to CA inhibitors. However, the potential of sACs as an antifungal drug target appears to be much greater than that of CAs. The use of quorum sensing molecules as novel antifungals also has merit, although proper investigation into the effects of the QSMs on the host microbiome need to be seen before these compounds can be considered as potential antifungal compounds. Antimicrobial peptides are hindered as antifungal compounds by the limited access to the cell membrane due to the impenetrable cell wall and in some cases capsule. If penetration can be improved then AMPs may become top priority for antifungal drug development.

In addition to the challenge of identifying a novel target, one issue in the development of antifungals lies with the pharmaceutical industry. Currently there are very few antifungal compounds in clinical trials. The time, financial investment and failure rate associated with drug development indicates that we are a long way off the introduction of a new antifungal drug to the market. Although the estimation of fatal fungal infections is high, many of the diseases are associated with developing countries and therefore do not promise substantial financial return for the large pharmaceutical companies. One alternative strategy is drug redeployment where approved compounds that have already passed clinical trials are tested for off target effects against fungi. This approach has the advantage that the time and cost of development of the drug are reduced. However, one drawback to this approach may be the lack of intellectual property rights to the compounds, which may prevent commercial companies from pursuing the compound.

In short, the rapid development of antifungal resistance coupled with the increase in patients receiving immunosuppressive therapy, and the lack of promising new antifungals in development is a cause for concern. A broad spectrum antifungal would be a welcome sight, but is still a long way off. Basic research into novel antifungal targets, as described here, are good but the potential of these targets for commercial development remains to be seen.

References

Abe, S., Tsunashima, R., Iijima, R., Yamada, T., Maruyama, N., Hisajima, T., Abe, Y., Oshima, H., and Yamazaki, M. (2009). Suppression of anti-*Candida* activity of macrophages by a quorum-sensing molecule, farnesol, through induction of oxidative stress. Microbiol. Immunol. *53*, 323–330.

Aguilera, J., Van Dijken, J.P., De Winde, J.H., and Pronk, J.T. (2005). Carbonic anhydrase (Nce103p): an essential biosynthetic enzyme for growth of *Saccharomyces cerevisiae* at atmospheric carbon dioxide pressure. Biochem. J. *391*, 311–316.

Alem, M.A.S., Oteef, M.D.Y., Flowers, T.H., and Douglas, L.J. (2006). Production of tyrosol by *Candida albicans* biofilms and its role in quorum sensing and biofilm development. Eukaryot. Cell *5*, 1770–1779.

Alspaugh, J.A., Pukkila-Worley, R., Harashima, T., Cavallo, L.M., Funnell, D., Cox, G.M., Perfect, J.R., Kronstad, J.W., and Heitman, J. (2002). Adenylyl cyclase functions downstream of the Gα protein Gpa1 and controls mating and pathogenicity of *Cryptococcus neoformans*. Eukaryot. Cell *1*, 75–84.

Amoroso, G., Morell-Avrahov, L., Muller, D., Klug, K., and Sultemeyer, D. (2005). The gene *NCE103* (YNL036w) from *Saccharomyces cerevisiae* encodes a functional carbonic anhydrase and its transcription is regulated by the concentration of inorganic carbon in the medium. Mol. Microbiol. *56*, 549–555.

Antonaroli, S., Bianco, A., Brufani, M., Cellai, L., Lo Baido, G., Potier, E., Bonomi, L., Perfetti, S., Fiaschi, A.I., and Segre, G. (1992). Acetazolamide-like carbonic anhydrase inhibitors with topical ocular hypotensive activity. J. Med. Chem. *35*, 2697–2703.

Arnusch, C.J., Ulm, H., Josten, M., Shadkchan, Y., Osherov, N., Sahl, H.-G., and Shai, Y. (2012). Ultrashort peptide bioconjugates are exclusively antifungal agents and synergize with cyclodextrin and amphotericin B. Antimicrob. Agents Chemother. *56*, 1–9.

Au-Young, J., and Robbins, P.W. (1990). Isolation of a chitin synthase gene (*CHS1*) from *Candida albicans* by expression in *Saccharomyces cerevisiae*. Mol. Microbiol. *4*, 197–207.

Bahn, Y.-S., Cox, G.M., Perfect, J.R., and Heitman, J. (2005). Carbonic anhydrase and CO_2 sensing during *Cryptococcus neoformans* growth, differentiation, and virulence. Curr. Biol. *15*, 2013–2020.

Baker, B.S. (2006). The role of microorganisms in atopic dermatitis. Clin. Exp. Immunol. *144*, 1–9.

Bassler, B.L. (1999). How bacteria talk to each other: regulation of gene expression by quorum sensing. Curr. Opin. Microbiol. *2*, 582–587.

Bates, S., Hughes, H.B., Munro, C.A., Thomas, W.P.H., MacCallum, D.M., Bertram, G., Atrih, A., Ferguson, M.A.J., Brown, A.J.P., Odds, F.C., *et al.* (2006). Outer chain *N*-glycans are required for cell wall integrity and virulence of *Candida albicans*. J. Biol. Chem. *281*, 90–98.

Boon, C., Deng, Y., Wang, L.-H., He, Y., Xu, J.-L., Fan, Y., Pan, S.Q., and Zhang, L.-H. (2007). A novel DSF-like signal from *Burkholderia cenocepacia* interferes with *Candida albicans* morphological transition. ISME J. *2*, 27–36.

vanden Bossche, H., Marichal, P., Odds, F.C., Le Jeune, L., and Coene, M.C. (1992). Characterization of an azole-resistant *Candida glabrata* isolate. Antimicrob. Agents Chemother. *36*, 2602–2610.

Brown, G.D., Denning, D.W., Gow, N.A.R., Levitz, S.M., Netea, M.G., and White, T.C. (2012). Hidden killers: human fungal infections. Sci.Transl. Med. *4*, 165rv113.

Bulawa, C.E., Miller, D.W., Henry, L.K., and Becker, J.M. (1995). Attenuated virulence of chitin-deficient mutants of *Candida albicans*. Proc. Natl. Acad. Sci. U.S.A. *92*, 10570–10574.

Burrows, L.L., Stark, M., Chan, C., Glukhov, E., Sinnadurai, S., and Deber, C.M. (2006). Activity of novel non-amphipathic cationic antimicrobial peptides against *Candida* species. J. Antimicrob. Chemother. *57*, 899–907.

Capasso, C., and Supuran, C.T. (2013). Anti-infective carbonic anhydrase inhibitors: a patent and literature review. Expert Opin. Ther. Pat. *23*, 693–704.

Carta, F., Innocenti, A., Hall, R.A., Mühlschlegel, F.A., Scozzafava, A., and Supuran, C.T. (2011). Carbonic anhydrase inhibitors. Inhibition of the β-class enzymes from the fungal pathogens *Candida albicans* and *Cryptococcus neoformans* with branched aliphatic/aromatic carboxylates and their derivatives. Bioorg. Med. Chem. Lett. *21*, 2521–2526.

Castelli, M.V., Butassi, E., Monteiro, M.C., Svetaz, L.A., Vicente, F., and Zacchino, S.A. (2014). Novel antifungal agents: a patent review (2011 – present). Expert Opin. Ther. Pat. *24*, 323–338.

Cerca, N., Gomes, F., Pereira, S., Teixeira, P., and Oliveira, R. (2012). Confocal laser scanning microscopy analysis of *S. epidermidis* biofilms exposed to farnesol, vancomycin and rifampicin. BMC Res. Notes *5*, 244.

Cerca, N., Gomes, F., Bento, J.C., Franca, A., Rolo, J., Miragaia, M., Teixeira, P., and Oliveira, R. (2013). Farnesol induces cell detachment from established *S. epidermidis* biofilms. J. Antibiot. *66*, 255–258.

Chang, R., Vo, T.-T., and Finney, N.S. (2006). Synthesis of the C1-phosphonate analog of UDP-GlcNAc. Carbohydr. Res. *341*, 1998–2004.

Chang, W., Li, Y., Zhang, L., Cheng, A., and Lou, H. (2012). Retigeric acid B attenuates the virulence of *Candida albicans* via inhibiting adenylyl cyclase activity targeted by enhanced farnesol production. PLoS One 7, e41624.

Chen, H., Fujita, M., Feng, Q., Clardy, J., and Fink, G.R. (2004). Tyrosol is a quorum-sensing molecule in *Candida albicans*. Proc. Natl. Acad. Sci. U.S.A. *101*, 5048–5052.

Chen, Y., Cann, M.J., Litvin, T.N., Iourgenko, V., Sinclair, M.L., Levin, L.R., and Buck, J. (2000). Soluble adenylyl cyclase as an evolutionarily conserved bicarbonate sensor. Science *289*, 625–628.

Cottier, F., Raymond, M., Kurzai, O., Bolstad, M., Leewattanapasuk, W., Jiménez-López, C., Lorenz, M.C., Sanglard, D., Váchová, L., Pavelka, N., *et al.* (2012). The bZIP transcription factor Rca1p is a central regulator of a novel CO_2 sensing pathway in yeast. PLoS Pathog. *8*, e1002485.

Cottier, F., Leewattanapasuk, W., Kemp, L.R., Murphy, M., Supuran, C.T., Kurzai, O., and Mühlschlegel, F.A. (2013). Carbonic anhydrase regulation and CO_2 sensing in the fungal pathogen *Candida glabrata* involves a novel Rca1p ortholog. Bioorg. Med. Chem. *21*, 1549–1554.

Csank, C., Schroppel, K., Leberer, E., Harcus, D., Mohamed, O., Meloche, S., Thomas, D.Y., and Whiteway, M. (1998). Roles of the *Candida albicans* mitogen-activated protein kinase homolog, Cek1p, in hyphal development and systemic candidiasis. Infect. Immun. *66*, 2713–2721.

Cugini, C., Calfee, M.W., Farrow, J.M., Morales, D.K., Pesci, E.C., and Hogan, D.A. (2007). Farnesol, a common sesquiterpene, inhibits PQS production in *Pseudomonas aeruginosa*. Mol. Microbiol. *65*, 896–906.

Davis, R.A., Hofmann, A., Osman, A., Hall, R.A., Mühlschlegel, F.A., Vullo, D., Innocenti, A., Supuran, C.T., and Poulsen, S.-A. (2011). Natural product-based phenols as novel probes for mycobacterial and fungal carbonic anhydrases. J. Med. Chem. *54*, 1682–1692.

Davis-Hanna, A., Piispanen, A.E., Stateva, L.I., and Hogan, D.A. (2008). Farnesol and dodecanol effects on the *Candida albicans* Ras1-cAMP signalling pathway and the regulation of morphogenesis. Mol. Microbiol. *67*, 47–62.

Décanis, N., Savignac, K., and Rouabhia, M. (2009). Farnesol promotes epithelial cell defense against *Candida albicans* through toll-like receptor 2 expression, interleukin-6 and human β-defensin 2 production. Cytokine 45, 132–140.

De Smet, K., Reekmans, R., and Contreras, R. (2004). Role of oxidative phosphorylation in histatin 5-induced cell death in *Saccharomyces cerevisiae*. Biotechnol. Lett. *26*, 1781–1785.

Dessauer, C.W., Tesmer, J.J., Sprang, S.R., and Gilman, A.G. (1999). Review the interactions of adenylate cyclases with P-site inhibitors. Trends Pharmacol. Sci. *20*, 205–210.

Deveau, A., Piispanen, A.E., Jackson, A.A., and Hogan, D.A. (2010). Farnesol induces hydrogen peroxide resistance in *Candida albicans* yeast by inhibiting the Ras-cyclic AMP signaling pathway. Eukaryot. Cell *9*, 569–577.

Díaz-Jiménez, D.F., Mora-Montes, H.M., Hernández-Cervantes, A., Luna-Arias, J.P., Gow, N.A.R., and Flores-Carreón, A. (2012). Biochemical characterization of recombinant *Candida albicans* mannosyltransferases Mnt1, Mnt2 and Mnt5 reveals new functions in *O*- and *N*-mannan biosynthesis. Biochem. Biophys. Res. Commun. *419* 77–82.

Drakulovski, P., Dunyach, C., Bertout, S., Reynes, J., and Mallié, M. (2011). A *Candida albicans* strain with high MIC for caspofungin and no *FKS1* mutations exhibits a high chitin content and mutations in two chitinase genes. Med. Mycol. *49*, 467–474.

Du, H., Guan, G., Xie, J., Cottier, F., Sun, Y., Jia, W., Mühlschlegel, F.A., and Huang, G. (2012). The transcription factor Flo8 mediates CO_2 sensing in the human fungal pathogen *Candida albicans*. Mol. Biol. Cell *23*, 2692–2701.

Elleuche, S., and Poggeler, S. (2009). Beta-carbonic anhydrases play a role in fruiting body development and ascospore germination in the filamentous fungus *Sordaria macrospora*. PLoS One *4*, e5177.

Fang, W., Du, T., Raimi, O.G., R., H.G., Mariño, K., Ibrahim, A.F., Albarbarawi, O., Ferguson, M.A., Jin, C., and van Aalten, D.M. (2013). Genetic and structural validation of *Aspergillus fumigatus N*-acetylphosphoglucosamine mutase as an antifungal target. Biosci. Rep. *33*, e00063.

Feldmesser, M., Kress, Y., Novikoff, P., and Casadevall, A. (2000). *Cryptococcus neoformans* is a facultative intracellular pathogen in murine pulmonary infection. Infect. Immun. *68*, 4225–4237.

Ganesan, L.T., Manavathu, E.K., Cutright, J.L., Alangaden, G.J., and Chandrasekar, P.H. (2004). *In vitro* activity of nikkomycin Z alone and in combination with polyenes, triazoles or echinocandins against *Aspergillus fumigatus*. Clin. Microbiol. Infect. *10*, 961–966.

Garcia-Effron, G., Lee, S., Park, S., Cleary, J.D., and Perlin, D.S. (2009). Effect of *Candida glabrata FKS1* and *FKS2* mutations on echinocandin sensitivity and kinetics of 1,3-β-D-glucan synthase: implication for the existing susceptibility breakpoint. Antimicrob. Agents Chemother. *53*, 3690–3699.

Gaughran, J.P., Lai, M.H., Kirsch, D.R., and Silverman, S.J. (1994). Nikkomycin Z is a specific inhibitor of *Saccharomyces cerevisiae* chitin synthase isozyme Chs3 *in vitro* and *in vivo*. J. Bacteriol. *176*, 5857–5860.

Gautier-Lefebvre, I., Behr, J.-B., Guillerm, G., and Muzard, M. (2005). Iminosugars as glycosyltransferase inhibitors: synthesis of polyhydroxypyrrolidines and their evaluation on chitin synthase activity. Eur. J. Med. Chem. *40*, 1255–1261.

Giorni, P., Battilani, P., Pietri, A., and Magan, N. (2008). Effect of aw and CO_2 level on *Aspergillus flavus* growth and aflatoxin production in high moisture maize postharvest. Int. J. Food Microbiol. *122*, 109–113.

Götz, R., Gnann, A., and Zimmermann, F.K. (1999). Deletion of the carbonic anhydrase-like gene *NCE103* of the yeast *Saccharomyces cerevisiae* causes an oxygen-sensitive growth defect. Yeast *15*, 855–864.

Gow, N.A., Robbins, P.W., Lester, J.W., Brown, A.J., Fonzi, W.A., Chapman, T., and Kinsman, O.S. (1994). A hyphal-specific chitin synthase gene (*CHS2*) is not essential for growth, dimorphism, or virulence of *Candida albicans*. Proc. Natl. Acad. Sci. U.S.A. *91*, 6216–6220.

Güzel, Ö., Maresca, A., Hall, R.A., Scozzafava, A., Mastrolorenzo, A., Mühlschlegel, F.A., and Supuran, C.T. (2010). Carbonic anhydrase inhibitors. The β-carbonic anhydrases from the fungal pathogens *Cryptococcus neoformans* and *Candida albicans* are strongly inhibited by substituted-phenyl-1H-indole-5-sulfonamides. Bioorg. Med. Chem. Lett. *20*, 2508–2511.

Hall, R.A., Turner, K.J., Chaloupka, J., Cottier, F., De Sordi, L., Sanglard, D., Levin, L.R., Buck, J., and Muhlschlegel, F.A. (2011). The quorum-sensing molecules farnesol/homoserine lactone and dodecanol operate via distinct modes of action in *Candida albicans*. Eukaryot. Cell *10*, 1034–1042.

Hall, R.A., and Gow, N.A.R. (2013). Mannosylation in *Candida albicans*: role in cell wall function and immune recognition. Mol. Microbiol. *90*, 1147–1161.

Han, K.H., Chun, Y.H., Figueiredo Bde, C., Soriani, F.M., Savold, i.M., Almeida, A., Rodrigues, F., Cairns, C.T., Bignell, E., Tobal, J.M., *et al.* (2010). The conserved and divergent roles of carbonic anhydrases in the filamentous fungi *Aspergillus fumigatus* and *Aspergillus nidulans*. Mol. Microbiol. *75*, 1372–1388.

Havlickova, B., Czaika, V.A., and Friedrich, M. (2008). Epidemiological trends in skin mycoses worldwide. Mycoses *51* (Suppl. 4), 2–15.

Hector, R.F., and Schaller, K. (1992). Positive interaction of nikkomycins and azoles against *Candida albicans in vitro* and *in vivo*. Antimicrob. Agents Chemother. *36*, 1284–1290.

Helmerhorst, E.J., Venuleo, C., Beri, A., and Oppenheim, F.G. (2005). *Candida glabrata* is unusual with respect to its resistance to cationic antifungal proteins. Yeast *22*, 705–714.

Hewitson, K.S., Vullo, D., Scozzafava, A., Mastrolorenzo, A., and Supuran, C.T. (2012). Molecular cloning, characterization, and inhibition studies of a β-carbonic anhydrase from *Malassezia globosa*, a potential anti-dandruff target. J. Med. Chem. *55*, 3513–3520.

Hisajima, T., Maruyama, N., Tanabe, Y., Ishibashi, H., Yamada, T., Makimura, K., Nishiyama, Y., Funakoshi, K., Oshima, H., and Abe, S. (2008). Protective effects of farnesol against oral candidiasis in mice. Microbiol. Immunol. *52*, 327–333.

Hitchcock, C.A., Pye, G.W., Troke, P.F., Johnson, E.M., and Warnock, D.W. (1993). Fluconazole resistance in *Candida glabrata*. Antimicrob. Agents Chemother. *37*, 1962–1965.

Hogan, D.A., and Kolter, R. (2002). *Pseudomonas–Candida* interactions: an ecological role for virulence factors. Science *296*, 2229–2232.

Hogan, D.A., Vik, A., and Kolter, R. (2004). A *Pseudomonas aeruginosa* quorum-sensing molecule influences *Candida albicans* morphology. Mol. Microbiol. *54*, 1212–1223.

Horfmann, M., Boles, E., and Zimmermann, F.K. (1994). Characterisation of the essential yeast gene encoding *N*-acetylglucosamine-phosphate mutase. Eur. J. Biochem. *221*, 741–747.

Hornby, J.M. (2001). Quorum sensing in the dimorphic fungus *Candida albicans* is mediated by farnesol. Appl. Environ. Microbiol. *67*, 2982–2992.

Hornby, J.M., Kebaara, B.W., and Nickerson, K.W. (2003). Farnesol biosynthesis in *Candida albicans*: cellular response to sterol inhibition by zaragozic acid B. Antimicrob. Agents Chemother. *47*, 2366–2369.

Huang, G., Srikantha, T., Sahni, N., Yi, S., and Soll, D.R. (2009). CO_2 regulates white-to-opaque switching in *Candida albicans*. Curr. Biol. *19*, 330–334.

Imtiaz, T., Lee, K.K., Munro, C.A., MacCallum, D.M., Shankland, G.S., Johnson, E.M., MacGregor, M.S., and Bal, A.M. (2012). Echinocandin resistance due to simultaneous FKS mutation and increased cell wall chitin in a *Candida albicans* bloodstream isolate following brief exposure to caspofungin. J. Med. Microbiol. *61*, 1330–1334.

Innocenti, A., Mühlschlegel, F.A., Hall, R.A., Steegborn, C., Scozzafava, A., and Supuran, C.T. (2008). Carbonic anhydrase inhibitors: Inhibition of the β-class enzymes from the fungal pathogens *Candida albicans* and *Cryptococcus neoformans* with simple anions. Bioorg. Med. Chem. Lett. *18*, 5066–5070.

Innocenti, A., Hall, R.A., Schlicker, C., Mühlschlegel, F.A., and Supuran, C.T. (2009a). Carbonic anhydrase inhibitors. Inhibition of the β-class enzymes from the fungal pathogens *Candida albicans* and *Cryptococcus neoformans* with aliphatic and aromatic carboxylates. Bioorg Med Chem *17*, 2654–2657.

Innocenti, A., Hall, R.A., Schlicker, C., Scozzafava, A., Steegborn, C., Mühlschlegel, F.A., and Supuran, C.T. (2009b). Carbonic anhydrase inhibitors. Inhibition and homology modeling studies of the fungal β-carbonic anhydrase from *Candida albicans* with sulfonamides. Bioorg. Med. Chem. *17*, 4503–4509.

Jain, P., Akula, I., and Edlind, T. (2003). Cyclic AMP signaling pathway modulates susceptibility of *Candida* species and *Saccharomyces cerevisiae* to antifungal azoles and other sterol biosynthesis inhibitors. Antimicrob. Agents Chemother. *47*, 3195–3201.

Jung, S.-I., Finkel, J.S., Solis, N.V., Chaili, S., Mitchell, A.P., Yeaman, M.R., and Filler, S.G. (2013). Bcr1 functions downstream of Ssd1 to mediate antimicrobial peptide resistance in *Candida albicans*. Eukaryot. Cell *12*, 411–419.

Kebaara, B.W., Langford, M.L., Navarathna, D.H.M.L.P., Dumitru, R., Nickerson, K.W., and Atkin, A.L. (2008). *Candida albicans* Tup1 is involved in farnesol-mediated inhibition of filamentous-growth induction. Eukaryot. Cell *7*, 980–987.

Kim, M.-K., Park, H.-S., Kim, C.-H., Park, H.-M., and Choi, W. (2002). Inhibitory effect of nikkomycin Z on chitin synthases in *Candida albicans*. Yeast *19*, 341–349.

Klengel, T., Liang, W.-J., Chaloupka, J., Ruoff, C., Schroppel, K., Naglik, J.R., Eckert, S.E., Mogensen, E.G., Haynes, K., Tuite, M.F., *et al.* (2005). Fungal adenylyl cyclase integrates CO_2 sensing with cAMP signaling and virulence. Curr. Biol. *15*, 2021–2026.

Klis, F., de Groot, P., and Hellingwerf, K. (2001). Molecular organization of the cell wall of *Candida albicans*. Med. Mycol. *39*, 1–8.

Kondoh, O., Tachibana, Y., Ohya, Y., Arisawa, M., and Watanabe, T. (1997). Cloning of the *RHO1* gene from *Candida albicans* and its regulation of beta-1,3-glucan synthesis. J. Bacteriol. *179*, 7734–7741.

Kruppa, M., Krom, B., Chauhan, N., Bambach, A., Cihlar, R., and Calderone, R. (2004). The two-component signal transduction protein Chk1p regulates quorum sensing in *Candida albicans*. Eukaryot. Cell *3*, 1062–1065.

Kusian, B., Sultemeyer, D., and Bowien, B. (2002). Carbonic anhydrase is essential for growth of *Ralstonia eutropha* at ambient CO_2 concentrations. J. Bacteriol. *184*, 5018–5026.

Lang-Yona, N., Levin, Y., Dannemiller, K., Yarden, O., Peccia, J., and Rudich, Y. (2013). Changes in atmospheric CO_2 influence the allergenicity of *Aspergillus fumigatus*. Glob. Chang. Biol. *19*, 2381–2388.

Latgé, J.P., Mouyna, I., Tekaia, F., Beauvais, A., Debeaupuis, J.P., and Nierman, W. (2005). Specific molecular features in the organization and biosynthesis of the cell wall of *Aspergillus fumigatus*. Med. Mycol. *43*, S15–S22.

Lee, K.K., MacCallum, D.M., Jacobsen, M.D., Walker, L.A., Odds, F.C., Gow, N.A.R., and Munro, C.A. (2012). Elevated cell wall chitin in *Candida albicans* confers echinocandin resistance *in vivo*. Antimicrob. Agents Chemother. *56*, 208–217.

Leippe, M. (1999). Antimicrobial and cytolytic polypeptides of amoeboid protozoa – effector molecules of primitive phagocytes. Dev. Comp. Immunol. *23*, 267–279.

Levy, D.E., Bao, M., Cherbavaz, D.B., Tomlinson, J.E., Sedlock, D.M., Homcy, C.J., and Scarborough, R.M. (2003). Metal coordination-based inhibitors of adenylyl cyclase: novel potent P-site antagonists. J. Med. Chem. *46*, 2177–2186.

Lewenza, S., Conway, B., Greenberg, E.P., and Sokol, P.A. (1999). Quorum sensing in *Burkholderia cepacia*: Identification of the LuxrI homologs CepRI. J. Bacteriol. *181*, 748–756.

Li, L., Redding, S., and Dongari-Bagtzoglou, A. (2007). *Candida glabrata*, an emerging oral opportunistic pathogen. J. Dent. Res. *86*, 204–215.

Lorek, J., Pöggeler, S., Weide, M.R., Breves, R., and Bockmühl, D.P. (2008). Influence of farnesol on the morphogenesis of *Aspergillus niger*. J. Basic Microbiol. *48*, 99–103.

Lorenz, M.C., Bender, J.A., and Fink, G.R. (2004). Transcriptional response of *Candida albicans* upon internalization by macrophages. Eukaryot. Cell *3*, 1076–1087.

Ma, H., Croudace, J.E., Lammas, D.A., and May, R.C. (2006). Expulsion of live pathogenic yeast by macrophages. Curr. Biol. *16*, 2156–2160.

Ma, H., Hagen, F., Stekel, D.J., Johnston, S.A., Sionov, E., Falk, R., Polacheck, I., Boekhout, T., and May, R.C. (2009). The fatal fungal outbreak on Vancouver Island is characterized by enhanced intracellular parasitism driven by mitochondrial regulation. Proc. Natl. Acad. Sci. *106*, 12980–12985.

Magellan, H., Boccara, M., Drujon, T., Soulié, M.-C., Guillou, C., Dubois, J., and Becker, H.F. (2013). Discovery of two new inhibitors of *Botrytis cinerea* chitin synthase by a chemical library screening. Bioorg. Med. Chem. *21*, 4997–5003.

Mania, D., Hilpert, K., Ruden, S., Fischer, R., and Takeshita, N. (2010). Screening for antifungal peptides and their modes of action in *Aspergillus nidulans*. Appl. Environ. Microbiol. *76*, 7102–7108.

Martins, M., Lazzell, A.L., Lopez-Ribot, J.L., Henriques, M., and Oliveira, R. (2012). Effect of exogenous administration of *Candida albicans* autoregulatory alcohols in a murine model of hematogenously disseminated candidiasis. J. Basic Microbiol. *52*, 487–491.

Masuda, S., and Ono, T.A. (2005). Adenylyl cyclase activity of *Cya1* from the *cyanobacterium synechocystis* sp. strain PCC 6803 is inhibited by bicarbonate. J. Bacteriol. *187*, 5032–5035.

Mazur, P., and Baginsky, W. (1996). *In vitro* activity of 1,3-beta-D-glucan synthase requires the GTP-binding protein Rho1. J. Biol. Chem. *271*, 14604–14609.

Meiller, T.F., Hube, B., Schild, L., Shirtliff, M.E., Scheper, M.A., Winkler, R., Ton, A., and Jabra-Rizk, M.A. (2009). A novel immune evasion strategy of *Candida albicans*: proteolytic cleavage of a salivary antimicrobial peptide. PLoS One *4*, e5039.

Merlin, C., Masters, M., McAteer, S., and Coulson, A. (2003). Why is carbonic anhydrase essential to *Escherichia coli*? J. Bacteriol. *185*, 6415–6424.

Mio, T., Adachi-Shimizu, M., Tachibana, Y., Tabuchi, H., Inoue, S.B., Yabe, T., Yamada-Okabe, T., Arisawa, M., Watanabe, T., and Yamada-Okabe, H. (1997). Cloning of the *Candida albicans* homolog of *Saccharomyces cerevisiae GSC1/FKS1* and its involvement in beta-1,3-glucan synthesis. J. Bacteriol. *179*, 4096–4105.

Mitsuhashi, S., Ohnishi, J., Hayashi, M., and M., I. (2004). A gene homologous to beta-type carbonic anhydrase is essential for the growth of *Corynebacterium glutamicum* under atmospheric conditions. Appl. Microbiol. Biotechnol. *63*, 592–601.

Mochon, A.B., and Liu, H. (2008). The antimicrobial peptide histatin-5 causes a spatially restricted disruption on the *Candida albicans* surface, allowing rapid entry of the peptide into the cytoplasm. PLoS Pathog. *4*, e1000190.

Mogensen, E.G., Janbon, G., Chaloupka, J., Steegborn, C., Fu, M.S., Moyrand, F., Klengel, T., Pearson, D.S., Geeves, M.A., Buck, J., *et al.* (2006). *Cryptococcus neoformans* senses CO_2 through the carbonic anhydrase Can2 and the adenylyl cyclase Cac1. Eukaryot. Cell *5*, 103–111.

Monti, S.M., Supuran, C.T., and De Simone, G. (2013). Anticancer carbonic anhydrase inhibitors: a patent review (2008 – 2013). Expert Opin. Ther. Pat. *23*, 737–749.

Moradi, S., Azerang, P., Khalaj, V., and Sardari, S. (2013). Antifungal indole and pyrrolidine-2,4-dione derivative peptidomimetic lead design based on *in silico* study of bioactive peptide families. Avicenna J. Med. Biotechnol. *5*, 42–53.

Mora-Montes, H.M., Bates, S., Netea, M.G., Diaz-Jimenez, D.F., Lopez-Romero, E., Zinker, S., Ponce-Noyola, P., Kullberg, B.J., Brown, A.J.P., Odds, F.C., *et al.* (2007). Endoplasmic Reticulum alpha-glycosidases of *Candida albicans* are required for *N*-glycosylation, cell wall integrity, and normal host–fungus interaction. Eukaryot. Cell *6*, 2184–2193.

Mora-Montes, H.M., López-Romero, E., Zinker, S., Ponce-Noyola, P., and Flores--Carreón, A. (2008). Heterologous expression and biochemical characterization of an beta1,2-mannosidase encoded by the *Candida albicans MNS1* gene. Mem. Inst. Oswaldo Cruz *103*, 724–730.

Mora-Montes, H.M., Bates, S., Netea, M.G., Castillo, L., Brand, A., Buurman, E.T., Diaz-Jimenez, D.F., Jan Kullberg, B., Brown, A.J.P., Odds, F.C., *et al.* (2010). A multifunctional mannosyltransferase family in *Candida albicans* determines cell wall mannan structure and host–fungus interactions. J. Biol. Chem. *285*, 12087–12095.

Munro, C.A. (2013). Chitin and glucan, the yin and yang of the fungal cell wall, implications for antifungal drug discovery and therapy. Adv. Appl. Microbiol. *83*, 145–172.

Munro, C.A., Winter, K., Buchan, A., Henry, K., Becker, J.M., Brown, A.J.P., Bulawa, C.E., and Gow, N.A.R. (2001). Chs1 of *Candida albicans* is an essential chitin synthase required for synthesis of the septum and for cell integrity. Mol. Microbiol. *39*, 1414–1426.

Munro, C.A., Whitton, R.K., Bleddyn Hughes, H., Rella, M., Selvaggini, S., and Gow, N.A.R. (2003). *CHS8*—a fourth chitin synthase gene of *Candida albicans* contributes to *in vitro* chitin synthase activity, but is dispensable for growth. Fungal Genet. Biol. *40*, 146–158.

Munro, C.A., Selvaggini, S., de Bruijn, I., Walker, L., Lenardon, M.D., Gerssen, B., Milne, S., Brown, A.J., and Gow, N.A. (2007). The PKC, HOG and Ca2+ signalling pathways co-ordinately regulate chitin synthesis in *Candida albicans*. Mol. Microbiol. *63*, 1399–1413.

Nealson, K.H., Platt, T., and Hastings, J.W. (1970). Cellular control of the synthesis and activity of the bacterial luminescent system. J. Bacteriol. *104*, 313–322.

Netea, M.G., Brown, G.D., Kullberg, B.J., and Gow, N.A.R. (2008). An integrated model of the recognition of *Candida albicans* by the innate immune system. Nat. Rev. Microbiol. *6*, 67–78.

Nishitani, Y., Maruyama, D., Nonaka, T., Kita, A., Fukami, T.A., Mio, T., Yamada-Okabe, H., Yamada-Okabe, T., and Miki, K. (2006). Purification, crystallization and preliminary X-ray diffraction studies of *N*-acetylglucosamine-phosphate mutase from *Candida albicans*. Acta Crystallogr. Sect. F Struct. Biol. Cryst. Commun. *62*, 419–421.

Passador, L., Cook, J.M., Gambello, M.J., Rust, L., and Iglewski, B.H. (1993). Expression of *Pseudomonas aeruginosa* virulence genes requires cell-to-cell communication. Science *260*, 1127–1130.

Pfaller, M.A., Boyken, L., Hollis, R.J., Messer, S.A., Tendolkar, S., and Diekema, D.J. (2006a). Global surveillance of *in vitro* activity of micafungin against *Candida*: a comparison with caspofungin by CLSI-recommended methods. J. Clin. Microbiol. *44*, 3533–3538.

Pfaller, M.A., Boyken, L., Hollis, R.J., Messer, S.A., Tendolkar, S., and Diekema, D.J. (2006b). *In vitro* susceptibilities of *Candida* spp. to caspofungin: four years of global surveillance. J. Clin. Microbiol. *44*, 760–763.

Pfaller, M.A., Diekema, D.J., Gibbs, D.L., Newell, V.A., Meis, J.F., Gould, I.M., Fu, W., Colombo, A.L., and Rodriguez-Noriega, E.; Global Antifungal Surveillance Study. (2007). Results from the ARTEMIS DISK global antifungal surveillance study, 1997 to 2005: an 8.5-year analysis of susceptibilities of *Candida* species and other yeast species to fluconazole and voriconazole determined by CLSI standardized disk diffusion testing. J. Clin. Microbiol. *45*, 1735–1745.

Prill, S.K.H., Klinkert, B., Timpel, C., Gale, C.A., Schröppel, K., and Ernst, J.F. (2005). PMT family of *Candida albicans*: five protein mannosyltransferase isoforms affect growth, morphogenesis and antifungal resistance. Mol. Microbiol. *55*, 546–560.

Rennemeier, C., Frambach, T., Hennicke, F., Dietl, J., and Staib, P. (2009). Microbial quorum-sensing molecules induce acrosome loss and cell death in human spermatozoa. Infect. Immun. *77*, 4990–4997.

Rocha, C.R.C., Schroppel, K., Harcus, D., Marcil, A., Dignard, D., Taylor, B.N., Thomas, D.Y., Whiteway, M., and Leberer, E. (2001). Signaling through adenylyl cyclase is essential for hyphal growth and virulence in the pathogenic fungus *Candida albicans*. Mol. Biol. Cell *12*, 3631–3643.

Rojas, F.J., Bruzzone, M.E., and Moretti-Rojas, I. (1992). Regulation of cyclic adenosine monophosphate synthesis in human ejaculated spermatozoa. II. The role of calcium and bicarbonate ions on the activation of adenylyl cyclase. Hum. Reprod. *7*, 1131–1135.

Román, E., Cottier, F., Ernst, J.F., and Pla, J. (2009). Msb2 signaling mucin controls activation of Cek1 mitogen-activated protein kinase in *Candida albicans*. Eukaryot. Cell *8*, 1235–1249.

Sato, T., Watanabe, T., Takeshi Mikami, and Matsumot, T. (2004). Farnesol, a morphogenetic autoregulatory substance in the dimorphic fungus *Candida albicans*, inhibits hyphae growth through suppression of a mitogen-activated protein kinase cascade. Biol. Pharm. Bull. *27*, 751–752.

Schlicker, C., Rauch, A., Hess, K.C., Kachholz, B., Levin, L.R., Buck, J., and Steegborn, C. (2008). Structure-based development of novel adenylyl cyclase inhibitors. J. Med. Chem. *51*, 4456–4464.

Schlicker, C., Hall, R.A., Vullo, D., Middelhaufe, S., Gertz, M., Supuran, C.T., Mühlschlegel, F.A., and Steegborn, C. (2009). Structure and inhibition of the CO_2-sensing carbonic anhydrase Can2 from the pathogenic fungus *Cryptococcus neoformans*. J. Mol. Biol. *385*, 1207–1220.

Scozzafava, A., Supuran, C.T., and Carta, F. (2013). Antiobesity carbonic anhydrase inhibitors: a literature and patent review. Expert Opin. Ther. Pat. *23*, 725–735.

Seider, K., Brunke, S., Schild, L., Jablonowski, N., Wilson, D., Majer, O., Barz, D., Haas, A., Kuchler, K., Schaller, M., *et al.* (2011). The facultative intracellular pathogen *Candida glabrata* subverts macrophage cytokine production and phagolysosome maturation. J. Immunol. *187*, 3072–3086.

Semighini, C.P., Hornby, J.M., Dumitru, R., Nickerson, K.W., and Harris, S.D. (2006). Farnesol-induced apoptosis in *Aspergillus nidulans* reveals a possible mechanism for antagonistic interactions between fungi. Mol. Microbiol. *59*, 753–764.

Sharma, M., and Prasad, R. (2011). The quorum-sensing molecule farnesol is a modulator of drug efflux mediated by ABC multidrug transporters and synergizes with drugs in *Candida albicans*. Antimicrob. Agents Chemother. *55*, 4834–4843.

Shchepin, R., Hornby, J.M., Burger, E., Niessen, T., Dussault, P., and Nickerson, K.W. (2003). Quorum sensing in *Candida albicans*: probing farnesol's mode of action with 40 natural and synthetic farnesol analogs. Chem. Biol. *10*, 743–750.

Shchepin, R., Dumitru, R., Nickerson, K.W., Lund, M., and Dussault, P.H. (2005). Biologically active fluorescent farnesol analogs. Chem. Biol. *12*, 639–641.

Sheth, C., Johnson, E., Baker, M.E., Haynes, K., and F.A., M. (2005). Phenotypic identification of *Candida albicans* by growth on chocolate agar. Med. Mycol. *43*, 735–378.

Shubitz, L.F., Roy, M.E., Nix, D.E., and Galgiani, J.N. (2013). Efficacy of Nikkomycin Z for respiratory coccidioidomycosis in naturally infected dogs. Med. Mycol. *0*, 1–8.

Sinclair, M.L., Wang, X.-Y., Mattia, M., Conti, M., Buck, J., Wolgemuth, D.J., and Levin, L.R. (2000). Specific expression of soluble adenylyl cyclase in male germ cells. Mol. Reprod. Dev. *56*, 6–11.

Smith, E.S., Fleischer, A.B., and Feldman, S.R.J. (1998). Nondermatologists are more likely than dermatologists to prescribe antifungal/corticosteroid products: an analysis of office visits for cutaneous fungal infections, 1990–1994. J. Am. Acad. Dermatol. *39*, 43–47.

Smith, L.M., and May, R.C. (2013). Mechanisms of microbial escape from phagocyte killing. Biochem. Soc. Trans. *41*, 475–490.

Steele, S., Brunton, J., Ziehr, B., Taft-Benz, S., Moorman, N., and Kawula, T. (2013). *Francisella tularensis* harvests nutrients derived via ATG5-independent autophagy to support intracellular growth. PLoS Pathog. *9*, e1003562.

Sudoh, M., Yamazaki, T., Masubuchi, K., Taniguchi, M., Shimma, N., Arisawa, M., and Yamada-Okabe, H. (2000). Identification of a novel inhibitor specific to the fungal chitin synthase: inhibition of chitin synthase 1 arrests the cell growth, but inhibition of chitin synthase 1 and 2 is leathal in the pathogenic fungus *Candida albicans* J. Biol. Chem. *275*, 32901–32905.

Szafranski-Schneider, E., Swidergall, M., Cottier, F., Tielker, D., Román, E., Pla, J., and Ernst, J.F. (2012). Msb2 shedding protects *Candida albicans* against antimicrobial peptides. PLoS Pathog. *8*, e1002501.

Tan, T.Y., Tan, A.L., Tee, N.W.S., Ng, L.S.Y., and Chee, C.W.J. (2010). The increased role of non-*albicans* species in candidaemia: results from a 3-year surveillance study. Mycoses *53*, 515–521.

Tashiro, M., Kimura, S., Tateda, K., Saga, T., Ohno, A., Ishii, Y., Izumikawa, K., Tashiro, T., Kohno, S., and Yamaguchi, K. (2012). Pravastatin inhibits farnesol production in *Candida albicans* and improves survival in a mouse model of systemic candidiasis. Med. Mycol. *50*, 353–360.

Tesmer, J.J., Dessauer, C.W., Sunahara, R.K., Murray, L.D., Johnson, R.A., Gilman, A.G., and Sprang, S.R. (2000). Molecular basis for P-site inhibition of adenylyl cyclase. Biochemistry (Mosc) *39*, 14464–14671.

Tian, J., Weng, L.-X., Zhang, Y.-Q., and Wang, L.-H. (2013). BDSF inhibits *Candida albicans* adherence to urinary catheters. Microb. Pathog. *64*, 33–38.

Tomita, M., Kikuchi, A., Kobayashi, M., Yamaguchi, M., Ifuku, S., Yamashoji, S., Ando, A., and Saito, A. (2013). Characterization of antifungal activity of the GH-46 subclass III chitosanase from *Bacillus circulans* MH-K1. Antonie Van Leeuwenhoek *104*, 737–748.

Topal, H., Fulcher, N.B., Bitterman, J., Salazar, E., Buck, J., Levin, L.R., Cann, M.J., Wolfgang, M.C., and Steegborn, C. (2012). Crystal structure and regulation mechanisms of the CyaB adenylyl cyclase from the human pathogen *Pseudomonas aeruginosa*. J. Mol. Biol. *416*, 271–286.

Tsai, P.-W., Yang, C.-Y., Chang, H.-T., and Lan, C.-Y. (2011). Human antimicrobial peptide LL-37 inhibits adhesion of *Candida albicans* by interacting with yeast cell-wall carbohydrates. PLoS One *6*, e17755.

Tucker, S.C., and Casadevall, A. (2002). Replication of *Cryptococcus neoformans* in macrophages is accompanied by phagosomal permeabilization and accumulation of vesicles containing polysaccharide in the cytoplasm. Proc. Natl. Acad. Sci. *99*, 3165–3170.

Vandeputte, P., Pradervand, S., Ischer, F., Coste, A.T., Ferrari, S., Harshman, K., and Sanglard, D. (2012). Identification and functional characterization of Rca1, a transcription factor involved in both antifungal susceptibility and host response in *Candida albicans*. Eukaryot. Cell *11*, 916–931.

Vílchez, R., Lemme, A., Ballhausen, B., Thiel, V., Schulz, S., Jansen, R., Sztajer, H., and Wagner-Döbler, I. (2010). *Streptococcus mutans* inhibits *Candida albicans* hyphal formation by the fatty acid signaling molecule trans-2-decenoic acid (SDSF). Chembiochem. *11*, 1552–1562.

Voelz, K., Ma, H., Phadke, S., Byrnes, E.J., Zhu, P., Mueller, O., Farrer, R.A., Henk, D.A., Lewit, Y., Hsueh, Y.-P., *et al.* (2013). Transmission of hypervirulence traits via sexual reproduction within and between lineages of the human fungal pathogen *Cryptococcus gattii*. PLoS Genet. 9, e1003771.

Walker, L.A., Munro, C.A., Bruijn, I.d., Lenardon, M.D., McKinnon, A., and Gow, N.A. (2008). Stimulation of chitin synthesis rescues *Candida albicans* from echinocandins. PLoS Pathog. *4*, e1000040.

Walker, L.A., Gow, N.A.R., and Munro, C.A. (2013). Elevated chitin content reduces the susceptibility of *Candida* Sspecies to caspofungin. Antimicrob. Agents Chemother. 57, 146–154.

Westwater, C., Balish, E., and Schofield, D. (2005). *Candida albicans*-conditioned medium protects yeast cells from oxidative stress: a possible link between quorum sensing and oxidative stress resistance. Eukaryot. Cell *4*, 1654–1661.

Yamaguchi, H., Uchida, K., Nagino, K., and Matsunaga, T. (2002). Usefulness of a colorimetric method for testing antifungal drug susceptibilities of *Aspergillus* species to voriconazole. J. Infect. Chemother. *8*, 374–377.

Zaragoza, O., Fries, B.C., and Casadevall, A. (2003). Induction of capsule growth in *Cryptococcus neoformans* by mammalian serum and CO_2. Infect. Immun. *71*, 6155–6164.

Zhou, X., Ma, Y., Fang, Y., gerile, W., Jaiseng, W., Yamada, Y., and Kuno, T. (2013). A genome-wide screening of potential target genes to enhance the antifungal activity of micafungin in *Schizosaccharomyces pombe*. PLoS One *8*, e65904.

Zou, H., Fang, H.-M., Zhu, Y., and Wang, Y. (2010). *Candida albicans* Cyr1, Cap1 and G-actin form a sensor/effector apparatus for activating cAMP synthesis in hyphal growth. Mol. Microbiol. 75, 579–591.

Strategies for the Identification of the Mode of Action of Antifungal Drug Candidates

9

Sadri Znaidi

Abstract

The characterization of the mode-of-action of antifungal drug candidates – also termed target deconvolution – is an integral part in the pharmaceutical drug discovery pipeline. The classical methods that were used during decades for deciphering antifungal drug mode-of-action relied specifically on classical microscopy, basic biochemical and biophysical approaches and classical forward genetics. With the era of functional genomics and high-throughput experimentation technologies, the implementation of large-scale target deconvolution strategies has drastically reshaped the field of drug discovery. In particular, the individual or combined use of chemogenomics, high-throughput drug–target interaction assays, transcriptomics and high-throughput imaging, together with powerful computer-based approaches for the integrative and predictive analyses from multiple data, offer a completely new vision on the potential and perspectives of exploring drug–target interactions in drug discovery and chemical biology. In this chapter, an overview of the classical approaches in antifungal drug mode-of-action studies is presented as a historical perspective, followed by description of the modern, currently used or likely to be more developed, approaches for conducting large-scale antifungal drug candidate target deconvolution studies.

Introduction

Antifungal drug discovery is the subject of extensive research and development strategies due to the increasing number of patients at risk for developing life-threatening fungal infections such as those with severely compromised immunity or critically ill patients (Pfaller and Diekema, 2010). In addition, recent reports raise concerns about emerging fungal diseases, including mucormycosis (Brown, 2005; Rammaert *et al.*, 2012), phaehyphomycosis (Chen *et al.*, 2011; Revankar and Sutton, 2010) and *Rasamsonia argillacea* mycosis (Doyon *et al.*, 2013; Giraud *et al.*, 2013) and many of their causative agents are intrinsically resistant to the currently used antifungals. The impact of drug resistance is also a major issue for clinicians (Kanafani and Perfect, 2008), as exemplified by the recent rise of azole resistance in *Aspergillus fumigatus* (Vermeulen *et al.*, 2013), reflecting the need for developing resistance reversal agents and new antifungal candidates with alternative target specificities. Unlike antibacterial drugs which target various cellular processes in bacteria such as DNA replication, RNA transcription, cell wall biogenesis and protein synthesis (Kohanski *et al.*, 2010), the current major systemic antifungals – polyenes, azoles and echinocandins – target components of the fungal cell membranes or the cell wall (Odds *et al.*, 2003). Because humans and fungi are more phylogenetically close to each other than either is to bacteria, antifungal drug candidates often carry high potential of toxic side-effects in patients; which may explain the relatively slow pace at which the development of clinically safe antifungals has been progressing.

Benzimidazole, the first azole drug, was brought to interest by Woolley in 1944 following the observation that it inhibited the growth of some fungi and bacteria (Woolley, 1944); and was, in fact, the starting molecule for the synthesis of a myriad of anti-infective and anticancer derivatives (Bansal and Silakari, 2012). Azole antifungals have been considerably developed during the 1970s and 1980s with the introduction of miconazole, econazole (both developed by Janssen Pharmaceutica, Belgium) and clotrimazole (Bayer, Germany) as topical agents and, most importantly, ketoconazole in 1979 as the first systemic azole (Janssen Pharmaceutica) (Heeres *et al.*, 1979). These were followed by the triazole antifungals fluconazole, itraconazole (first-generation azoles), then voriconazole and posaconazole (second-generation azoles) (Krcmery and Kalavsky, 2007). The extensive development of azole antifungals was the result of sequential manipulations and improvements of the chemical structures and properties of the conazole family in particular – that is miconazole and its congeners (Heeres *et al.*, 2010). Most azoles exert a fungistatic action on fungi, which impacts on the development of azole drug resistance (Kanafani and Perfect, 2008; Lupetti *et al.*, 2002; Morschhauser, 2002). Azoles bind to fungal CYP51/Erg11 (lanosterol 14α-demethylase) with high affinity (Warrilow *et al.*, 2013) and inhibit demethylation of the ergosterol precursor lanosterol leading to sterol depletion and accumulation of toxic methylated sterol derivatives (Lamb *et al.*, 1999). This perturbs both the structure and function of fungal membranes, causing inhibition of cell growth and proliferation. One exception is miconazole which inhibits peroxidases in addition to targeting ergosterol biosynthesis, leading to peroxide accumulation and cell death (Barasch and Griffin, 2008). In fact, early studies already pointed to an effect of miconazole on the function of oxidative and peroxidative enzymes in *S. cerevisiae* and *C. albicans* (De Nollin and Borgers, 1977; De Nollin *et al.*, 1977).

In 1948 was discovered the first polyene macrolide antifungal nystatin by Hazen and Brown, then rapidly marketed by Squibb Laboratories in 1954 (Hazen and Brown, 1951; Swain *et al.*, 2003). Polyenes are a class of bacterial secondary metabolites from *Streptomyces*, an aerobic actinomycete obtained from soil, and include over 200 macrolide molecules (Lewis, 2010; Zotchev, 2003). Among these, only amphotericin B has been successfully developed into a highly efficient systemic antifungal agent (Fungizone™, deoxycholate conjugate by Squibb Laboratories), whereas nystatin was used as a topical agent (Lewis, 2010; Zotchev, 2003). The particularly potent antifungal properties of polyenes are reflected through their strong fungicidal action and their broad-spectrum potential (Lewis, 2010; Zotchev, 2003). However, strong toxic side-effects – nephrotoxicity and hepatotoxicity in particular – have limited the use of amphotericin B, which underlined the need for developing amphotericin B formulations with decreased toxicity (e.g. liposomal formulations and colloidal dispersion) (Carrillo-Munoz *et al.*, 2006; Ostrosky-Zeichner *et al.*, 2003). Polyenes directly target ergosterol, the major fungal membrane sterol (Odds *et al.*, 2003; Palacios *et al.*, 2011). It was thought for years that the fungicidal action of polyenes was primarily due to the induction of ion channel formation upon polyene–ergosterol interaction in fungal membranes leading to leakage of cellular content and cell death, until the recent demonstration that the simple direct interaction of amphotericin B and natamycin with ergosterol was sufficient for fungicidal activity (Gray *et al.*, 2012; Palacios *et al.*, 2011; te Welscher *et al.*, 2008). The potent fungicidal action of polyenes explains the relative rarity of occurrence of clinical polyene resistance. One recent study suggested that mutations conferring resistance to amphotericin B in *C. albicans* and *C. tropicalis* clinical isolates evolve with a dramatic fitness cost, including the inability to respond to various stresses encountered in the host and decreased virulence, which prevents the emergence of resistance (Vincent *et al.*, 2013). Nevertheless, mutations in *ERG6* (encoding delta 24-sterol methyltransferase) conferring polyene resistance from *C. glabrata* clinical isolates were identified and characterized (Vandeputte *et al.*, 2007, 2008).

The discovery of the echinocandins (caspofungin, micafungin, anidulafungin) was a milestone in antifungal development, owing to their highly efficient antifungal action against *Candida* and *Aspergillus* infections in particular (Morris and Villmann, 2006a,b). Echinocandins are large cyclic lipopeptides with an amphiphilic hexapeptide nucleus into which is inserted an acyl lipidic chain (Debono and Gordee, 1994). They exert their antifungal effect through inhibition of cell wall synthesis by targeting β-(1,3)-D-glucan synthase, encoded by the *FKS1* gene in *C. albicans*, leading to depletion of glucans and alteration of cell wall integrity (Denning, 1997, 2002). Echinocandins are rapidly fungicidal against *Candida* spp., and fungistatic against *Aspergillus* spp., but are inefficient against *Cryptococcus neoformans* and some moulds (Denning, 2002). In 2001, caspofungin (Cancidas™, by Merck laboratories) was the first echinocandin to receive approval for clinical use (Deresinski and Stevens, 2003; Letscher-Bru and Herbrecht, 2003). It is a semi-synthesized derivative of pneumocandin B_0, a fermentation product from the filamentous fungus *Glarea lozoyensis* (Masurekar *et al.*, 2012). Later were developed micafungin (Fugisawa Pharmaceutical) and anidulafungin (originally identified by Eli Lilly and subsequently licensed to Versicor, Inc.) which, together with the azole drug posaconazole, constitute the newest members of the antifungal armamentarium currently available for treating fungal infections (Stevens, 2012).

Most of the currently available antifungals exert toxic side-effects on patients, typically through cross-inhibiting the human homologue of the fungal target molecule. For instance, both human CYP51 enzyme variants bound clotrimazole, itraconazole, and ketoconazole with high affinity (Warrilow *et al.*, 2013). Similarly, amphotericin B binds to cholesterol, even though with lower affinity than ergosterol, and this accounts for a large part of its cellular toxicity and hepatotoxicity (Carrillo-Munoz *et al.*, 2006; Lewis, 2010). Griseofulvin, the first developed antifungal drug (Oxford *et al.*, 1939), equally affects mammalian and fungal microtubule function and was repositioned as an anti-cancer drug (Panda *et al.*, 2005). Identifying the mode-of-action of antifungal candidates (also termed target deconvolution) is therefore crucial for (i) predicting their impact on the metabolism of both the fungal pathogen and the host cell, (ii) assessing their potential toxic side-effects, (iii) optimizing their use and finding new antimicrobial or therapeutic properties, and (iv) better understanding the mechanisms of drug resistance. Mode-of-action studies are consequently an integral part in the classical drug discovery pipeline. High-throughput technologies are currently cornerstone in identifying the mode-of-action of antifungal candidates, owing to the development of screening technologies but also to the availability of biological resources for performing such screens. For instance, sequencing and annotation of the *C. albicans* genome (Braun *et al.*, 2005; Jones *et al.*, 2004) has been of a paramount importance to antifungal discovery, as this allows to mine the genome for potential drug target candidates (Roemer *et al.*, 2003) or perform drug mode-of-action studies directly in the fungal pathogen using high-throughput reverse genetics approaches (see below). But prior to the development of the current technologies, such as functional genomics, computational biology and high-throughput imaging, antifungal mode-of-action studies extensively relied on the use of classical microscopy, biochemical or biophysical approaches in addition to classical forward genetics.

Mode-of-action studies of currently used antifungal agents: a historical perspective

The first and significant step in antifungal development was achieved in the late 1930s with the discovery of the antifungal properties of griseofulvin (Oxford *et al.*, 1939), a metabolite from the mould *Penicillium griseofulvum*. Griseofulvin is still currently used for the treatment of superficial fungal infections and is proposed as an anticancer drug owing to its antimitotic activity (Panda *et al.*, 2005). Using the fungus *Basidiobolus ranarum* as a model and through microscopic examination of *B. ranarum* large nuclei, it was shown for the first time that griseofulvin inhibits fungal mitosis (Gull and Trinci, 1973). Subsequent mode-of-action studies were conflicting on whether or not fungal mitosis inhibition occurred through direct interference with microtubule organization. While one report showed that griseofulvin exerts its effect without disrupting microtubules (Grisham *et al.*, 1973), combination of immunofluorescence microscopy, light-scattering and electron microscopy suggested that griseofulvin inhibits tubulin polymerization both *in vivo* and *in vitro* (Weber *et al.*, 1976; Wehland *et al.*, 1977). Further analyses through gel filtration chromatography (Sloboda *et al.*, 1982) and measurement of tubulin-tryptophan fluorescence quenching by griseofulvin (Chaudhuri and Luduena, 1996) pointed to a direct interaction of griseofulvin with tubulin. Live cell imaging of β-tubulin fused to enhanced yeast fluorescent protein in the filamentous fungus *Magnaporthe grisea* revealed extensive microtubule depolymerization and aggregate formation upon treatment with griseofulvin (Czymmek *et al.*, 2005). Quantitative polymerase-chain reaction analyses showed that griseofulvin treatment also led to a dose-dependent decrease of β-tubulin transcript levels in the dermatophyte *Trichophyton rubrum* (Zomorodian *et al.*, 2007). Collectively, these studies support a model in which, griseofulvin interferes with the microtubule mitotic spindle, through direct interaction with tubulin and/or alteration of tubulin expression, leading to inhibition of mitosis.

Major steps in elucidating the mode-of-action of azoles were achieved during the 1970s owing to studies aimed at clarifying the mechanism of miconazole and clotrimazole actions (Henry and Sisler, 1979; Iwata *et al.*, 1973a,b; Sud and Feingold, 1981; Van Den Bossche *et al.*, 1975, 1978). Electronic microscopy examinations (Iwata *et al.*, 1973a,b) and measurement of the release of ^{32}P-labelled cellular constituents (through incorporation of ^{32}P-ortho-phosphoric acid) (Sreedhara Swamy *et al.*, 1974) initially proposed an effect of miconazole on cell membrane or cell wall integrity, as prompt and extensive cell content leakage was observed upon treatment of *Candida albicans* with miconazole. But the link between azole mode-of-action and ergosterol biosynthesis was hypothesized following Van Tuyl's observation that cross-resistance invariably occurred between fenarimol, an ergosterol biosynthesis inhibitor, and imazalil, a miconazole analogue, in *Aspergillus nidulans* mutants (Van Tuyl, 1977). Van Den Bossche *et al.* (1978) tested this hypothesis in *C. albicans* by measuring ^{14}C-labelled acetate incorporation into ergosterol and observed a time- and dose-dependent inhibition of ^{14}C-acetate incorporation with a concomitant increase in 14α-methyl sterols, which was suggestive of potent inhibition of one of the metabolic steps involved in the demethylation at C-14. Thin layer chromatography followed by gas–liquid chromatography and mass spectroscopy also confirmed miconazole-mediated inhibition of C-14 sterol demethylation in *Ustilago maydis* (Henry and Sisler, 1979). Importantly, the inhibitory effect of ketoconazole, the first orally active azole drug, on ergosterol biosynthesis was characterized for the first time *in vivo* in a rat model of vaginal candidiasis, by measuring ^{14}C-labelled mevalonate incorporation into the ergosterol biosynthetic pathway following intraperitoneal injection of ^{14}C-mevalonate (Van den Bossche *et al.*, 1980). This, combined with the observation that azoles affect the function of cytochrome P-450 in *S. cerevisiae* microsomes, allowed to emit the hypothesis that azoles interact with cytochrome P-450, which inhibits ergosterol biosynthesis and leads to accumulation of toxic methylsterols (Van den Bossche *et al.*, 1983). Indeed, a strong interaction was demonstrated between azole derivatives and purified *S. cerevisiae* cytochrome P-450 lanosterol 14α demethylase and structural analyses of the interaction were performed (Vanden Bossche *et al.*, 1995; Yoshida and Aoyama, 1987, 1991). Genetic studies allowed a better characterization of azole drug mode-of-action. An *S. cerevisiae* genomic library in a multicopy plasmid was screened for resistance to ketoconazole, which allowed the isolation of plasmids carrying the sequence of the same gene, *ERG11*, encoding the azole target P-450 lanosterol 14α demethylase (Kalb *et al.*, 1986, 1987). Importantly, isolation of *ERG11* from *C. albicans* was achieved through hybridizing an *S. cerevisiae ERG11* probe to a phage lambda library of *C. albicans* nucleotide sequences and overexpression of *C. albicans ERG11* conferred azole resistance (Kirsch *et al.*, 1988). In fact, the increasing cases of clinical azole resistance in the late eighties, especially during the first decade of the AIDS epidemic, led to the isolation and characterization of various *ERG11* mutations conferring reduced susceptibility to azoles in *C. albicans* in particular (Marichal *et al.*, 1999; Marichal and Vanden Bossche, 1995; Sanglard, 2002; Sanglard *et al.*, 1995; Sanglard and Odds, 2002). The characterization of additional *ERG11* mutations from additional pathogenic fungi further increased our understanding of the mechanism-of-action of (or resistance to) azoles as well as their toxic side-effects (Maertens, 2004).

The characterization of the mode-of-action of polyenes has been critical for understanding their impact on the patient's tolerance. Nystatin was initially thought to exert its action on *C. albicans* primarily by depleting energy reserves of the fungal cell as a consequence of depressing glucose uptake (Harman and Masterson, 1957). The same study also reported alteration of membrane permeability and decreased cell volume (Harman and Masterson, 1957). The work of Lampen and colleagues in particular provided a series of biochemical data pointing to a strong link between polyenes and fungal sterols (Lampen *et al.*, 1962). Based on the observation that addition of sterols prevented the antifungal action of nystatin (Gottlieb *et al.*, 1958) and that only sterol-containing organisms were sensitive to polyenes (Lampen and Arnow, 1959; Lampen *et al.*, 1959), they showed that ^{14}C-labelled nystatin irreversibly bound to both intact and protoplasted *S. cerevisiae* cells as well as to protoplast membranes; in addition to showing efficient extraction of protoplast membrane-bound ^{14}C-nystatin with the sterol-complexing agent digitonin (Lampen *et al.*, 1962). The distribution of radioactive nystatin within multiple fractions from sonicated cells pre-incubated with ^{14}C-nystatin revealed strong enrichment of radioactive cell membranes containing high proportions of ergosterol (Lampen *et al.*, 1962), thus reinforcing the link between polyene antifungal action and membrane sterols. Similar observations were made by Kinsky in *Neurospora crassa* using both nystatin and amphotericin B (Kinsky, 1962a,b). Additional experiments showed alteration of cell permeability in *N. crassa* upon treatment with polyene derivatives via direct action on the cell membrane (Kinsky, 1961a,b). During the same period, another study by Kinsky also pointed to a lytic effect of polyenes on mammalian erythrocytes (Kinsky *et al.*, 1962), which was an early indication of polyene potential toxic side-effects. Using artificial spherules of sterol-free or sterol-containing phospholipids encapsulating anions or glucose, polyene action on membrane integrity as well as the requirement for sterol presence were further confirmed by measuring the release of ions or glucose upon nystatin and amphotericin B treatment (Weissmann and Sessa, 1967). By measuring ion permeability of thin lipid membrane layers, Finkelstein and colleagues provided, for the first time, biophysical data that strongly suggested the creation of aqueous pores in lipid membranes following treatment with nystatin or amphotericin B, in addition to demonstrating that the observed effect required the presence of sterols (Cass *et al.*, 1970; Holz and Finkelstein, 1970). Using *Acholesplasma laidlawii* as a model, an organism that does not require sterol for growth but is capable of incorporating in its membranes sterols introduced in the medium, the molecular structure of the amphotericin B–sterol complex was described in a series of studies by de Kruijff *et al.* and visualized as a circular arrangement of 8 amphotericin B molecules interdigitated by eight cholesterol molecules (de Kruijff and Demel, 1974; de Kruijff *et al.*, 1974a,b). Many studies aimed at elucidating the mode-of-action of polyenes also provided data supporting the higher affinity of polyenes to ergosterol as compared to cholesterol (Hsuchen and Feingold, 1973a,b; Kotler-Brajtburg *et al.*, 1974), but this selectivity is weak which explains polyene toxic side-effects. Years after these initial key mode-of-action studies of nystatin and amphotericin B, the precise molecular mechanism of polyene fungicidal action is still unclear, but several biochemical, biophysical and computational chemistry analyses are in agreement with a model in which polyenes exert their direct action on sterol-containing membranes and that this action requires the presence of sterols through generating both aqueous and non-aqueous pores of polyene–sterol complexes, leading to leakage of cell content and cell death (Baginski *et al.*, 2005; Cohen, 2010; Neumann *et al.*, 2010). One exception to this rule is the polyene drug natamycin which binds specifically to ergosterol without permeabilizing the membrane (te Welscher *et al.*, 2008). Interestingly, this observation is consistent with an important recent study showing that amphotericin B can exert its fungicidal action simply by directly binding ergosterol, without necessarily inducing membrane ion channel formation (Gray *et al.*, 2012). Another report by te Welscher *et al.* (2012) suggests that natamycin has an effect on nutrient-transport membrane proteins, suggesting the existence of alternative mechanisms of polyene fungicidal action.

The earliest echinocandin mode-of-action studies were performed using the first discovered members of the echinocandin family, echinocandin B, acealucin A and papulacandin isolated from *Aspergillus nidulans* (Nyfeler and Keller-Schierlein, 1974), *Aspergillus aculeatus* (Mizuno *et al.*, 1977) and the deuteromycete *Arthrinium sphaerospermum* (Traxler *et al.*, 1977), respectively. Using *S. cerevisiae* as a model, a significant reduction in U-^{14}C-labelled D-glucose incorporation into cell wall glucans was observed upon treatment with acealucin A and the effect on cell wall integrity was further examined by electronic microscopy (Mizoguchi *et al.*, 1977). Similarly, incorporation of ^{14}C-glucose into the cell wall glucan was strongly altered in both *S. cerevisiae* and *C. albicans* intact cells treated with papulacandin as well as in yeast protoplasts monitored during regeneration of cell wall polysaccharides in the presence of either papulacandin or echinocandin B (Baguley *et al.*, 1979). The inhibition of glucan cell wall regeneration in *C. albicans* and *S. pombe* protoplasts by acealucin A was also documented by electron microscopy (Osumi *et al.*, 1989; Yamaguchi *et al.*, 1987). The effect of echinocandin B on the β-(1,3)-D-glucan synthase activity from *C. albicans* mixed-membrane preparations showed a strong inhibitory action (Sawistowska-Schroder *et al.*, 1984). Other studies characterized with similar approaches the antifungal action of cilofungin (LY121019), an analogue of echinocandin B (Gordee *et al.*, 1984), and found that it specifically inhibited the activity of β-(1,3)-D-glucan synthase from *N. crassa* (Taft and Selitrennikoff, 1988) and *C. albicans* (Taft *et al.*, 1988). Additional key mode-of-action studies used forward genetics approaches, through generating *S. cerevisiae* or *S. pombe* echinocandin-resistant mutants and isolating alleles conferring echinocandin resistance (Castro *et al.*, 1995; Douglas *et al.*, 1994; el-Sherbeini and Clemas, 1995; Eng *et al.*, 1994; Ram *et al.*, 1995). These independent studies isolated the same gene, *ETG1* (Echinocandin Target Gene 1, currently named *FKS1*), encoding the catalytic subunit of β-(1,3)-D-glucan synthase, the target of echinocandins (Douglas *et al.*, 1994). Biochemical and gene deletion analyses allowed to identify *S. cerevisiae FKS2*, encoding functionally redundant paralogue of *FKS1* (Inoue *et al.*, 1995; Mazur *et al.*, 1995). In a key study by Douglas *et al.*, (1997) *C. albicans FKS1* has been identified by Southern blot analyses using a portion of *S. cerevisiae FKS1* as a probe. Characterization of the resulting DNA fragments allowed cloning of *C. albicans FKS1* and generation of homozygous knock-out mutants which displayed hypersusceptibility to pneumocandin (Douglas *et al.*, 1997). The characterization of *C. albicans* echinocandin-resistant mutants in the same study combined with virulence assays *in vivo* in a mouse model allowed to provide strong indication that *FKS1* is the target of echinocandins (Douglas *et al.*, 1997). During many years, the characterization of echinocandin-resistant clinical isolates of *C. albicans*, *C. glabrata* and other fungal pathogens has reinforced the assumption that the Fks proteins are the primary target of echinocandins (Balashov *et al.*, 2006; Garcia-Effron *et al.*, 2009; Park *et al.*, 2005; Perlin, 2007; Singh-Babak *et al.*, 2012). However, the difficulty of solubilizing and purifying the membrane-spanning glucan synthase complex prevented clear binding assays from being performed. Partial purification of the *S. cerevisiae FKS* complex by glucan product entrapment was achieved, but binding assays were not performed (Inoue *et al.*, 1995). A photoaffinity analogue of the echinocandin antifungal LY303366 was used as a probe for direct binding to proteins in crude *C. albicans* microsomal membrane preparations carrying glucan synthase activity, but revealed association with two proteins that were different from Fks1p (Radding *et al.*, 1998).

Chemical genetics and chemogenomics strategies for the identification of the mode-of-action of antifungal drug candidates: a high-throughput reverse-genetics approach

Genetics is a powerful tool for deciphering pathways and circuits involved in response to treatment with chemicals and small compounds. The mode-of-action of the currently used antifungal agents was initially characterized, in part, through forward genetics studies. For instance, initial studies characterizing the mechanism of action of azoles or the echinocandins relied on screening *S. cerevisiae* genomic libraries to isolate mutated alleles conferring drug resistance, as stated above. In fact, the use of multiple fungal models for such studies revealed the remarkable conservation of the drug mode-of-action across species, which makes complex screens amenable to genetically flexible and attractive organisms such as *S. cerevisiae* and *S. pombe* (Kapitzky *et al.*, 2010). Furthermore, combining data from different model organisms leads to more accurate prediction of the drug's mechanism of action (Kapitzky *et al.*, 2010).

The significant improvements in whole genome sequencing approaches, the development of the next-generation sequencing technologies and the creation of highly efficient genome annotation algorithms drastically reshaped the perspectives and potential of functional genetics. The tedious and long process of screening genomic fragment libraries to identify point mutations in genes that may encode the drug target using the genetically flexible fungal model *S. cerevisiae* is now gradually replaced by relatively straightforward, rapid and powerful whole genome sequencing of drug-resistant fungal pathogens. In addition, the availability of complete genome sequences of multiple fungal species, including *C. albicans*, the major fungal pathogen in humans (Braun *et al.*, 2005; Jones *et al.*, 2004), allows the creation of resources for high-throughput phenotyping and therefore establishing large-scale phenotypic assays by testing in parallel a wide array of single gene-deletion mutants or gene-overexpression strain collections. Importantly, such achievements have been cornerstone in implementing powerful drug mode-of-action screens relying on chemical genetics or chemogenomics (Lokey, 2003; Roemer *et al.*, 2012). Chemical genetics is a discipline that explores interactions between bioactive chemicals or small compounds and biological pathways. Consequently, establishing assays linking a given chemical bioactivity to perturbation of gene function/dosage is an attractive strategy for defining an antifungal drug candidate mode-of-action. One of the most comprehensive and powerful resources for performing large-scale chemical genetics (also termed chemogenomics) assays has been the yeast open reading frame (ORF) deletion library (Giaever *et al.*, 2002; Winzeler *et al.*, 1999). The *S. cerevisiae* ORFs were systematically replaced, from the start to stop codon, with a Kanamycin resistance marker (*KanMX*) that confers resistance to geneticin (G418) (Giaever *et al.*, 2002; Winzeler *et al.*, 1999). In addition, every *KanMX* marker contains two unique 20-nucleotide sequences (and thus are specific to each gene deletion mutation) serving as signature tags, or barcodes, for the identification and quantification of each deletion strain in one single culture of a mixed mutant-population (Giaever *et al.*, 2002; Winzeler *et al.*, 1999). The 20-bp molecular tags are flanked by universal sequences allowing amplification by polymerase chain reaction and quantification using microarrays or high-throughput sequencing (Bar-Seq) (Shoemaker *et al.*, 1996; Smith *et al.*, 2012; Xu *et al.*, 2009b). One crucial aspect in barcode design is that tag sequences should be highly divergent to facilitate simultaneous analysis on oligonucleotide microarrays or using Bar-Seq. Powerful algorithms can filter barcode sequences to exclude, for instance, those carrying restriction sites that are reserved for library cloning or those not matching certain criteria, such as a range of melting temperatures or GC content (Shoemaker *et al.*, 1996; Xu *et al.*, 2009b). Barcode libraries can be produced using *in situ* microarray DNA synthesis, instead of the relatively expensive and time-consuming conventional methods for generating complex libraries of defined nucleic acids (e.g. synthesized oligonucleotides) (Cleary *et al.*, 2004).

Similar deletion collections were established in *S. pombe* (Kim *et al.*, 2010; Spirek *et al.*, 2010) (http://pombe.bioneer.co.kr/technic_infomation/construction.jsp), *C. albicans* (Oh *et al.*, 2010b; Roemer *et al.*, 2011; Vandeputte *et al.*, 2011; Xu *et al.*, 2007) and *C. glabrata* (Rai *et al.*, 2012; Seider *et al.*, 2013). These resources combined with the available read-out technologies such as microarrays and high-throughput sequencing represent powerful approaches for drug target discovery and antifungal candidate mode-of-action studies. A complement to the gene deletion approach is gene overexpression or multicopy suppression (Prelich, 2012). Chemogenomic profiling by gene overexpression or multicopy suppression relies on the creation of barcoded

ORF libraries for the highly parallel screening of genes whose increased dosage confers resistance to a given compound or drug candidate. Such collections have been created in *S. cerevisiae* (Butcher *et al.*, 2006; Douglas *et al.*, 2012; Ho *et al.*, 2009) and are presently developed in *C. albicans* (Cabral *et al.*, 2012; Chauvel *et al.*, 2012). They extend the repertoire of chemogenomics approaches for conducting antifungal candidate mode-of-action screens. I invite the reader to examine various detailed protocols for strain-pool construction and the setting up of competitive growth assays for the identification of the mode-of-action of drugs or exploring gene function through the use of either gene deletion or gene overexpression/multicopy suppression coupled to microarray or high-throughput sequencing analyses (Butcher and Schreiber, 2006; Ericson *et al.*, 2010; Pierce *et al.*, 2007; Smith *et al.*, 2011, 2012).

Gene deletion approaches

Heterozygous deletion screens

Chemogenomics strategies have been first developed in *S. cerevisiae*. They rely on the generation of an array of heterozygous single-gene deletion mutants carrying signature tags (also referred to as barcodes) followed by pooling of the mutants in one single culture and screening for growth alterations in the presence or absence of growth subinhibitory concentrations of a given chemical (typically ranging from $IC_{10\%}$ to $IC_{50\%}$) (Giaever *et al.*, 1999, 2004). The sensitized mutants are thus depleted from the pool over time and the relative abundance of the corresponding strains is measured by quantification of barcode abundance (Giaever *et al.*, 1999, 2004). The screen can also be performed through individual growth rate analysis, either by quantifying strain growth in liquid medium in the presence or absence of the drug (Baetz *et al.*, 2004) or by measuring colony size (Carroll *et al.*, 2009). Knowing that the lack of sufficient material (e.g. natural product extracts) is a limiting factor in the industrial settings, one disadvantage of the individual growth assay is that higher amounts of the tested compound is needed for preparing the solid or liquid media used for testing growth of the array of mutants individually. The principle of the heterozygous deletion screen is based on the assumption that chemical inhibition of the drug target induces haploinsufficiency phenotype in the corresponding mutant – in other words, the heterozygous strain is sensitized to the compound that inhibits the product of the heterozygous locus (Fig. 9.1A). More specifically, pools of equally represented strains are grown in the presence or absence of the antifungal drug candidate during an optimal number of generations (typically up to 20 generations), followed by genomic DNA extraction, barcode amplification by PCR using common primers and quantification of barcode abundance by microarrays (Fig. 9.1A) or high-throughput sequencing (Ericson *et al.*, 2010; Pierce *et al.*, 2007; Smith *et al.*, 2011, 2012). If a given ORF is important for cell growth, the decreased gene dosage in the heterozygous context combined with chemical inhibition of the product of the remaining allele will confer drug sensitivity. In the extreme case where an antifungal drug candidate pharmacologically inhibits the function of an essential gene, the haploinsufficiency profiling approach still allows to identify the direct drug target owing to viability of the heterozygous mutation, unlike a homozygous deletion collection (see below). Therefore, theoretically the mutant displaying the strongest drug sensitivity phenotype would be the strain carrying the heterozygous deletion of the drug target. The *S. cerevisiae* signature-tagged mutant collection presents the advantage of including the complete *S. cerevisiae* ORFeome (Giaever *et al.*, 2004). As a proof-of-principle, the screen identified the targets of miconazole and fluconazole (*ERG11*) and suggested that *ERG24*, not *ERG2*, was more likely the target for fenpropimorph (Giaever *et al.*, 2004). Importantly, the same screen identified *PDR5*, encoding a drug-efflux pump, whose heterozygous deletion conferred the strongest sensitivity to azoles, reflecting the notion that the haploinsufficiency chemogenomic profiling also detects indirect effects such as drug detoxification mechanisms (Giaever *et al.*, 2004). Another similar study tested the effect of 78 compounds on pools of 3503 barcoded *S. cerevisiae* heterozygous deletion mutants originating from the same collection used by Giaever *et al.* and the direct targets of known antifungal agents were also resolved,

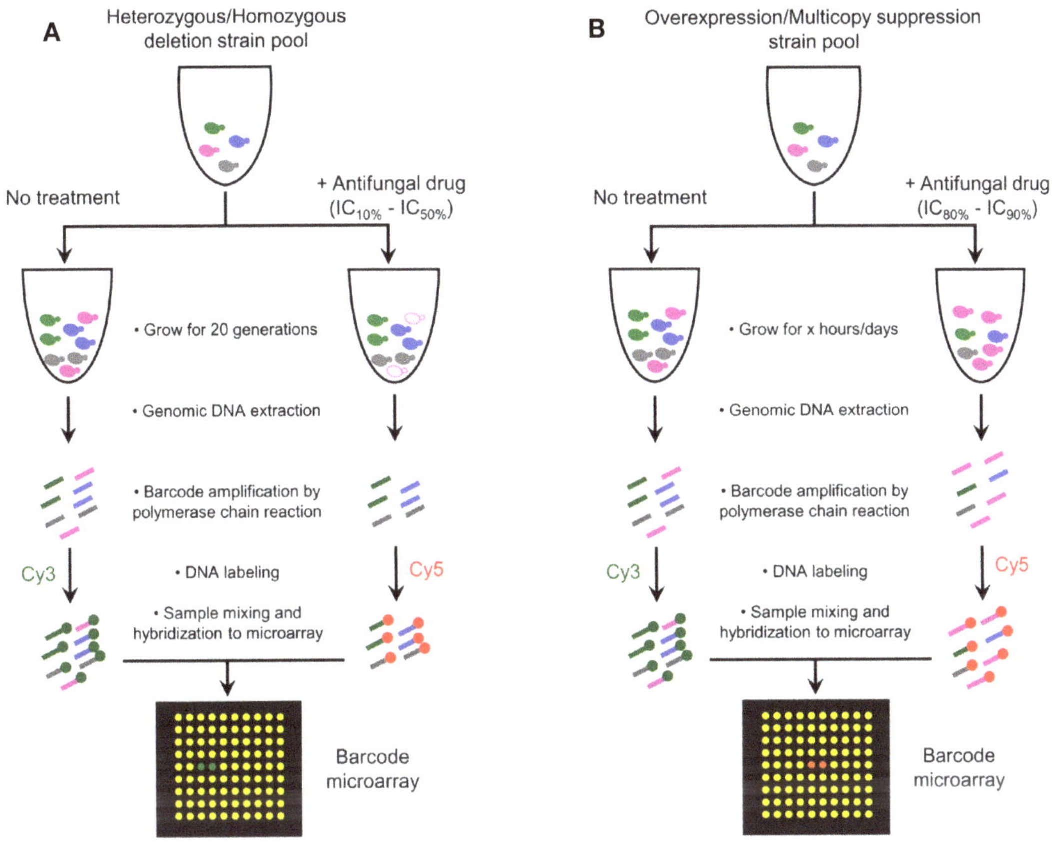

Figure 9.1 Chemogenomic profiling for the identification of protein targets or pathways involved in the mechanism of action of antifungal drug candidates. (A) Heterozygous/homozygous deletion screens. Pools of equally represented strains from a heterozygous or homozygous deletion collection (every mutant is depicted with a different colour) carrying specific barcodes for each deletion mutation are grown in the absence (No treatment) or presence (+ Antifungal drug) of an antifungal drug candidate at inhibitory concentrations ranging from 10% to 50% ($IC_{10\%}$ – $IC_{50\%}$) during 20 generations. The strain carrying a heterozygous deletion of the drug target or a homozygous deletion of a gene required for growth in the presence of the drug (magenta strain) is depleted from the pool over time when the drug is added to the medium (pink dashed strain). Strains are collected and genomic DNA of the strain mixture is extracted. The molecular barcodes are amplified by PCR such that modified nucleotides allowing coupling, post-PCR, of cyanine dyes are incorporated and the resulting amplified barcodes are differentially labelled with the Cy dyes (Cy5 for the drug-treated samples, red spheres; Cy3 for control, green spheres). The labelled DNA samples are mixed and hybridized to a barcode microarray carrying oligos that are complementary to each barcode sequence of the mutant collection. Microarray slides are scanned, followed by acquisition of fluorescence intensities of each channel and normalization. The Cy5 signal from the depleted, sensitized, strain is decreased relative to the Cy3 signal from same untreated strain in the control sample (green spots on microarray depiction), which identifies the strain carrying a heterozygous deletion of the drug target. (B) Overexpression/Multicopy suppression screens. Pooled strains from a collection of overexpression or multicopy suppression clones, each overexpressing one given ORF and carrying one specific barcode (every overexpression/multicopy suppression strain is indicated with a different colour) are competitively grown in the absence (No treatment) or presence of high concentrations of an antifungal drug candidate (+Antifungal drug; typically $IC_{80\%}$ to $IC_{90\%}$). The strain overexpressing the drug target (magenta strain) is enriched over time relative to the remaining strains. Following genomic DNA extraction, barcode amplification (coloured bars), DNA labelling (Cy5 for treated sample, red spheres; Cy3 for untreated control, green spheres), sample mixing and hybridization to barcode microarrays, the Cy3 and Cy5 signals are normalized and relative intensities are quantified. The enriched, drug-resistant, strain overexpressing the antifungal drug target displays Cy5 signals that are higher than the Cy3 signals from the same untreated strain (red spots on the barcode microarray depiction).

including terbinafine (*ERG1*), fenpropimorph (*ERG2* or *ERG24*) and clotrimazole (*ERG11*) (Lum *et al.*, 2004). The first compendium of chemical genetic profiles in *S. pombe* was recently established, reporting the interaction of 47 drugs with a collection of 4836 heterozygous diploid deletion mutants (Han *et al.*, 2013). An analogous collection of 2868 heterozygous (~45% genome coverage), double barcoded, mutants was constructed in *C. albicans* and allowed to perform mode-of-action assays with various antifungal compounds of natural product or synthetic origin (Xu *et al.*, 2007), including various azoles, caspofungin, amphotericin B and 5-fluorocytosine. The azole drugs showed similar chemogenomic profiles, with the *ERG11* heterozygote displaying the highest sensitivity score followed by strain carrying the heterozygous deletions of *NCP1*, encoding an accessory protein required for Erg11p-mediated sterol 14 α-demethylation and *CDR1*, encoding a drug-efflux pump that is important for conferring clinical azole resistance (Xu *et al.*, 2007). Strains heterozygous for *RHO1*, encoding the GTP-binding protein required for the function of the β-(1,3)-D-glucan synthase complex, followed by *FKS1*, the target of echinocandins, were the highest sensitive mutants among the strains tested for altered growth in the presence of caspofungin (Xu *et al.*, 2007). No clear target was resolved in the amphotericin B chemogenomic screen, consistent with the mode-of-action of polyene antifungals which do not act through specific protein targets (Xu *et al.*, 2007), although one expects buffering interactions to show up such as a resistance phenotype in certain *erg* mutants but such mutants might not be part of the set of 2868 mutants used in the study. A screen with the same collection allowed the discovery, isolation and determination of the mode-of-action of parnafungin (Poly(A) RNA fungin), a natural product derived from a culture broth of *Fusarium larvarum* with a potent antifungal activity both *in vitro* and *in vivo* in a mouse model of disseminated candidiasis (Jiang *et al.*, 2008). The mechanism of antifungal action of parnafungin is unique, as it inhibits polyadenylation of fungal mRNA by acting on the product of the essential gene *PAPa*. Consistently, the haploinsufficiency chemogenomic profile displayed by parnafungin was similar to that of cordycepin (3′-deoxyadenosine), a known 3′ mRNA processing inhibitor (Jiang *et al.*, 2008). Both compounds caused hypersensitivity of strains that are heterozygous for genes encoding multiple components of the mRNA cleavage and polyadenylation complex (e.g. *CLP1*, *PCF11*, *YSH1*, *PTA1*, *CFT1*) (Jiang *et al.*, 2008). Another essential gene, *OLE1*, encoding fatty acid Δ9 desaturase, displayed specific sensitivity to two synthetic compounds, ECC145 and ECC188 (Xu *et al.*, 2009a) which suggested that both compounds affect unsaturated fatty acid synthesis. The genetic compromise of *OLE1* function, via tetracycline-induced repression of *OLE1* expression, phenocopied chemical inhibition with ECC145 and ECC188 and displayed avirulence in a mouse model of systemic infection, which makes *OLE1* an attractive antifungal target (Xu *et al.*, 2009a). The *C. albicans* barcoded heterozygous collection has been recently expanded to include ~5400 double-barcoded heterozygotes (~90% genome coverage) (Roemer *et al.*, 2011), which represents a valuable resource for performing target deconvolution studies directly in the pathogen. This is reflected through an extensive screening campaign, achieved by the Merck laboratories, for characterizing the mode-of-action of a myriad of natural products with growth-inhibitory activities against *C. albicans* totalling 1801 natural extracts (1231 fungal and 570 actinomycete extracts) (Roemer *et al.*, 2011). One variant of the heterozygous gene knock-out collection is the heterozygous transposon insertion mutant collection, recently constructed in *C. albicans* (Oh *et al.*, 2010b) and covering 3633 ORFs (~60% of the genome). The creation of this set of mutants was made by exploiting the flexibility of the Gateway™ technology, allowing the recombination-mediated transfer of 4280 double signature-tagged modules from entry vectors to any compatible genetic system, including those intended for transposon-mediated mutagenesis (Oh *et al.*, 2010a). Screening of 57 chemicals in the resulting *C. albicans* insertion mutants identified two synthetic, previously uncharacterized, molecules that inhibited vesicle transport in *C. albicans* in addition to revealing that the antimicrobial nigericin, which affects ion gradients across lipid membranes, interacted with pathways associated with the GTP exchange factor *SEC7*, involved in vesicle transport and target of brefeldin A (Oh *et al.*, 2010b). An alternative haploinsufficiency profiling strategy is the use of DAmP, decreased abundance by mRNA perturbation (Yan *et al.*, 2008), in which disruption of the 3′ untranslated region of an essential gene yields a hypomorphic allele (i.e. a gene knock-down effect), reducing mRNA levels 4- to 10-fold. This approach is advantageous in case a heterozygous context is not sufficient to observe phenotypic differences, and was successfully tested in proof-of-principle target-deconvolution experiments (Yan *et al.*, 2009). The *S. cerevisiae* barcoded DAmP collection includes 958 essential and 444 nonessential 'slow-growing' alleles, for a total of 1402 DAmPed ORFs (Yan *et al.*, 2008).

Homozygous deletion screens

The homozygous deletion screen relies on the same principle than the heterozygous, haploinsufficiency, profiling (Fig. 9.1A) except that both copies of a given gene are deleted (or a single copy is deleted in each haploid mutant strain). Consequently, essential genes are not included in the screen. Additionally, the assay does not identify the direct drug target, because the gene encoding the target is absent. However, this approach is relevant for the identification of gene sets and pathways that modulate drug sensitivity – i.e. displaying buffering interactions – thus required for growth in the presence of the antifungal candidate. It also mimics a double-deletion mutant context, because one gene is deleted and the function of the second is altered through chemical inhibition. For those drugs that do not target proteins (e.g. polyene antifungals amphotericin B or nystatin), the homozygous deletion screen identifies protein-encoding genes involved in the synthesis or metabolism of the direct drug target. For instance, Parsons *et al.* (2006) used ~5000 viable barcoded yeast haploid deletion mutants in screens for identifying genes whose deletion confers hypersensitivity to a myriad of compounds. They particularly identified the membrane lipid phosphatidylserine as a target for papuamide B, a lipopeptide with antifungal and anti-HIV activity. Notably, such screens allow clustering of the chemogenomic profiles of different compounds, including those with a known mode-of-action, to get insight into the target pathways for new, uncharacterized, bioactive compounds (Parsons *et al.*, 2006). Hillenmeyer *et al.* (2008) surveyed the effect of over 400 compounds on the complete homozygous and heterozygous *S. cerevisiae* mutants and found that nearly all of the deletion strains displayed a phenotype in at least one condition. Clustering of all chemogenic profiling data allowed to propose mechanisms of action for drugs with unknown/unclear target (e.g. alverine citrate, dyclonine, dihydromotuporamine C) and, importantly, identify sets of genes whose deletion conferred invariable sensitivity to various compounds, including detoxification and pleiotropic drug resistance genes (e.g. *PDR5*, *PDR1*). Consequently, the homozygous deletion screen is complementary to the heterozygous assay, in that it detects indirect targets as well as pathways and circuits involved in drug detoxification and resistance. Data integration from both homozygous and heterozygous chemogenomic profiling allows, therefore, to better resolve mechanism-of-action studies. For instance, the mode-of-action of the synergistic effect of combining colistin, an antibacterial agent, with the echinocandin antifungal aminocandin was interrogated through the combination of *S. cerevisiae* haploinsufficiency assay and haploid deletion collection assay to suggest that echinocandin-mediated alteration of the cell wall might facilitate colistin's access to and perturbation of fungal membranes (Zeidler *et al.*, 2013). Effort in generating comprehensive homozygous deletion collections in various fungal models, including pathogenic fungi, is currently being made. Chen *et al.* (2012) created a double-barcoded haploid fission yeast insertion mutant collection, which provides an additional tool for performing screens in *S. pombe*. Noble *et al.* (2010) used a set of 48 distinct barcodes to construct a *C. albicans* homozygous full gene knock-out subcollections covering 674 genes (~11% of the genome). Vandeputte *et al.* (2011, 2012) created a set of transcription factor gene knock-out strains, including a subset of barcoded mutants that were successfully used in virulence screens *in vivo*. A set of barcoded *Candida glabrata* mutants for 619 genes has also been developed by the groups of Karl Kuchler (Vienna, Austria),

Ken Haynes (Exeter, UK) and Brendan Cormack (Philadelphia, USA) (Seider *et al.*, 2013).

Gene overexpression approaches

Chemogenomic profiling by gene overexpression leads to identification of the drug's mode-of-action based on the assumption that increasing drug-target dosage confers a drug resistance phenotype. Consequently, a strain overexpressing the drug target will be gradually enriched over time within a mixed population (Fig. 9.1B). Gene overexpression or multicopy suppression profiling are complementary approaches to gene deletion screens. The advantages of overexpression include the possibility of performing screens with essential genes or detecting phenotypes for functionally redundant genes (Prelich, 2012). Gene overexpression also mimics gain-of-function mutations and leads to pathway activation (Sopko *et al.*, 2006). However, unlike phenotypic profiling through a gene deletion approach, which relies on an all-or-none effect, the expected subtle or even significant phenotypic differences might be difficult to resolve through gene overexpression. This is because overexpression levels can significantly vary from one strain to another, depending on the strength of the promoter that is used for driving overexpression, the stabilities of the overproduced protein or the overexpressed RNA and their endogenous basal expression level. Therefore, an antifungal-drug candidate screen is performed at high inhibitory concentrations, usually approaching $IC_{90\%}$, for the phenotypes to be efficiently resolved. An elegant overexpression screen in yeast for assaying drug mode-of-action used a genomic DNA library constructed in the multicopy YEp13 plasmid (with an average insert size of ~5 kb) (Luesch *et al.*, 2005). A total of 7296 individual randomly picked *S. cerevisiae* transformants were arrayed on 384-well format agar plates supplemented with a growth inhibitory concentration of a compound that shares a phenylaminopyrimidine (PAP) scaffold with some anticancer kinase inhibitors (Luesch *et al.*, 2005). Genes conferring increased resistance to the compound included detoxification genes (e.g. *PDR3*) in addition to various genes acting downstream of the *PKC1-MPK1* MAP kinase pathway (the essential gene *PKC1*, *BCK1* and *MKK2*) or exhibiting genetic interactions with the Pkc1 cascade (Luesch *et al.*, 2005). Complementary drug-induced haploinsufficiency assay of the heterozygous diploid *pkc1Δ/PKC1* mutant followed by target validation using affinity chromatography and *in vitro* kinase assay confirmed that Pkc1p was the target of the PAP compound (Luesch *et al.*, 2005). This early high-throughput individual-growth overexpression assay already reflected the powerful use of overexpression (i) to reveal the mechanism of action of drugs targeting the product of essential genes and (ii) as a strategy that is complementary to the gene deletion assay. Another approach used galactose-inducible overexpression of ~3900 *S. cerevisiae* ORFs (~60% genome coverage) from centromeric (low copy) plasmids in a competitive assay to identify genes conferring resistance to rapamycin (Butcher *et al.*, 2006). But instead of using barcodes, the authors used the coding sequence of the overexpressed ORF as a strain identifier and measured the relative strain abundance on a microarray of spotted yeast genes (Butcher *et al.*, 2006; Butcher and Schreiber, 2006). As expected, strain overexpressing the *TOR1* (target of rapamycin) gene was among the top 10 most enriched ones, including those overexpressing general drug resistance genes (Butcher *et al.*, 2006). An important application of the overexpression approach in yeast is the ability to complement recessive resistance mutations in drug targets as an alternative, elegant, way of assaying drug mode-of-action (Butcher *et al.*, 2006; Ho *et al.*, 2009). This strategy could be considered as a combination of forward and reverse genetics. First, resistant mutants are isolated from growth plates supplemented with high concentration of the drug. Second, instead of isolating and sequencing the mutant gene, the identity of the recessive mutant allele can be revealed through complementation by transforming the mutant with a library of gene overexpression/multicopy gene-expression plasmids (Butcher *et al.*, 2006) or barcoded low-copy vectors carrying ORFs with their native promoter and terminator (also known as the MoBY-ORF resource) (Ho *et al.*, 2009; Xu and Roemer, 2009). Again, the advantage here is that the assay is performed in a single liquid culture of mixed population, which reduces the amount of the compound required for the assay. One drawback is that the drug-resistant mutations need to be recessive or, at the extreme case, partially dominant, for the phenotype to be complemented. Butcher *et al.* used their galactose-inducible ORF overexpression library in a proof-of-concept experiment to identify a rapamycin-resistant mutant carrying a mutation in *FPR1* (Butcher *et al.*, 2006). Ho *et al.* (2009) used the MoBY-ORF collection to complement resistance to cycloheximide (C623A mutation in *RPL28*) and rapamycin (29-bp deletion in *FPR1*), followed by confirmation of the drug-resistance mutations by single-nucleotide polymorphism tiling microarrays (Gresham *et al.*, 2006). Importantly, using a combination of homozygous deletion chemogenomic profile clustering, MoBY-ORF drug resistance complementation and additional genetic and biochemical analyses, they identified the same target, ergosterol, for two structurally dissimilar natural products with antifungal activity: theopalauamide (isolated from a bacterial symbiont of marine sponge) and stichloroside (isolated from a marine sea cucumber) (Ho *et al.*, 2009). Owing to the highly flexible genetics of *S. cerevisiae*, overexpression/multicopy suppression chemogenomic resources have been extensively developed, as reflected by the recent creation of an additional barcoded *S. cerevisiae* overexpression collection, termed barFLEX (barcoded full-length EXpression-ready, ~5100 strains), which places every yeast ORF under the control of the *GAL1* promoter (Douglas *et al.*, 2012). The generation of ORFeome libraries in various yeast models and ultimately in fungal pathogens will be a great addition to the repertoire of chemogenomics resources for performing target deconvolution studies. Matsuyama *et al.* created an *S. pombe* overexpression collection carrying 4910 ORFs (99.2% genome coverage) under the control of the inducible strong *nmt1*$^+$ promoter (Matsuyama *et al.*, 2006). The mode-of-action of theonellamide F, a derivative of theopalauamide, was assayed through individually screening the *S. pombe* collection in parallel for altered viability in the presence of theonellamide F and additional compounds, using a tetrazolium salt colorimetric assay as a readout (Nishimura *et al.*, 2010). The assay revealed that theonellamide F counteracted the mode-of-action of the echinocandin antifungal micafungin in addition to targeting ergosterol (Nishimura *et al.*, 2010), which confirms theopalauamide mode-of-action using the *S. cerevisiae* MoBY-ORF library (Ho *et al.*, 2009). The two studies are therefore a nice example of how combination of drug mode-of-action analyses in various fungal models could better resolve the targets of antifungal drug candidates. The goal of the Wellcome Trust-funded *C. albicans* ORFeome project, led by Drs Carol Munro (University of Aberdeen, UK) and Christophe d'Enfert (Institut Pasteur, France), is to create a Gateway™ entry vector library covering each of the 6214 *C. albicans* ORFs (Cabral *et al.*, 2012). From these entry vectors were created partial overexpression collections, including a set of 302 barcoded overexpression strains with a tetracycline derivative-inducible gene expression system (Chauvel *et al.*, 2012). The collection is currently being expanded to cover the whole genome, which will provide a highly valuable resource for performing chemogenomic overexpression screens directly in the pathogen.

Biochemical and pharmacological approaches exploring drug–target physical associations

During decades, assaying a drug candidate mode-of-action involved testing the effect of the compound directly on the predicted target(s), such as purified proteins or protein complexes or using binding affinity methods with total protein extracts (Terstappen *et al.*, 2007). With the development of functional genomics resources and high-throughput screening technologies, most of the current drug mode-of-action study campaigns are gradually reorienting towards cell-based phenotypic assays that have the advantage of preserving the intracellular environment necessary for protein function. Nevertheless, the use of purified proteins is still cornerstone in clearly assessing the direct effect of small molecules or drug candidates on their hypothetical targets, through performing direct binding assays and/or engaging in direct enzyme activity inhibition testing. One of the high-throughput technologies that could serve in identifying the direct target of drug candidate are protein chips (Zhu *et al.*, 2001;

Zhu and Snyder, 2003). In its simplest form, the technology relies on attaching purified proteins on a solid substrate; usually a glass slide covered with nitrocellulose or poly-L-lysine, followed by incubation with a fluorescent or biotin-labelled drug candidate, washing and detection of bound molecules by scanning with a microarray scanner (Fig. 9.2A). The biotin-labelled compound necessitates subsequent coupling with streptavidin fluorescent conjugate for scanning, e.g. cyanine dyes (Cy3 or Cy5). Huang *et al.* (2004) probed a yeast proteome chip carrying 5800 purified proteins from a collection of gene-overexpression yeast strains (Zhu *et al.*, 2001) with two small molecules that suppressed rapamycin's anti-proliferative effect (Huang *et al.*, 2004). They found 38 binding hits, among them Tep1p (PTEN mammalian tumour suppressor protein homologue) showing the strongest binding with one of the two compounds, and Nir1p, a new modulator of the TOR pathway as reflected through the subsequent chemical genetics assays that the authors performed (Huang *et al.*, 2004). Because the array is covered with purified proteins, one of the advantages of the protein chip technology is that it allows measuring direct drug–target associations. Another advantage is that it can detect drug-target binding events for proteins that are expressed at low levels, because the proteome is equally represented on the array. The drawbacks include the difficulty of producing and purifying eukaryotic proteins, especially those with a high molecular weight. Another limitation is that the attached proteins are randomly oriented on the matrix, which may alter their native conformation and prevent drug binding.

Affinity chromatography is another valuable method for target identification. The principle is to immobilize the drug on a chromatography solid support (i.e. matrix, usually agarose beads), followed by incubation with total protein extracts and washing to remove unbound proteins (Burdine and Kodadek, 2004; Terstappen *et al.*, 2007) (Fig. 9.2B). The bound protein is eluted using buffers that dissociate the drug–target interaction. Then mass spectrometry analyses are performed to identify the eluted protein. Attachment of the drug candidate to the matrix requires chemical modification – called derivatization – of the drug molecule by adding a linker, e.g. biotin or ethylene glycol, without altering its activity (Burdine and Kodadek, 2004; Terstappen *et al.*, 2007). An example is the successful synthesis of a biotinylated derivative of phorboxazole A, a natural product isolated from the ocean sponge species *Phorbas* that exhibits potent antifungal action against *C. albicans* (Hansen *et al.*, 2003). A procedure that circumvents small molecule derivatization is the use of photo-cross-linked small molecule affinity chromatography (Takayama *et al.*, 2012), in which aryl diazirine groups covalently attached to the chromatography matrix are transformed into highly reactive carbenes that immobilize intact small molecules upon irradiation with UV light. Some of the most popular studies reflecting the successful use of affinity chromatography for target deconvolution are the identification of the target proteins for FK506, the immunosuppressant macrolide acting as cyclosporine A inhibitor, and trapoxin, a microbially derived cyclotetrapeptide that inhibits histone deacetylation *in vivo* and used as an anticancer drug (Harding *et al.*, 1989; Taunton *et al.*, 1996). The technique was also used to identify tubulin and actin as targets for an antifungal natural product, fusarielin A, isolated from *Fusarium* and inducing 'curling phenotype' in filamentous fungi in a manner similar to the mode-of-action of griseofulvin (Noguchi-Yachide *et al.*, 2010). Derivatized cyclosporin A was similarly used in affinity chromatography experiments to pull down the interacting partners cyclophilins PfCyP-19A and PfCyP-19B from the human malarial intracellular parasite *Plasmodium falciparum* (Gavigan *et al.*, 2003). One limitation of using affinity chromatography in target deconvolution is that the drug–target associations might be indirect. Another issue is that low affinity interactions are most of the time, if not always, undetected (Burdine and Kodadek, 2004), in addition to the highly likely presence of unspecific binding (Chan *et al.*, 2010). Chemical proteomics is a newly developed technology that combines affinity chromatography with high-resolution mass spectrometry for a more efficient and sensitive identification of drug targets (Rix and Superti-Furga, 2009). For instance, Ong *et al.* (2009) used metabolic labelling (light- vs. heavy-isotope cell labelling) in combination with affinity

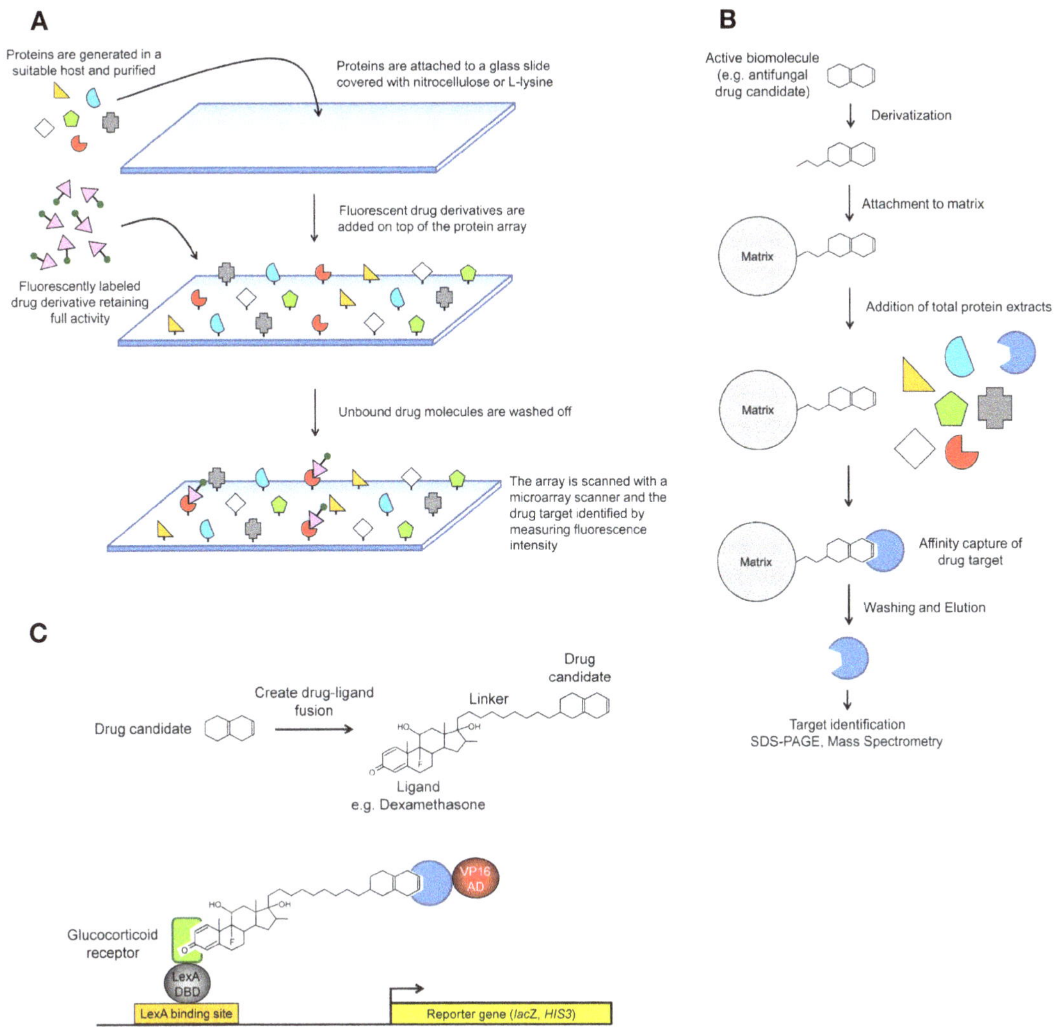

Figure 9.2 Biochemical and pharmacological methods exploring drug-target physical associations. (A) Protein chip array. Proteins are individually generated in a suitable host organism then purified (the different proteins are depicted with different coloured shapes) and attached to a microarray glass slide covered with nitrocellulose or l-lysine. A fluorescently labelled (green sphere) drug candidate (pink triangle) is incubated on top of the protein array. The unbound molecules are washed off and the specific direct binding to the target protein (red symbol) is identified by scanning the slide with a fluorescence scanner. (B) Affinity Chromatography. An antifungal drug candidate is chemically modified (Derivatization) to allow subsequent attachment to matrix (grey sphere). Total protein extracts from a fungal pathogen are then added and incubated to allow affinity capture of the drug target (blue sphere) by the immobilized drug candidate. Following extensive washing and elution, the drug target is identified by mass spectrometry analyses. (C) Yeast three-hybrid assay. A drug candidate molecule is covalently linked to a ligand molecule (shown is a drug–dexamethasone fusion) to allow indirect anchoring of the drug to bacterial LexA DNA binding domain (LexA DBD, grey sphere) via a ligand binding protein (Glucocorticoid receptor, green rectangle) expressed as a fusion with LexA DBD in a genetically engineered yeast strain optimized for the assay (carrying LexA binding site, orange rectangle, a reporter gene, yellow rectangle). The overall complex, consisting of the drug-ligand derivative anchored to the ligand-binding domain that is fused to LexA DBD, constitute the bait. The prey is a library of VP16 activation domain-tagged ORFs to test for interaction with the drug candidate (e.g. ORFeome prey collection for *C. albicans*, or any other relevant fungal pathogen model). Interaction of the small molecule with the protein target allows the reconstitution of a functionally active transcription factor that induces the expression of a reporter gene (*lacZ* or *HIS3*).

chromatography and quantitative mass spectrometry to assess the direct, indirect and unspecific interactions between small molecules and their targets (Ong *et al.*, 2009). Affinity chromatography can also be coupled to a phage-display library screen (Morohashi *et al.*, 2006). Proteins or peptides can be expressed as a fusion with the phage coat protein and consequently displayed at the surface of the bacteriophage for screening (Casey *et al.*, 2008). The DNA encoding the displayed protein is contained within the phage particle, which establishes a physical association between the genotype (target identity) and phenotype (binding). Phages displaying the small molecule target are enriched during the binding assay by affinity capture and subsequently eluted, amplified and sequenced for drug target identification (Morohashi *et al.*, 2006; Terstappen *et al.*, 2007).

The yeast three-hybrid technology has been one of the approaches used for identifying drug-interacting proteins (Becker *et al.*, 2004; Chidley *et al.*, 2011; Lefurgy and Cornish, 2004; Moser and Johnsson, 2013). Although it is considered as a genetics method, the principle still relies on direct physical interaction between a ligand and a target molecule. The classical powerful two-hybrid approach was modified such that it carries a third interaction partner: a small molecule or drug candidate (Fig. 9.2C) (Licitra and Liu, 1996). The three components comprise (1) the yeast *GAL4* or bacterial LexA DNA binding domain fused to, for example, the mammalian glucocorticoid receptor or the ligand-binding domain of dihydrofolate reductase (DHFR); (2) the small molecule (i.e. drug candidate) covalently linked to the respective ligand, glucocorticoid or methotrexate; and (3) a *GAL4* or viral VP16 activation domain is fused to a library of ORFs to screen for interaction with the drug candidate (e.g. ORFeome 'prey' collection for *C. albicans*, or any other relevant fungal pathogen model). By analogy to the two-hybrid assay, interaction of the small molecule with the protein target reconstitutes a functionally active transcription factor that induces the expression of a reporter gene. Two reporters are classically used; one that allows to monitor interaction through an enzymatic assay, such as the *lacZ*/β-galactosidase colorimetric assay (Fig. 9.2C). The second allows performing growth-deficiency complementation assays on media lacking certain nutrients – e.g. *his3*-auxotrophy complementation by induction of *HIS3* expression conferring histidine prototrophy. The recent development of a *C. albicans* strain optimized for two-hybrid assay (Stynen *et al.*, 2010) opens up the possibility of performing such screens directly in the pathogen, knowing that the CUG codon usage in *C. albicans* prevents such assays from being performed in *S. cerevisiae*. The great advantage of this technology is that the interaction between the drug candidate and its target occurs *in vivo*. One major disadvantage is that false positive hits are frequent, resulting from activation of reporter expression independently of the small-molecule hybrid component.

Clues from transcriptomics analyses

Transcriptomics is a powerful technology for getting insights into gene function or pathways and circuits activated upon exerting environmental perturbations or cell insults. Chemical stress induces transcriptional changes reflecting both general (e.g. detoxification pathways) and specific responses of the organism to alteration of one or more biological pathways that are affected by treatment with the chemical. Consequently, transcriptional signatures originating from the specific inhibition of the direct target of a small molecule or drug candidate could be detected, if a series of transcriptional profiles from cells treated with various unrelated compounds are analysed. This allows to discriminate the specific responses from the general ones and to hypothesize on the drug's mechanism of action based on transcriptional signatures. When these transcript signatures are compared to comprehensive transcript profiling data of gene-deletion, drug–gene associations can be established and the drug mode of action inferred. One nice example reflecting this concept is the study by Hughes *et al.* in which gene expression profiles of yeast cells treated with both known and unknown drugs were compared with a compendium of transcript profiles from an array of yeast deletion mutants (Hughes *et al.*, 2000). The study particularly identified the mode-of-action of dyclonine, a topical anaesthetic with antimicrobial properties (Hughes *et al.*, 2000).

A similar approach exploring transcript profiling combines a 'virtual gene technique', where the drug is regarded as a 'virtual gene', with a method based on Bayesian network model to construct gene networks and infer drug targets (Imoto *et al.*, 2003, 2007). Savoie *et al.* (2003) used this strategy by examining expression profiling data for over 500 yeast strains, each carrying a single gene disruption, in combination with dose- and time-course expression experiments with griseofulvin antifungal agent. They identified *CIK1*, encoding a kinesin-associated protein required for both karyogamy and mitotic spindle organization, as an important griseofulvin-affected target gene in line with the known role of griseofulvin in altering mitotic spindle formation (Savoie *et al.*, 2003). *CIK1* expression was indeed highly induced at most of the time-points and drug doses that were tested (Savoie *et al.*, 2003). One can therefore speculate that inhibition of the function of a drug target leads to increased expression of the genes that function in a common pathway with the drug target, as a result of a compensatory transcriptional response due to reduced activity of the target. Clear examples reflecting this hypothesis is the study conducted by Rogers and coworkers, which characterized the transcriptional profiles of *C. albicans* cells treated with different antifungal agents at $IC_{50\%}$, including azoles, polyenes, echinocandins (Liu *et al.*, 2005) and ciclopirox olamine (Lee *et al.*, 2005). Strong induction of most of the *ERG* gene family members involved in ergosterol biosynthesis was observed upon treatment with ketoconazole, including *ERG11* encoding the target of azoles (Liu *et al.*, 2005). Similarly, cell-wall maintenance-encoding genes were induced upon caspofungin treatment, and, interestingly, down-regulation of *ERG3* and *ERG11* were detected following incubation with amphotericin B (Liu *et al.*, 2005). Consistent with the notion that ciclopirox olamine acts as an iron chelator, alteration of the expression of iron uptake genes was observed (Lee *et al.*, 2005), in agreement with data from a similar study that used qRT-PCR (Niewerth *et al.*, 2003). One recent report aimed at characterizing the mode-of-action of the polyene antifungal natamycin examined the transcriptional response of germinating conidia of the fungus *Aspergillus niger* to treatment with natamycin and found that the majority of sugar and amino acid transporters displayed a clear tendency of up-regulation in the presence of the drug (te Welscher *et al.*, 2012). Subsequent validation experiments demonstrated that natamycin inhibited membrane transport proteins through binding to ergosterol (te Welscher *et al.*, 2012). As stated above, transcriptomics can definitely provide valuable information on the mechanism of action of antifungal agents and data originating from gene expression signatures that connect small molecules to their drug targets are presently under extensive mining owing to the development of algorithms and computational approaches with powerful integrative analytical abilities (see concluding remarks section).

High-throughput imaging

The development of sophisticated high-throughput image capture technologies and automated microscopy, also referred to as 'High Content Screening' (HCS), has added a great value to the current arsenal of approaches for studying cell biology, cellular phenotypes (Rausch, 2006) and host–pathogen interactions (Brodin and Christophe, 2011). In fact, HCS has originally been specifically created for the purpose of drug discovery and drug mode-of-action studies in the pharmaceutical settings (Giuliano *et al.*, 1997). The technology initially used fluorescent proteins to image over time and space, before and after drug treatment, the intracellular molecular processes affected by the drug; which gives clues on the drug mode-of-action (Giuliano and Taylor, 1998). Nowadays, HCS is amenable to automation for large screening campaigns and involves both hardware and image analysis software improvements as well as significant development of fluorescent antibodies and probes, allowing multiplexed assays to be performed (Taylor, 2007; Taylor and Giuliano, 2005; Zanella *et al.*, 2010). The inventor of HCS, D. Lansing Taylor, defines the technology as follows:

> Platform and methods, including instruments, biological application software, reagents, assays, and informatics software used to automatically screen and analyze

> arrays of cells to define the temporal and spatial activities and functions of cells, and their constituents, on a cell-by-cell basis, including subcellular features.
>
> (Taylor, 2007)

Such definition reflects the complexity of the process as a whole, which probably explains why the spread of the HCS technology and the expertise behind it throughout the academic and industrial settings has been relatively slow since the late 1990s. Nevertheless, major studies exploring target deconvolution approaches were successful at proving the value of the HCS technology. An example is the work by Perlman *et al.*, in which automated fluorescence microscopy images of up to ~8000 HeLa (human cancer) cells/well cultured in 384-well plates, stained with 11 fluorescent probes and treated with 100 compounds (including 90 with known mode-of-action) were captured and analysed (Perlman *et al.*, 2004). The 11 fluorescent probes were specific to various cell components and metabolic pathways (e.g. antibodies raised against α-tubulin, actin; various phosphoproteins, etc.) (Perlman *et al.*, 2004). A set of descriptors was measured such as cell size and shape, relative fluorescence intensity of each probe, which gave both quantitative and qualitative information on several processes affected by drug treatment (Perlman *et al.*, 2004). The authors were able to show that drugs with common mode-of-action and diverse chemical structures often display similar profiles that were distinguished from drugs with different mechanisms; and that the affected descriptors were consistent with the known mode-of-action of the tested compounds (Perlman *et al.*, 2004). An elegant study reflecting the powerful use of HCS to study gene function in yeast used a collection of 787 yeast strains expressing temperature-sensitive (ts) alleles of essential genes to explore the phenotypes caused by essential gene mutation through combination of synthetic genetic arrays and HCS (Li *et al.*, 2011). A panel of strains expressing fluorescent markers representing six fundamental subcellular compartments or structures were crossed to 762 ts mutants to create double mutants expressing a subcellular marker and a ts allele, followed by screening through HCS, at 26°C and 32°C, for alterations in cell shape, budding as well as intracellular features such as organelle density and 85 reporter-specific parameters (Li *et al.*, 2011). An online database, PhenoM, was created for mining the data collected from the experiment (Jin *et al.*, 2011). Strains of the complete European *Saccharomyces cerevisiae* Archive for Functional Analysis (EUROSCARF) haploid deletion collection, covering 4786 of non-essential genes, were individually triple-stained and used to screen for genes affecting morphology via HCS (Ohya *et al.*, 2005). Extensive data mining established relationships between morphology and gene function (Ohya *et al.*, 2005). The use of the HCS technology to decipher the mode-of-action of chemicals and antifungal agents in fungal species has not been extensively explored. But the availability of various genetic resources in *S. cerevisiae* allowed some functional and phenotypic studies with chemicals and antifungal drugs to be performed. For instance, recently, Mazumder *et al.* (2013) exploited the *S. cerevisiae* GFP-tagged library to examine by HCS the effect of treating cells with the DNA damaging agent methyl methanesulfonate (MMS). Various groups of induced, repressed, nucleus- and cytoplasm-enriched proteins were identified and data mining detected processes that were significantly enriched by MMS treatment; in particular the enrichment of target proteins of major stress-induced transcription factors (Mazumder *et al.*, 2013). Using HCS in yeast cells stained with various markers, including the cell wall (FITC-Concanavalin A), actin cytoskeleton (rhodamin-phalliodin) and nuclear DNA (DAPI), Ohnuki *et al.* (2010) inferred the mechanism of action of four compounds, namely hydroxyurea, concanavalin A, lovastatin and the antifungal echinocandin B, by comparing morphologic changes induced by the compounds with data originating from the HCS-resolved morphological phenotypes of the EUROSCARF collection (Ohya *et al.*, 2005). Okada *et al.* (2013) recently gained further insight into the relationship between the cell wall and morphogenesis by quantitatively analysing the morphological changes induced by echinocandin B, tunicamycin, and nikkomycin Z in yeast using HCS, and particularly revealed that echinocandin B impacted on two related and important functions of its target Fks1, namely, β-(1,3)-D-glucan synthesis and cell polarity. Such examples reflect the high potential of the HCS technology in characterizing the mode-of-action of antifungal drug candidates and stress the need for developing HCS-optimized biological resources in fungal pathogen models.

Concluding remarks

An integrative strategy should be considered in order to infer drug's mode-of-action. As stated above, each of the technologies has advantages and limitations. Therefore the use of multiple approaches allows cross-validation and better resolving the mechanism of action of a given drug. Chemogenomics approaches are powerful for screening compound targets; however they can detect secondary, indirect, mechanisms of action and off-target effects. In fact, examination of the yeast knock out collection revealed the presence of secondary mutations of functional consequence, including mutations affecting nutrient responses and/or heat-stress-induced cell death, which might complicate reverse genetics analyses (Teng *et al.*, 2013). Similarly, the biochemical approaches suffer from some limitations, including the generation of artefacts during affinity purification. They are also limited by the highly variable affinity of chemicals to targets and the availability of linker sites for conjugation on the affinity matrix without affecting the activity. Transcriptomics are powerful at detecting drug target transcriptional signatures, but subtle signals are often hidden by additional, indirect, transcriptional responses, batch effects and other parameters (Lamb *et al.*, 2006; Leek *et al.*, 2010). HCS is based on cellular imaging, which might be quite descriptive and certainly requires extensively diverse resources, highly specialized expertise and skilled personnel (Lang *et al.*, 2006). Consequently, an integrative approach allows complementing the limitation of each technology with the advantageous aspects of the alternative ones. Computational analysis is essential for integrating high-throughput data. For instance, it was used for combining chemogenomics with genetic interaction analyses to better linking bioactive compounds to their targets (Heiskanen and Aittokallio, 2013; Parsons *et al.*, 2004) or for mining data resulting from the integration of the three genome-wide gene-dosage assays, namely homozygous, heterozygous and multicopy suppression assays, which improved the sensitivity and specificity of drug-target identification (Hoon *et al.*, 2008). A highly integrative computational re-analysis of the yeast heterozygous and homozygous deletion chemical profiling data allowed building of an algorithm that predicts drug targets, followed by verification of its accuracy with experimental testing (Hillenmeyer *et al.*, 2010). Powerful algorithms currently analyse, predict and more precisely assess the mechanism of drug action (Hopkins, 2009; Keiser *et al.*, 2009). Now, through the analysis of the chemical structure of a given drug, it is possible to predict with a high level of accuracy its biological target (Lounkine *et al.*, 2012). This is reflected through the recent creation of a public database, www.drugable.com, aimed at storing data from the

> largest computational docking ever done by mankind.
>
> (Reardon, 2013)

With the growing number of biological resources and the accelerated development of high-throughput screening, high-throughput sequencing and computational data analysis technologies, highly integrative systems chemical biology-driven strategies in drug discovery and drug target deconvolution are expected to be implemented in the near future.

Acknowledgements

I would like to thank Dr Christophe d'Enfert for critically reading the manuscript and Dr Marie-Elisabeth Bougnoux for her insight on some aspects of clinical mycology. I apologize to authors whose work could not be cited in this chapter due to space constraints. My work at the Institut Pasteur Biologie et Pathogénicité Fongiques Research Unit was supported by grants from the European commission (FinSysB PITN-GA-2008-214004), the Agence Nationale de la Recherche (KANJI, ANR-08-MIE-033-01) and a BIOASTER-Sanofi-Alliance pour les Sciences de la Vie et de la Santé (AVIESAN) joint programme to Dr Christophe d'Enfert.

References

Baetz, K., McHardy, L., Gable, K., Tarling, T., Reberioux, D., Bryan, J., Andersen, R.J., Dunn, T., Hieter, P., and Roberge, M. (2004). Yeast genome-wide drug-induced haploinsufficiency screen to determine drug mode of action. Proc. Natl. Acad. Sci. U.S.A. *101*, 4525–4530.

Baginski, M., Sternal, K., Czub, J., and Borowski, E. (2005). Molecular modelling of membrane activity of amphotericin B, a polyene macrolide antifungal antibiotic. Acta Biochim. Pol. *52*, 655–658.

Baguley, B.C., Rommele, G., Gruner, J., and Wehrli, W. (1979). Papulacandin B: an inhibitor of glucan synthesis in yeast spheroplasts. Eur. J. Biochem. *97*, 345–351.

Balashov, S.V., Park, S., and Perlin, D.S. (2006). Assessing resistance to the echinocandin antifungal drug caspofungin in *Candida albicans* by profiling mutations in FKS1. Antimicrob. Agents Chemother. *50*, 2058–2063.

Bansal, Y., and Silakari, O. (2012). The therapeutic journey of benzimidazoles: a review. Bioorg. Med. Chem. *20*, 6208–6236.

Barasch, A., and Griffin, A.V. (2008). Miconazole revisited: new evidence of antifungal efficacy from laboratory and clinical trials. Future Microbiol. *3*, 265–269.

Becker, F., Murthi, K., Smith, C., Come, J., Costa-Roldan, N., Kaufmann, C., Hanke, U., Degenhart, C., Baumann, S., Wallner, W., *et al.* (2004). A three-hybrid approach to scanning the proteome for targets of small molecule kinase inhibitors. Chem. Biol. *11*, 211–223.

Braun, B.R., van Het Hoog, M., d'Enfert, C., Martchenko, M., Dungan, J., Kuo, A., Inglis, D.O., Uhl, M.A., Hogues, H., Berriman, M., *et al.* (2005). A human-curated annotation of the *Candida albicans* genome. PLoS Genet. *1*, 36–57.

Brodin, P., and Christophe, T. (2011). High-content screening in infectious diseases. Curr. Opin. Chem. Biol. *15*, 534–539.

Brown, J. (2005). Zygomycosis: an emerging fungal infection. Am. J. Health Syst. Pharm. *62*, 2593–2596.

Burdine, L., and Kodadek, T. (2004). Target identification in chemical genetics: the (often) missing link. Chem. Biol. *11*, 593–597.

Butcher, R.A., and Schreiber, S.L. (2006). A microarray-based protocol for monitoring the growth of yeast overexpression strains. Nat. Protoc. *1*, 569–576.

Butcher, R.A., Bhullar, B.S., Perlstein, E.O., Marsischky, G., LaBaer, J., and Schreiber, S.L. (2006). Microarray-based method for monitoring yeast overexpression strains reveals small-molecule targets in TOR pathway. Nat. Chem. Biol. *2*, 103–109.

Cabral, V., Chauvel, M., Firon, A., Legrand, M., Nesseir, A., Bachellier-Bassi, S., Chaudhari, Y., Munro, C.A., and d'Enfert, C. (2012). Modular gene over-expression strategies for *Candida albicans*. Methods Mol. Biol. *845*, 227–244.

Carrillo-Munoz, A.J., Giusiano, G., Ezkurra, P.A., and Quindos, G. (2006). Antifungal agents: mode of action in yeast cells. Rev. Esp. Quimioter. *19*, 130–139.

Carroll, S.Y., Stirling, P.C., Stimpson, H.E., Giesselmann, E., Schmitt, M.J., and Drubin, D.G. (2009). A yeast killer toxin screen provides insights into a/b toxin entry, trafficking, and killing mechanisms. Dev. Cell *17*, 552–560.

Casey, J.L., Coley, A.M., and Foley, M. (2008). Phage display of peptides in ligand selection for use in affinity chromatography. Methods Mol. Biol. *421*, 111–124.

Cass, A., Finkelstein, A., and Krespi, V. (1970). The ion permeability induced in thin lipid membranes by the polyene antibiotics nystatin and amphotericin B. J. Gen. Physiol. *56*, 100–124.

Castro, C., Ribas, J.C., Valdivieso, M.H., Varona, R., del Rey, F., and Duran, A. (1995). Papulacandin B resistance in budding and fission yeasts: isolation and characterization of a gene involved in (1,3)beta-D-glucan synthesis in *Saccharomyces cerevisiae*. J. Bacteriol. *177*, 5732–5739.

Chan, J.N., Nislow, C., and Emili, A. (2010). Recent advances and method development for drug target identification. Trends Pharmacol. Sci. *31*, 82–88.

Chaudhuri, A.R., and Luduena, R.F. (1996). Griseofulvin: a novel interaction with bovine brain tubulin. Biochem. Pharmacol. *51*, 903–909.

Chauvel, M., Nesseir, A., Cabral, V., Znaidi, S., Goyard, S., Bachellier-Bassi, S., Firon, A., Legrand, M., Diogo, D., Naulleau, C., *et al.* (2012). A versatile overexpression strategy in the pathogenic yeast *Candida albicans*: identification of regulators of morphogenesis and fitness. PLoS One *7*, e45912.

Chen, B.R., Hale, D.C., Ciolek, P.J., and Runge, K.W. (2012). Generation and analysis of a barcode-tagged insertion mutant library in the fission yeast *Schizosaccharomyces pombe*. BMC Genomics *13*, 161.

Chen, S.C., Blyth, C.C., Sorrell, T.C., and Slavin, M.A. (2011). Pneumonia and lung infections due to emerging and unusual fungal pathogens. Semin. Respir. Crit. Care Med. *32*, 703–716.

Chidley, C., Haruki, H., Pedersen, M.G., Muller, E., and Johnsson, K. (2011). A yeast-based screen reveals that sulfasalazine inhibits tetrahydrobiopterin biosynthesis. Nat. Chem. Biol. *7*, 375–383.

Cleary, M.A., Kilian, K., Wang, Y., Bradshaw, J., Cavet, G., Ge, W., Kulkarni, A., Paddison, P.J., Chang, K., Sheth, N., *et al.* (2004). Production of complex nucleic acid libraries using highly parallel *in situ* oligonucleotide synthesis. Nat. Methods *1*, 241–248.

Cohen, B.E. (2010). Amphotericin B membrane action: role for two types of ion channels in eliciting cell survival and lethal effects. J. Membr. Biol. *238*, 1–20.

Czymmek, K.J., Bourett, T.M., Shao, Y., DeZwaan, T.M., Sweigard, J.A., and Howard, R.J. (2005). Live-cell imaging of tubulin in the filamentous fungus *Magnaporthe grisea* treated with anti-microtubule and anti-microfilament agents. Protoplasma *225*, 23–32.

Debono, M., and Gordee, R.S. (1994). Antibiotics that inhibit fungal cell wall development. Annu. Rev. Microbiol. *48*, 471–497.

Denning, D.W. (1997). Echinocandins and pneumocandins--a new antifungal class with a novel mode of action. J. Antimicrob. Chemother. *40*, 611–614.

Denning, D.W. (2002). Echinocandins: a new class of antifungal. J. Antimicrob. Chemother. *49*, 889–891.

De Nollin, S., and Borgers, M. (1977). The effects of miconazole on the ultrastructure of *Candida albicans*. Proc. R. Soc. Med. *70* (Suppl. 1), 9–12.

De Nollin, S., Van Belle, H., Goossens, F., Thone, F., and Borgers, M. (1977). Cytochemical and biochemical studies of yeasts after *in vitro* exposure to miconazole. Antimicrob. Agents Chemother. *11*, 500–513.

Deresinski, S.C., and Stevens, D.A. (2003). Caspofungin. Clin Infect Dis *36*, 1445–1457.

Douglas, A.C., Smith, A.M., Sharifpoor, S., Yan, Z., Durbic, T., Heisler, L.E., Lee, A.Y., Ryan, O., Gottert, H., Surendra, A., *et al.* (2012). Functional analysis with a barcoder yeast gene overexpression system. G3 (Bethesda) *2*, 1279–1289.

Douglas, C.M., Foor, F., Marrinan, J.A., Morin, N., Nielsen, J.B., Dahl, A.M., Mazur, P., Baginsky, W., Li, W., el-Sherbeini, M., *et al.* (1994). The *Saccharomyces cerevisiae* FKS1 (ETG1) gene encodes an integral membrane protein which is a subunit of 1,3-beta-D-glucan synthase. Proc. Natl. Acad. Sci. U.S.A. *91*, 12907–12911.

Douglas, C.M., D'Ippolito, J.A., Shei, G.J., Meinz, M., Onishi, J., Marrinan, J.A., Li, W., Abruzzo, G.K., Flattery, A., Bartizal, K., *et al.* (1997). Identification of the FKS1 gene of *Candida albicans* as the essential target of 1,3-beta-D-glucan synthase inhibitors. Antimicrob. Agents Chemother. *41*, 2471–2479.

Doyon, J.B., Sutton, D.A., Theodore, P., Dhillon, G., Jones, K.D., Thompson, E.H., Fu, J., Wickes, B.L., Koehler, J.E., and Schwartz, B.S. (2013). Rasamsonia argillacea pulmonary and aortic graft infection in an immune-competent patient. J. Clin. Microbiol. *51*, 719–722.

Eng, W.K., Faucette, L., McLaughlin, M.M., Cafferkey, R., Koltin, Y., Morris, R.A., Young, P.R., Johnson, R.K., and Livi, G.P. (1994). The yeast FKS1 gene encodes a novel membrane protein, mutations in which confer FK506 and cyclosporin A hypersensitivity and calcineurin-dependent growth. Gene *151*, 61–71.

Ericson, E., Hoon, S., St Onge, R.P., Giaever, G., and Nislow, C. (2010). Exploring gene function and drug action using chemogenomic dosage assays. Methods Enzymol. *470*, 233–255.

Garcia-Effron, G., Lee, S., Park, S., Cleary, J.D., and Perlin, D.S. (2009). Effect of Candida glabrata FKS1 and FKS2 mutations on echinocandin sensitivity and kinetics of 1,3-beta-D-glucan synthase: implication for the existing susceptibility breakpoint. Antimicrob. Agents Chemother. *53*, 3690–3699.

Gavigan, C.S., Kiely, S.P., Hirtzlin, J., and Bell, A. (2003). Cyclosporin-binding proteins of *Plasmodium falciparum*. Int. J. Parasitol. 33, 987–996.

Giaever, G., Shoemaker, D.D., Jones, T.W., Liang, H., Winzeler, E.A., Astromoff, A., and Davis, R.W. (1999). Genomic profiling of drug sensitivities via induced haploinsufficiency. Nat. Genet. *21*, 278–283.

Giaever, G., Chu, A.M., Ni, L., Connelly, C., Riles, L., Veronneau, S., Dow, S., Lucau-Danila, A., Anderson, K., Andre, B., *et al.* (2002). Functional profiling of the *Saccharomyces cerevisiae* genome. Nature *418*, 387–391.

Giaever, G., Flaherty, P., Kumm, J., Proctor, M., Nislow, C., Jaramillo, D.F., Chu, A.M., Jordan, M.I., Arkin, A.P., and Davis, R.W. (2004). Chemogenomic profiling: identifying the functional interactions of small molecules in yeast. Proc. Natl. Acad. Sci. U.S.A. *101*, 793–798.

Giraud, S., Favennec, L., Bougnoux, M.E., and Bouchara, J.P. (2013). Rasamsonia argillacea species complex: taxonomy, pathogenesis and clinical relevance. Future Microbiol. *8*, 967–978.

Giuliano, K.A., and Taylor, D.L. (1998). Fluorescent-protein biosensors: new tools for drug discovery. Trends Biotechnol. *16*, 135–140.

Giuliano, K.A., DeBiasio, R.L., Dunlay, R.T., Gough, A., Volosky, J.M., Zock, J., Pavlakis, G.N., and Taylor, D.L. (1997). High-content screening: a new approach to easing key bottlenecks in the drug discovery process. J. Biomol. Screen. 2, 249–259.

Gordee, R.S., Zeckner, D.J., Ellis, L.F., Thakkar, A.L., and Howard, L.C. (1984). *In vitro* and *in vivo* anti-*Candida* activity and toxicology of LY121019. J. Antibiot. (Tokyo) 37, 1054–1065.

Gottlieb, D., Carter, H.E., Sloneker, J.H., and Ammann, A. (1958). Protection of fungi against polyene antibiotics by sterols. Science *128*, 361.

Gray, K.C., Palacios, D.S., Dailey, I., Endo, M.M., Uno, B.E., Wilcock, B.C., and Burke, M.D. (2012). Amphotericin primarily kills yeast by simply binding ergosterol. Proc. Natl. Acad. Sci. U.S.A. *109*, 2234–2239.

Gresham, D., Ruderfer, D.M., Pratt, S.C., Schacherer, J., Dunham, M.J., Botstein, D., and Kruglyak, L. (2006). Genome-wide detection of polymorphisms at nucleotide resolution with a single DNA microarray. Science *311*, 1932–1936.

Grisham, L.M., Wilson, L., and Bensch, K.G. (1973). Antimitotic action of griseofulvin does not involve disruption of microtubules. Nature *244*, 294–296.

Gull, K., and Trinci, A.P. (1973). Griseofulvin inhibits fungal mitosis. Nature *244*, 292–294.

Han, S., Lee, M., Chang, H., Nam, M., Park, H.O., Kwak, Y.S., Ha, H.J., Kim, D., Hwang, S.O., Hoe, K.L., *et al.* (2013). Construction of the first compendium of chemical-genetic profiles in the fission yeast *Schizosaccharomyces pombe* and comparative compendium approach. Biochem. Biophys. Res. Commun. *436*, 613–618.

Hansen, T.M., Engler, M.M., and Forsyth, C.J. (2003). Total synthesis of a biotinylated derivative of phorboxazole A via sonogashira coupling. Bioorg. Med. Chem. Lett. *13*, 2127–2130.

Harding, M.W., Galat, A., Uehling, D.E., and Schreiber, S.L. (1989). A receptor for the immunosuppressant FK506 is a cis–trans peptidyl-prolyl isomerase. Nature *341*, 758–760.

Harman, J.W., and Masterson, J.G. (1957). The mechanism of nystatin action. Ir. J. Med. Sci., 249–253.

Hazen, E.L., and Brown, R. (1951). Fungicidin, an antibiotic produced by a soil actinomycete. Proc. Soc. Exp. Biol. Med. *76*, 93–97.

Heeres, J., Backx, L.J., Mostmans, J.H., and Van Cutsem, J. (1979). Antimycotic imidazoles. Part 4. Synthesis and antifungal activity of ketoconazole, a new potent orally active broad-spectrum antifungal agent. J. Med. Chem. *22*, 1003–1005.

Heeres, J., Meerpoel, L., and Lewi, P. (2010). Conazoles. Molecules *15*, 4129–4188.

Heiskanen, M.A., and Aittokallio, T. (2013). Predicting drug–target interactions through integrative analysis of chemogenetic assays in yeast. Mol. Biosyst. *9*, 768–779.

Henry, M.J., and Sisler, H.D. (1979). Effects of miconazole and dodecylimidazole on sterol biosynthesis in Ustilago maydis. Antimicrob. Agents Chemother. *15*, 603–607.

Hillenmeyer, M.E., Fung, E., Wildenhain, J., Pierce, S.E., Hoon, S., Lee, W., Proctor, M., St Onge, R.P., Tyers, M., Koller, D., *et al.* (2008). The chemical genomic portrait of yeast: uncovering a phenotype for all genes. Science *320*, 362–365.

Hillenmeyer, M.E., Ericson, E., Davis, R.W., Nislow, C., Koller, D., and Giaever, G. (2010). Systematic analysis of genome-wide fitness data in yeast reveals novel gene function and drug action. Genome Biol. *11*, R30.

Ho, C.H., Magtanong, L., Barker, S.L., Gresham, D., Nishimura, S., Natarajan, P., Koh, J.L., Porter, J., Gray, C.A., Andersen, R.J., *et al.* (2009). A molecular barcoded yeast ORF library enables mode-of-action analysis of bioactive compounds. Nat. Biotechnol. *27*, 369–377.

Holz, R., and Finkelstein, A. (1970). The water and nonelectrolyte permeability induced in thin lipid membranes by the polyene antibiotics nystatin and amphotericin B. J. Gen. Physiol. *56*, 125–145.

Hoon, S., Smith, A.M., Wallace, I.M., Suresh, S., Miranda, M., Fung, E., Proctor, M., Shokat, K.M., Zhang, C., Davis, R.W., *et al.* (2008). An integrated platform of genomic assays reveals small-molecule bioactivities. Nat. Chem. Biol. *4*, 498–506.

Hopkins, A.L. (2009). Drug discovery: Predicting promiscuity. Nature *462*, 167–168.

Hsuchen, C.C., and Feingold, D.S. (1973a). Selective membrane toxicity of the polyene antibiotics: studies on lecithin membrane models (liposomes). Antimicrob. Agents Chemother. *4*, 309–315.

Hsuchen, C.C., and Feingold, D.S. (1973b). Selective membrane toxicity of the polyene antibiotics: studies on natural membranes. Antimicrob. Agents Chemother. *4*, 316–319.

Huang, J., Zhu, H., Haggarty, S.J., Spring, D.R., Hwang, H., Jin, F., Snyder, M., and Schreiber, S.L. (2004). Finding new components of the target of rapamycin (TOR) signaling network through chemical genetics and proteome chips. Proc. Natl. Acad. Sci. U.S.A. *101*, 16594–16599.

Hughes, T.R., Marton, M.J., Jones, A.R., Roberts, C.J., Stoughton, R., Armour, C.D., Bennett, H.A., Coffey, E., Dai, H., He, Y.D., *et al.* (2000). Functional discovery via a compendium of expression profiles. Cell *102*, 109–126.

Imoto, S., Savoie, C.J., Aburatani, S., Kim, S., Tashiro, K., Kuhara, S., and Miyano, S. (2003). Use of gene networks for identifying and validating drug targets. J. Bioinform. Comput. Biol. *1*, 459–474.

Imoto, S., Tamada, Y., Savoie, C.J., and Miyano, S. (2007). Analysis of gene networks for drug target discovery and validation. Methods Mol. Biol. *360*, 33–56.

Inoue, S.B., Takewaki, N., Takasuka, T., Mio, T., Adachi, M., Fujii, Y., Miyamoto, C., Arisawa, M., Furuichi, Y., and Watanabe, T. (1995). Characterization and gene cloning of 1,3-beta-D-glucan synthase from *Saccharomyces cerevisiae*. Eur. J. Biochem. *231*, 845–854.

Iwata, K., Kanda, Y., Yamaguchi, H., and Osumi, M. (1973a). Electron microscopic studies on the mechanism of action of clotrimazole on *Candida albicans*. Sabouraudia *11*, 205–209.

Iwata, K., Yamaguchi, H., and Hiratani, T. (1973b). Mode of action of clotrimazole. Sabouraudia *11*, 158–166.

Jiang, B., Xu, D., Allocco, J., Parish, C., Davison, J., Veillette, K., Sillaots, S., Hu, W., Rodriguez-Suarez, R., Trosok, S., *et al.* (2008). PAP inhibitor with *in vivo* efficacy identified by *Candida albicans* genetic profiling of natural products. Chem. Biol. *15*, 363–374.

Jin, K., Li, J., Vizeacoumar, F.S., Li, Z., Min, R., Zamparo, L., Vizeacoumar, F.J., Datti, A., Andrews, B., Boone, C., *et al.* (2011). PhenoM: a database of morphological phenotypes caused by mutation of essential genes in *Saccharomyces cerevisiae*. Nucleic Acids Res. *40*, D687–694.

Jones, T., Federspiel, N.A., Chibana, H., Dungan, J., Kalman, S., Magee, B.B., Newport, G., Thorstenson, Y.R., Agabian, N., Magee, P.T., *et al.* (2004). The diploid genome sequence of *Candida albicans*. Proc. Natl. Acad. Sci. U.S.A. *101*, 7329–7334.

Kalb, V.F., Loper, J.C., Dey, C.R., Woods, C.W., and Sutter, T.R. (1986). Isolation of a cytochrome P-450 structural gene from *Saccharomyces cerevisiae*. Gene *45*, 237–245.

Kalb, V.F., Woods, C.W., Turi, T.G., Dey, C.R., Sutter, T.R., and Loper, J.C. (1987). Primary structure of the P450 lanosterol demethylase gene from *Saccharomyces cerevisiae*. DNA *6*, 529–537.

Kanafani, Z.A., and Perfect, J.R. (2008). Antimicrobial resistance: resistance to antifungal agents: mechanisms and clinical impact. Clin. Infect. Dis. *46*, 120–128.

Kapitzky, L., Beltrao, P., Berens, T.J., Gassner, N., Zhou, C., Wuster, A., Wu, J., Babu, M.M., Elledge, S.J., Toczyski, D., *et al.* (2010). Cross-species chemogenomic profiling reveals evolutionarily conserved drug mode of action. Mol. Syst. Biol. *6*, 451.

Keiser, M.J., Setola, V., Irwin, J.J., Laggner, C., Abbas, A.I., Hufeisen, S.J., Jensen, N.H., Kuijer, M.B., Matos, R.C., Tran, T.B., *et al.* (2009). Predicting new molecular targets for known drugs. Nature *462*, 175–181.

Kim, D.U., Hayles, J., Kim, D., Wood, V., Park, H.O., Won, M., Yoo, H.S., Duhig, T., Nam, M., Palmer, G., *et al.* (2010). Analysis of a genome-wide set of gene deletions in the fission yeast *Schizosaccharomyces pombe*. Nat. Biotechnol. *28*, 617–623.

Kinsky, S.C. (1961a). Alterations in the permeability of *Neurospora crassa* due to polyene antibiotics. J. Bacteriol. *82*, 889–897.

Kinsky, S.C. (1961b). The effect of polyene antibiotics on permeability in *Neurospora crassa*. Biochem. Biophys. Res. Commun. *4*, 353–357.

Kinsky, S.C. (1962a). Effect of polyene antibiotics on protoplasts of *Neurospora crassa*. J Bacteriol *83*, 351–358.

Kinsky, S.C. (1962b). Nystatin binding by protoplasts and a particulate fraction of *Neurospora crassa*, and a basis for the selective toxicity of polyene antifungal antibiotics. Proc. Natl. Acad. Sci. U.S.A. *48*, 1049–1056.

Kinsky, S.C., Avruch, J., Permutt, M., and Rogers, H.B. (1962). The lytic effect of polyene antifungal antibiotics on mammalian erythrocytes. Biochem. Biophys. Res. Commun. *9*, 503–507.

Kirsch, D.R., Lai, M.H., and O'Sullivan, J. (1988). Isolation of the gene for cytochrome P450L1A1 (lanosterol 14 alpha-demethylase) from *Candida albicans*. Gene *68*, 229–237.

Kohanski, M.A., Dwyer, D.J., and Collins, J.J. (2010). How antibiotics kill bacteria: from targets to networks. Nat. Rev. Microbiol. *8*, 423–435.

Kotler-Brajtburg, J., Price, H.D., Medoff, G., Schlessinger, D., and Kobayashi, G.S. (1974). Molecular basis for the selective toxicity of amphotericin B for yeast and filipin for animal cells. Antimicrob. Agents Chemother. *5*, 377–382.

Krcmery, V., and Kalavsky, E. (2007). Antifungal drug discovery, six new molecules patented after 10 years of feast: why do we need new patented drugs apart from new strategies? Recent Pat. Antiinfect. Drug Discov. *2*, 182–187.

de Kruijff, B., and Demel, R.A. (1974). Polyene antibiotic–sterol interactions in membranes of *Acholeplasma laidlawii* cells and lecithin liposomes. 3. Molecular structure of the polyene antibiotic–cholesterol complexes. Biochim. Biophys. Acta *339*, 57–70.

de Kruijff, B., Gerritsen, W.J., Oerlemans, A., Demel, R.A., and van Deenen, L.L. (1974a). Polyene antibiotic–sterol interactions in membranes of *Acholeplasma laidlawii* cells and lecithin liposomes. I. Specificity of the membrane permeability changes induced by the polyene antibiotics. Biochim. Biophys. Acta *339*, 30–43.

de Kruijff, B., Gerritsen, W.J., Oerlemans, A., van Dijck, P.W., Demel, R.A., and van Deenen, L.L. (1974b). Polyene antibiotic–sterol interactions in membranes of *Acholesplasma laidlawii* cells and lecithin liposomes. II. Temperature dependence of the polyene antibiotic–sterol complex formation. Biochim. Biophys. Acta *339*, 44–56.

Lamb, D., Kelly, D., and Kelly, S. (1999). Molecular aspects of azole antifungal action and resistance. Drug Resist. Updat. *2*, 390–402.

Lamb, J., Crawford, E.D., Peck, D., Modell, J.W., Blat, I.C., Wrobel, M.J., Lerner, J., Brunet, J.P., Subramanian, A., Ross, K.N., *et al.* (2006). The Connectivity Map: using gene-expression signatures to connect small molecules, genes, and disease. Science *313*, 1929–1935.

Lampen, J.O., and Arnow, P. (1959). Significance of nystatin uptake for its antifungal action. Proc. Soc. Exp. Biol. Med. *101*, 792–797.

Lampen, J.O., Morgan, E.R., Slocum, A., and Arnow, P. (1959). Absorption of nystatin by microorganisms. J. Bacteriol. *78*, 282–289.

Lampen, J.O., Arnow, P.M., Borowska, Z., and Laskin, A.I. (1962). Location and role of sterol at nystatin-binding sites. J. Bacteriol. *84*, 1152–1160.

Lang, P., Yeow, K., Nichols, A., and Scheer, A. (2006). Cellular imaging in drug discovery. Nat. Rev. Drug Discov. *5*, 343–356.

Lee, R.E., Liu, T.T., Barker, K.S., and Rogers, P.D. (2005). Genome-wide expression profiling of the response to ciclopirox olamine in *Candida albicans*. J. Antimicrob. Chemother. *55*, 655–662.

Leek, J.T., Scharpf, R.B., Bravo, H.C., Simcha, D., Langmead, B., Johnson, W.E., Geman, D., Baggerly, K., and Irizarry, R.A. (2010). Tackling the widespread and critical impact of batch effects in high-throughput data. Nat. Rev. Genet. *11*, 733–739.

Lefurgy, S., and Cornish, V. (2004). Finding Cinderella after the ball: a three-hybrid approach to drug target identification. Chem. Biol. *11*, 151–153.

Letscher-Bru, V., and Herbrecht, R. (2003). Caspofungin: the first representative of a new antifungal class. J. Antimicrob. Chemother. *51*, 513–521.

Lewis, R.E. (2010). Polyene antifungal agents. In Aspergillosis: From Diagnosis to Prevention (Springer, Dordrecht, the Netherlands), pp. 281–305.

Li, Z., Vizeacoumar, F.J., Bahr, S., Li, J., Warringer, J., Vizeacoumar, F.S., Min, R., Vandersluis, B., Bellay, J., Devit, M., *et al.* (2011). Systematic exploration of essential yeast gene function with temperature-sensitive mutants. Nat. Biotechnol. *29*, 361–367.

Licitra, E.J., and Liu, J.O. (1996). A three-hybrid system for detecting small ligand-protein receptor interactions. Proc. Natl. Acad. Sci. U.S.A. *93*, 12817–12821.

Liu, T.T., Lee, R.E., Barker, K.S., Wei, L., Homayouni, R., and Rogers, P.D. (2005). Genome-wide expression profiling of the response to azole, polyene, echinocandin, and pyrimidine antifungal agents in *Candida albicans*. Antimicrob. Agents Chemother. *49*, 2226–2236.

Lokey, R.S. (2003). Forward chemical genetics: progress and obstacles on the path to a new pharmacopoeia. Curr. Opin. Chem. Biol. *7*, 91–96.

Lounkine, E., Keiser, M.J., Whitebread, S., Mikhailov, D., Hamon, J., Jenkins, J.L., Lavan, P., Weber, E., Doak, A.K., Cote, S., *et al.* (2012). Large-scale prediction and testing of drug activity on side-effect targets. Nature *486*, 361–367.

Luesch, H., Wu, T.Y., Ren, P., Gray, N.S., Schultz, P.G., and Supek, F. (2005). A genome-wide overexpression screen in yeast for small-molecule target identification. Chem. Biol. *12*, 55–63.

Lum, P.Y., Armour, C.D., Stepaniants, S.B., Cavet, G., Wolf, M.K., Butler, J.S., Hinshaw, J.C., Garnier, P., Prestwich, G.D., Leonardson, A., *et al.* (2004). Discovering modes of action for therapeutic compounds using a genome-wide screen of yeast heterozygotes. Cell *116*, 121–137.

Lupetti, A., Danesi, R., Campa, M., Del Tacca, M., and Kelly, S. (2002). Molecular basis of resistance to azole antifungals. Trends Mol. Med. *8*, 76–81.

Maertens, J.A. (2004). History of the development of azole derivatives. Clin. Microbiol. Infect. *10* (Suppl. 1), 1–10.

Marichal, P., and Vanden Bossche, H. (1995). Mechanisms of resistance to azole antifungals. Acta Biochim. Pol. *42*, 509–516.

Marichal, P., Koymans, L., Willemsens, S., Bellens, D., Verhasselt, P., Luyten, W., Borgers, M., Ramaekers, F.C., Odds, F.C., and Bossche, H.V. (1999). Contribution of mutations in the cytochrome P450 14α-demethylase (Erg11p, Cyp51p) to azole resistance in *Candida albicans*. Microbiology *145*, 2701–2713.

Masurekar, P.S., Fountoulakis, J.M., Hallada, T.C., Sosa, M.S., and Kaplan, L. (2012). Fermentation development of Pneumocandin B0: the ultimate lead for making Caspofungin. Planta Medica 78, IL24.

Matsuyama, A., Arai, R., Yashiroda, Y., Shirai, A., Kamata, A., Sekido, S., Kobayashi, Y., Hashimoto, A., Hamamoto, M., Hiraoka, Y., *et al.* (2006). ORFeome cloning and global analysis of protein localization in the fission yeast *Schizosaccharomyces pombe*. Nat. Biotechnol. *24*, 841–847.

Mazumder, A., Pesudo, L.Q., McRee, S., Bathe, M., and Samson, L.D. (2013). Genome-wide single-cell-level screen for protein abundance and localization changes in response to DNA damage in *S. cerevisiae*. Nucleic Acids Res. *41*, 9310–9324.

Mazur, P., Morin, N., Baginsky, W., el-Sherbeini, M., Clemas, J.A., Nielsen, J.B., and Foor, F. (1995). Differential expression and function of two homologous subunits of yeast 1,3-beta-D-glucan synthase. Mol. Cell Biol. *15*, 5671–5681.

Mizoguchi, J., Saito, T., Mizuno, K., and Hayano, K. (1977). On the mode of action of a new antifungal antibiotic, aculeacin A: inhibition of cell wall synthesis in *Saccharomyces cerevisiae*. J. Antibiot. (Tokyo) *30*, 308–313.

Mizuno, K., Yagi, A., Satoi, S., Takada, M., and Hayashi, M. (1977). Studies on aculeacin. I. Isolation and characterization of aculeacin A. J. Antibiot. (Tokyo) *30*, 297–302.

Morohashi, K., Arai, T., Saito, S., Watanabe, M., Sakaguchi, K., and Sugawara, F. (2006). A high-throughput phage display screening method using a combination of real-time PCR and affinity chromatography. Comb. Chem. High Throughput Screen *9*, 55–61.

Morris, M.I., and Villmann, M. (2006a). Echinocandins in the management of invasive fungal infections, Part 1. Am. J. Health Syst. Pharm. *63*, 1693–1703.

Morris, M.I., and Villmann, M. (2006b). Echinocandins in the management of invasive fungal infections, Part 2. Am. J. Health Syst. Pharm. *63*, 1813–1820.

Morschhauser, J. (2002). The genetic basis of fluconazole resistance development in *Candida albicans*. Biochim. Biophys. Acta *1587*, 240–248.

Moser, S., and Johnsson, K. (2013). Yeast three-hybrid screening for identifying anti-tuberculosis drug targets. Chembiochem. *14*, 2239–2242.

Neumann, A., Baginski, M., and Czub, J. (2010). How do sterols determine the antifungal activity of amphotericin B? Free energy of binding between the drug and its membrane targets. J. Am. Chem. Soc. *132*, 18266–18272.

Niewerth, M., Kunze, D., Seibold, M., Schaller, M., Korting, H.C., and Hube, B. (2003). Ciclopirox olamine treatment affects the expression pattern of *Candida albicans* genes encoding virulence factors, iron metabolism proteins, and drug resistance factors. Antimicrob. Agents Chemother. *47*, 1805–1817.

Nishimura, S., Arita, Y., Honda, M., Iwamoto, K., Matsuyama, A., Shirai, A., Kawasaki, H., Kakeya, H., Kobayashi, T., Matsunaga, S., *et al.* (2010). Marine antifungal theonellamides target 3β-hydroxysterol to activate Rho1 signaling. Nat. Chem. Biol. *6*, 519–526.

Noble, S.M., French, S., Kohn, L.A., Chen, V., and Johnson, A.D. (2010). Systematic screens of a *Candida albicans* homozygous deletion library decouple morphogenetic switching and pathogenicity. Nat. Genet. *42*, 590–598.

Noguchi-Yachide, T., Dodo, K., Aoyama, H., Fujimoto, H., Hori, M., Hashimoto, Y., and Kobayashi, H. (2010). Identification of binding proteins of fusarielin A as actin and tubulin. Chem. Pharm. Bull. (Tokyo) *58*, 129–134.

Nyfeler, R., and Keller-Schierlein, W. (1974). [Metabolites of microorganisms. 143. Echinocandin B, a novel polypeptide-antibiotic from Aspergillus nidulans var. echinulatus: isolation and structural components]. Helv. Chim. Acta *57*, 2459–2477.

Odds, F.C., Brown, A.J., and Gow, N.A. (2003). Antifungal agents: mechanisms of action. Trends Microbiol. *11*, 272–279.

Oh, J., Fung, E., Price, M.N., Dehal, P.S., Davis, R.W., Giaever, G., Nislow, C., Arkin, A.P., and Deutschbauer, A. (2010a). A universal TagModule collection for parallel genetic analysis of microorganisms. Nucleic Acids Res. *38*, e146.

Oh, J., Fung, E., Schlecht, U., Davis, R.W., Giaever, G., St Onge, R.P., Deutschbauer, A., and Nislow, C. (2010b). Gene annotation and drug target discovery in *Candida albicans* with a tagged transposon mutant collection. PLoS Pathog. *6*, e1001140.

Ohnuki, S., Oka, S., Nogami, S., and Ohya, Y. (2010). High-content, image-based screening for drug targets in yeast. PLoS One *5*, e10177.

Ohya, Y., Sese, J., Yukawa, M., Sano, F., Nakatani, Y., Saito, T.L., Saka, A., Fukuda, T., Ishihara, S., Oka, S., *et al.* (2005). High-dimensional and large-scale phenotyping of yeast mutants. Proc. Natl. Acad. Sci. U.S.A. *102*, 19015–19020.

Okada, H., Ohnuki, S., Roncero, C., Konopka, J.B., and Ohya, Y. (2013). Distinct roles of cell wall biogenesis in yeast morphogenesis as revealed by multivariate analysis of high-dimensional morphometric data. Mol. Biol. Cell *25*, 222–233.

Ong, S.E., Schenone, M., Margolin, A.A., Li, X., Do, K., Doud, M.K., Mani, D.R., Kuai, L., Wang, X., Wood, J.L., *et al.* (2009). Identifying the proteins to which small-molecule probes and drugs bind in cells. Proc. Natl. Acad. Sci. U.S.A. *106*, 4617–4622.

Ostrosky-Zeichner, L., Marr, K.A., Rex, J.H., and Cohen, S.H. (2003). Amphotericin B: time for a new "gold standard". Clin. Infect. Dis. 37, 415–425.

Osumi, M., Yamada, N., Kobori, H., Taki, A., Naito, N., Baba, M., and Nagatani, T. (1989). Cell wall formation in regenerating protoplasts of *Schizosaccharomyces pombe*: study by high resolution, low voltage scanning electron microscopy. J. Electron Microsc. (Tokyo) *38*, 457–468.

Oxford, A.E., Raistrick, H., and Simonart, P. (1939). Studies in the biochemistry of micro-organisms: Griseofulvin, C(17)H(17)O(6)Cl, a metabolic product of Penicillium griseo-fulvum Dierckx. Biochem. J. *33*, 240–248.

Palacios, D.S., Dailey, I., Siebert, D.M., Wilcock, B.C., and Burke, M.D. (2011). Synthesis-enabled functional group deletions reveal key underpinnings of amphotericin B ion channel and antifungal activities. Proc. Natl. Acad. Sci. U.S.A. *108*, 6733–6738.

Panda, D., Rathinasamy, K., Santra, M.K., and Wilson, L. (2005). Kinetic suppression of microtubule dynamic instability by griseofulvin: implications for its possible use in the treatment of cancer. Proc. Natl. Acad. Sci. U.S.A. *102*, 9878–9883.

Park, S., Kelly, R., Kahn, J.N., Robles, J., Hsu, M.J., Register, E., Li, W., Vyas, V., Fan, H., Abruzzo, G., *et al.* (2005). Specific substitutions in the echinocandin target Fks1p account for reduced susceptibility of rare laboratory and clinical *Candida* sp. isolates. Antimicrob. Agents Chemother. *49*, 3264–3273.

Parsons, A.B., Brost, R.L., Ding, H., Li, Z., Zhang, C., Sheikh, B., Brown, G.W., Kane, P.M., Hughes, T.R., and Boone, C. (2004). Integration of chemical-genetic and genetic interaction data links bioactive compounds to cellular target pathways. Nat. Biotechnol. *22*, 62–69.

Parsons, A.B., Lopez, A., Givoni, I.E., Williams, D.E., Gray, C.A., Porter, J., Chua, G., Sopko, R., Brost, R.L., Ho, C.H., *et al.* (2006). Exploring the mode-of-action of bioactive compounds by chemical-genetic profiling in yeast. Cell *126*, 611–625.

Perlin, D.S. (2007). Resistance to echinocandin-class antifungal drugs. Drug Resist. Updat. *10*, 121–130.

Perlman, Z.E., Slack, M.D., Feng, Y., Mitchison, T.J., Wu, L.F., and Altschuler, S.J. (2004). Multidimensional drug profiling by automated microscopy. Science *306*, 1194–1198.

Pfaller, M.A., and Diekema, D.J. (2010). Epidemiology of invasive mycoses in North America. Crit. Rev. Microbiol. *36*, 1–53.

Pierce, S.E., Davis, R.W., Nislow, C., and Giaever, G. (2007). Genome-wide analysis of barcoded *Saccharomyces cerevisiae* gene-deletion mutants in pooled cultures. Nat. Protoc. *2*, 2958–2974.

Prelich, G. (2012). Gene overexpression: uses, mechanisms, and interpretation. Genetics *190*, 841–854.

Radding, J.A., Heidler, S.A., and Turner, W.W. (1998). Photoaffinity analog of the semisynthetic echinocandin LY303366: identification of echinocandin targets in *Candida albicans*. Antimicrob. Agents Chemother. *42*, 1187–1194.

Rai, M.N., Balusu, S., Gorityala, N., Dandu, L., and Kaur, R. (2012). Functional genomic analysis of *Candida glabrata*–macrophage interaction: role of chromatin remodeling in virulence. PLoS Pathog. *8*, e1002863.

Ram, A.F., Brekelmans, S.S., Oehlen, L.J., and Klis, F.M. (1995). Identification of two cell cycle regulated genes affecting the β1,3-glucan content of cell walls in *Saccharomyces cerevisiae*. FEBS Lett. *358*, 165–170.

Rammaert, B., Lanternier, F., Zahar, J.R., Dannaoui, E., Bougnoux, M.E., Lecuit, M., and Lortholary, O. (2012). Healthcare-associated mucormycosis. Clin. Infect. Dis. *54* (Suppl. 1), S44–54.

Rausch, O. (2006). High content cellular screening. Curr. Opin. Chem. Biol. *10*, 316–320.

Reardon, S. (2013). Project ranks billions of drug interactions. Nature *503*, 449–450.

Revankar, S.G., and Sutton, D.A. (2010). Melanized fungi in human disease. Clin. Microbiol. Rev. *23*, 884–928.

Rix, U., and Superti-Furga, G. (2009). Target profiling of small molecules by chemical proteomics. Nat. Chem. Biol. *5*, 616–624.

Roemer, T., Jiang, B., Davison, J., Ketela, T., Veillette, K., Breton, A., Tandia, F., Linteau, A., Sillaots, S., Marta, C., *et al.* (2003). Large-scale essential gene identification in *Candida albicans* and applications to antifungal drug discovery. Mol. Microbiol. *50*, 167–181.

Roemer, T., Xu, D., Singh, S.B., Parish, C.A., Harris, G., Wang, H., Davies, J.E., and Bills, G.F. (2011). Confronting the challenges of natural product-based antifungal discovery. Chem. Biol. *18*, 148–164.

Roemer, T., Davies, J., Giaever, G., and Nislow, C. (2012). Bugs, drugs and chemical genomics. Nat. Chem. Biol. *8*, 46–56.

Sanglard, D. (2002). Resistance of human fungal pathogens to antifungal drugs. Curr. Opin. Microbiol. *5*, 379–385.

Sanglard, D., and Odds, F.C. (2002). Resistance of *Candida* species to antifungal agents: molecular mechanisms and clinical consequences. Lancet Infect. Dis. *2*, 73–85.

Sanglard, D., Kuchler, K., Ischer, F., Pagani, J.L., Monod, M., and Bille, J. (1995). Mechanisms of resistance to azole antifungal agents in *Candida albicans* isolates from AIDS patients involve specific multidrug transporters. Antimicrob. Agents Chemother. *39*, 2378–2386.

Savoie, C.J., Aburatani, S., Watanabe, S., Eguchi, Y., Muta, S., Imoto, S., Miyano, S., Kuhara, S., and Tashiro, K. (2003). Use of gene networks from full genome microarray libraries to identify functionally relevant drug-affected genes and gene regulation cascades. DNA Res. *10*, 19–25.

Sawistowska-Schroder, E.T., Kerridge, D., and Perry, H. (1984). Echinocandin inhibition of 1,3-beta-D-glucan synthase from *Candida albicans*. FEBS Lett. *173*, 134–138.

Seider, K., Gerwien, F., Kasper, L., Allert, S., Brunke, S., Jablonowski, N., Schwarzmuller, T., Barz, D., Rupp, S., Kuchler, K., *et al.* (2013). Immune evasion, stress resistance and efficient nutrient acquisition are crucial for intracellular survival of *Candida glabrata* within macrophages. Eukaryot. Cell *13*, 170–183.

el-Sherbeini, M., and Clemas, J.A. (1995). Nikkomycin Z supersensitivity of an echinocandin-resistant mutant of *Saccharomyces cerevisiae*. Antimicrob. Agents Chemother. *39*, 200–207.

Shoemaker, D.D., Lashkari, D.A., Morris, D., Mittmann, M., and Davis, R.W. (1996). Quantitative phenotypic analysis of yeast deletion mutants using a highly parallel molecular bar-coding strategy. Nat. Genet. *14*, 450–456.

Singh-Babak, S.D., Babak, T., Diezmann, S., Hill, J.A., Xie, J.L., Chen, Y.L., Poutanen, S.M., Rennie, R.P., Heitman, J., and Cowen, L.E. (2012). Global analysis of the evolution and mechanism of echinocandin resistance in *Candida glabrata*. PLoS Pathog. *8*, e1002718.

Sloboda, R.D., Van Blaricom, G., Creasey, W.A., Rosenbaum, J.L., and Malawista, S.E. (1982). Griseofulvin: association with tubulin and inhibition of *in vitro* microtubule assembly. Biochem. Biophys. Res. Commun. *105*, 882–888.

Smith, A.M., Durbic, T., Oh, J., Urbanus, M., Proctor, M., Heisler, L.E., Giaever, G., and Nislow, C. (2011). Competitive genomic screens of barcoded yeast libraries. J Vis Exp. *Aug 11* (54), 2864.

Smith, A.M., Durbic, T., Kittanakom, S., Giaever, G., and Nislow, C. (2012). Barcode sequencing for understanding drug–gene interactions. Methods Mol. Biol. *910*, 55–69.

Sopko, R., Huang, D., Preston, N., Chua, G., Papp, B., Kafadar, K., Snyder, M., Oliver, S.G., Cyert, M., Hughes, T.R., *et al.* (2006). Mapping pathways and phenotypes by systematic gene overexpression. Mol. Cell *21*, 319–330.

Spirek, M., Benko, Z., Carnecka, M., Rumpf, C., Cipak, L., Batova, M., Marova, I., Nam, M., Kim, D.U., Park, H.O., *et al.* (2010). *S. pombe* genome deletion project: an update. Cell Cycle *9*, 2399–2402.

Sreedhara Swamy, K.H., Sirsi, M., and Ramananda Rao, G.R. (1974). Studies on the mechanism of action of miconazole: effect of miconazole on respiration and cell permeability of *Candida albicans*. Antimicrob. Agents Chemother. *5*, 420–425.

Stevens, D.A. (2012). Advances in systemic antifungal therapy. Clin. Dermatol. *30*, 657–661.

Stynen, B., Van Dijck, P., and Tournu, H. (2010). A CUG codon adapted two-hybrid system for the pathogenic fungus *Candida albicans*. Nucleic Acids Res. *38*, e184.

Sud, I.J., and Feingold, D.S. (1981). Mechanisms of action of the antimycotic imidazoles. J. Invest. Dermatol. *76*, 438–441.

Swain, M.H., Payne, E.A., and Spruill, M.J. (2003). Mississippi women: their histories, their lives, Vol 1 (University of Georgia Press, Athens, GA, USA).

Taft, C.S., and Selitrennikoff, C.P. (1988). LY121019 inhibits *Neurospora crassa* growth and (1–3)-beta-D-glucan synthase. J. Antibiot. (Tokyo) *41*, 697–701.

Taft, C.S., Stark, T., and Selitrennikoff, C.P. (1988). Cilofungin (LY121019) inhibits *Candida albicans* (1-3)-beta-D-glucan synthase activity. Antimicrob. Agents Chemother. *32*, 1901–1903.

Takayama, H., Moriya, T., and Kanoh, N. (2012). Preparation of photo-cross-linked small molecule affinity matrices for affinity selection of protein targets for biologically active small molecules. Methods Mol. Biol. *800*, 75–83.

Taunton, J., Hassig, C.A., and Schreiber, S.L. (1996). A mammalian histone deacetylase related to the yeast transcriptional regulator Rpd3p. Science *272*, 408–411.

Taylor, D.L. (2007). Past, present, and future of high content screening and the field of cellomics. Methods Mol. Biol. *356*, 3–18.

Taylor, D.L., and Giuliano, K.A. (2005). Multiplexed high content screening assays create a systems cell biology approach to drug discovery. Drug Discov. Today *2005*, (Suppl.), 13–18.

te Welscher, Y.M., ten Napel, H.H., Balague, M.M., Souza, C.M., Riezman, H., de Kruijff, B., and Breukink, E. (2008). Natamycin blocks fungal growth by binding specifically to ergosterol without permeabilizing the membrane. J. Biol. Chem. *283*, 6393–6401.

te Welscher, Y.M., van Leeuwen, M.R., de Kruijff, B., Dijksterhuis, J., and Breukink, E. (2012). Polyene antibiotic that inhibits membrane transport proteins. Proc. Natl. Acad. Sci. U.S.A. *109*, 11156–11159.

Teng, X., Dayhoff-Brannigan, M., Cheng, W.C., Gilbert, C.E., Sing, C.N., Diny, N.L., Wheelan, S.J., Dunham, M.J., Boeke, J.D., Pineda, F.J., *et al.* (2013). Genome-wide consequences of deleting any single gene. Mol Cell *52*, 485–494.

Terstappen, G.C., Schlupen, C., Raggiaschi, R., and Gaviraghi, G. (2007). Target deconvolution strategies in drug discovery. Nat. Rev. Drug Discov. *6*, 891–903.

Traxler, P., Gruner, J., and Auden, J.A. (1977). Papulacandins, a new family of antibiotics with antifungal activity, I. Fermentation, isolation, chemical and biological characterization of papulacandins A, B, C, D and E. J. Antibiot. (Tokyo) *30*, 289–296.

Van Den Bossche, H., Willemsens, G., and Van Cutsem, J.M. (1975). The action of miconazole on the growth of *Candida albicans*. Sabouraudia *13* (Pt 1), 63–73.

Van Den Bossche, H., Willemsens, G., Cools, W., Lauwers, W.F., and Le Jeune, L. (1978). Biochemical effects of miconazole on fungi. II. Inhibition of ergosterol biosynthesis in *Candida albicans*. Chem. Biol. Interact. *21*, 59–78.

Van den Bossche, H., Willemsens, G., Cools, W., Cornelissen, F., Lauwers, W.F., and van Cutsem, J.M. (1980). *In vitro* and *in vivo* effects of the antimycotic drug ketoconazole on sterol synthesis. Antimicrob. Agents Chemother. *17*, 922–928.

Van den Bossche, H., Willemsens, G., Cools, W., Marichal, P., and Lauwers, W. (1983). Hypothesis on the molecular basis of the antifungal activity of N-substituted imidazoles and triazoles. Biochem. Soc. Trans. *11*, 665–667.

Vanden Bossche, H., Koymans, L., and Moereels, H. (1995). P450 inhibitors of use in medical treatment: focus on mechanisms of action. Pharmacol. Ther. *67*, 79–100.

Vandeputte, P., Tronchin, G., Berges, T., Hennequin, C., Chabasse, D., and Bouchara, J.P. (2007). Reduced susceptibility to polyenes associated with a missense mutation in the ERG6 gene in a clinical isolate of *Candida glabrata* with pseudohyphal growth. Antimicrob. Agents Chemother. *51*, 982–990.

Vandeputte, P., Tronchin, G., Larcher, G., Ernoult, E., Berges, T., Chabasse, D., and Bouchara, J.P. (2008). A nonsense mutation in the ERG6 gene leads to reduced susceptibility to polyenes in a clinical isolate of *Candida glabrata*. Antimicrob. Agents Chemother. *52*, 3701–3709.

Vandeputte, P., Ischer, F., Sanglard, D., and Coste, A.T. (2011). *In vivo* systematic analysis of *Candida albicans* Zn2-Cys6 transcription factors mutants for mice organ colonization. PLoS One *6*, e26962.

Vandeputte, P., Pradervand, S., Ischer, F., Coste, A.T., Ferrari, S., Harshman, K., and Sanglard, D. (2012). Identification and functional characterization of Rca1, a transcription factor involved in both antifungal susceptibility and host response in *Candida albicans*. Eukaryot. Cell *11*, 916–931.

Van Tuyl, J.M. (1977). Genetic aspects of resistance to imazalil in *Aspergillus nidulans*. Neth. J. Plant Pathol. *83*, 169–176.

Vermeulen, E., Lagrou, K., and Verweij, P.E. (2013). Azole resistance in *Aspergillus fumigatus*: a growing public health concern. Curr. Opin. Infect Dis. *26*, 493–500.

Vincent, B.M., Lancaster, A.K., Scherz-Shouval, R., Whitesell, L., and Lindquist, S. (2013). Fitness trade-offs restrict the evolution of resistance to amphotericin B. PLoS Biol. *11*, e1001692.

Warrilow, A.G., Parker, J.E., Kelly, D.E., and Kelly, S.L. (2013). Azole affinity of sterol 14alpha-demethylase (CYP51) enzymes from *Candida albicans* and *Homo sapiens*. Antimicrob. Agents Chemother. *57*, 1352–1360.

Weber, K., Wehland, J., and Herzog, W. (1976). Griseofulvin interacts with microtubules both *in vivo* and *in vitro*. J. Mol. Biol. *102*, 817–829.

Wehland, J., Herzog, W., and Weber, K. (1977). Interaction of griseofulvin with microtubules, microtubule protein and tubulin. J. Mol. Biol. *111*, 329–342.

Weissmann, G., and Sessa, G. (1967). The action of polyene antibiotics on phospholipid-cholesterol structures. J. Biol. Chem. *242*, 616–625.

Winzeler, E.A., Shoemaker, D.D., Astromoff, A., Liang, H., Anderson, K., Andre, B., Bangham, R., Benito, R., Boeke, J.D., Bussey, H., *et al.* (1999). Functional characterization of the *S. cerevisiae* genome by gene deletion and parallel analysis. Science *285*, 901–906.

Woolley, D.W. (1944). Some biological effects produced by benzimidazole and their reversal by purines. J. Biol. Chem. *152*, 225–232.

Xu, D., and Roemer, T. (2009). A whale of a library. Nat. Biotechnol. *27*, 342–344.

Xu, D., Jiang, B., Ketela, T., Lemieux, S., Veillette, K., Martel, N., Davison, J., Sillaots, S., Trosok, S., Bachewich, C., *et al.* (2007). Genome-wide fitness test and mechanism-of-action studies of inhibitory compounds in *Candida albicans*. PLoS Pathog. 3, e92.

Xu, D., Sillaots, S., Davison, J., Hu, W., Jiang, B., Kauffman, S., Martel, N., Ocampo, P., Oh, C., Trosok, S., *et al.* (2009a). Chemical genetic profiling and characterization of small-molecule compounds that affect the biosynthesis of unsaturated fatty acids in *Candida albicans*. J. Biol. Chem. *284*, 19754–19764.

Xu, Q., Schlabach, M.R., Hannon, G.J., and Elledge, S.J. (2009b). Design of 240,000 orthogonal 25mer DNA barcode probes. Proc. Natl. Acad. Sci. U.S.A. *106*, 2289–2294.

Yamaguchi, H., Hiratani, T., Baba, M., and Osumi, M. (1987). Effect of aculeacin A on reverting protoplasts of *Candida albicans*. Microbiol. Immunol. *31*, 625–638.

Yan, Z., Costanzo, M., Heisler, L.E., Paw, J., Kaper, F., Andrews, B.J., Boone, C., Giaever, G., and Nislow, C. (2008). Yeast barcoders: a chemogenomic application of a universal donor-strain collection carrying barcode identifiers. Nat. Methods *5*, 719–725.

Yan, Z., Berbenetz, N.M., Giaever, G., and Nislow, C. (2009). Precise gene-dose alleles for chemical genetics. Genetics *182*, 623–626.

Yoshida, Y., and Aoyama, Y. (1987). Interaction of azole antifungal agents with cytochrome P-45014DM purified from *Saccharomyces* cerevisiae microsomes. Biochem. Pharmacol. *36*, 229–235.

Yoshida, Y., and Aoyama, Y. (1991). Sterol 14 alpha-demethylase and its inhibition: structural considerations on interaction of azole antifungal agents with lanosterol 14 alpha-demethylase (P-450(14DM)) of yeast. Biochem. Soc. Trans. *19*, 778–782.

Zanella, F., Lorens, J.B., and Link, W. (2010). High content screening: seeing is believing. Trends Biotechnol. *28*, 237–245.

Zeidler, U., Bougnoux, M.E., Lupan, A., Helynck, O., Doyen, A., Garcia, Z., Sertour, N., Clavaud, C., Munier-Lehmann, H., Saveanu, C., *et al.* (2013). Synergy of the antibiotic colistin with echinocandin antifungals in *Candida* species. J. Antimicrob. Chemother. *68*, 1285–1296.

Zhu, H., and Snyder, M. (2003). Protein chip technology. Curr. Opin. Chem. Biol. *7*, 55–63.

Zhu, H., Bilgin, M., Bangham, R., Hall, D., Casamayor, A., Bertone, P., Lan, N., Jansen, R., Bidlingmaier, S., Houfek, T., *et al.* (2001). Global analysis of protein activities using proteome chips. Science *293*, 2101–2105.

Zomorodian, K., Uthman, U., Tarazooie, B., and Rezaie, S. (2007). The effect of griseofulvin on the gene regulation of beta-tubulin in the dermatophyte pathogen *Trichophyton rubrum*. J. Infect. Chemother. *13*, 373–379.

Zotchev, S.B. (2003). Polyene macrolide antibiotics and their applications in human therapy. Curr. Med. Chem. *10*, 211–223.

Genome Integrity: Mechanisms and Contribution to Antifungal Resistance

10

Raphaël Loll-Krippleber, Adeline Feri, Christophe d'Enfert and Mélanie Legrand

Abstract

Along with a growing population of immunocompromised patients, a steady increase in the number of invasive fungal infections has been observed over the last decades. Among human fungal pathogens, *Candida albicans* is one of the most prevalent and can cause life-threatening infections in humans. Although antifungal treatments are available and routinely used against fungal infections, one of the major issues in the treatment of these infections is the appearance of antifungal-resistant cells. Over the last decade, genome plasticity of fungal pathogens in general and *C. albicans* in particular has emerged to be an important factor that contributes to their ability to acquire resistance to antifungal drugs. In this chapter, we provide an overview of the most important genome rearrangements that have been shown to occur upon antifungal exposure and/or lead to antifungal resistance. We also review recent advances in another important aspect in the field of genome stability and antifungal resistance, which is the characterization of the molecular mechanisms that can drive or prevent genome rearrangements in fungal pathogens. Finally, we address the effect of such large-scale genome changes on the fitness of antifungal-resistant cells.

Introduction

Maintaining genome integrity is central to the survival of every living cell and the mechanisms ensuring it are similar in all organisms. Despite a sophisticated network of molecular mechanisms to maintain DNA integrity and stability, genome rearrangements are still observed and poorly tolerated. For instance, gross chromosomal rearrangements, such as aneuploidies or chromosomal translocations, are often associated with negative consequences, such as embryonic lethality, developmental defects or cancer in humans (Colnaghi *et al.*, 2011). In some instances, however, somatic genome rearrangements can be beneficial. This is the case of recombination-mediated gene rearrangements of the variable, diverse and joining gene segments (V(D)J recombination) involved in the generation of a highly diverse antibody repertoire (Bogue and Roth, 1996; Grawunder *et al.*, 1998).

Microbes seem more tolerant to genome changes than their pluricellular counterparts, as illustrated by many instances of genome rearrangements conferring a fitness advantage. From an evolutionary perspective, exposure of pluricellular organisms to abrupt environmental changes is minimal as compared to the frequent environmental challenges encountered by microbial pathogens. As a consequence, tolerance to genome rearrangements could constitute a strategy for microbial pathogens to generate genetic diversity and to rapidly adapt to a new environment (Hoffman and Parsons, 1991). Indeed, there have been several reports of functional DNA rearrangements in the DNA of viruses, parasites and bacteria that allow the 'pathogen' to express new variant surface proteins and to escape the defences of the infected host (Chen and Deng, 2009; Stockdale *et al.*, 2008; Wisniewski-Dye and Vial, 2008).

As a result of modern medicines, the number of immunocompromised patients has increased worldwide, favouring the incidence of fungal infections to rise sharply, to the point of becoming a significant cause of morbidity and mortality (Pfaller and Diekema, 2010). *Candida albicans, Cryptococcus neoformans, Aspergillus fumigatus* and *Candida glabrata* are the leading causes of opportunistic fungal infections amongst the immunocompromised. Electrophoretic karyotype analyses have highlighted the remarkable capacity of some of these organisms to rapidly undergo gross chromosomal rearrangements. In several instances, the observation of genome rearrangements has been correlated with the acquisition of new phenotypic traits such as the acquisition of antifungal resistance, both *in vivo* and *in vitro*. Hence, the purpose of this chapter is to provide an update on the genome rearrangements that alter antifungal drug susceptibility and to highlight the different molecular mechanisms that cells have evolved to prevent genome instability. We will also briefly discuss the impact of the acquisition of antifungal resistance on cell fitness.

Genomic alterations involved in the acquisition of antifungal resistance

Various forms of genetic alterations have been involved in the acquisition of antifungal resistance. Point mutations in genes encoding the targets of antifungals, drug efflux pumps and transcription factors controlling those genes are at the core of antifungal resistance. These mutations can lead to (i) decreased affinity of the target for the drug, (ii) increased dosage of the target, (iii) increased drug efflux, (iv) enzymatic modification and (v) modification of the membrane content that decreases access to the antifungal target, and are reviewed in the preceding chapters of this book. While these mutations provide the ground for antifungal resistance, it is increasingly evident that further genomic rearrangements contribute to increasing antifungal resistance, such as aneuploidies in haploid and diploid fungal pathogens and loss of heterozygosity (LOH) in diploid species.

Aneuploidy

Deviations from multiples of the haploid genome by either gain or loss of entire chromosomes or chromosome fragments are defined as aneuploidies. Aneuploidies are deleterious in humans and mice, resulting in embryonic lethality or heavy mental defects when viable at birth. Aneuploidy has also been associated with developmental defects and lethality in a variety of other organisms (Hodgkin *et al.*, 1979; Lindsley *et al.*, 1972; McClintock, 1929). In contrast, aneuploidies can be an advantage in fungi, providing a rapid ability of adaptation.

Segmental aneuploidies

Segmental aneuploidies are the result of truncated chromosomes or extra copies of a chromosome fragment. The latter category is often referred to as supernumerary chromosomes or isochromosomes. Interestingly, both truncated chromosomes and isochromosomes have been found in *C. albicans* isolates resistant to azole antifungals. *C. albicans* isochromosomes often derive from chromosome 5 (Ch5) and can form quickly after fluconazole exposure (Coste *et al.*, 2007; Selmecki *et al.*, 2006, 2008, 2009). Formation of an isochromosome i(5L) (two left arms of Ch5 separated by a copy of the Ch5 centromere) was shown to occur in 19% of the isolates evolved *in vitro* in the presence of fluconazole (Selmecki *et al.*, 2009) (Fig. 10.1). i(5L) can be present in one or multiple copies. Other types of isochromosomes have been described in azole-resistant *C. albicans* isolates, such as i(5L) fused to a fragment of chromosome 3 (Ch3) right arm (named i(5L)+3R) or i(5L) attached to an intact Ch5 (named att-i(5L)) (Fig. 10.1). It has been suggested that the formation of i(5L)+3R was the result of a non-reciprocal recombination event (Selmecki *et al.*, 2009). Alternatively, the formation of att-i(5L) has been described as the result of telomeric fusion between the left arm of Ch5 and i(5L) (Fig. 10.1). Interestingly, the dynamics of formation and loss of att-i(5L) was shown to be rapid, suggesting that the acquisition of an isochromosome can be a transient event for rapid adaptation to antifungal pressure (Selmecki *et al.*, 2009).

In addition to isochromosomes, truncated chromosomes have also been observed in azole-resistant *C. albicans* isolates. Selmecki *et al.* (2008) detected a truncation on Ch5 in

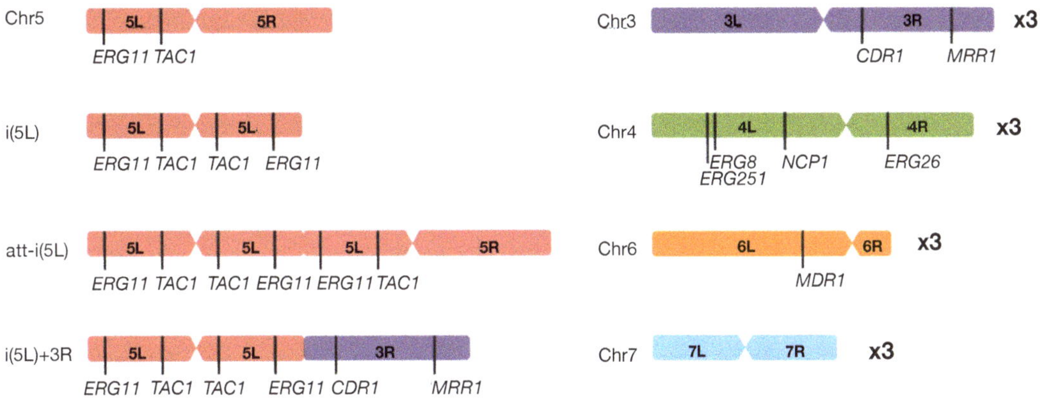

Figure 10.1 Different types of aneuploidies have been described in azole-resistant *C. albicans* isolates. Isochromosomes often derive from chromosome 5 (Ch5), a chromosome that carries important genes involved in azole resistance such as *TAC1* and *ERG11.* Isochromosome i(5)L consists of two left arms of Ch5 separated by Ch5 centromeric region (*CEN5*). Isochromosome att-i(5)L is a multicentric chromosome and consists of two Ch5 fused together via their telomeric region. Isochromosome i(5)L+3R is composed of an isochromosome i(5)L attached to the right arm of chromosome 3 (Ch3). Ch3 also contains important genes involved in azole resistance such as *CDR1* and *MRR1.* Whole-chromosome aneuploidies (mostly trisomies) have also been described for Ch3, Ch4, Ch6 and Ch7. Similarly to Ch5, these chromosomes carry genes implicated in antifungal resistance, except Ch7 that does not seem to contain any azole resistance gene.

a fluconazole-resistant clinical isolate (named FH8). This chromosome truncation implied the loss of a 200 kb region located on the left arm of Ch5 and occurred at the *GIT* locus carrying three genes (*GIT1*, *GIT2* and *GIT4*) that display almost 75% sequence identity. Repetitive sequences have been shown to form fragile sites where DNA breakage may occur (Lemoine *et al.*, 2005). The FH8 clinical isolate was also shown to carry i(5L). Interestingly, fluconazole resistance was greater in another isolate carrying only i(5L) but no truncation on Ch5, suggesting that deletion of one or more genes due to the truncation resulted in a decrease in fluconazole resistance.

It is notable that *C. albicans* Ch5 left arm carries genes implicated in azole resistance, namely *TAC1* and *ERG11* (Fig. 10.1), as described in Chapter 1. Briefly, *TAC1* encodes a transcription activator for the *CDR1* and *CDR2* Candida Drug Resistance genes that encode ATP-binding cassette (ABC) transporters (Coste *et al.*, 2004). ABC transporters are responsible for antifungal efflux and the increased expression of the corresponding genes has been shown to contribute to antifungal resistance. *ERG11* encodes a lanosterol demethylase, an enzyme from the P450 cytochrome family involved in ergosterol synthesis and which is the target of azole antifungals (Kontoyiannis and Lewis, 2002). Thus, formation of isochromosomes deriving from Ch5 leads to an increase in *TAC1* and *ERG11* copy number and consequently overproduction of antifungal efflux pumps or the target of azoles, allowing cells to become more tolerant to the presence of antifungals. In the case of i(5L)+3R formation, the *MRR1* copy number is also increased since this gene is located on the right arm of Ch3 attached to i(5L) (Fig. 10.1). *MRR1* encodes a transcription activator of *MDR1* whose gene product is a major facilitator efflux pump and whose overexpression has been shown to increase fluconazole resistance (Dunkel *et al.*, 2008a).

Segmental aneuploidies have also been observed in other pathogenic fungi. For instance, Comparative Genome Hybridization (CGH) analysis revealed that chromosomes 1, 2 and 13 can have elevated copy numbers for specific chromosomal segments in *Cryptococcus gattii* cells that have acquired drug resistance (D'Souza *et al.*, 2011).

There have also been reports of generation of novel small chromosomes as a virulence mechanism in *C. glabrata* (Ahmad *et al.*, 2013; Polakova *et al.*, 2009). These small chromosomes can originate through a segmental duplication encompassing the centromeric region. In several instances, these extra minichromosomes have been shown to carry transporters of the ATP-binding cassette family and strongly correlate with increased azole resistance in the patients. Indeed, when grown in absence of antifungals, these minichromosomes tend to be lost, resulting in a decrease in drug resistance.

Whole-chromosome aneuploidies

Acquisition of whole-chromosome aneuploidies in *C. albicans* has been well documented, in particular in cells exposed to various kinds of stress and growth conditions. For instance, Selmecki *et al.* (2009) described whole-chromosome aneuploidies of *C. albicans* chromosomes 3, 4, 6 and 7, representing up to 20% of extra-genetic material, in fluconazole-resistant strains. Interestingly, aneuploidies on these chromosomes result in copy number increase of some antifungal resistance-related genes such as *CDR1* and *MRR1* (Ch3) or *MDR1* (Ch6). Ch4 carries *NCP1* that encodes a NADPH-cytochrome P450 reductase, a cofactor of Erg11p, and other genes involved in ergosterol biosynthesis such as *ERG8*, *ERG26* and *ERG251* (Fig. 10.1). Ch7 does not carry any known genes implicated in antifungal resistance, but it has been suggested that Ch7 trisomy may improve the ability of *C. albicans* cells to tolerate gene imbalance due to other chromosome alterations and may provide a selective advantage under specific stress conditions (Selmecki *et al.*, 2009). Indeed, Ch7 trisomy has been described in some clinical isolates that had undergone homozygosity at the mating-type-like (*MTL*) locus on Ch5 (Legrand *et al.*, 2004). Generation of aneuploid chromosomes in *C. albicans* has also been shown to allow cells to metabolize specific carbon sources such as sorbose or arabinose (Janbon *et al.*, 1998; Rustchenko *et al.*, 1994).

Acquisition of azole-resistance associated with whole-chromosome aneuploidies has also been described in the haploid fungi *C. glabrata* and *Cryptococcus species.* In *C. glabrata*, molecular characterization of an azole-resistant isolate revealed duplication of the entire chromosome E. Here again, ChE carries the gene coding for Erg11p, the azole target (Marichal *et al.*, 1997). Whole-chromosome aneuploidies have also been described in *C. neoformans* and *C. gattii*, and associated with increased resistance to fluconazole (D'Souza *et al.*, 2011; Sionov *et al.*, 2010). This is the case for *C. neoformans* Ch1 that carries *ERG11* and *AFR1*, a gene that encodes the major transporter of azoles in *C. neoformans.* Interestingly, acquisition of other disomies in addition to Ch1 has been shown to influence the level of drug resistance (Sionov *et al.*, 2010) and the virulence of *C. neoformans* (Hu *et al.*, 2011).

Whole-genome ploidy changes have been described in *C. albicans* both *in vivo* and *in vitro.* Indeed, *C. albicans* cells of opposite mating type can mate, generating tetraploid cells (Hull *et al.*, 2000; Magee and Magee, 2000). Very recently, Hickman *et al.* isolated haploid cells from patients and from specific *in vitro* culture conditions including growth in presence of fluconazole (Hickman *et al.*, 2013). However, if fungus virulence is modified, there is no evidence for changes in antifungal susceptibility between the different ploidy phases.

In *C. neoformans*, formation of titan cells has been described. They are 5 to 10 times larger than the normal yeast cells of *C. neoformans* and are polyploid (tetraploid or octoploid). Titan cells display an increased resistance to oxidative and nitrosative stresses (Shapiro *et al.*, 2011) but, to our knowledge, change in antifungal sensitivity has not been reported. On the other hand, there has been a report of heteroresistance to fluconazole in isolates of the closely related species *C. gattii* associated with changes in ploidy (D'Souza *et al.*, 2011). In one of those isolates whose baseline ploidy appears to be diploid, Ch2, 9 and 10 have copy number above 2N while Ch14 is present in a single copy. Segmental aneuploidies on Ch1 and 13 are also present in this strain (D'Souza *et al.*, 2011). This isolate illustrates the sophisticated genotypes that can be obtained as a consequence of multiple genome alterations and resulting in a high level of genetic diversity within a population.

Based on these studies, the appearance of aneuploidy seems to be an efficient strategy adopted by pathogenic fungi to rapidly modify gene expression in response to environmental stresses.

Loss of heterozygosity

Loss of heterozygosity (LOH) results from chromosome missegregation or DNA double-strand breaks (DSBs) in diploid organisms. DNA DSBs can be repaired via recombination-dependent mechanisms such as gene conversion, break-induced replication and mitotic crossover. Inability to repair correctly a DSB results in the loss or truncation of the damaged chromosome or in lethality.

Generally, LOH has a positive effect on the adaptation speed in asexual diploid populations (Mandegar and Otto, 2007). Indeed, LOH allows to pass from a heterozygous situation in which one allele having acquired a beneficial mutation is competing with the wild-type allele, to a homozygous situation in which only the mutated allele is expressed. In *C. albicans*, LOH events are frequent and have been observed during *C. albicans* commensal carriage, superficial infection in humans and systemic infection in mice. Various stresses mimicking the conditions encountered by *C. albicans* within the host during colonization and infection, such as heat shock, oxidative stress and presence of antifungals, increase the LOH frequency (Bouchonville *et al.*, 2009; Diogo *et al.*, 2009; Forche *et al.*, 2005, 2009, 2011). Importantly, LOH has been shown to play an important role in the acquisition of antifungal resistance and to occur via Break-induced replication (BIR)/Mitotic recombination (MR), gene conversion (GC) or chromosome loss and reduplication (Coste *et al.*, 2007).

A first example of LOH-mediated antifungal resistance refers to the control of drug export from the cells by the ABC transporters Cdr1 and Cdr2 (Coste *et al.*, 2004, 2006, 2007). Mutated versions of *TAC1*, also known as hyperactive alleles, are associated with increased *CDR1* and *CDR2* expression and fluconazole resistance. Interestingly, hyperactive *TAC1* alleles are semi-dominant, allowing only a slight increase in *CDR1* and *CDR2* expression and azole resistance when co-expressed with wild-type *TAC1* allele (Coste *et al.*, 2006). Resistance to high concentrations of fluconazole (>128 μg/ml) is observed when hyperactive alleles are present in a homozygous state, as a result of LOH. *TAC1* location on Ch5, the chromosome that also carries the *MTL* locus, probably explains the strong correlation observed between *MTL* homozygosity and azole resistance (Rustad *et al.*, 2002).

The *ERG11* gene encoding the azole target lanosterol demethylase is similarly located on Ch5, and numerous mutations in this gene, conferring resistance to azoles, have been described (reviewed in (Morio *et al.*, 2010)). Notably, Coste *et al.* (2007) could show that increased azoles resistance was accompanied by LOH events resulting in homozygosity of *ERG11* hyperactive alleles and that these LOH events could also involve hyperactive alleles at the *TAC1* locus. In this respect, Morio *et al.* (2010) found that *ERG11* hyperactive alleles are present in a homozygous state in 77% of the cases (88/123 isolates) suggesting that azole resistance due to hyperactive *ERG11* alleles is often associated with LOH events.

In *C. albicans*, the transcription factor Upc2 has been shown to control *ERG11* up-regulation in response to azoles. Here again, gain-of-function mutations in *UPC2* followed by LOH have been observed (Dunkel *et al.*, 2008b; Heilmann *et al.*, 2010; Hoot *et al.*, 2011a). Antifungals can also be exported from the cells by major facilitator transporters. A mechanism involving the acquisition of mutations in *MRR1*, the transcriptional activator of the major facilitator transporter *MDR1*, followed by a LOH has been described and allows greater tolerance to fluconazole (Morschhauser *et al.*, 2007).

Homozygosity has also been observed for genes involved in the resistance to other classes of antifungal agents. Ergosterol is the target of polyene antifungals such as amphotericin B. Missense mutations in *C. albicans ERG3*, involved in ergosterol biosynthesis, result in decreased amount of ergosterol within the cell wall and thus decrease cell susceptibility to amphotericin B (Morio *et al.*, 2012). Mutations in *ERG3* in *C. albicans* have been found in homozygous state. Echinocandin resistance has been linked to mutations in the *FKS1* gene, encoding a β-1-3-glucan synthase subunit resistant to echinocandins, and these are found in a homozygous state (Niimi *et al.*, 2010; Park *et al.*, 2005).

In conclusion, these studies clearly demonstrated that LOH contributes to elevate drug resistance in *C. albicans* by increasing the copy number of mutant alleles of drug resistance genes.

Other genomic rearrangements

In *A. fumigatus*, decreased azoles susceptibility has always been linked to point mutations in the *cyp51A* gene, encoding the target of azole antifungals (Diaz-Guerra *et al.*, 2003). Sequence analysis of *cyp51A* in clinical isolates of *A. fumigatus* exhibiting cross-resistance to azoles has uncovered a new resistance mechanism, involving the presence of a non-synonymous substitution in the coding sequence, in combination with the duplication in tandem of a 34 bp sequence in the promoter region (Mellado *et al.*, 2007). Interestingly, the authors showed that each alteration was not sufficient to confer azole resistance on its own and that combination of both modifications is needed for the azole-resistant phenotype.

Also, the new minichromosomes presented in the section 'Segmental aneuploidies' can also originate from translocation of a centromere-less arm to another chromosome in *C. glabrata*. Unlike the new chromosomes resulting from segmental duplication, these extra chromosomes tend to be more stable since the genes on the extra chromosome are present in only one copy in the genome (Ahmad *et al.*, 2013).

Molecular mechanisms that suppress genome instability

As illustrated in the section 'Genomic alterations involved in the acquisition of antifungal resistance', various genome changes have been presented as major contributors to antifungal resistance. Research efforts are now being made to study genome dynamics in the context of drug resistance acquisition. Here, we describe the molecular mechanisms that cells have evolved to maintain genome integrity and that might be at fault in cells that have acquired drug resistance, namely checkpoint and DNA repair pathways.

C. albicans genes with a known and characterized function in genome integrity maintenance are listed in Table 10.1.

Cell cycle checkpoints

Cell cycle checkpoints or CCCPs are essential for the maintenance of genome integrity as they monitor DNA integrity along the cell cycle and ensure that critical stages of the cell cycle are accurately completed before progression into the following stage. Upon CCCPs activation, signalling pathways delay cell cycle progression to allow recovery of DNA integrity before reentering the cell cycle. In addition to their action on the cell cycle, CCCPs directly activate repair mechanisms, regulate the recruitment of DNA repair proteins to the site of damage and activate cell death programmes if the DNA lesion is not reparable (Finn *et al.*, 2012; Zhou and Elledge, 2000).

CCCPs have been well studied in the non-pathogenic fungi *S. cerevisiae* and *A. nidulans*. By contrast, little is known about cell cycle checkpoints in pathogenic fungi. We will briefly present the work that has been done in *C. albicans* to better understand the role of CCCPs proteins in genome stability and antifungal resistance.

Overall organisation

The best-characterized CCCPs pathways are the DNA damage and DNA replication checkpoints, as well as the spindle-assembly checkpoint, each involved in a different step of the cell cycle (Finn *et al.*, 2012).

The DNA damage checkpoint (DDCP) recognizes and responds to DNA damage. The DDCP can occur (i) at the G1/S transition thus stopping the cell cycle before entry into S phase by regulating cell budding and spindle pole body duplication, allowing cells to repair damaged DNA before replication, (ii) during the S phase (the intra-S phase checkpoint), thus slowing replication when damaged DNA is present and allowing coordination between repair mechanisms and the cell cycle, ensuring that replication is complete before mitosis or (iii) at the G2/M transition thus stopping the cell cycle prior to entry into anaphase (Finn *et al.*, 2012). The DNA replication checkpoint monitors the fidelity of copying DNA and prevents progression into mitosis when DNA replication is incomplete. Finally, during mitosis, the spindle-assembly checkpoint prevents anaphase onset until all chromosomes are properly attached to the spindle.

Table 10.1 *C. albicans* genes involved in genome integrity maintenance

Gene in *C. albicans*	Function	Genetics approach for gene function characterization	Mutant phenotype	Reference
MEC1	DNA damage checkpoint	Knock-out	Growth defect	Legrand *et al.* (2011)
			Sensitivity to genotoxic stresses (i.e. UV radiations (slight sensitivity), hydroxyurea (HU), ethyl methanesulfonate (EMS) and methyl methanesulfonate (MMS))	
			Increased genome instability (increase in BIR-mediated LOH frequency)	
			Impairment in genotoxic stress-induced polarized growth	
RAD9	DNA damage checkpoint	Knock-out	Sensitivity to genotoxic stresses (i.e. MMS)	Shi *et al.* (2007)
			Defect in cell cycle arrest upon MMS treatment	
			Defect in genotoxic stress (i.e. UV and MMS)-induced polarized growth	
MRC1	DNA damage checkpoint	Knock-out	Sensitivity to genotoxic stresses (i.e. HU and MMS)	Shi *et al.* (2007)
			Elongation and cell separation defects	
			Slow DNA synthesis	
			Alteration in cell morphology (filament formation)	
RAD53	DNA damage checkpoint	Knock-out	Growth defect	Loll-Krippleber *et al.* (2013); Shi *et al.* (2007)
			Sensitivity to genotoxic stresses (i.e. UV radiations, HU and MMS)	
			Impairment in genotoxic stress-induced polarized growth	
			Defect in genotoxic stress-induced cell cycle arrest	
			Increased genome instability (increase in LOH frequency by BIR and chromosome loss; frequent aneuploidies)	
		Overexpression	Growth defect	R.L.-K., A.F., C.d'E. and M.L. (unpublished)
			Genome instability (increase in LOH frequency)	
			Alteration in cell morphology (filament formation)	
DUN1	DNA damage checkpoint	Knock-out	Slight growth defect	Loll-Krippleber *et al.* (2013)
			Moderate sensitivity to genotoxic stresses (i.e. UV radiations, HU and MMS)	
			Alteration in cell morphology (filament formation)	
CDC20	DNA damage checkpoint	Knock-out	Cell cycle defect	Chou *et al.* (2011)
		Overexpression	Genome instability (increase in LOH frequency)	R.L.-K., A.F., C.d'E. and M.L. (unpublished)

Table 10.1 continued

Gene in *C. albicans*	Function	Genetics approach for gene function characterization	Mutant phenotype	Reference
PPH3/ PSY2	DNA damage checkpoint	Knock-out	Sensitivity to genotoxic stresses (i.e. UV radiations, MMS and cisplatin)	Sun *et al.* (2011)
			Defect in cell cycle regulation in presence of genotoxic stress (i.e. MMS)	
			Formation of longer filaments upon genotoxic stress (i.e. MMS) as compared to wild-type	
RNR2	Ribonucleotide synthesis	Knock-out	Sensitive to HU and amphidicolin, dramatic cell elongation	Shi *et al.* (2007)
SGS1	Homologous recombination/ DNA damage checkpoint	Knock-out	Growth defect	Legrand *et al.* (2011)
			High sensitivity to oxidative stress (i.e. tert-butyl hydroperoxide (TBHP))	
			High sensitivity to genotoxic stresses (i.e. UV radiations, camptothecin, HU, EMS and MMS)	
			Alteration in cell morphology (filament formation)	
MRE11–RAD50	Homologous recombination	Knock-out	Growth defect	Legrand *et al.* (2007)
			Sensitivity to oxidative stress (i.e. hydrogen peroxide (H_2O_2) and TBHP	
			Sensitivity to genotoxic stresses (i.e. UV radiations, camptothecin, EMS and MMS)	
			Increased genome instability (increase in BIR-mediated LOH frequency)	
			Alteration in cell morphology (filament formation)	
RAD51	Homologous recombination	Knock-out	Growth defect	Garcia-Prieto *et al.* (2010)
			Moderate sensitivity to oxidative stress (i.e. TBHP)	
			Moderate sensitivity to genotoxic stresses (i.e. MMS, bleomycin)	
			Alteration in cell morphology (filament formation)	
		Overexpression	Growth defect	R.L.-K., A.F., C.d'E. and M.L. (unpublished)
			Genome instability (increase in LOH frequency)	
			Alteration in cell morphology (filament formation)	

Table 10.1 continued

Gene in *C. albicans*	Function	Genetics approach for gene function characterization	Mutant phenotype	Reference
RAD52	Homologous recombination	Knock-out	Growth defect Sensitivity to oxidative stress (i.e. H_2O_2 and TBHP) Sensitivity to genotoxic stresses (i.e. UV radiations, camptothecin, EMS and MMS) Increased genome instability (increase in LOH frequency by chromosome loss and truncation; frequent aneuploidies) Alteration in cell morphology (filament formation) Decreased virulence lig4/rad52 double mutant avirulent in a murine model of hematogenously disseminated candidiasis	Andaluz *et al*. (2011); Chauhan *et al*. (2005); Ciudad *et al*. (2004); Legrand *et al*. (2007); Takagi *et al*. (2008)
RAD54	Homologous recombination	Knock-out	Growth defect High sensitivity to genotoxic stress (i.e. MMS) Moderate sensitivity to oxidative stress (i.e. menadione) Alteration in cell morphology (filament formation)	Hoot *et al*. (2011b)
RDH54	Homologous recombination	Knock-out	No growth defect No sensitivity to oxidative and genotoxic stresses	Hoot *et al*. (2011)
RAD59	Homologous recombination	Knock-out	High sensitivity to genotoxic stresses (i.e. UV radiations)	Garcia-Prieto *et al*. (2010)
YKU70	Non-homologous end joining	Knock-out	Sensitivity to genotoxic stresses (i.e. MMS) Increased telomere length	Chico *et al*. (2011)
YKU80	Non-homologous end joining	Knock-out	No phenotype (no growth defect, no sensitivity to DNA damaging agents, no LOH increase and no increase in antifungal drug resistance)	Legrand *et al*. (2007)
LIG4	Non-homologous end joining	Knock-out	Attenuated virulence in a murine model of systemic candidiasis Defect in filamentation	Andaluz *et al*. (2001, 2002)
RTT109	Non-homologous end joining	Knock-out	Growth defect Sensitivity to oxidative and genotoxic stresses (i.e. H_2O_2, MMS and camptothecin) Attenuated virulence in a murine systemic candidiasis model Alteration in cell morphology (filament formation) Constitutive expression of DNA repair gene Defective in white-to-opaque switching	Lopes da Rosa *et al*. (2010); Stevenson and Liu (2011); Wurtele *et al*. (2010)

Table 10.1 continued

Gene in *C. albicans*	Function	Genetics approach for gene function characterization	Mutant phenotype	Reference
MSH2	Mismatch repair	Knock-out	Increased in frequency of appearance of fluconazole-resistant colonies	Legrand *et al.* (2007)
PMS1				
RAD2	Nucleotide excision repair	Knock-out	Highly sensitive to UV radiations	Legrand *et al.* (2008)
RAD10	Nucleotide excision repair			
RAD3	Nucleotide excision repair	Protein purification and biochemical characterization	Possess a helicase activity similar to *S. cerevisiae* Rad53 Complement the defects in both nucleotide excision pathway and transcription of *S. cerevisiae* rad3ΔΔ mutant	Seong *et al.* (2011)
APN1	Base excision repair	Knock-out	No phenotype (no growth defect, no sensitivity to DNA damaging agents, no LOH increase and no increase in antifungal drug resistance)	Legrand *et al.* (2008)
NTG1	Base excision repair			
OGG1	Base excision repair			
BIM1	Chromosome segregation	Overexpression	Growth defect and genome instability (increase in LOH frequency)	R.L.-K., A.F., C.d'E. and M.L. (unpublished)

Organization of the different CCCPs in *S. cerevisiae* is very similar and requires the contribution of sensors, mediators, and effectors (Zhou and Elledge, 2000). Sensors detect abnormal structures in the DNA and convey this information via mediators to effectors. In turn, checkpoint effector kinases amplify and transduce the signal to a large panel of target proteins that act directly on cell cycle and DNA repair mechanisms. Signal transduction mainly occurs through phosphorylation/dephosphorylation of CCCP proteins. It is important to note that the roles of each actor in the spindle-assembly checkpoint are not always as precisely defined as they are in the DNA damage and replication checkpoints. Fig. 10.2 shows the major players involved in cell cycle checkpoints in *S. cerevisiae* and *C. albicans*.

DNA damage and replication checkpoints

Sensors

Activation of the DNA damage and replication checkpoints is mainly triggered by proteins of the PIKK (phosphoinositide 3-kinase-related kinase) family. These kinases have been conserved during evolution and orthologues are found in most eukaryotic species (ATR and ATM in humans and Mec1 and Tel1 in *S. cerevisiae*). ATM/Tel1 is involved in telomere maintenance and mainly responds to DSBs, while ATR/Mec1 is activated in response to replicative stress and is especially recruited via the RPA protein that protects the single-stranded DNA exposed when a replication fork is stalled. They both function as central regulators of DNA damage response as they modulate cell cycle progression by phosphorylating multiple substrates.

In contrast with the essential role of *MEC1* in several organisms (*S. cerevisiae, Caenorhabditis elegans* and *Drosophila melanogaster*) (reviewed in (Schultz *et al.*, 2000)), *C. albicans MEC1* and human ATR are not essential for viability, indicating that *MEC1*-dependent signalling in *C. albicans* may be closer to that in humans than in other model organisms (Legrand *et al.*, 2011; Schultz *et al.*, 2000). In *S. cerevisiae, MEC1* deletion results in lethality while hypomorphic mutants display

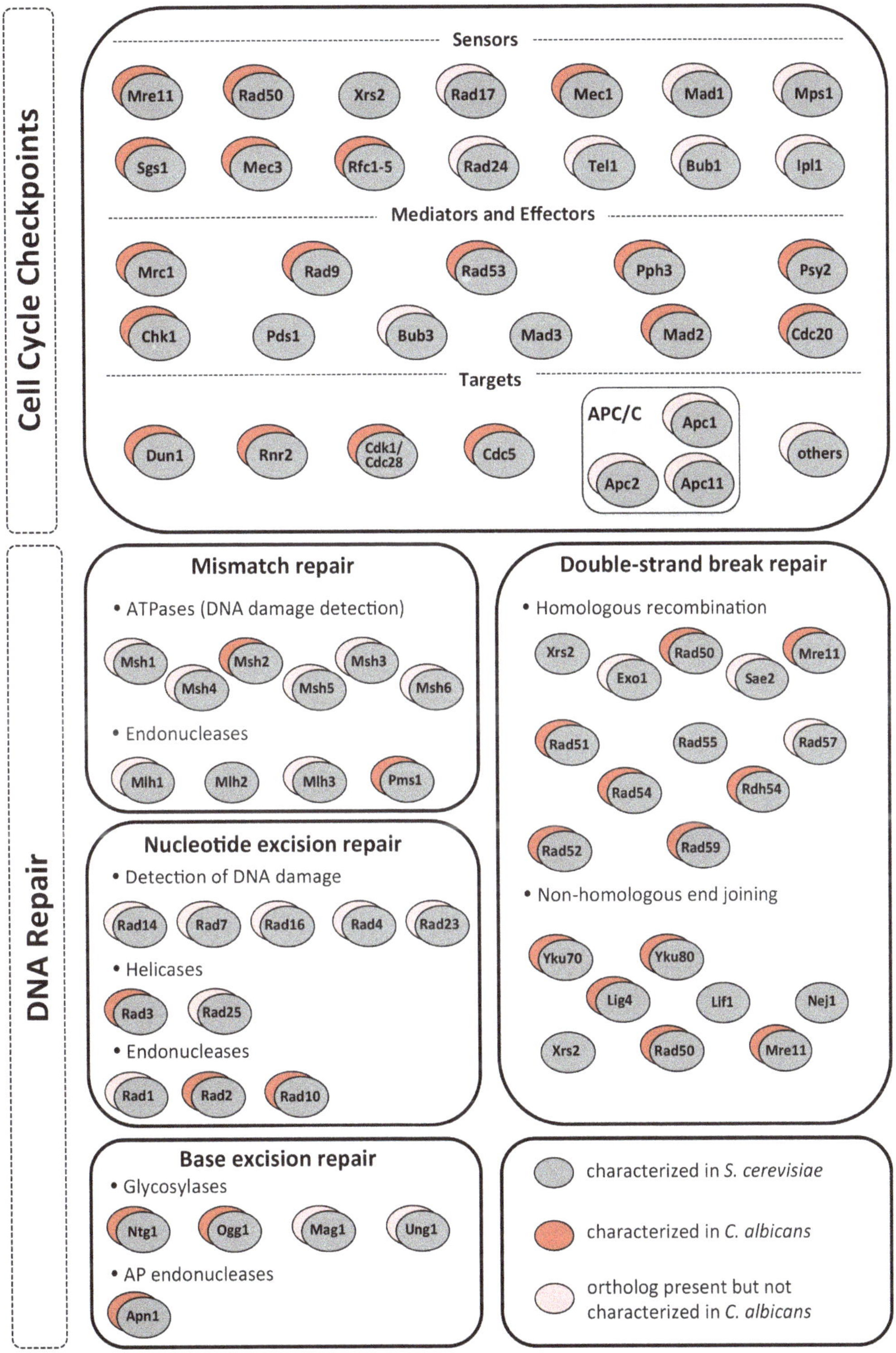

Figure 10.2 Major players of the genome stability maintenance network in *S. cerevisiae* and *C. albicans*. Most of these proteins have been identified and characterized using the budding yeast *S. cerevisiae* as a model organism and are depicted as a grey circle. Some of these proteins have orthologues in *C. albicans* that have been studied (dark red circle) or not (light red circle).

increased sensitivity to DNA-damaging agents and genome instability. Lethality can be rescued by increasing cellular deoxyribonucleotide (dNTP) levels. This can be achieved by overexpression of *RNR3*, which encodes a subunit of the ribonucleotide reductase (RNR) or inactivation of *SML1*, which encodes an inhibitor of RNR. A putative orthologue of *SML1* in *C. albicans*, namely *ORF19.5642*, has been identified but its sequence similarity with ScSml1p is quite low (e-value 1.1×10^{-5}) (Legrand *et al.*, 2011). It could therefore be hypothesized that there is no functional orthologue of *SML1* in *C. albicans*, which could explain why *MEC1* is not essential in this organism. Similar to hypomorphic *Scmec1* mutants, *C. albicans mec1* null mutants are highly sensitive to genotoxic stresses (Legrand *et al.*, 2011). In addition, LOH monitoring at the *GAL1* locus revealed an increase in genomic instability in these mutants (Legrand *et al.*, 2011), suggesting that the *C. albicans* Mec1 protein is likely to have the same function as the one observed in *S. cerevisiae*. Finally, the *Camec1* null mutants do not appear to have any alteration in their overall sensitivity to fluconazole and no increase in the appearance of drug-resistant colonies was observed in the *mec1* mutants. These data suggest either that the genome changes leading to acquisition of drug resistance might still be properly repaired in *mec1* null mutants or that they could be too detrimental for survival and trigger cell death. An orthologue of *TEL1* is present in the genome of *C. albicans* but has not been studied yet.

The MRX complex (Mre11/Rad50/Xrs2) is another important sensor in the DNA damage and replication checkpoints as it binds DSBs and recruits Tel1 (Finn *et al.*, 2012). Interestingly, the MRX complex is also involved in the initial processing of double-strand DNA breaks prior to repair by Non-Homologous End Joining (NHEJ) or Homologous Recombination (HR), particularly by stabilizing the broken site, tethering broken ends prior to repair. In *C. albicans*, *RAD50* and *MRE11* genes were identified and their contribution to repair mechanisms was evaluated (Legrand *et al.*, 2007). *rad50* and *mre11* deletion mutants were shown to be particularly sensitive to DNA-damaging agents, and to display an increased LOH frequency at the *GAL1* locus (Legrand *et al.*, 2007). Interestingly, *rad50* null mutations gave rise to fluconazole-resistant isolates at a much higher frequency than the parental strain (Legrand *et al.*, 2007), suggesting that the loss of this protein results in a higher frequency of genome rearrangements that favour acquisition of antifungal resistance.

Mec1 and Tel1 kinases are considered as central actors of the DNA damage and replication checkpoints because they have the ability to bind the site of DNA damage and activate the checkpoints. However, it is important to note that these kinases work together with other factors. Most of the orthologues of these proteins are present in *C. albicans*, but their functions in genome instability-acquired drug resistance have not been characterized yet. Among them are Mec3 and the Rfc1-5 proteins. Mec3 is a subunit of a PCNA-like ring that functions as a clamp involved in DNA damage checkpoint control. Similarly, the Rfc proteins are part of the RFC–Rad24 clamp loader complex that plays a role during the early stages of the checkpoint control response. Interestingly, transcriptomic analysis has revealed differential expression of these genes in the context of antifungal treatments. *MEC3* has been shown to be up-regulated upon treatment with the fungicide benomyl, a drug that also induces up-regulation of the major facilitator gene *MDR1* (Karababa *et al.*, 2004). *RFC4* has been shown to be up-regulated in response to 5-FC treatment, and, along with *RFC5*, to be down-regulated in fluconazole-resistant isolates (Liu *et al.*, 2005; Yan *et al.*, 2008). These data suggest a potential role of these checkpoints proteins in the response to antifungals.

Additional sensor proteins can bind DNA lesions and activate the S phase DNA checkpoint, either in response to a replicative stress (replication checkpoint) or damaged DNA (intra-S phase checkpoint). The intra-S phase checkpoint is characterized by the occurrence of two partially redundant pathways that differ by the set of sensor proteins detecting the stress or lesion: a *RAD17* and *RAD24*-dependent pathway and a *SGS1*-dependent pathway (Kolodner *et al.*, 2002). Sgs1p, the orthologue of BLM, the Bloom's syndrome gene, is a RecQ-related DNA helicase;

cells from BLM patients are characterized by an increase in genome instability. *C. albicans* strains bearing deletions in *SGS1* are viable but slow-growing, producing wrinkled colonies with pseudohyphal-like cells. The *sgs1* null mutants exhibit a high sensitivity to all DNA-damaging agents tested. In an assay monitoring chromosome 1 integrity, no change was observed in the *sgs1* deletion mutants as compared to a wild-type strain, possibly because of the redundant branch depending on Rad17/Rad24 or because genomic rearrangements are lethal in these cells. Finally, loss of Sgs1 leads to an increased susceptibility to fluconazole but no increase in the level of antifungal drug resistance acquisition (Legrand *et al.*, 2011). Although orthologues are present in the *C. albicans* genome, the *RAD17/RAD24* branch has not been characterized.

Mediators and effectors

After Mec1 or Tel1 recruitment at the injury site and the formation of protein complexes required for the activation of these two kinases, mediators Rad9 and Mrc1 are required to bring Mec1 and the effector protein Rad53 together, allowing Rad53 activation by Mec1. In *C. albicans*, orthologues of *RAD9*, *MRC1* and *RAD53* are present and their roles in CCCP seem conserved (Loll-Krippleber *et al.*, 2013; Shi *et al.*, 2007). An additional role for these proteins in polarized growth in response to genotoxic stress has also been demonstrated in *C. albicans* (Loll-Krippleber *et al.*, 2013; Shi *et al.*, 2007). It should be noted that this response is independent of classical filamentation pathways involving cAMP/PKA and MAP kinases, and instead requires cell cycle regulators such as Cln3, Clb4 and Clb2 cyclins (Bachewich and Whiteway, 2005; Bensen *et al.*, 2005). Loll-Krippeleber *et al.* (2013) also conducted a site-directed mutagenesis study on Rad53 phospho-sites and demonstrated that phosphorylation of specific Rad53 sites were required for maintaining genome integrity and for other various cellular processes such as cell cycle regulation, genotoxic stress resistance and genotoxic stress-induced polarized growth (Loll-Krippleber *et al.*, 2013). Upon DNA damage recovery, CaRad53 dephosphorylation is regulated by the phosphatase Pph3 and its regulatory subunit Psy2 (Sun *et al.*, 2011).

Targets

The recent development of high-definition methods for quantitative phosphoproteomics has resulted in the identification of numerous targets of Mec1, Tel1 and Rad53 kinases (Albuquerque *et al.*, 2008; Chen *et al.*, 2010; Smolka *et al.*, 2005; Smolka *et al.*, 2007). The significant accumulation of data about these phosphorylation networks allows a better understanding of the DNA damage and replication checkpoints organization. However, the regulatory mechanisms between checkpoint kinases and their targets, and their impact on the cell physiology remain in many cases to be elucidated. The development of phosphoproteomics approaches to identify targets of kinases involved in DNA-damage response is still pending in pathogenic fungi.

S phase

An important target of Rad53 during S phase is the Dun1 protein (Finn *et al.*, 2012). The functioning of the Rad53/Dun1 couple has been well studied in *S. cerevisiae*, where Dun1 binds the phosphorylated form of Rad53 via its FHA domain (Chen *et al.*, 2007). Dun1 regulates the transcription of many genes induced during the CCCP and especially genes coding for ribonucleotide reductase (RNR) subunits, which is involved in dNTPs synthesis. In *C. albicans*, deletion of *DUN1* resulted in an increase in genome instability and sensitivity to methyl methanesulfonate (MMS) and hydroxyurea (HU), identifying *CaDUN1* as part of the DNA damage response. Unlike Rad53p, Dun1p is not required for genotoxic-stress induced polarized growth (Loll-Krippleber *et al.*, 2013), indicating that a different cascade, downstream of Rad53 and Dun1-independent, is involved in genotoxic-stress induced polarized growth.

G2/M phase

The checkpoint of the G2/M phase is critical in eukaryotic cells because it prevents the cells from entering mitosis in presence of DNA lesions, thus avoiding the transmission of incomplete or damaged genetic material to the next generation. In most eukaryotic cells, activation of the G2/M checkpoint is regulated by phosphorylation of the CDKs (cyclin-dependent kinases) family (Finn *et al.*, 2012). In human cells, this regulation is done via the inhibition of *CDK1* complex phosphorylation. In *S. cerevisiae,* there is no such mechanism for cell cycle arrest in G2. Instead, a different mechanism acts later during mitosis and prevents the entry of cells into anaphase. This mechanism is controlled by Pds1, a protein that inhibits the separation of sister chromatids and therefore, the entry of cells into anaphase (Ciosk *et al.*, 1998; Yamamoto *et al.*, 1996a, b). Upon genotoxic stress, Pds1 is phosphorylated, preventing its degradation and blocking the entry into anaphase (Agarwal *et al.*, 2003; Sanchez *et al.*, 1999; Wang *et al.*, 2001). In *C. albicans,* no orthologue of *PDS1* has been identified. However, there is an orthologue of Cdk1 (encoded by *CDC28*), which suggests that there is potentially a checkpoint-dependent regulation of the entry of cells into mitosis similar to that in human cells. The role of *CDC28* in morphogenesis has been studied in *C. albicans* but its function in the regulation of the G2 phase checkpoint is unknown (Finn *et al.*, 2012; Gutierrez-Escribano *et al.*, 2011; Umeyama *et al.*, 2005).

Spindle checkpoint and mitotic exit network pathway (MEN)

The spindle checkpoint operates during early mitosis and prevents anaphase onset until all chromosomes are properly attached to the spindle. To achieve proper segregation, the sister chromatids are attached to microtubules via their centromeres through a structure called kinetochore. Several proteins are involved in monitoring kinetochore assembly during mitosis: these include Mad and Bub proteins (Mad1, Mad2, Mad3 and Bub1, Bub3) but also Mps1 (May and Hardwick, 2006) and the Aurora B kinase (Ipl1). The activation of the spindle checkpoint through Mad, Bub, Mps1 and Ipl1 proteins leads to delay in the entry of the cells into anaphase. This regulation occurs in particular through the inactivation of Cdc20p. *CDC20* encodes an activator of the anaphase-promoting complex/cyclosome (APC/C), a multiprotein complex responsible for the ubiquitination of many cell cycle regulators, leading to their degradation by the 26S proteasome (May and Hardwick, 2006).

Conditional *CDC20* deletion mutants have been characterized in *C. albicans* by Chou *et al.* (Chou *et al.*, 2011). Their work suggested that, similarly to *ScCDC20, CaCDC20* was important for the metaphase-to-anaphase transition and mitotic exit. Interestingly, *CDC20* has a distinct role in morphogenesis in *C. albicans,* which is not the case in *S. cerevisiae*. In addition, an overexpression screen performed in *C. albicans* identified *CDC20* as the gene whose overexpression triggers a 35-fold increase ($P<0.0001$) in LOH frequency (R.L.-K., A.F., C.d'E. and M.L., unpublished). Several studies have shown that degradation of Cdc20p seems to be an essential and conserved mechanism to maintain the spindle assembly checkpoint (Stevenson *et al.*, 2001; Yu *et al.*, 2006) and therefore, one can understand why the sustained presence of the Cdc20 protein in an overexpression mutant could perturb genome integrity. The acquisition of drug resistance was not investigated in this mutant.

A *MAD2* orthologue has been identified and inactivated in *C. albicans* (Bai *et al.*, 2002). In this study, *mad2* null mutants exhibited an increase in chromosome loss frequency, monitored as an increased number of sorbose-utilizing (Sou+) colonies when plated on sorbose plates as compared to the wild-type strain. Wild-type *C. albicans* cells can not utilize sorbose and therefore can not grow on sorbose plates. Only cells that have become monosomic for Ch5 are able to grow on sorbose (Janbon *et al.*, 1998; Magee and Magee, 2000; Rustchenko *et al.*, 1994). This defect in chromosome mis-segregation is consistent with a defective spindle checkpoint. Also, the *Camad2* null mutants displayed an increased sensitivity to oxidative stress, which can explain the virulence defect of the *mad2* mutants in a mouse systemic candidiasis model. The acquisition of drug resistance was not investigated in these mutants.

In an effort to better understand the mode-of-action of inhibitory compounds, Xu *et al.* (2007) performed a chemogenomic assay by screening a collection of 2868 *C. albicans* heterozygous deletion strains for growth defect in presence of diverse inhibitory compounds and identified *BUB1* heterozygous mutants as highly sensitive to microtubule inhibitors. This study confirmed the role of *CaBUB1* in monitoring kinetochore and spindle integrity.

Because of the critical functions of the DNA checkpoint proteins, new therapeutic research strategies are being developed to identify compounds that would specifically inhibit these proteins in *C. albicans*. Tsuda *et al.* (2011) identified the molecule LY83583 as a specific inhibitor of CaMps1p with no effect on human Mps1p activity. Human Mps1p is already considered as a target of choice, and small molecule inhibitors against it have been reported to be anticancer drug candidates. Indeed, cells that display the level of genome instability observed in cancer cells or antifungal-resistant cells are likely to be highly sensitive to inhibition of Mps1. For this reason, CaMps1p is an ideal target molecule for antifungal therapy.

In *S. cerevisiae*, the mitotic exit network (MEN) pathway is regulated by the Cdc5 kinase that triggers the degradation process of specific mitotic cyclins. Cdc5 is a target for phosphorylation by Rad53. However, the mechanism by which Rad53 inhibits the MEN pathway via its action on Cdc5 is unknown (Finn *et al.*, 2012). Similar to *CDC28*, Cdc5 protein has a role in morphogenesis in *C. albicans* and is also important to regulate the progression of cells into anaphase (Bachewich *et al.*, 2003). However, the link between Cdc5 and Rad53 has not been studied.

Overall, we have only started to scratch the surface of the complex role of DNA damage checkpoint proteins in generating genome rearrangements associated with antifungals resistance and their potential as antifungal targets.

DNA repair mechanisms

Fig. 10.2 shows the major players involved in DNA repair in *S. cerevisiae* and *C. albicans*.

Chemically altered nucleotides

Exposure to various environmental agents (UV radiations) or chemicals such as alkylating agents or reactive oxygen species leads to chemical modifications of nucleotide bases. These alterations lead to changes in the structure of the DNA double helix that are recognized and repaired by two mechanisms called Base Excision Repair (BER) and Nucleotide Excision Repair (NER).

BER

Chemical modifications of bases as a result of alkylation, deamination and oxidation are specifically recognized and repaired by the BER machinery. DNA glycosylases are particularly important in this process as they allow the recognition and removal of the damaged base by cleaving the N-glycosidic bond (Memisoglu and Samson, 2000), leaving the sugar-phosphate backbone intact and creating an apurinic/apyrimidinic site or AP site. AP sites are then processed by AP endonucleases, excising a 1 to 10 nucleotide(s) fragment. After ssDNA formation, particular polymerases and ligases intervene to synthesize a new damage-free DNA strand (Memisoglu and Samson, 2000).

In *C. albicans*, single deletion mutants for genes involved in BER (*NTG1* and *APN1*) did not display increased sensitivity to various DNA-damaging agents, including oxidants (Legrand *et al.*, 2008). In *S. cerevisiae*, inactivation of a single DNA glycosylase does not lead to increased sensitivity to genotoxic agents; only the subsequent inactivation of several DNA glycosylases leads to the appearance of a phenotype, which can be explained by the redundancy of these enzymes. Similarly, in *C. albicans*, only the *ogg1/ngt1* double mutant displayed an increased sensitivity to oxidizing agents (Legrand *et al.*, 2008). Finally, LOH frequency at the *GAL1/gal1::URA3* locus on Ch1 was not affected in BER mutants and no change in antifungal sensibility was observed (Legrand *et al.*, 2008), suggesting that the BER pathway is not involved in suppressing genome alterations causing antifungal resistance acquisition. There are additional *C. albicans* BER genes, encoding other DNA glycosylases, such as *MAG1* and *UNG1*, but the effect of the inactivation of these genes on cell physiology and response to genotoxic agents has not yet been tested.

NER

Nucleotide excision repair (NER) primarily removes DNA-distorting lesions including UV-light induced thymine dimers, as well as substrates induced by DNA cross-linking agents. The importance of this repair mechanism is illustrated by the severe human diseases that result from mutations in NER proteins. Patients diagnosed with Xeroderma pigmentosum or Cockayne's syndrome are prone to develop cancer. Defects in the NER pathway were implied by an abnormal sensitivity of these patients to sunlight (Deschavanne *et al.*, 1984; Setlow *et al.*, 1969).

From *E. coli* to *S. cerevisiae*, NER requires the action of proteins that associate to form a complex and whose role is to detect the damaged site (Rad14, Rad7/Rad16 and Rad4/Rad23). After detection of the lesion site, helicases (Rad3 and Rad25) open up the DNA helix to allow resection of DNA in both 5′ and 3′ side of the lesion by endonucleases. After the fragment excision, polymerases and ligases synthesize an intact complementary strand. Along with Rad1p, Rad2p and Rad10p are two of the endonucleases involved in resection of the DNA fragment containing the lesion. Legrand *et al.* have shown that *C. albicans rad2* and *rad10* null mutants were highly sensitive to UV radiations (Legrand *et al.*, 2008). Interestingly, the level of UV sensitivity of *C. albicans rad2* and *rad10* mutants was higher than *S. cerevisiae* mutants. In addition, the *S. cerevisiae* mutants showed increased recovery after UV treatment as compared to the *C. albicans* mutants. This difference could be explained by the specific environmental conditions in which the two yeasts grow. *S. cerevisiae* is commonly found on leaf and fruit surfaces, where the cells would be directly exposed to the UV light found in sunlight, whereas the normal habitat of *C. albicans* is the mucosal membranes of humans and other warm-blooded animals, and therefore would be unlikely to be exposed to significant amounts of UV light. As for the BER mutants, Ch1 stability and antifungal sensitivity were not affected in the NER mutants (Legrand *et al.*, 2008). The *RAD3* gene has been identified in *C. albicans* but only its physicochemical properties have been reported, as well as its ability to complement the viability and DNA repair defects of ScRad3 mutant cells (Seong *et al.*, 2011).

DNA double-strand break (DSB) repair

A double-strand break (DSB), where both strands of the double helix are severed, is the most serious type of DNA damage. Indeed, failure to repair a DSB leads to loss of the fragment lacking a centromere, while improper repair of a DSB can lead to gross chromosomal rearrangements such as translocations, deletions and inversions. Two main pathways in DSBs repair have been described: the homologous recombination (HR) pathway and the non-homologous end joining (NHEJ) pathway. HR is favoured in diploid cells as a second intact copy of the broken DNA is present and can serve as a template for repairing the damaged DNA molecule. NHEJ can be used in haploid cells by ligating the ends of the DNA molecule without the need for an intact DNA template.

After DNA DSB detection, the first step in the setting of DSB repair is DNA resection. Some of the DDCPs sensors, such as the MRX complex (Mre11/Rad50/Xrs2), are involved in DNA resection (Aylon and Kupiec, 2004) and are required for both HR and NHEJ (Deng *et al.*, 2009; Dinkelmann *et al.*, 2009; Williams *et al.*, 2008). The characterization of the MRX complex in *C. albicans* has been described in the section on 'DNA damage and replication checkpoints'.

Homologous recombination (HR)

Homologous recombination refers to an exchange of genetic information between two DNA molecules. HR requires the interaction of DNA sequences that show perfect or significant homology over several tens or hundreds bases (Krogh and Symington, 2004). Several conservative homology-dependent pathways have been characterized: DSBR (double-strand break repair), SDSA (synthesis-dependent strand annealing) and BIR (break-induced replication) pathways (Krogh and Symington, 2004) (Fig. 10.3). In all cases, the repair process is triggered by the occurrence of a DSB and begins with the resection of the DNA at the break site in order to generate 3′ ssDNA overhangs. Strand invasion by these 3′ ssDNA overhangs into a sequence of perfect or high homology is followed by DNA synthesis at the invading end. The second DSB end can then be captured to form an intermediate with two Holliday junctions that can be solved through a process involving or not reciprocal exchanges (DSBR). Alternatively, the reaction can proceed to SDSA by strand displacement, annealing of the extended single-strand end to the ssDNA on the other break end, followed by gap-filling DNA synthesis and ligation. The repair product from SDSA is always non-crossover. Finally, BIR is initiated when one of the DSB end is lost. In this case, a

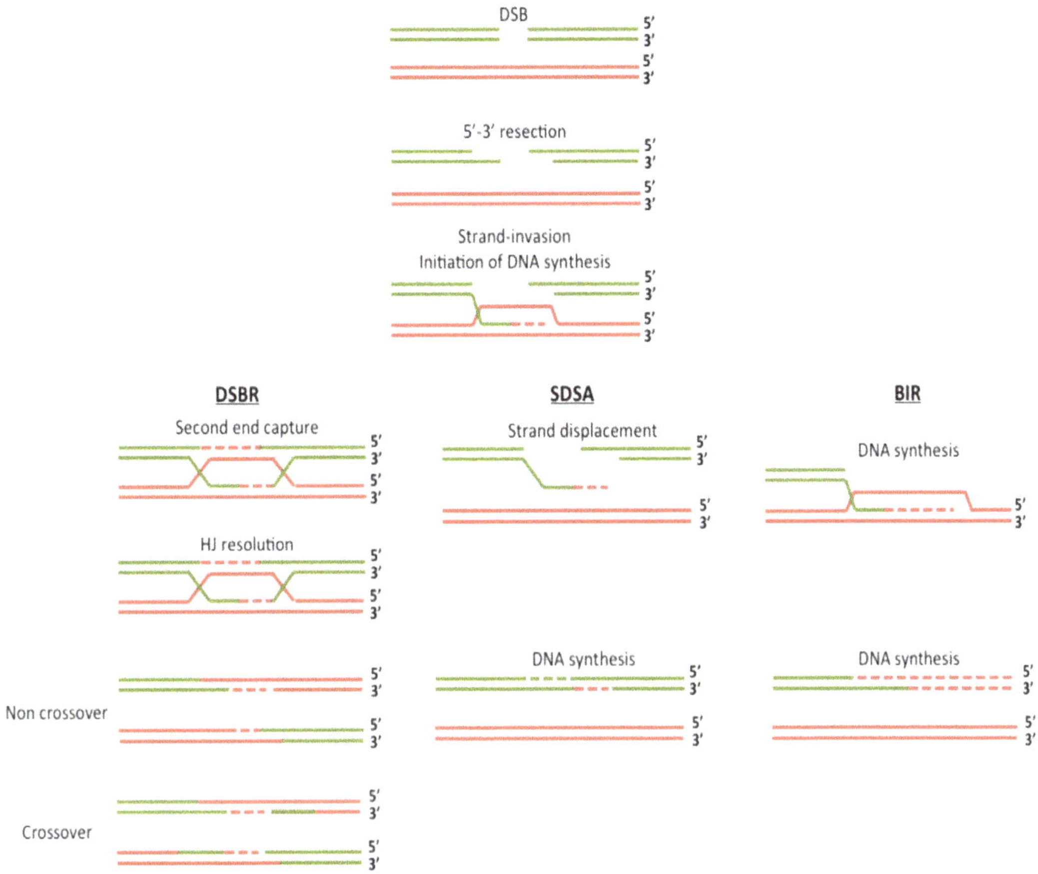

Figure 10.3 DNA double strand breaks (DSB) can be repaired by homologous recombination through different ways. When a DSB occurs, the homologous recombination machinery is recruited to the site of damage. First, the DNA ends are resected to generate 3′ ssDNA ends. Second, one of the 3′ ssDNA end invades a homologous template and DNA is newly synthesized using this homologous template (dashed line). Third, three major pathways can then be used. In the DSBR (double-strand break repair) pathway, the first 3′ ssDNA end capture is followed by the capture of the remaining 3′ ssDNA end and the synthesis of new DNA using the homologous DNA as template resulting in the formation of a Holliday Junction that can be resolved by nucleases into crossover or non-crossover. In the SDSA pathway (synthesis dependent strand annealing), after the first 3′ ssDNA end capture and the synthesis of the DNA using the homologous DNA as template, the newly synthesized fragment can be unwound by helicase and can anneal with the second resected 3′ ssDNA end. The BIR (break-induced replication) pathway occurs when one DNA resected end is lost. In this case, the remaining 3′ ssDNA end invades a homologous DNA region and DNA is synthesized using the homologous template to the end of the chromosome.

single stranded tail invades a homologous duplex DNA molecule and prime DNA replication to the telomeric end of the intact chromosome. These recombination mechanisms result in the appearance of LOH. In the case of DSBR without crossover (gene conversion), the LOH is limited to a small region, while in the DSBR pathway with crossover, LOH may spread out over wider region. SDSA is a source of gene conversions and leads to short-range LOH whereas BIR is a source of LOH that can span large regions. In addition, DSBs that occur between two repeat sequences can be repaired by Single Strand Annealing (SSA), a non-conservative homology-dependent pathway resulting in deletions of genetic material.

A number of proteins are required to process DSB repair via HR in yeast. After detection of the DSB and nucleolytic processing of DNA ends by the MRX complex and nucleases Sae2 and Exo1, the newly formed ssDNA is immediately coated with the RPA protein, triggering the subsequent recruitment of the Rad51 recombinase, through the action of the mediator proteins Rad55 and Rad57. Rad51 plays a key role in the homology search and is also important in strand exchange between DNA molecules (Krogh and Symington, 2004). Along other functions, chromatin remodellers Rad54 and Rdh54 both facilitate and stabilize Rad51 binding to ssDNA, stimulating homologous DNA pairing (Krogh and Symington, 2004). Rad59 stimulates the activity of Rad52, the central component of the HR machinery, in single-strand annealing. This Rad51-dependent recombination pathway is the most effective way to repair DSBs and the most frequently used by mitotic cells (Krogh and Symington, 2004).

To assess the role of HR in the biology of *C. albicans*, several groups have characterized deletion mutants of genes involved in HR (*RAD52, RAD51, RAD59, RAD54* and *RDH54*). The work by Andaluz *et al.* (2011) showed an increase in LOH frequency at the *HIS4* locus (Ch4) in *rad52* mutants, resulting from chromosome loss and truncation; while LOH at this locus arises through BIR in wild-type cells. However, other studies reported no change in LOH frequency on Ch1 in *rad52* mutants and the requirement of Rad52 for UV-induced LOH to occur on Ch3 (Legrand *et al.*, 2007; Takagi *et al.*, 2008). These discrepancies could be explained by strain or chromosome/homologue specificities. It has also been shown that *C. albicans* cells that are lacking *RAD52* exhibit a very high sensitivity to genotoxic agents and are unable to incorporate exogenous linear DNA (Ciudad *et al.*, 2004; Legrand *et al.*, 2007). Together, these results suggest that Rad52 has also a central role in recombination in *C. albicans*. Interestingly, it is worth noting that *RAD52* inactivation renders *C. albicans* avirulent in a disseminated infection model in mice (Chauhan *et al.*, 2005). In comparison, inactivation of *RAD51* or *RAD59* resulted in milder phenotypes, with *rad51* mutants displaying a moderate sensitivity to genotoxic agents and *rad59* mutants being sensitive only to high UV (Garcia-Prieto *et al.*, 2010). Also, Rad54 appears to be important for MMS damage repair, unlike Rdh54 (Hoot *et al.*, 2011b). The recent work by Mitra *et al.* also demonstrated the importance of Rad51 and Rad52 in proper assembly of the kinetochore proteins (Mitra *et al.*, 2014). Regarding antifungal sensitivity, while *rad54* and *rdh54* mutants do not exhibit altered susceptibility to antifungals (Hoot *et al.*, 2011b), *rad52* mutants are more sensitive to fluconazole than wild-type cells (Legrand *et al.*, 2007). Compounds that specifically inhibit *RAD52*, or possibly HR, could be used as companion drugs to improve treatment for *Candida* infections, as they would increase susceptibility of the cells to the antifungal drug without increasing the frequency of appearance of antifungal drug-resistant cells.

Non-homologous end joining (NHEJ)

The NHEJ pathway can repair DSBs by a simple process of religation between two DNA ends without searching a homologous sequence. In *S. cerevisiae*, NHEJ is primarily used during G1 phase in haploid cells and is inhibited in diploid cells by a mechanism controlled by the heterozygosity status of the *MAT* locus (Krogh and Symington, 2004). NHEJ requires the action of a protein complex composed of yKu70, yKu80 and the DNA ligase IV complex, whose catalytic subunit Lig4 mediates interactions with its critical partners proteins Lif1 and Nej1 (Deshpande and Wilson, 2007; Krogh and Symington, 2004). In absence of DNA damage, yKu70 and yKu80 proteins are located at the telomeres and are important for the protection of these regions. When DNA damage occurs, these proteins are relocated to the break site, along with the DNA ligase IV complex, via interactions with proteins of the MRX complex.

In *C. albicans, YKU70, YKU80* and *LIG4* have been partially characterized. Mutations in *CaYKU80* do not affect the sensitivity of the mutants to genotoxic agents or antifungal drugs (fluconazole), nor do they affect genome stability at the *GAL1* locus (Legrand *et al.*, 2007). Similarly, *CaLIG4* inactivation appears to have a minor effect on DNA repair and genome stability (Andaluz *et al.*, 2002). Nevertheless, virulence of *lig4* mutants in a murine model of candidiasis is attenuated, possibly linked to impairment in filamentation (Andaluz *et al.*, 2001).

Similarly to *S. cerevisiae*, *C. albicans* seems to utilize HR over NHEJ to repair DNA damages. Despite the diploid status of *C. albicans*, in which NHEJ is likely to be inhibited similarly to what is observed in *S. cerevisiae* diploid cells, NHEJ genes have been conserved, suggesting additional functions. This is the case for *YKU70*, whose role in telomere maintenance has been demonstrated in *C. albicans*, *C. glabrata* and *S. cerevisiae* (Chico *et al.*, 2011; Rosas-Hernandez *et al.*, 2008; Stellwagen *et al.*, 2003). In agreement with the haploid state of *C. glabrata*, the work by Corrigan *et al.* (2013) showed that NHEJ is more prevalent in this organism (Corrigan *et al.*, 2013). Repair of DSBs by the more error-prone NHEJ pathway in haploid organisms such as *C. glabrata* may contribute to genomic instability and a higher incidence of antifungal resistance acquisition.

Mismatch repair

DNA replication is a faithful mechanism, but as any biological process, it can generate errors such as misincorporations of bases by the DNA polymerases during DNA synthesis. Base mismatches can be repaired by the Methyl Mismatch Repair (MMR) pathway, initially characterized in *E. coli*, in which three major proteins are involved: MutS, MutH and MutL (Harfe and Jinks-Robertson, 2000).

DNA mismatches result in the distortion of the double helix structure and are recognized by the MMR machinery. After detection of the nucleotide mismatch by ATPase proteins (MutS and MutS homologues), endonucleases (MutL and MutL homologues) are recruited at the damaged site. These endonucleases limit their actions to the newly synthesized strand either by monitoring DNA methylation in bacteria or by detecting the 5′ and 3′ ends at the Okazaki fragments formed during replication in eukaryotes (Harfe and Jinks-Robertson, 2000). After cleavage of DNA at the mismatch site, DNA can be degraded by various exonucleases. DNA is then resynthesized with the requirement of helicases. Inactivation of the MMR pathway proteins leads to an increase in mutation rate in both *E. coli* and *S. cerevisiae* (Drotschmann *et al.*, 1999; Funchain *et al.*, 2000).

In *C. albicans*, the role of *PMS1* and *MSH2* in DNA damage response and genome integrity has been studied. Legrand *et al.* have shown that *C. albicans pms1* and *msh2* mutants display no increased sensitivity to genotoxic stresses (UV, MMS, tert-Butyl hydroperoxide or TBHP, camptothecin) (Legrand *et al.*, 2007). The frequency of LOH at the *GAL1/gal1::URA3* locus has also been measured and is similar to wild-type cells. However, the analysis of 2-DG^R (*gal1*$^{-/-}$ cells) and 5-FOA^R (*ura3*$^{-/-}$ cells) colonies in *pms1* and *msh2* mutants revealed an increase in point mutations in the *GAL1* and *URA3* genes. Mutations causing *GAL1* and *URA3* inactivation were predominantly insertions/deletions at repeated regions in the coding sequence resulting in a shift of the reading frame. This suggests that, similarly to what has been observed in *E. coli* and *S. cerevisiae*, inactivation of the MMR pathway in *C. albicans* leads to an increased mutation frequency. Interestingly, an increase in the appearance of fluconazole-resistant cells has been observed with *C. albicans pms1* and *msh2* mutants, which can possibly be connected to the mutator phenotype of these strains. However, it should be noted that the mutator phenotype of *C. albicans* is likely to be less pronounced because of the diploid nature of this organism that buffers the impact of point mutations.

Defects in maintenance of genome integrity might cause the kind of large-scale genome rearrangements that have been shown to occur in fungi under various *in vitro* and *in vivo* stresses. The consequence of the genome modifications leading to acquisition of antifungal resistance on the overall fitness of the cell is crucial in determining the persistence of the resistant population in absence of the stress.

The effect of drug resistance acquisition on fitness

The persistence of drug-resistant organisms in nature depends on their fitness relative to their drug-sensitive counterparts in the absence of the drug. Even though exceptions are known, antibiotic resistance mutations are generally accompanied by a fitness cost in bacteria, manifesting as reduced growth rate or virulence relative to antibiotic sensitive strains. Indeed, drug resistance usually generates a reduction in biological fitness because antibiotics target basic bacterial functions, important in maintaining the cell wall, transcription and protein synthesis, and regulating chromosome supercoiling. Nevertheless, costs of resistance are expected to fade away during subsequent evolution, as natural selection will favour compensatory mutations that will reduce the fitness cost but maintain the level of resistance. The evolution of drug resistance can be studied by following patient isolates during the course of treatment or by long-term laboratory evolution experiments. Unlike bacteria, fungi are more able to accommodate resistance mutations without a devastating loss of fitness in the absence of the drug. In many instances, fluconazole-resistant strains of *C. albicans* selected *in vitro* from susceptible clinical isolates even exhibited increased virulence (Angiolella *et al.*, 2008; Fekete-Forgacs *et al.*, 2000). Similarly, gain-of-function mutations in Pdr1, which cause deregulated gene expression, improved the *in vivo* fitness of drug-resistant *C. glabrata* strains (Ferrari *et al.*, 2009, 2011).

On the other hand, Cowen *et al.* observed both increased and decreased competitive fitness in the absence of the drug in independent resistant populations that evolved from the same drug susceptible progenitor (Cowen *et al.*, 2000, 2001).

Other investigators observed decreased fitness and virulence of fluconazole or echinocandin-resistant isolates of *C. albicans* (Ben-Ami *et al.*, 2011; Graybill *et al.*, 1998; Hayama *et al.*, 2012; Sasse *et al.*, 2012; Schulz *et al.*, 2011).

Therefore, although the evolved resistant *C. albicans* cells commonly achieve equal or superior fitness relative to that of the parental drug-sensitive strain even in the absence of drug, some mechanisms of drug resistance can also be associated with a fitness cost. A novel strategy for the design of antifungal agents would be to identify targets for which the associated resistance mechanism has the most negative effect on fitness.

Conclusions and perspectives

Aneuploidies in human cells are known to largely affect their ability to proliferate. However, there are some tissues in which a significant proportion of human cells naturally carry aneuploid genomes. For example, about 10% of the adult brain cells are aneuploid (carrying trisomies or monosomies). Nevertheless, the absence of detrimental effects of such aneuploidies can be attributed to the fact that brain cells are differentiated and therefore, no longer divide (Pfau and Amon, 2012). Apart from these few exceptions, aneuploidies in humans may cause deleterious effects and are thought to play a role in cancer development. Indeed, genome instability, in particular the occurrence of aneuploidies, point mutations and LOH, is a shared feature of cancer cells. For example, individuals with Down syndrome, also known as trisomy 21, have an increased risk of developing cancer during childhood, and acquired trisomy 21 has been associated with cancer development in non-Down syndrome patients (Hasle *et al.*, 2000; Larsson *et al.*, 2012). Another example showing the importance of aneuploidy in oncogenesis is the effect of CENP-E protein misexpression in mice. CENP-E is a centromere-associated protein and is important in the process of chromosome segregation during mitosis. Mice under-expressing CENP-E have a higher rate of aneuploidy and frequently develop tumours (Weaver *et al.*, 2007).

From a clinical perspective, it is worth noting that the appearance of aneuploidy is also observed in tumour cells that have acquired resistance to anticancer treatment. Among other instances, amplification of the thymidylate synthetase gene in aneuploid cancer cells leads to resistance to 5-fluorouracil, an anti-metabolite drug that acts as a thymidylate synthetase inhibitor (Wang *et al.*, 2004). This scenario is reminiscent of the azole resistance observed in cells that have a copy number increase of the azole target Erg11 as a result of isochromosome formation. In several different cancers, drug resistance has also been associated with increased expression level of efflux pumps, as a result of gene rearrangements. Interestingly, increased drug efflux from fungal cells has also been described as a mechanism of antifungal drug resistance. With an increasing number of patients suffering from cancers, the challenge encountered by modern cancer therapies is the development of resistance to chemotherapeutic agents.

S. cerevisiae was the first eukaryotic model organism for studying aneuploidy. However, a disadvantage of using this model is that aneuploidy acquisition is often associated with a drastic reduction of cell fitness (Torres *et al.*, 2007), which is obviously not the case in cancer cells. In contrast, *C. albicans* is much more tolerant to the presence of aneuploidies. Overall, *C. albicans*'s ability to acquire antifungal resistance through genome rearrangements could serve as a model to study acquisition of resistance to chemotherapeutic agents in cancer cells. Similarly to the work undertaken by anti-cancer drug manufacturers, a new class of resistance reversal drugs could be molecules that would block the synthesis of the efflux pumps or alter their function (Coley, 2010). In addition, the characterization of the molecular mechanisms underlying genome rearrangement-mediated resistance will identify potential new targets for a future class of drugs. New compounds could be effective as companion drugs during treatment for *Candida* infections, as they would increase susceptibility to the antifungal drug without increasing the frequency of appearance of antifungal drug-resistant cells.

Acknowledgements

Our work has received funding from the French Government's Investissement d'Avenir program, Laboratoire d'Excellence 'Integrative Biology of Emerging Infectious Diseases' (Grant # ANR-10-LABX-62-IBEID). RL and AF are the recipients of a PhD grant funded by INRA and Institut Pasteur. We are grateful to other members of the Fungal Biology and Pathogenicity Unit for helpful discussions and apologize to colleagues whose work we could not cite due to space limitations.

References

Agarwal, R., Tang, Z., Yu, H., and Cohen-Fix, O. (2003). Two distinct pathways for inhibiting pds1 ubiquitination in response to DNA damage. J. Biol. Chem. *278*, 45027–45033.

Ahmad, K.M., Ishchuk, O.P., Hellborg, L., Jorgensen, G., Skvarc, M., Stenderup, J., Jorck-Ramberg, D., Polakova, S., and Piskur, J. (2013). Small chromosomes among Danish *Candida glabrata* isolates originated through different mechanisms. Antonie Van Leeuwenhoek *104*, 111–122.

Albuquerque, C.P., Smolka, M.B., Payne, S.H., Bafna, V., Eng, J., and Zhou, H. (2008). A multidimensional chromatography technology for in-depth phosphoproteome analysis. Mol. Cell. Proteomics *7*, 1389–1396.

Andaluz, E., Calderone, R., Reyes, G., and Larriba, G. (2001). Phenotypic analysis and virulence of *Candida albicans* LIG4 mutants. Infect. Immun. *69*, 137–147.

Andaluz, E., Ciudad, T., and Larriba, G. (2002). An evaluation of the role of LIG4 in genomic instability and adaptive mutagenesis in *Candida albicans*. FEMS Yeast Res. *2*, 341–348.

Andaluz, E., Bellido, A., Gomez-Raja, J., Selmecki, A., Bouchonville, K., Calderone, R., Berman, J., and Larriba, G. (2011). Rad52 function prevents chromosome loss and truncation in *Candida albicans*. Mol. Microbiol. *79*, 1462–1482.

Angiolella, L., Stringaro, A.R., De Bernardis, F., Posteraro, B., Bonito, M., Toccacieli, L., Torosantucci, A., Colone, M., Sanguinetti, M., Cassone, A., *et al.* (2008). Increase of virulence and its phenotypic traits in drug-resistant strains of *Candida albicans*. Antimicrob. Agents Chemother. *52*, 927–936.

Aylon, Y., and Kupiec, M. (2004). DSB repair: the yeast paradigm. DNA Repair *3*, 797–815.

Bachewich, C., and Whiteway, M. (2005). Cyclin Cln3p links G1 progression to hyphal and pseudohyphal development in *Candida albicans*. Eukaryot. Cell *4*, 95–102.

Bachewich, C., Thomas, D.Y., and Whiteway, M. (2003). Depletion of a polo-like kinase in *Candida albicans* activates cyclase-dependent hyphal-like growth. Mol. Biol. Cell *14*, 2163–2180.

Bai, C., Ramanan, N., Wang, Y.M., and Wang, Y. (2002). Spindle assembly checkpoint component CaMad2p is indispensable for *Candida albicans* survival and virulence in mice. Mol. Microbiol. *45*, 31–44.

Ben-Ami, R., Garcia-Effron, G., Lewis, R.E., Gamarra, S., Leventakos, K., Perlin, D.S., and Kontoyiannis, D.P. (2011). Fitness and virulence costs of *Candida albicans* FKS1 hot spot mutations associated with echinocandin resistance. J. Infect. Dis. *204*, 626–635.

Bensen, E.S., Clemente-Blanco, A., Finley, K.R., Correa-Bordes, J., and Berman, J. (2005). The mitotic cyclins Clb2p and Clb4p affect morphogenesis in *Candida albicans*. Mol. Biol. Cell *16*, 3387–3400.

Bogue, M., and Roth, D.B. (1996). Mechanism of V(D)J recombination. Curr. Opin. Immunol. *8*, 175–180.

Bouchonville, K., Forche, A., Tang, K.E., Selmecki, A., and Berman, J. (2009). Aneuploid chromosomes are highly unstable during DNA transformation of *Candida albicans*. Eukaryot. Cell *8*, 1554–1566.

Chauhan, N., Ciudad, T., Rodriguez-Alejandre, A., Larriba, G., Calderone, R., and Andaluz, E. (2005). Virulence and karyotype analyses of rad52 mutants of *Candida albicans*: regeneration of a truncated chromosome of a reintegrant strain (rad52/RAD52) in the host. Infect. Immun. *73*, 8069–8078.

Chen, J., and Deng, Y.M. (2009). Influenza virus antigenic variation, host antibody production and new approach to control epidemics. Virol. J. *6*, 30.

Chen, S.H., Smolka, M.B., and Zhou, H. (2007). Mechanism of Dun1 activation by Rad53 phosphorylation in *Saccharomyces cerevisiae*. J. Biol. Chem. *282*, 986–995.

Chen, S.H., Albuquerque, C.P., Liang, J., Suhandynata, R.T., and Zhou, H. (2010). A proteome-wide analysis of kinase–substrate network in the DNA damage response. J. Biol. Chem. *285*, 12803–12812.

Chico, L., Ciudad, T., Hsu, M., Lue, N.F., and Larriba, G. (2011). The *Candida albicans* Ku70 modulates telomere length and structure by regulating both telomerase and recombination. PloS One *6*, e23732.

Chou, H., Glory, A., and Bachewich, C. (2011). Orthologues of the anaphase-promoting complex/cyclosome coactivators Cdc20p and Cdh1p are important for mitotic progression and morphogenesis in *Candida albicans*. Eukaryot. Cell *10*, 696–709.

Ciosk, R., Zachariae, W., Michaelis, C., Shevchenko, A., Mann, M., and Nasmyth, K. (1998). An ESP1/PDS1 complex regulates loss of sister chromatid cohesion at the metaphase to anaphase transition in yeast. Cell 93, 1067–1076.

Ciudad, T., Andaluz, E., Steinberg-Neifach, O., Lue, N.F., Gow, N.A., Calderone, R.A., and Larriba, G. (2004). Homologous recombination in *Candida albicans*: role of CaRad52p in DNA repair, integration of linear DNA fragments and telomere length. Mol. Microbiol. 53, 1177–1194.

Coley, H.M. (2010). Overcoming multidrug resistance in cancer: clinical studies of p-glycoprotein inhibitors. Methods Mol. Biol. *596*, 341–358.

Colnaghi, R., Carpenter, G., Volker, M., and O'Driscoll, M. (2011). The consequences of structural genomic alterations in humans: genomic disorders, genomic instability and cancer. Semin. Cell Dev. Biol. *22*, 875–885.

Corrigan, M.W., Kerwin-Iosue, C.L., Kuczmarski, A.S., Amin, K.B., and Wykoff, D.D. (2013). The fate of linear DNA in *Saccharomyces cerevisiae* and *Candida glabrata*: the role of homologous and non-homologous end joining. PloS One *8*, e69628.

Coste, A.T., Karababa, M., Ischer, F., Bille, J., and Sanglard, D. (2004). TAC1, transcriptional activator of CDR genes, is a new transcription factor involved in the regulation of *Candida albicans* ABC transporters CDR1 and CDR2. Eukaryot. Cell 3, 1639–1652.

Coste, A., Turner, V., Ischer, F., Morschhauser, J., Forche, A., Selmecki, A., Berman, J., Bille, J., and Sanglard, D. (2006). A mutation in Tac1p, a transcription factor regulating CDR1 and CDR2, is coupled with loss of heterozygosity at chromosome 5 to mediate antifungal resistance in *Candida albicans*. Genetics *172*, 2139–2156.

Coste, A., Selmecki, A., Forche, A., Diogo, D., Bougnoux, M.E., d'Enfert, C., Berman, J., and Sanglard, D. (2007). Genotypic evolution of azole resistance mechanisms in sequential *Candida albicans* isolates. Eukaryot. Cell *6*, 1889–1904.

Cowen, L.E., Sanglard, D., Calabrese, D., Sirjusingh, C., Anderson, J.B., and Kohn, L.M. (2000). Evolution of drug resistance in experimental populations of *Candida albicans*. J. Bacteriol. *182*, 1515–1522.

Cowen, L.E., Kohn, L.M., and Anderson, J.B. (2001). Divergence in fitness and evolution of drug resistance in experimental populations of *Candida albicans*. J. Bacteriol. *183*, 2971–2978.

Deng, Y., Guo, X., Ferguson, D.O., and Chang, S. (2009). Multiple roles for MRE11 at uncapped telomeres. Nature *460*, 914–918.

Deschavanne, P.J., Chavaudra, N., Fertil, B., and Malaise, E.P. (1984). Abnormal sensitivity of some Cockayne's syndrome cell strains to UV- and gamma-rays. Association with a reduced ability to repair potentially lethal damage. Mutat. Res. *131*, 61–70.

Deshpande, R.A., and Wilson, T.E. (2007). Modes of interaction among yeast Nej1, Lif1 and Dnl4 proteins and comparison to human XLF, XRCC4 and Lig4. DNA Repair *6*, 1507–1516.

Diaz-Guerra, T.M., Mellado, E., Cuenca-Estrella, M., and Rodriguez-Tudela, J.L. (2003). A point mutation in the 14α-sterol demethylase gene cyp51A contributes to itraconazole resistance in *Aspergillus fumigatus*. Antimicrob. Agents Chemother. *47*, 1120–1124.

Dinkelmann, M., Spehalski, E., Stoneham, T., Buis, J., Wu, Y., Sekiguchi, J.M., and Ferguson, D.O. (2009). Multiple functions of MRN in end-joining pathways during isotype class switching. Nature Struct. Mol. Biol. *16*, 808–813.

Diogo, D., Bouchier, C., d'Enfert, C., and Bougnoux, M.E. (2009). Loss of heterozygosity in commensal isolates of the asexual diploid yeast *Candida albicans*. Fungal Genet. Biol. *46*, 159–168.

Drotschmann, K., Clark, A.B., and Kunkel, T.A. (1999). Mutator phenotypes of common polymorphisms and missense mutations in MSH2. Curr. Biol. *9*, 907–910.

D'Souza, C.A., Kronstad, J.W., Taylor, G., Warren, R., Yuen, M., Hu, G., Jung, W.H., Sham, A., Kidd, S.E., Tangen, K., *et al.* (2011). Genome variation in *Cryptococcus gattii*, an emerging pathogen of immunocompetent hosts. MBio *2*, e00342–10.

Dunkel, N., Blass, J., Rogers, P.D., and Morschhauser, J. (2008a). Mutations in the multi-drug resistance regulator MRR1, followed by loss of heterozygosity, are the main cause of MDR1 overexpression in fluconazole-resistant *Candida albicans* strains. Mol. Microbiol. *69*, 827–840.

Dunkel, N., Liu, T.T., Barker, K.S., Homayouni, R., Morschhauser, J., and Rogers, P.D. (2008b). A gain-of-function mutation in the transcription factor Upc2p causes upregulation of ergosterol biosynthesis genes and increased fluconazole resistance in a clinical *Candida albicans* isolate. Eukaryot. Cell 7, 1180–1190.

Fekete-Forgacs, K., Gyure, L., and Lenkey, B. (2000). Changes of virulence factors accompanying the phenomenon of induced fluconazole resistance in *Candida albicans*. Mycoses 43, 273–279.

Ferrari, S., Ischer, F., Calabrese, D., Posteraro, B., Sanguinetti, M., Fadda, G., Rohde, B., Bauser, C., Bader, O., and Sanglard, D. (2009). Gain of function mutations in CgPDR1 of *Candida glabrata* not only mediate antifungal resistance but also enhance virulence. PLoS Pathog. *5*, e1000268.

Ferrari, S., Sanguinetti, M., Torelli, R., Posteraro, B., and Sanglard, D. (2011). Contribution of CgPDR1-regulated genes in enhanced virulence of azole-resistant *Candida glabrata*. PloS One *6*, e17589.

Finn, K., Lowndes, N.F., and Grenon, M. (2012). Eukaryotic DNA damage checkpoint activation in response to double-strand breaks. Cell. Mol. Life Sci. *69*, 1447–1473.

Forche, A., May, G., and Magee, P.T. (2005). Demonstration of loss of heterozygosity by single-nucleotide polymorphism microarray analysis and alterations in strain morphology in *Candida albicans* strains during infection. Eukaryot. Cell 4, 156–165.

Forche, A., Magee, P.T., Selmecki, A., Berman, J., and May, G. (2009). Evolution in *Candida albicans* populations during a single passage through a mouse host. Genetics *182*, 799–811.

Forche, A., Abbey, D., Pisithkul, T., Weinzierl, M.A., Ringstrom, T., Bruck, D., Petersen, K., and Berman, J. (2011). Stress alters rates and types of loss of heterozygosity in *Candida albicans*. MBio *2*, e00129–11.

Funchain, P., Yeung, A., Stewart, J.L., Lin, R., Slupska, M.M., and Miller, J.H. (2000). The consequences of growth of a mutator strain of *Escherichia coli* as measured by loss of function among multiple gene targets and loss of fitness. Genetics *154*, 959–970.

Garcia-Prieto, F., Gomez-Raja, J., Andaluz, E., Calderone, R., and Larriba, G. (2010). Role of the homologous recombination genes RAD51 and RAD59 in the resistance of *Candida albicans* to UV light, radiomimetic and anti-tumor compounds and oxidizing agents. Fungal Genet. Biol. *47*, 433–445.

Grawunder, U., West, R.B., and Lieber, M.R. (1998). Antigen receptor gene rearrangement. Curr. Opin. Immunol. *10*, 172–180.

Graybill, J.R., Montalbo, E., Kirkpatrick, W.R., Luther, M.F., Revankar, S.G., and Patterson, T.F. (1998). Fluconazole versus *Candida albicans:* a complex relationship. Antimicrob. Agents Chemother. *42*, 2938–2942.

Gutierrez-Escribano, P., Gonzalez-Novo, A., Suarez, M.B., Li, C.R., Wang, Y., de Aldana, C.R., and Correa-Bordes, J. (2011). CDK-dependent phosphorylation of Mob2 is essential for hyphal development in *Candida albicans*. Mol. Biol. Cell 22, 2458–2469.

Harfe, B.D., and Jinks-Robertson, S. (2000). DNA mismatch repair and genetic instability. Annu. Rev. Genet. *34*, 359–399.

Hasle, H., Clemmensen, I.H., and Mikkelsen, M. (2000). Risks of leukaemia and solid tumours in individuals with Down's syndrome. Lancet 355, 165–169.

Hayama, K., Ishibashi, H., Ishijima, S.A., Niimi, K., Tansho, S., Ono, Y., Monk, B.C., Holmes, A.R., Harding, D.R., Cannon, R.D., *et al.* (2012). A D-octapeptide drug efflux pump inhibitor acts synergistically with azoles in a murine oral candidiasis infection model. FEMS Microbiol. Lett. *328*, 130–137.

Heilmann, C.J., Schneider, S., Barker, K.S., Rogers, P.D., and Morschhauser, J. (2010). An A643T mutation in the transcription factor Upc2p causes constitutive ERG11 upregulation and increased fluconazole resistance in *Candida albicans*. Antimicrob. Agents Chemother. *54*, 353–359.

Hickman, M.A., Zeng, G., Forche, A., Hirakawa, M.P., Abbey, D., Harrison, B.D., Wang, Y.M., Su, C.H., Bennett, R.J., Wang, Y., *et al.* (2013). The 'obligate diploid' *Candida albicans* forms mating-competent haploids. Nature *494*, 55–59.

Hodgkin, J., Horvitz, H.R., and Brenner, S. (1979). Nondisjunction mutants of the nematode *Caenorhabditis elegans*. Genetics *91*, 67–94.

Hoffmann, A.A., and Parsons, P.A. (1991). Evolutionary Genetics and Environmental Stress (Oxford University Press, Oxford, UK).

Hoot, S.J., Smith, A.R., Brown, R.P., and White, T.C. (2011a). An A643V amino acid substitution in Upc2p contributes to azole resistance in well-characterized clinical isolates of *Candida albicans*. Antimicrob. Agents Chemother. *55*, 940–942.

Hoot, S.J., Zheng, X., Potenski, C.J., White, T.C., and Klein, H.L. (2011b). The role of *Candida albicans* homologous recombination factors Rad54 and Rdh54 in DNA damage sensitivity. BMC Microbiol. *11*, 214.

Hu, G., Wang, J., Choi, J., Jung, W.H., Liu, I., Litvintseva, A.P., Bicanic, T., Aurora, R., Mitchell, T.G., Perfect, J.R., *et al.* (2011). Variation in chromosome copy number influences the virulence of *Cryptococcus neoformans* and occurs in isolates from AIDS patients. BMC Genomics *12*, 526.

Hull, C.M., Raisner, R.M., and Johnson, A.D. (2000). Evidence for mating of the "asexual" yeast *Candida albicans* in a mammalian host. Science *289*, 307–310.

Janbon, G., Sherman, F., and Rustchenko, E. (1998). Monosomy of a specific chromosome determines L-sorbose utilization: a novel regulatory mechanism in Candida albicans. Proc. Natl. Acad. Sci. U.S.A. *95*, 5150–5155.

Karababa, M., Coste, A.T., Rognon, B., Bille, J., and Sanglard, D. (2004). Comparison of gene expression profiles of *Candida albicans* azole-resistant clinical isolates and laboratory strains exposed to drugs inducing multidrug transporters. Antimicrob. Agents Chemother. *48*, 3064–3079.

Kolodner, R.D., Putnam, C.D., and Myung, K. (2002). Maintenance of genome stability in *Saccharomyces cerevisiae*. Science *297*, 552–557.

Kontoyiannis, D.P., and Lewis, R.E. (2002). Antifungal drug resistance of pathogenic fungi. Lancet *359*, 1135–1144.

Krogh, B.O., and Symington, L.S. (2004). Recombination proteins in yeast. Annu. Rev. Genet. *38*, 233–271.

Larsson, N., Lilljebjorn, H., Lassen, C., Johansson, B., and Fioretos, T. (2012). Myeloid malignancies with acquired trisomy 21 as the sole cytogenetic change are clinically highly variable and display a heterogeneous pattern of copy number alterations and mutations. Eur. J. Haematol. *88*, 136–143.

Legrand, M., Lephart, P., Forche, A., Mueller, F.M., Walsh, T., Magee, P.T., and Magee, B.B. (2004). Homozygosity at the MTL locus in clinical strains of *Candida albicans*: karyotypic rearrangements and tetraploid formation. Mol. Microbiol. *52*, 1451–1462.

Legrand, M., Chan, C.L., Jauert, P.A., and Kirkpatrick, D.T. (2007). Role of DNA mismatch repair and double-strand break repair in genome stability and antifungal drug resistance in *Candida albicans*. Eukaryot. Cell *6*, 2194–2205.

Legrand, M., Chan, C.L., Jauert, P.A., and Kirkpatrick, D.T. (2008). Analysis of base excision and nucleotide excision repair in *Candida albicans*. Microbiology *154*, 2446–2456.

Legrand, M., Chan, C.L., Jauert, P.A., and Kirkpatrick, D.T. (2011). The contribution of the S-phase checkpoint genes MEC1 and SGS1 to genome stability maintenance in *Candida albicans*. Fungal Genet. Biol. *48*, 823–830.

Lemoine, F.J., Degtyareva, N.P., Lobachev, K., and Petes, T.D. (2005). Chromosomal translocations in yeast induced by low levels of DNA polymerase a model for chromosome fragile sites. Cell *120*, 587–598.

Lindsley, D.L., Sandler, L., Baker, B.S., Carpenter, A.T., Denell, R.E., Hall, J.C., Jacobs, P.A., Miklos, G.L., Davis, B.K., Gethmann, R.C., *et al.* (1972). Segmental aneuploidy and the genetic gross structure of the *Drosophila* genome. Genetics *71*, 157–184.

Liu, T.T., Lee, R.E., Barker, K.S., Lee, R.E., Wei, L., Homayouni, R., and Rogers, P.D. (2005). Genome-wide expression profiling of the response to azole, polyene, echinocandin, and pyrimidine antifungal agents in *Candida albicans*. Antimicrob. Agents Chemother. *49*, 2226–2236.

Loll-Krippleber, R., d'Enfert, C., Feri, A., Diogo, D., Perin, A., Marcet-Houben, M., Bougnoux, M.E., and Legrand, M. (2013). A study of the DNA damage checkpoint in *Candida albicans*: uncoupling of the functions of Rad53 in DNA repair, cell cycle regulation and genotoxic stress-induced polarized growth. Mol. Microbiol. *91*, 452–471.

Lopes da Rosa, J., Boyartchuk, V.L., Zhu, L.J., and Kaufman, P.D. (2010). Histone acetyltransferase Rtt109 is required for *Candida albicans* pathogenesis. Proc. Natl. Acad. Sci. U.S.A. *107*, 1594–1599.

McClintock, B. (1929). A cytological and genetical study of triploid maize. Genetics *14*, 180–222.

Magee, B.B., and Magee, P.T. (2000). Induction of mating in *Candida albicans* by construction of MTLa and MTLalpha strains. Science *289*, 310–313.

Mandegar, M.A., and Otto, S.P. (2007). Mitotic recombination counteracts the benefits of genetic segregation. Proc. Biol. Sci. *274*, 1301–1307.

Marichal, P., Vanden Bossche, H., Odds, F.C., Nobels, G., Warnock, D.W., Timmerman, V., Van Broeckhoven, C., Fay, S., and Mose-Larsen, P. (1997). Molecular biological characterization of an azole-resistant *Candida glabrata* isolate. Antimicrob. Agents Chemother. *41*, 2229–2237.

May, K.M., and Hardwick, K.G. (2006). The spindle checkpoint. J. Cell Sci. *119*, 4139–4142.

Mellado, E., Garcia-Effron, G., Alcazar-Fuoli, L., Melchers, W.J., Verweij, P.E., Cuenca-Estrella, M., and Rodriguez-Tudela, J.L. (2007). A new *Aspergillus fumigatus* resistance mechanism conferring *in vitro* cross-resistance to azole antifungals involves a combination of cyp51A alterations. Antimicrob. Agents Chemother. *51*, 1897–1904.

Memisoglu, A., and Samson, L. (2000). Base excision repair in yeast and mammals. Mutat. Res. *451*, 39–51.

Mitra, S., Gomez-Raja, J., Larriba, G., Dubey, D.D., and Sanyal, K. (2014). Rad51–Rad52 mediated maintenance of centromeric chromatin in *Candida albicans*. PLoS Genet. *10*, e1004344.

Morio, F., Loge, C., Besse, B., Hennequin, C., and Le Pape, P. (2010). Screening for amino acid substitutions in the *Candida albicans* Erg11 protein of azole-susceptible and azole-resistant clinical isolates: new substitutions and a review of the literature. Diagn. Microbiol. Infect. Dis. *66*, 373–384.

Morio, F., Pagniez, F., Lacroix, C., Miegeville, M., and Le Pape, P. (2012). Amino acid substitutions in the *Candida albicans* sterol Delta5,6-desaturase (Erg3p) confer azole resistance: characterization of two novel mutants with impaired virulence. J. Antimicrob. Chemother. *67*, 2131–2138.

Morschhauser, J., Barker, K.S., Liu, T.T., Bla, B.W.J., Homayouni, R., and Rogers, P.D. (2007). The transcription factor Mrr1p controls expression of the MDR1 efflux pump and mediates multidrug resistance in *Candida albicans*. PLoS Pathog. *3*, e164.

Niimi, K., Monk, B.C., Hirai, A., Hatakenaka, K., Umeyama, T., Lamping, E., Maki, K., Tanabe, K., Kamimura, T., Ikeda, F., *et al.* (2010). Clinically significant micafungin resistance in *Candida albicans* involves modification of a glucan synthase catalytic subunit GSC1 (FKS1) allele followed by loss of heterozygosity. J. Antimicrob. Chemother. *65*, 842–852.

Park, S., Kelly, R., Kahn, J.N., Robles, J., Hsu, M.J., Register, E., Li, W., Vyas, V., Fan, H., Abruzzo, G., *et al.* (2005). Specific substitutions in the echinocandin target Fks1p account for reduced susceptibility of rare laboratory and clinical *Candida* sp. isolates. Antimicrob. Agents Chemother. *49*, 3264–3273.

Pfaller, M.A., and Diekema, D.J. (2010). Epidemiology of invasive mycoses in North America. Crit. Rev. Microbiol. *36*, 1–53.

Pfau, S.J., and Amon, A. (2012). Chromosomal instability and aneuploidy in cancer: from yeast to man. EMBO Rep. *13*, 515–527.

Polakova, S., Blume, C., Zarate, J.A., Mentel, M., Jorck-Ramberg, D., Stenderup, J., and Piskur, J. (2009). Formation of new chromosomes as a virulence mechanism in yeast *Candida glabrata*. Proc. Natl. Acad. Sci. U.S.A. *106*, 2688–2693.

Rosas-Hernandez, L.L., Juarez-Reyes, A., Arroyo-Helguera, O.E., De Las Penas, A., Pan, S.J., Cormack, B.P., and Castano, I. (2008). yKu70/yKu80 and Rif1 regulate silencing differentially at telomeres in *Candida glabrata*. Eukaryot. Cell 7, 2168–2178.

Rustad, T.R., Stevens, D.A., Pfaller, M.A., and White, T.C. (2002). Homozygosity at the *Candida albicans* MTL locus associated with azole resistance. Microbiology *148*, 1061–1072.

Rustchenko, E.P., Howard, D.H., and Sherman, F. (1994). Chromosomal alterations of *Candida albicans* are associated with the gain and loss of assimilating functions. J. Bacteriol. *176*, 3231–3241.

Sanchez, Y., Bachant, J., Wang, H., Hu, F., Liu, D., Tetzlaff, M., and Elledge, S.J. (1999). Control of the DNA damage checkpoint by chk1 and rad53 protein kinases through distinct mechanisms. Science *286*, 1166–1171.

Sasse, C., Dunkel, N., Schafer, T., Schneider, S., Dierolf, F., Ohlsen, K., and Morschhauser, J. (2012). The stepwise acquisition of fluconazole resistance mutations causes a gradual loss of fitness in *Candida albicans*. Mol. Microbiol. *86*, 539–556.

Schultz, L.B., Chehab, N.H., Malikzay, A., DiTullio, R.A., Jr., Stavridi, E.S., and Halazonetis, T.D. (2000). The DNA damage checkpoint and human cancer. Cold Spring Harb. Symp. Quant. Biol. *65*, 489–498.

Schulz, B., Weber, K., Schmidt, A., Borg-von Zepelin, M., and Ruhnke, M. (2011). Difference in virulence between fluconazole-susceptible and fluconazole-resistant *Candida albicans* in a mouse model. Mycoses *54*, e522–530.

Selmecki, A., Forche, A., and Berman, J. (2006). Aneuploidy and isochromosome formation in drug-resistant *Candida albicans*. Science *313*, 367–370.

Selmecki, A., Gerami-Nejad, M., Paulson, C., Forche, A., and Berman, J. (2008). An isochromosome confers drug resistance *in vivo* by amplification of two genes, ERG11 and TAC1. Mol. Microbiol. *68*, 624–641.

Selmecki, A.M., Dulmage, K., Cowen, L.E., Anderson, J.B., and Berman, J. (2009). Acquisition of aneuploidy provides increased fitness during the evolution of antifungal drug resistance. PLoS Genet. *5*, e1000705.

Seong, K.M., Lee, S.H., Kim, H.D., Lee, C.H., Youn, H., Youn, B., and Kim, J. (2011). Expression, purification, and characterization of putative *Candida albicans* Rad3, the product of orf19.7119. Biochemistry (Mosc) *76*, 666–676.

Setlow, R.B., Regan, J.D., German, J., and Carrier, W.L. (1969). Evidence that xeroderma pigmentosum cells do not perform the first step in the repair of ultraviolet damage to their DNA. Proc. Natl. Acad. Sci. U.S.A. *64*, 1035–1041.

Shapiro, R.S., Robbins, N., and Cowen, L.E. (2011). Regulatory circuitry governing fungal development, drug resistance, and disease. Microbiol. Mol. Biol. Rev. *75*, 213–267.

Shi, Q.M., Wang, Y.M., Zheng, X.D., Lee, R.T., and Wang, Y. (2007). Critical role of DNA checkpoints in mediating genotoxic-stress-induced filamentous growth in *Candida albicans*. Mol. Biol. Cell *18*, 815–826.

Sionov, E., Lee, H., Chang, Y.C., and Kwon-Chung, K.J. (2010). *Cryptococcus neoformans* overcomes stress of azole drugs by formation of disomy in specific multiple chromosomes. PLoS Pathog. *6*, e1000848.

Smolka, M.B., Albuquerque, C.P., Chen, S.H., Schmidt, K.H., Wei, X.X., Kolodner, R.D., and Zhou, H. (2005). Dynamic changes in protein–protein interaction and protein phosphorylation probed with amine-reactive isotope tag. Mol. Cell. Proteomics *4*, 1358–1369.

Smolka, M.B., Albuquerque, C.P., Chen, S.H., and Zhou, H. (2007). Proteome-wide identification of *in vivo* targets of DNA damage checkpoint kinases. Proc. Natl. Acad. Sci. U.S.A. *104*, 10364–10369.

Stellwagen, A.E., Haimberger, Z.W., Veatch, J.R., and Gottschling, D.E. (2003). Ku interacts with telomerase RNA to promote telomere addition at native and broken chromosome ends. Genes Dev. *17*, 2384–2395.

Stevenson, L.F., Kennedy, B.K., and Harlow, E. (2001). A large-scale overexpression screen in *Saccharomyces cerevisiae* identifies previously uncharacterized cell cycle genes. Proc. Natl. Acad. Sci. U.S.A. *98*, 3946–3951.

Stevenson, J.S., and Liu, H. (2011). Regulation of white and opaque cell-type formation in *Candida albicans* by Rtt109 and Hst3. Mol. Microbiol. *81*, 1078–1091.

Stockdale, C., Swiderski, M.R., Barry, J.D., and McCulloch, R. (2008). Antigenic variation in *Trypanosoma brucei*: joining the DOTs. PLoS Biol. *6*, e185.

Sun, L.L., Li, W.J., Wang, H.T., Chen, J., Deng, P., Wang, Y., and Sang, J.L. (2011). Protein phosphatase Pph3 and its regulatory subunit Psy2 regulate Rad53 dephosphorylation and cell morphogenesis during recovery from DNA damage in *Candida albicans*. Eukaryot. Cell *10*, 1565–1573.

Takagi, Y., Akada, R., Kumagai, H., Yamamoto, K., and Tamaki, H. (2008). Loss of heterozygosity is induced in *Candida albicans* by ultraviolet irradiation. Appl. Microbiol. Biotechnol. *77*, 1073–1082.

Torres, E.M., Sokolsky, T., Tucker, C.M., Chan, L.Y., Boselli, M., Dunham, M.J., and Amon, A. (2007). Effects of aneuploidy on cellular physiology and cell division in haploid yeast. Science *317*, 916–924.

Tsuda, K., Nishiya, N., Umeyama, T., and Uehara, Y. (2011). Identification of LY83583 as a specific inhibitor of *Candida albicans* MPS1 protein kinase. Biochem. Biophys. Res. Commun. *409*, 418–423.

Umeyama, T., Kaneko, A., Nagai, Y., Hanaoka, N., Tanabe, K., Takano, Y., Niimi, M., and Uehara, Y. (2005). *Candida albicans* protein kinase CaHsl1p regulates cell elongation and virulence. Mol. Microbiol. *55*, 381–395.

Wang, H., Liu, D., Wang, Y., Qin, J., and Elledge, S.J. (2001). Pds1 phosphorylation in response to DNA damage is essential for its DNA damage checkpoint function. Genes Dev. *15*, 1361–1372.

Wang, T.L., Diaz, L.A., Jr., Romans, K., Bardelli, A., Saha, S., Galizia, G., Choti, M., Donehower, R., Parmigiani, G., Shih Ie, M., *et al.* (2004). Digital karyotyping identifies thymidylate synthase amplification as a mechanism of resistance to 5-fluorouracil in metastatic colorectal cancer patients. Proc. Natl. Acad. Sci. U.S.A. *101*, 3089–3094.

Weaver, B.A., Silk, A.D., Montagna, C., Verdier-Pinard, P., and Cleveland, D.W. (2007). Aneuploidy acts both oncogenically and as a tumor suppressor. Cancer Cell *11*, 25–36.

Williams, R.S., Moncalian, G., Williams, J.S., Yamada, Y., Limbo, O., Shin, D.S., Groocock, L.M., Cahill, D., Hitomi, C., Guenther, G., *et al.* (2008). Mre11 dimers coordinate DNA end bridging and nuclease processing in double-strand-break repair. Cell *135*, 97–109.

Wisniewski-Dye, F., and Vial, L. (2008). Phase and antigenic variation mediated by genome modifications. Antonie Van Leeuwenhoek *94*, 493–515.

Wurtele, H., Tsao, S., Lepine, G., Mullick, A., Tremblay, J., Drogaris, P., Lee, E.H., Thibault, P., Verreault, A., and Raymond, M. (2010). Modulation of histone H3 lysine 56 acetylation as an antifungal therapeutic strategy. Nat. Med. *16*, 774–780.

Xu, D., Jiang, B., Ketela, T., Lemieux, S., Veillette, K., Martel, N., Davison, J., Sillaots, S., Trosok, S., Bachewich, C., *et al.* (2007). Genome-wide fitness test and mechanism-of-action studies of inhibitory compounds in *Candida albicans*. PLoS Pathog. *3*, e92.

Yamamoto, A., Guacci, V., and Koshland, D. (1996a). Pds1p is required for faithful execution of anaphase in the yeast, *Saccharomyces cerevisiae*. J. Cell Biol. *133*, 85–97.

Yamamoto, A., Guacci, V., and Koshland, D. (1996b). Pds1p, an inhibitor of anaphase in budding yeast, plays a critical role in the APC and checkpoint pathway(s). J. Cell Biol. *133*, 99–110.

Yan, L., Zhang, J., Li, M., Cao, Y., Xu, Z., Cao, Y., Gao, P., Wang, Y., and Jiang, Y. (2008). DNA microarray analysis of fluconazole resistance in a laboratory *Candida albicans* strain. Acta Biochim. Biophys. Sin. *40*, 1048–1060.

Yu, L., Pena Castillo, L., Mnaimneh, S., Hughes, T.R., and Brown, G.W. (2006). A survey of essential gene function in the yeast cell division cycle. Mol. Biol. Cell *17*, 4736–4747.

Zhou, B.B., and Elledge, S.J. (2000). The DNA damage response: putting checkpoints in perspective. Nature *408*, 433–439.

Modulation of the Host Response to Control Invasive Fungal Infections

11

Flavie Courjol, Thierry Jouault and Chantal Fradin

Abstract

Disruption of the host immune defence is one of the most important risk factors for invasive fungal infections (IFIs). Knowledge of both the immune factors involved in defence against fungi and the immune risk factors for IFIs is important to target specific immune functions in order to regulate them in patients at risk. *Candida albicans*, *Aspergillus fumigatus* and *Cryptococcus neoformans* are the most common fungal pathogens of humans. They are opportunistic pathogens, which can disseminate in hosts with acquired defects in immune homeostasis. Risk factors for IFIs caused by these three fungi are well-known. Different studies have reported immune defence against their invasion, highlighting putative immune targets for immunotherapy. Some pre-clinical and clinical trials using non-specific immunotherapies (leucocytes transfusion, growth factors and cytokines) have been tested. Although an antifungal immunotherapy has not been discovered yet, some of the assays have given promising results.

Introduction

Humans are exposed to and tolerate numerous fungi that are either environmental or commensal. However, some fungi can become pathogenic when not properly controlled (Domer and Kobayashi, 2004). Most of these fungal pathogens cause recurring infections, mainly superficial infections of the skin and mucosal tissues, that can be easily diagnosed and effectively treated (Balkus *et al.*, 2011; Del Rosso and Kircik, 2013; Miele *et al.*, 2002; Schwartz, 2004). Invasive infections that occur following superficial fungal infections or direct exposure to fungi are less common although their incidence has risen dramatically in the past decade (Enoch *et al.*, 2006; Kriengkauykiat *et al.*, 2011; Tuite and Lacey, 2013). Invasive fungal infections (IFIs) may be caused either by opportunistic fungi that take advantage of a weakened immune system or by true pathogens whose resulting infections generally worsen in immunocompromised patients (Ascioglu *et al.*, 2002; Pfaller and Diekema, 2010). There is a clear balance between immune tolerance to colonization and immune defence against invasion in healthy subjects (Romani, 2011; Sales-Campos *et al.*, 2013). This balance is disrupted in immunocompromised patients, who then become susceptible to IFIs due to opportunistic (environmental or commensal) or true pathogenic fungi.

On infection, fungi are rapidly recognized by cells of the innate immune system which constitute the first line of defence against most microorganisms (Mariscalco, 2006). Innate immune cells express different receptors able to recognize self-molecules that opsonize microorganisms or non-self-molecules at the microbial surface (Bourgeois *et al.*, 2010; Cheng *et al.*, 2012; Cunha *et al.*, 2012; Faro-Trindade *et al.*, 2012; Gresnigt *et al.*, 2012; Jouault *et al.*, 2009; Netea *et al.*, 2004). These latter molecules include fungal cell wall components, mainly glycans, which can be ubiquitous or restricted to certain species. Fungal molecules can be detected by different immune cell types including professional antigen-presenting cells (APCs) such as dendritic cells

(DCs) and macrophages, other leucocytes such as neutrophils and natural killer (NK) cells, and endothelial and epithelial cells. Effective control of fungal invasion also involves an adaptive response, humoral or cellular, towards the fungus.

IFIs are difficult to treat because of the lack of sensitive and specific methods for their early diagnosis, which is crucial to the initiation of appropriate and effective treatments (Cuenca-Estrella *et al.*, 2011; Gould, 2011; Gullo, 2009; Perfect, 2013; Tuite and Lacey, 2013). IFIs then become life-threatening if not diagnosed and treated early. Antifungal treatments against invasive fungi are available but limited, can be toxic to the host, and can induce resistance (Kriengkauykiat *et al.*, 2011; Petrikkos and Skiada, 2007; Shao *et al.*, 2007). Direct or indirect disruption of host immune homeostasis is the main risk factor for IFIs and their associated high morbidity and mortality (Muskett *et al.*, 2011; Pagano *et al.*, 2011; Pfaller and Diekema, 2010). Restoring immune function, possibly together with antifungal treatment, in at risk-patients could contribute to the recognition of invasive fungi and their clearance from the host. Special care should be taken to avoid impairment of treatment of the primary disease by the chosen immune therapy. Analysis of the effective systemic immune response against fungal pathogens can help to predict endogenous immunomodulators or specialized immunoadjuvants able to restore, alone or in combination with antifungal treatment, immune attributes to limit fungal invasion.

Risk factors for invasive fungal infections

Fungi are common type of human pathogen. They are the source of numerous superficial infections, but some of them, mainly in clinical settings, are a leading cause of life-threatening invasive infections (Garber, 2001; Walsh and Dixon, 1996). Most fungal pathogens, either commensal or environmental, become invasive when host mucosal and immune barriers are compromised. Depending on the nature of the fungus, different risk factors are involved and different types of infection occur.

Commensal fungi and disruption of the microbial flora

Candida albicans is the most documented of the different commensal and opportunistic fungal pathogens. This yeast is the leading cause of invasive mycosis (Poulain, 2013). This yeast is part of the human microbial flora, mainly in the gut and female genital tract (Angebault *et al.*, 2013; Giraldo *et al.*, 2000). As a commensal, *C. albicans* is well adapted to the human mucosal environment and is well tolerated by its host. *C. albicans* in the gut can be considered as a time bomb because, in the majority of cases, invasive candidiasis is due to the patient's own endogenous mucosal strain(s) (Bonnin, 2012; Eggimann and Pittet, 2006). Tolerance towards this yeast in healthy subjects is mainly due to the development of mucosal immunity after stimulation by *C. albicans* itself, but also by other microorganisms of the intestinal microbiome, particularly bacteria (Belkaid and Artis, 2013; Hernandez-Santos and Gaffen, 2012; Moyes and Naglik, 2011; Owen and Mohamadzadeh, 2013; Yan *et al.*, 2013). There are several benefits of a healthy balanced microbiome. In addition to the digestion of food and the synthesis of essential nutrients for the host, the intestinal microbiome is important in the defence against invading pathogens by competing for space, essential nutrients, and attachment sites on the epithelium in order to prevent gut colonization by pathogenic microorganisms (Kamada *et al.*, 2013). Toxins are also secreted by the intestinal microbiota, which inhibit the growth of other microorganisms including pathogens. The intestinal microbiota is essential to prevent *C. albicans* colonization as germ-free animals are more susceptible to mucosal yeast infections (Naglik *et al.*, 2008). Interaction of *C. albicans* with bacteria from the intestinal microbiome helps to inhibit yeast adhesion, colonization, and dissemination from the gastrointestinal tract (Kennedy and Volz, 1985; Van der Waaij and Van der Waaij, 1990). Inhibition of *C. albicans* colonization by the gut microbiota is only partial as healthy subjects can be naturally colonized by the yeast, with close control of the cells attached to the epithelium. Other non-bacterial microorganisms, which do not belong to host microbiome can also limit

C. albicans growth and colonization. The non-pathogenic yeast *Saccharomyces boulardii*, which is very similar to *S. cerevisiae*, can compete with *C. albicans* for adherence and colonization of the intestinal mucosa (Jawhara and Poulain, 2007). *S. boulardii* can affect *C. albicans* colonization and possible dissemination by secreting a molecule, identified as capric acid, which acts as a quorum sensor that can inhibit *C. albicans* growth, hyphal formation, and adhesion (Krasowska *et al.*, 2009; Murzyn *et al.*, 2010). The interest of this yeast as a probiotic, especially when homeostasis of the gut microbiota is disturbed, to decrease the risk of *C. albicans* infection is discussed in the paragraph 'Interplay between fungi and host immune cells' later in this chapter.

A healthy balanced microbiome and its beneficial effect can be disrupted by antibiotic therapy which eliminates members of the microbial flora. Antibiotic treatment results in long-lasting alterations and overgrowth of opportunistic pathogens which can then become invasive (Fig. 11.1). Furthermore, antibiotics disturb epithelial permeability allowing colonization and invasion by both endogenous and exogenous microorganisms. Broad-spectrum antimicrobial therapy is one of the major risk factors for invasive *Candida* infection (Morace and Borghi, 2010; Singla *et al.*, 2012) and also favours non-*albicans* species, including non-commensal species, that can then colonize the gut and invade the host (Cornistein *et al.*, 2013). These other species are emerging

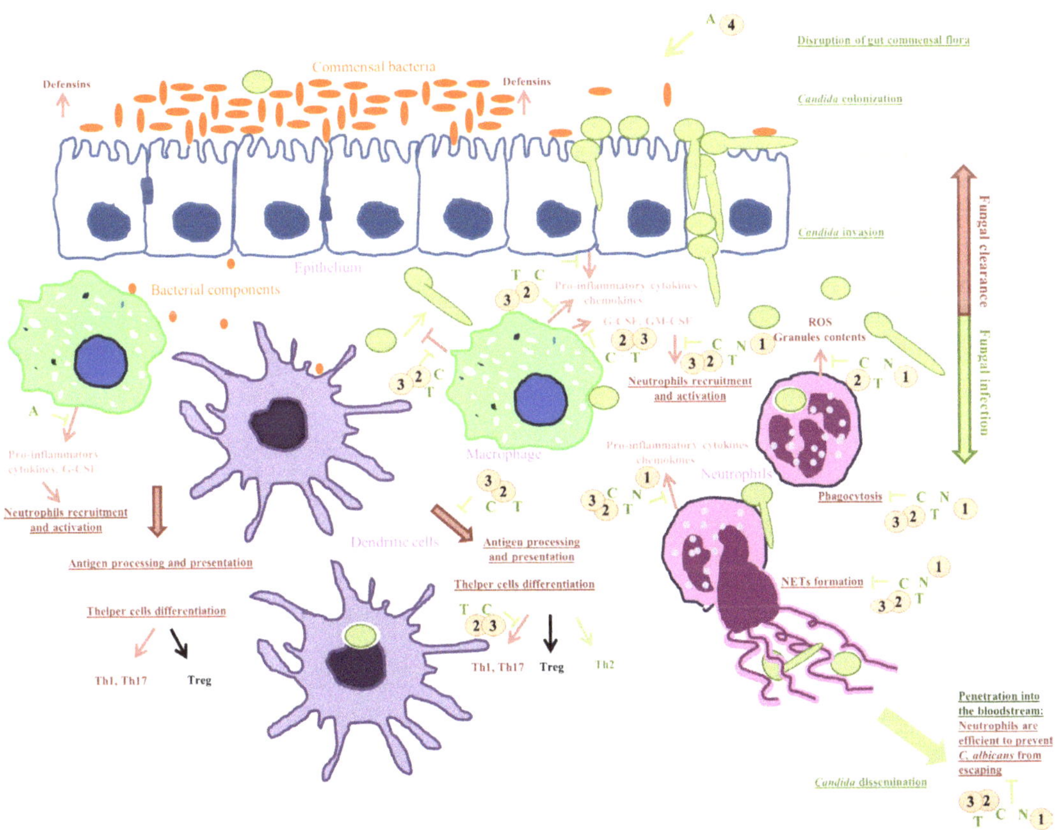

Figure 11.1 *Candida albicans* colonization and dissemination from the gastrointestinal tract. Shown are (1) host cells involved in the defence against the yeast colonization, invasion and dissemination; (2) factors, which predisposed to *C. albicans* infection (A, wide spectrum antibiotic treatment; C, corticosteroids; N, neutropenia (cancer or chemotherapy); and T, immunosuppressive drugs after organ transplantation); and (3) immunotherapies to recover immune function (1, leucocytes transfusion possibly with growth factors or antifungal treatments; 2, growth factors; and 3, inflammatory cytokines).

pathogens and are being implicated more frequently in invasive candidiasis, although *C. albicans* remains the predominant cause (Echeverria *et al.*, 2011; Giri and Kindo, 2012; Pfaller *et al.*, 2011). Emergence of these species is mainly due to their intrinsic poor susceptibility to existing antifungals as compared to *C. albicans* (Collin *et al.*, 1999; Krcmery and Barnes, 2002).

Commensal fungi and disruption of the immune system

Antibiotic treatment can also alter homeostasis of systemic immunity maintained by the normal microbial flora. The gut microbiota can stimulate the mucosal immune system, but components released by these organisms can also pass through the mucosal barrier and challenge the immune cells involved in adaptive immunity and prime systemic immunity, allowing the host to rapidly and effectively prevent microbial invasion (Guarner, 2005; Kamada *et al.*, 2013; Lamouse-Smith *et al.*, 2011). A weakened immune system is a major risk factor for invasive *Candida* infections (Pfaller and Diekema, 2010) and patients with acquired immunodeficiencies are predisposed to these infections. Immunosuppressors can suppress cell-mediated as well as humoral immunity, which reduces the host defence against fungal pathogens. Secondary immunodeficiencies can have different origins: (i) medications such as corticosteroids, chemotherapy, or immunosuppressive drugs after organ transplantation (Cupps and Fauci, 1982; Fauci *et al.*, 1976; Fedor and Rubinstein, 2006; Mahmud *et al.*, 2010; Rasmussen and Arvin, 1982); or (ii) diseases such as different types of cancer or HIV infection (Jaffe *et al.*, 1996; Rosenberg and Fauci, 1989). Depending on the type of immunodeficiency, specific immune functions will be altered with a distinct impact on susceptibility to fungal invasion and dissemination (Fig. 11.1). Although acquired immunosuppression by drugs such as corticosteroids predisposes patients to invasive infections (Nucci and Colombo, 2002), HIV infection is rarely associated with *Candida* dissemination but rather with oral and oesophageal candidiasis (Meyer and Schuster, 2002). Chemotherapy and some types of cancer induce neutropenia, which is frequently associated with invasive candidiasis (Karabinis *et al.*, 1988). Immunosuppressive agents can have additional effects on susceptibility to *Candida* infections. For example, dexamethasone, a glucocorticoid, has been reported to enhance the growth of the yeast *in vitro* and in the murine upper intestinal tract (Gupta *et al.*, 1982; Uematsu and Fujimoto, 2010). The mechanism involved is not known, but *C. albicans* strains, like other *Candida* species and other types of fungi, possess cell wall corticosteroid-binding proteins (Loose *et al.*, 1983). Cell wall-incorporated corticosteroids can then bind to corticosteroid receptors on host cells and tissues, promoting adherence of *C. albicans* to epithelial cells and colonization of the epithelium by the yeast (Ghannoum and Elteen, 1987). Some other glucocorticoids like cortisone acetate or hydrocortisone can, however, block adherence of the yeast to epithelial cells (Ghannoum and Elteen, 1987).

Although invasive candidiasis is usually caused by endogenous yeasts, invasive surgical devices such as indwelling venous catheters are putative sources of exogenous *C. albicans* (Blot, 2003; Bouza *et al.*, 2004; Ruzicka *et al.*, 2012; Talaat *et al.*, 2010). Like bacteria, *C. albicans* can form biofilms on plastic devices/supports which are resistant to antifungals and can be a constant source of invading microorganisms as patients with catheters are usually receiving treatments that facilitate this invasion (see Chapter 4) (Al-Fattani and Douglas, 2006; Andes *et al.*, 2004; Hawser and Douglas, 1994, 1995; Kuhn and Ghannoum, 2004; Ricicova *et al.*, 2010). Yeast cells are delivered directly into the bloodstream from which they can easily disseminate.

Environmental fungi

Environmental fungi have invasive properties that are less dependent on the microbial flora than commensal fungi. Some of them, like *Coccidioides immitis* and its relative *C. posadasii*, *Blastomyces dermatitidis* and *Histoplasma capsulatum*, are infectious in immunocompetent subjects without specific risk factors (Bradsher *et al.*, 2003; Olivere *et al.*, 1999; Wheat and Kauffman, 2003). These primary pathogens are dimorphic fungi with a mycelial saprophytic phase and a yeast pathogenic phase. They are usually endemic pathogens found in soil (*C. immitis*, *C. posadasii* and *B. dermatitidis*) or in bird and bat droppings (*H. capsulatum*) with spores inhaled by humans, delivering the fungi into the lungs. In immunocompetent patients, the resulting infections usually resolve without specific treatments due to effective cellular immunity that controls fungal development and dissemination. Resulting invasive mycosis can be observed mainly in predisposed patients with altered cellular immunity due to diseases such as HIV infection or some cancers, or to medical therapies such as corticosteroid treatment or chemotherapy (Pfaller and Diekema, 2010). Alteration of the immune defences also favours infections by opportunistic environmental fungal pathogens. *Aspergillus fumigatus* and *Cryptococcus neoformans* are the most common of these pathogens. By contrast with the true pathogens mentioned above, these fungi are ubiquitous in the environment. Most humans inhale *Aspergillus* spores every day but have an effective immune defence that controls the fate of the fungus in the respiratory tract. Depending on various risk factors, this mould can cause different forms of aspergillosis ranging from allergic bronchopulmonary aspergillosis (predisposing factor: asthma or cystic fibrosis), pulmonary aspergilloma (predisposing factor: preexisting lung cavity), and invasive aspergillosis (predisposing factor: immunosuppression). The latter is the most serious form of the disease and has a high mortality rate. *A. fumigatus* is the main cause of aspergillosis in humans as, by contrast with other *Aspergillus* species such as *A. flavus* (Krishnan *et al.*, 2009) and more rarely *A. niger* (Person *et al.*, 2010), it is well adapted to the human body temperature (McCormick *et al.*, 2010b). Whatever the species, immunodeficiency is the main condition required to develop invasive aspergillosis, with invasive pulmonary aspergillosis being the predominant form of the disease (Fig. 11.2). The most important risk factor is neutropenia but there are other factors such as haematopoietic stem-cell or solid-organ transplants and their associated medications, prolonged treatment with corticosteroids, advanced HIV infection, or chronic granulomatous disease (Baddley, 2011; Gangneux *et al.*, 2010; Li *et al.*, 2012b; Maschmeyer *et al.*, 2007). Advanced HIV infection and organ transplantation are also risk factors for cryptococcal meningitis, which occurs after dissemination of *C. neoformans* from a pulmonary infection (Baddley *et al.*, 2008; Pfaller and Diekema, 2010; Pippard *et al.*, 1986). Disseminated cryptococcosis particularly affects people who have defects in their cell-mediated immunity. Factors such as lymphomas, liver cirrhosis and long-term corticosteroid therapy also predispose patients to such infection (Baddley *et al.*, 2008). The fungus can be found worldwide in different soils contaminated with bird droppings. Its spores can be inhaled by humans and cause three types of infections: wound or cutaneous cryptococcosis, pulmonary cryptococcosis, and the invasive form, cryptococcal meningitis, which is severe and often fatal (Perfect and Casadevall, 2002). *C. neoformans* is the main species responsible for invasive cryptococcosis but *C. gattii* can also cause this disease (Oliveira Fde *et al.*, 2007; Rolston, 2013; Santos *et al.*, 2008; Springer *et al.*, 2012). By contrast to *C. neoformans*, *C. gattii* can be infectious in non-immunocompromised patients but, like the primary pathogens, this species is endemic to specific regions of the world, decreasing its incidence. It is unknown if the low incidence of cryptococcosis caused by *C. gattii* is due to rare exposure to this fungus or to unidentified immune defects predisposing to its dissemination.

Interplay between fungi and host immune cells

Immunocompetent individuals are rarely affected by IFIs as their immune system sets up an efficient defence to prevent fungal development and invasion (Ruping *et al.*, 2008). Several studies have focused on innate and adaptive immunity essential to limit the spread of commensal fungi in the gut or colonization and invasion by environmental fungi. Several questions have been addressed to identify the protective host response and its mechanisms against fungi: (i) Which *in vitro*, *ex vivo*, and *in vivo* models are relevant to study the immune response to fungal infection?; (ii) Which immune cells are effective against the fungi?; (iii) Which fungal molecules and host cells trigger or down-regulate the protective immune response?; and (iv) Which fungal cell morphotypes modulate the immune response? These studies and studies to come are crucial to define putative

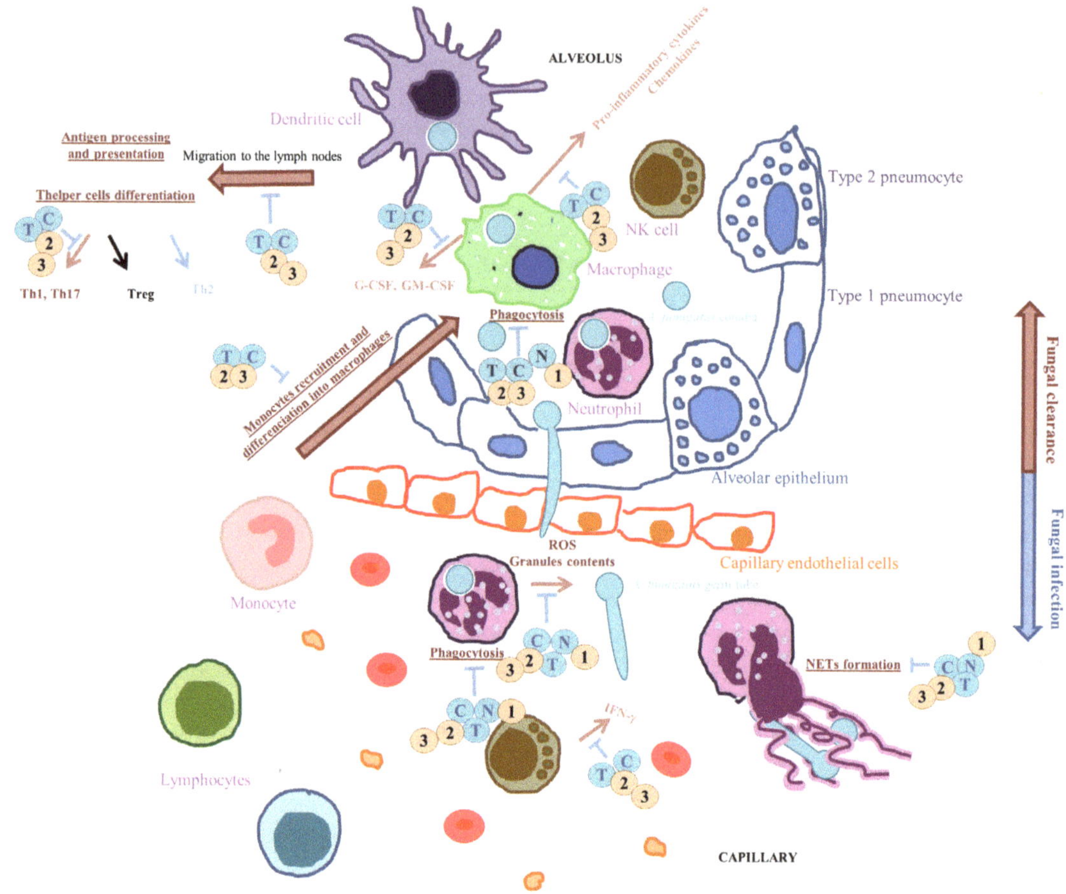

Figure 11.2 *Aspergillus fumigatus* dissemination from the alveolus. Shown are (1) host cells involved in the defence against the yeast colonization, invasion and dissemination; (2) factors, which predisposed to *A. fumigatus* invasion (C, corticosteroids; N, neutropenia (cancer or chemotherapy); and T, immunosuppressive drugs after organ transplantation); and (3) immunotherapies to recover immune function (1, leucocytes transfusion possibly with growth factors or antifungal treatments; 2, growth factors; and 3, inflammatory cytokines).

immunomodulators able to counteract specific immunodeficiencies induced by chemical agents or disease. Key immune factors that are crucial to stopping fungi from disseminating have already been identified.

In vitro and *ex vivo* approaches to analyse the host responses against fungi

The epithelium is a natural barrier to microorganisms. Epithelial cells can sense microorganisms and secrete antimicrobial peptides such as defensins and cathelecidins (Gallo and Hooper, 2012; Ganz, 2003; Schneider *et al.*, 2005) (Fig. 11.1) that directly kill or limit the growth of microorganisms like fungi, and inflammatory cytokines and chemokines that attract phagocytes to the sites of infection and orientate mucosal immunity against the pathogens (Stadnyk, 2002). This barrier is very important to control the natural microbial flora of the host and to tolerate the different species within it. It is also the surface to which commensal fungi such as *C. albicans* are well adapted and adhere to in order to colonize their host. Studies on the interaction between fungi and the epithelium have focused on fungal adherence and the release of antimicrobial peptides, cytokines and chemokines on interaction with fungi. *In vitro* analyses involve different epithelial cell lines depending on the fungus being studied. For *C. albicans*, oral, intestinal, or vaginal epithelial cells (Dalle *et al.*, 2010; Dongari-Bagtzoglou and Kashleva, 2003; Holmes *et al.*, 2002; Steele and Fidel, 2002) are preferred whereas pulmonary epithelial cells are appropriate for *A. fumigatus* and *C. neoformans* (Bellanger *et al.*, 2009; Merkel and Scofield, 2000; Tomee *et al.*, 1997). *In vitro* models based on reconstituted human oral, oesophageal, and vaginal epithelium (RHE) that mimic more closely the complexity of mucosal infections are often used to study mucosal candidiasis (Jayatilake *et al.*, 2007; Lu *et al.*, 2006; Schaller *et al.*, 2008; Wagener *et al.*, 2012). Moreover, *ex vivo* assays can be performed using human epithelial cells taken from the buccal cavity, cervix, or inferior turbinate of volunteers (Amitani *et al.*, 1995; Steele and Fidel, 2002; Wiechula *et al.*, 2010; Yano *et al.*, 2005). The latter are attractive as they allow comparison between cells from healthy donors and cells from patients (Fortis *et al.*, 1998). They have, however, never been widely used because only one type of cell is challenged at once. *In vivo* conditions are not limited to one cell type and more relevant *in vitro* and *ex-vivo* assays should include different cell types. There are different types of epithelium: one cell thick epithelium (simple epithelium), or two or more cells thick epithelium (stratified epithelium) as well as different types of epithelial cells that constitute one epithelium. The principle cell types of the intestinal epithelium (enterocytes, enteroendocrine cells, goblet cells and paneth cells) may have distinct roles in recognition of commensal microorganisms and maintenance of immune homeostasis in the gut (Artis, 2008). Mucosal immunity is also dependent on other cells that act together with epithelial cells to eliminate invasive microorganisms (Weindl *et al.*, 2010, 2011). In some studies, phagocytes such as neutrophils were added to RHE and were shown to have a clear impact on the epithelial response and damage due to fungal infection (Schaller *et al.*, 2004; Weindl *et al.*, 2007).

Leucocytes are essential cells of innate immunity and some of them (monocytes), which can differentiate into APCs (macrophages and dendritic cells), are also required for adaptive immunity (Fig. 11.3). Phagocytosis and

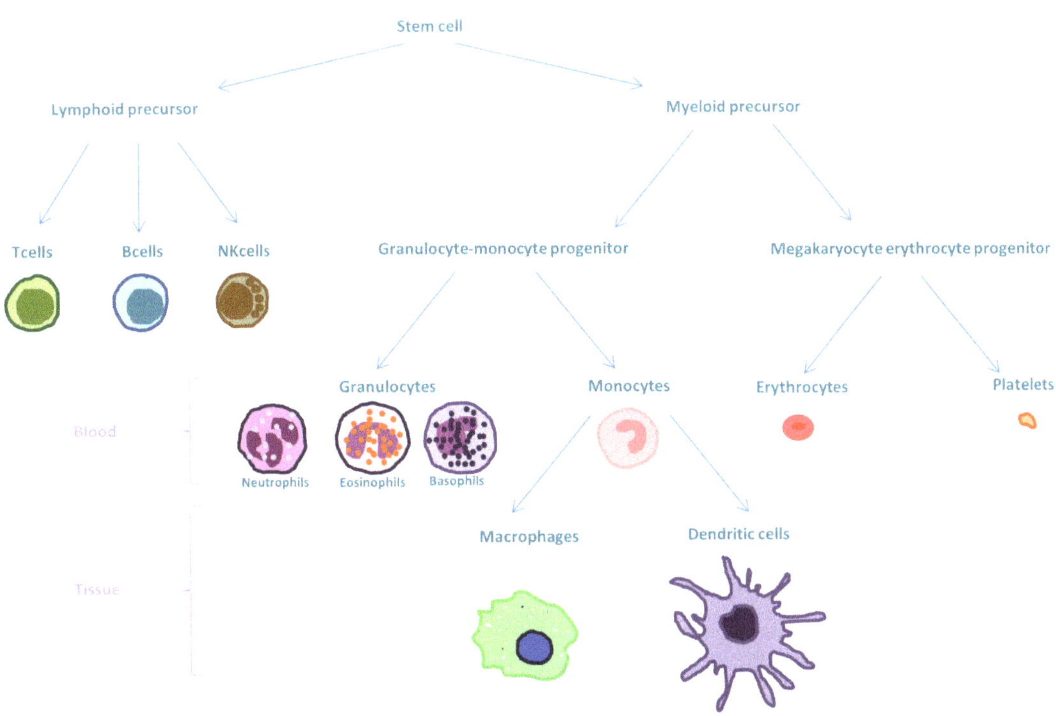

Figure 11.3 Key immune system cells from the myeloid and lymphoid lineages.

efficient killing of fungi, as well as cytokine secretion (inflammatory versus anti-inflammatory response) after stimulation by fungi and their components, can be evaluated.

Cells of macrophage lineage are most often studied. Macrophages, together with neutrophils, are very efficient phagocytes of the innate immunity, but have also a relevant role in adaptive immunity as they are APCs and can orientate the T-cell response through cytokine secretion. Preliminary experiments are usually performed with different cell lines (Fradin *et al.*, 1996; Ibrahim-Granet *et al.*, 2003; Martinez-Solano *et al.*, 2009; Mukherjee *et al.*, 1996; Sasada and Johnston, 1980; Wasylnka and Moore, 2002; Xiao *et al.*, 2008). Mononuclear cells can be differentiated into macrophages using IFN-γ or a phorbol ester, phorbol-12-myristate-13-acetate (PMA) (Jouault *et al.*, 1994, 2006). This last model is interesting as the response of both monocytes and macrophages can be monitored. Differentiated macrophages can express different receptors, secrete distinct cytokines and are more efficient at killing fungi than monocytes. Macrophages from different organs (e.g. Kupffer cells in the liver, alveolar macrophages in the pulmonary alveolus, etc.) can display specific traits and functions. It is therefore important to use cells that are relevant to the pathophysiology of the fungal infection. A human promyelocytic cell line, HL-60, can be differentiated into two distinct cell types: (i) either macrophages with plasminogen activator; or (ii) neutrophils with dimethyl formamide (Baquir *et al.*, 2012; Mullick *et al.*, 2004). Differentiation of cell lines can be associated with a background level of cytokine production that might interfere with the study and differentiated cells can be less efficient (phagocytosis, cytokines production) than mature native cells (Pullen and Hosking, 1985). Cells taken directly from animals or humans might be preferred. Human peripheral blood mononuclear cells or animal bone marrow-derived monocytes can be differentiated into macrophages with GM-CSF (Granulocytes-macrophages colony-stimulating factors) or into dendritic cells with GM-CSF plus IL-4 (Kelly *et al.*, 2005; Marr *et al.*, 2003). Macrophages can also be harvested from the peritoneal cavity of animals or by bronchoalveolar lavage of humans or animals (Alvarez *et al.*, 2008; Brummer *et al.*, 1994, 2002; Cameron *et al.*, 1990; Gaziri *et al.*, 1999; Kamberi *et al.*, 2002).

Animal or human neutrophils taken from blood have been widely used to analyse their interaction with fungi (Fradin *et al.*, 2005, 2007; Kozel *et al.*, 1987; Thywissen *et al.*, 2011; Urban *et al.*, 2006). Blood is a predominant source of immune cells, mononuclear cells (monocytes and lymphocytes), and polymorphonuclear cells (neutrophils, eosinophils, and basophils) (Fig. 11.3), that can be purified or used in selected combinations. Mixed cells can diminish the response (killing, modulation of microbial fitness, and immune modulation) of a specific cell type within the mixture (Fradin *et al.*, 2005, 2007).

Host receptors involved in the interaction with fungi, either fungal uptake or triggering of the immune response, can be identified either by affinity techniques using targeted ligands (Fradin *et al.*, 2000), using monoclonal antibodies or agonists that block the receptor–ligand interaction (Maneu *et al.*, 2011; Netea *et al.*, 2002), or using knock-out animals that no longer express specific receptors (Netea *et al.*, 2008).

Fungi often have different cellular morphotypes that trigger distinct immune responses and may have different susceptibilities to phagocytes (d'Ostiani *et al.*, 2000; Hohl *et al.*, 2005; Moyes *et al.*, 2010). Furthermore, their wall is composed of different molecules, mainly glycans, which are able to interact with host cells (Bourgeois *et al.*, 2010; Jouault *et al.*, 2009; Netea *et al.*, 2006b). They may be targets of the immune response or its modulators. Fungal mutants as well as purified molecules enable analysis of the immune response triggered by specific interactions (Netea *et al.*, 2006b).

In vivo approaches to analyse the host response against fungi

Additional *in vivo* assays are necessary to validate the relevance of data obtained with *in vitro* and *ex vivo* assays. Different animal models allow the analysis of fungal colonization (Bendel *et al.*, 2003; Clemons and Stevens, 2005, 2012; Jawhara and Poulain, 2007; Koh *et al.*, 2008; Rahman *et al.*, 2007; Takahashi *et al.*, 2003; Zwolinska-Wcislo *et al.*, 2007). These models are very valuable for *C. albicans* as invasive candidiasis is mainly due to endogenous strains that disseminate in high-risk patients. Whereas models of oral and vaginal colonization are very valuable to analyse superficial candidiasis, such as oral thrush and vaginitis, intestinal colonization is more relevant for invasive candidiasis as treatment of colonized mice with immunosuppressors results in dissemination of the yeast (Szabo and MacCallum, 2011). However, most studies are performed with mice, whose gut is not a natural host of *C. albicans*, generating a putative bias in these assays. Murine colonization can be established with antibiotic treatment prior to oral yeast inoculation. This treatment eliminates part of the bacterial flora and favours *C. albicans* colonization. However, it also modifies mucosal immunity, making it difficult to study normal protective mucosal immunity against yeast colonization and growth. Alteration of immune homeostasis can influence *C. albicans* colonization as observed in a murine model of inflammatory bowel disease (IBD). The gut of mice treated with a chemical agent that injures epithelial cells and promotes inflammation (dextran-sulfate sodium), prior to oral inoculation of *C. albicans*, was colonized with the yeast (Jawhara *et al.*, 2008). In this model, a pre-treatment with the yeast *S. boulardii* or with yeast compounds such as β-glucans inhibits inflammation and then *C. albicans* gastrointestinal tract colonization (Jawhara *et al.*, 2012; Jawhara and Poulain, 2007). This model is interesting to analyse modulation of inflammation by *C. albicans* in conditions similar to IBD, even though the clinical link between this yeast and IBD is unclear. However, the model has more limited value for the analysis of invasive candidiasis.

The relevance of *in vitro*/*ex vivo* models has also been also assessed using *in vivo* models of systemic infection, which are numerous for the different pathogenic fungi (Carroll *et al.*, 2007; Clemons and Stevens, 2005; Maccallum, 2012). To establish these infections, some of the models require immunosuppression induced by agents such as corticosteroids (dexamethasone) or chemotherapeutic agents (cyclophosphamide), which are risk factors for these diseases. Immunosuppressive treatment is necessary to allow *C. albicans* and *A. fumigatus* to disseminate from the gastrointestinal tract and to invade the lungs, respectively (Bendel *et al.*, 2003; Richie *et al.*, 2011). However, although they are relevant to study the establishment of IFIs immunosuppressive treatments are not recommended to analyse the host protective response against fungal invasion. Depending on the fungus studied, different routes of infection in animals are chosen. To deliver *C. neoformans* and *A. fumigatus* strains to the lungs of animals, these fungi are inoculated via the intratracheal route (Carroll *et al.*, 2012; Deak *et al.*, 2011). To mimic its dissemination to the brain, *C. neoformans* can also be delivered intracerebrally or intravenously to animals (Van Cutsem, 1993; Zhou *et al.*, 2007). Systemic *C. albicans* infection is usually established by intravenous challenge of the yeast via the murine caudal vein (Szabo and MacCallum, 2011). These animal models enable the identification of (i) fungal determinants essential for invasive infection when the virulence of deleted strains is investigated; (ii) the protective immune response to fungal invasion; (iii) host determinants essential to establish this response when modulators and knock-out animals are used; and (iv) risk factors for invasive infection when animals are treated with immunosuppressors. These last three points will be discussed later in this review.

Innate immunity

Cells of innate immunity can distinguish self- from non-self elements with the help of different receptors expressed at their surface. These receptors, called pattern recognition receptors (PRRs), recognize different motifs at the microbial surface identified as pathogen-associated molecular patterns (PAMPs) (Filler, 2006). As discussed below, the interaction of these receptors with their ligands defines the outcome of innate immunity such as clearance of fungi by phagocytes but also the control of acquired immunity against these fungi. Several reviews have described fungal PAMPs and their corresponding PRRs (Bourgeois *et al.*, 2010; Cheng *et al.*, 2012; Cunha *et al.*, 2012; Jouault *et al.*, 2009; Netea *et al.*, 2006b; Romani, 2011; Sorrell and Chen, 2009), which are listed in Table 11.1.

Phagocytic cells are able to control and, for some fungi, to clear fungal infection. Among these cells are neutrophils, which are very efficient for

Table 11.1 Host pattern recognition receptors (PRRs) and their corresponding fungal pathogen-associated molecular patterns (PAMPs)

PRRs		PAMPs	Outcome	References
C-type lectins	Mannose receptor (MR)	α-Mannosides	Phagocytosis, inflammation	Bennett *et al.* (1987), Cambi *et al.* (2008), Porcaro *et al.* (2003), van de Veerdonk *et al.* (2009), Yamamoto *et al.* (1997)
	Dectin-1	β-Glucans	Phagocytosis, signalling through TLR2, TLR2/6 and TLR4	Brown *et al.* (2002), Ferwerda *et al.* (2008), Gersuk *et al.* (2006), Leal *et al.* (2010), Roeder *et al.*, (2004), Tsoni and Brown (2008), Viriyakosol *et al.* (2005)
	Dectin-2	High mannose, α-mannan, α-i,3-glucans	Signalling through FcγR, inflammation	Bittencourt *et al.* (2006), McGreal *et al.* (2006), Sato *et al.* (2006)
	Mincle	α-Mannose	Signalling through FcγR	Lee *et al.* (2011), Wells *et al.* (2008), Yamasaki *et al.* (2008)
	DCSIGN/ SIGNR1	High mannose, fucose	Anti-inflammatory response	Cambi *et al.* (2008), Serrano-Gomez *et al.* (2005), Svajger *et al.* (2010)
	Mannose binding lectin (MBL), soluble	α-Mannosides	Complement activation, phagocytosis	Bidula *et al.*, (2013), Ip and Lau (2004), Li *et al.* (2012a); Neth *et al.* (2000), Panepinto *et al.* (2007); van Asbeck *et al.* (2008)
S-type lectin	Galectin-3 (Gal-3), soluble	β-1,2-mannosides	Antimicrobial activity, phagocytosis, signalling through TLR2	Fradin *et al.* (2000); Jouault *et al.* (2003, 2006), Kohatsu *et al.* (2006), Linden *et al.* (2013)
TLRs	TLR2	β-glucans, phospholipomannan (PLM), chitin?	Pro-/anti-inflammatory responses, fungal clearance	Da Silva *et al.* (2008), Jouault *et al.* (2003), Luther *et al.* (2007), Netea *et al.* (2006a)
	TLR4	O-mannans	Pro-inflammatory response, fungal clearance	Netea *et al.* (2006b)
	TLR6	Zymosan	Pro-inflammatory response (heterodimer TLR2/TLR6)	Gil and Gozalbo (2009), Roeder *et al.* (2004)
	TLR9	CpG DNA	Pro/anti-inflammatory response, fungal clearance	Edwards *et al.* (2005), Kasperkovitz *et al.* (2010), Kasperkovitz *et al.* (2011), Miyazato *et al.* (2009), Nakamura *et al.* (2008), Ramirez-Ortiz *et al.* (2008)
Unknown	Unknown	Chitin	Possible signalling through TLR2, inflammation	Da Silva *et al.* (2008), Da Silva *et al.* (2010), Lee *et al.* (2008)

Host PRRs and fungal PAMPs can be ubiquitous molecules found on different cell types/species or specific to one cell type (e.g. DCSIGN expressed by cells of the dendritic cell/macrophage lineage) or some fungal species (e.g. β-1,2 mannosides expressed only by some *Candida* species and *Pichia pastoris*).

host defence against candidiasis, either mucosal or disseminated (Richardson and Rautemaa, 2009; van de Veerdonk *et al.*, 2010). These cells are recruited at the infection site by the haematopoietic growth factor G-CSF, and different proinflammatory cytokines such as IL-8, IL-6, TNF-α and IL-17 (Balish *et al.*, 1999; Basu *et al.*, 2008; Kullberg *et al.*, 1999; Romani *et al.*, 1996; Saunus *et al.*, 2010). These molecules, which are produced by different immune cells, are also able to activate neutrophil function. Neutrophils phagocytose and kill *C. albicans* yeast cells with antimicrobial peptides, hydrolytic enzymes, and reactive oxygen species (ROS) (Fig. 11.1).

They inhibit yeast to hyphal transition and once activated, release chromatin and extracellular granules that form neutrophil extracellular traps (NETs) able to kill both yeast and hyphal cells (Fradin *et al.*, 2005; Peltroche-Llacsahuanga *et al.*, 2000; Urban *et al.*, 2006). NETs contain the antifungal peptide calprotectin, which mediates zinc chelation and is crucial for clearance of *C. albicans* infection (Urban *et al.*, 2009). Interaction between *C. albicans* and neutrophils is mediated through β-glucans and β-1,2-oligomannosides that bind to the C-type lectin dectin-1 and the S-type lectin galectin-3, respectively (Fradin *et al.*, 2000; Kennedy *et al.*, 2007; Li *et al.*, 2011; Linden *et al.*, 2013). Furthermore, soluble β-glucans can promote neutrophil motility (Harler and Reichner, 2001). These glycans are also essential for the recruitment of neutrophils at the *A. fumigatus* infection site (Gessner *et al.*, 2012; Hohl *et al.*, 2005; Werner *et al.*, 2009). Conidia are phagocytosed by neutrophils but are also trapped like hyphae in NETs, which have a fungistatic effect to prevent the fungus from spreading (Bruns *et al.*, 2010; McCormick *et al.*, 2010a). Calprotectin, which is contained in the NETs, is essential for the anti-*Aspergillus* defence (Bianchi *et al.*, 2011) as well as ROS release by neutrophils to kill hyphae (Boyle *et al.*, 2011). Neutrophils are able to phagocytose and kill *C. neoformans* independently of ROS as the fungus releases mannitol, which scavenges oxygen intermediates (Chaturvedi *et al.*, 1996; Mambula *et al.*, 2000; Miller and Mitchell, 1991). Interaction of neutrophils with fungi involves different PAMPs and PRRs (Table 11.1) that may lead to similar, complementary, or antagonistic effects that are beneficial or not to the host.

Mononuclear phagocytes (monocytes/macrophages) have a key role in preventing *Cryptococcus* and *Aspergillus* infections. The recruitment of monocytes that are activated to become macrophages is an essential step for an effective pulmonary immune response against *C. neoformans* infection (Osterholzer *et al.*, 2011). In addition to neutrophils, resident macrophages can ingest and kill the opsonized (complement) or unopsonized (via CD14) fungus, limiting its growth, but this is not sufficient to completely clear the infection (Brummer and Stevens, 1994; Lipovsky *et al.*, 1997). Macrophages, like dendritic cells, produce cytokines that control the development of the adaptive cellular immune response: the level and types of T-cells and the types of cytokines they produce (Guillot *et al.*, 2008; Osterholzer *et al.*, 2009), leading, among other effects, to recruitment and activation of monocytes. Activated macrophages produce TNF-α and fungicidal nitric oxide (NO) (Alspaugh and Granger, 1991; Osterholzer *et al.*, 2011). Orientation of the adaptive immune response is also very important for the outcome of *Aspergillus* and *Candida* infections (Figs. 11.1 and 11.2). Macrophages are not very powerful phagocytes towards *C. albicans*, mainly because the yeast can escape the phagosome after germ tube production (Lo *et al.*, 1997). However, sensitized macrophages are more efficient at killing the yeast and inhibiting yeast-to-hypha transition (Evron, 1980), highlighting the importance of a primed immune response (Fig. 11.1). *A. fumigatus* conidia but not hyphae can be efficiently cleared by macrophages (Brakhage *et al.*, 2010). This phagocytosis is mediated by dectin-1 and Toll-like receptor (TLR) 2 (Luther *et al.*, 2007). As for neutrophils, the interaction of macrophages with fungi involves different PRRs and corresponding PAMPs (Table 11.1).

NK cells, innate lymphocytes with potent cytotoxic activity, display anti-cryptococcal activity *in vitro* through the expression and release of a granular cytolytic protein, perforin (Ma *et al.*, 2004; Murphy and McDaniel, 1982; Wiseman *et al.*, 2007). They also enhance clearance of *C. neoformans* from tissues *in vivo* (Balkus *et al.*, 2011; Hidore and Murphy, 1986; Salkowski and Balish, 1991). NK cells are very important effector cells in invasive aspergillosis (Park *et al.*, 2009). They are recruited to the lungs during early infection via CC chemokine ligand-2 (MCP-1/CCL2) to eliminate the fungus and enhance host survival (Morrison *et al.*, 2003). The direct activity of NK cells on *A. fumigatus* hyphae is independent of granular cytolytic proteins but dependent on IFN-γ release which acts directly on fungal cells (Bouzani *et al.*, 2011) (Fig. 11.2). The involvement of NK cells in the immune response against disseminated candidiasis has not been widely studied but it seems that these cells could act on *C. albicans* clearance via perforin (Voigt *et al.*, 2013).

Adaptive immunity

A successful outcome against IFIs often relies on the development of adaptive immunity. This antigen-specific immunity will be the topic of the Chapter 12. Cell-mediated immunity will be briefly discussed here. APCs such as macrophages and dendritic cells are essential to establish this immunity through phagocytosis of the pathogenic fungi and modulation of inflammation via specific fungal determinants, which promote T-cell differentiation. Different PRRs and their corresponding PAMPs are involved in phagocytosis and also signalling in the immune response (Table 11.1).

The effects of immune polarization are not limited to the T-helper (Th) 1/Th2 balance as Th17 and regulatory T-cells (Treg) also have a role during fungal infection (Figs. 11.1 and 11.2). CD4+ Th1 immunity with T-cells producing IFN-γ, IL-2, and TNFα cytokines is acknowledged as a protective response against IFIs whereas CD4+ Th2 immunity with T-cells producing IL-4, IL-5, IL-6, IL-10, and IL-13 cytokines is believed to exacerbate these infections (Cenci *et al.*, 1997a,b, 1998; Clemons *et al.*, 2000; Koguchi and Kawakami, 2002; Schop, 2007; Stuehler *et al.*, 2011; Wormley *et al.*, 2007). By contrast to the Th2 response, which is associated more with the humoral response and allergy, the Th1 response is associated with macrophages and cytotoxic T-cell activation and the inflammatory response. There is a balance between these two responses as the Th2 response has anti-inflammatory properties that can inhibit the Th1 response, as the development of Th1 or Th2 cells is driven by IL-12/IFN-γ and IL-4, respectively. Tregs can also modulate the Th1 response with IL-10 production to avoid the harmful consequences of a long-term Th1 response (Montagnoli *et al.*, 2006). Th17 polarization, usually associated with auto-immune and auto-inflammatory diseases (Marwaha *et al.*, 2012), is observed during fungal infections (Vautier *et al.*, 2010). Th17 cells that can act together with or repress Th1, have controversial roles in the outcome of IFIs (Sales-Campos *et al.*, 2013) with both protective (Lin *et al.*, 2009; Qiu *et al.*, 2012) and impaired (Zelante *et al.*, 2007, 2009) immune responses reported. The Th response clearly depends on the fungal molecule types (Bozza *et al.*, 2009; Jouault *et al.*, 2009) and fungal morphotypes (d'Ostiani *et al.*, 2000).

Restoration of the immune system

Immunotherapy can either be non-specific (leucocytes, cytokines, and growth factors) or specific (primed DCs, T-cells, and antibodies). This paragraph will focus on non-specific immunotherapy of fungal infections, for which the pre-clinical and clinical trials are summarized in Table 11.2. Specific immunotherapy is developed further in Chapter 12. Some interesting observations have been made after pre-clinical and clinical trials of immunotherapy to overcome IFIs. The promise and disappointment of immunotherapies for fungal infections have been the focus of different reviews (Antachopoulos *et al.*, 2012; Antachopoulos and Walsh, 2012; Casadevall and Pirofski, 2001; Hamad, 2008; Katragkou and Roilides, 2012; Lehrnbecher *et al.*, 2011; Nosanchuk and Taborda, 2013; Pikman and Ben-Ami, 2012; Safdar, 2013; Segal *et al.*, 2006).

Neutropenia recovery

Neutropenia is an important risk factor for fungal infections such as candidiasis and aspergillosis. Trials to overcome neutropenia by leucocyte transfusion were first performed several decades ago with limited success due to marginal efficacy, low neutrophil harvests, and association with toxic reactions and alloimmunization (Chanock and Gorlin, 1996; Wright, 1984). To increase the yield of granulocytes, donors were administered granulocyte colony-stimulating factors (G-CSFs), with dexamethasone or prednisolone. G-CSF, alone or together with pro-inflammatory cytokines such as IFN-γ, can stimulate antimicrobial functions of neutrophils (Dale *et al.*, 1998). Results of different clinical studies failed to demonstrate a clear benefit of granulocyte transfusion on IFIs. Whereas some authors reported a beneficial effect of the administration of G-CSF-primed leucocytes in patients with *Aspergillus* and *Candida* infections (Mousset *et al.*, 2005; Peters *et al.*, 1999), other studies showed only a limited or no effect (Bhatia *et al.*, 1994; Lee *et al.*, 2001; Seidel *et al.*, 2008). Furthermore, there appeared to be no benefit in

Table 11.2 Summary of immunotherapeutic studies on invasive fungal infections

Risk factor	Restoration of immune response	Immunotherapeutic studies	References
Neutropenia	Leucocyte transfusion	Not successful due to marginal efficacy/ low neutrophil harvest/toxic reactions/ alloimmunization	Chanock and Gorlin (1996), Wright (1984)
	G-CSF primed leucocyte transfusion	No clear benefit: no correlation between transfused dose and response to therapy (PS); no benefit (RS, RaPIIIS)	Bhatia *et al.* (1994), Lee *et al.* (2001), Seidel *et al.* (2008)
		Clearance of the infection independent on G-CSF (PI/IIS); prevention of infection-related mortality (PNRaS)	Mousset *et al.* (2005), Peters *et al.* (1999)
	Leucocyte transfusion + antifungal treatment	No effect on survival rate (RS) + possible neutrophil aggregation problems	Boxer *et al.* (1981), Safdar *et al.* (2004), Wright *et al.* (1981)
Depression of leucocyte and macrophage function	**CSFs**		
	G-CSF	No protective activity alone (AS and clinical prophylactic use)	Giles (1998), Sionov *et al.* (2005)
		Enhanced activity of antifungal drugs (AS)	Kullberg *et al.* (1999), Sionov *et al.* (2005)
		Induced GVDH and impaired immune recovery (allogenic stem-cell transplant recipients)	Ringden *et al.* (2004), Volpi *et al.* (2001)
	M-CSF	Reversed immunosuppression, induced protection against candidiasis and aspergillosis and enhanced anti-*Candida* activity of amphotericin B (AS)	Gonzalez *et al.* (2001), Kuhara *et al.* (2000)
		Beneficial combination with antifungal treatment for stem-cell transplant recipients (F-US) protection against	Nemunaitis *et al.* (1993)
	GM-CSF	Restored fungicidal activity of alveolar macrophages against *A. fumigatus* conidia (AS)	Brummer *et al.* (2001)
		Prophylactic use reduced fungal infections associated with high mortality for patients with acute myeloid leukaemia	Giles (1998)
Cellular immune deficiency	**Pro-inflammatory cytokines**		
	IFN-γ	No reproducible or significant efficiency against disseminated candidiasis (AS)	Balish *et al.* (1998), Kaposzta *et al.* (1998), Lavigne *et al.* (1998)
		Prevented cryptococcosis and aspergillosis and potentiated antifungal activity of drugs against the two corresponding pathogens (AS)	Clemons *et al.* (2001), Naik *et al.* (2011)
		Prevented cryptococcal meningitis in clinical trials of adjunct immunotherapy in HIV-infected patients (RaS, PIIDBP-CS)	Jarvis *et al.* (2012), Pappas *et al.* (2004)
		Enhanced clearance of *C. neoformans* from cerebrospinal fluid of HIV-patients with cryptococcal meningitis by standard antifungal treatment (PS)	Bicanic *et al.* (2009)
		Successful treatment of invasive aspergillosis when used in combination with antifungal drugs (CR)	Al-Tawfiq and Al-Abdely (2010), Kelleher *et al.* (2006), Saulsbury (2001)

Table 11.2 continued

Risk factor	Restoration of immune response	Immunotherapeutic studies	References
		Does not fully prevent invasive aspergillosis in CGD patients when used as prophylactic treatment	Lee *et al.* (2010), Segal and Walsh (2006)
	IL-12	Protected neutropenic mice from disseminated candidiasis but induced septic shock in non-neutropenic mice	Romani *et al.* (1995)
		Enhanced antifungal activity of monocytes against *A. fumigatus* hyphae and prevented pulmonary and invasive cryptococcosis (*in vitro* and AS)	Antachopoulos and Walsh (2012), Roilides *et al.* (1999)
		Increased susceptibility of patients receiving autologous stem-cell transplants to fungal infections	Toren *et al.* (1997)
	IL-18, IL-15	Involved in the protective effect of IL-12 against murine pulmonary cryptococcosis (IL-18) and enhanced antifungal activity of granulocytes and NK cytotoxicity (IL-15)	Kawakami *et al.* (2000), Ma *et al.* (2002), Mody *et al.* (1998), Musso *et al.* (1998), Tran *et al.* (2003), Vazquez *et al.* (1998), Winn *et al.* (2003)
	TNF-α	Resistance to *A. fumigatus* and *C. neoformans* enhanced by its agonists	Kawakami *et al.* (1996), Mehrad *et al.* (1999)
		Linked to inflammatory diseases	Bradley (2008)

PS, prospective studies; RS, retrospective studies; RaPIIIS, randomized phase III studies; PI/IIS, phase I and II studies; PNRaS, prospective non-randomized studies; F-US, follow-up studies; RaS, randomized studies; AS, animal studies; PIIDBP-CS, phase II double-blind, placebo-controlled studies; CR, case reports.

terms of the survival rate of patients with candidaemia (Safdar *et al.*, 2004). Neutrophils are very important to control *C. albicans* growth at the beginning of the infection. However, one of the main problems of *Candida* infections is making an early diagnosis so that it can be treated effectively. It is often difficult to establish an early diagnosis and treatment is therefore started too late to eliminate the fungus. Once the disease is diagnosed, the fungus has started to develop and disseminate, neutrophils might not be efficient enough to overcome the disease. A combination of antifungal treatment with leucocyte transfusion could then be a good alternative. Unfortunately, this combination has failed to give promising results. These disappointing results are attributed to the limited number of neutrophils administered and/or to patient heterogeneity as, as for leucocyte transfusion alone (Seidel *et al.*, 2008), the treatment is successful in some patients but not in others (Safdar *et al.*, 2004). Furthermore, some studies have reported that antifungal treatment also had an impact on leucocyte transfusion with increased obstruction of capillaries by neutrophils and/or neutrophil aggregations (Boxer *et al.*, 1981; Wright *et al.*, 1981). Neutrophils are not the only innate immune cells important to limit fungal development. For aspergillosis, mononuclear cells are also important in the initial control of the disease. Reports on immunity against the different invasive fungi clearly show that this immunity is very specific to the fungal species and the time of infection. It is not surprising then that leucocyte transfusion is not indicated for all fungal infections.

Growth factors

Antifungal immunotherapies using growth factors that can differentiate granulocytes (G-CSF) or macrophages (M-CSF) or both (GM-CSF) and modulate the antifungal activity of the differentiated cells (neutrophils, monocytes, and macrophages) have been investigated (Kuhara *et al.*, 2000; Kullberg *et al.*, 1999). In animal models, these CSFs have shown promising results to protect against lethal fungal infection and to potentiate antifungal drugs. Combination of recombinant human M-CSF with amphotericin B or recombinant G-CSF with fluconazole prolonged mice survival against disseminated *Candida* infection (Kuhara *et al.*, 2000; Kullberg *et al.*, 1999). A benefit of recombinant human M-CSF could be also observed when used alone, because it reversed the immunosuppressive treatment of mice and prevented yeast dissemination. Macrophages, but not neutrophils, seem important to overcome invasive aspergillosis in animal models. Although G-CSF enhanced the antifungal properties of caspofungin, amphotericin B, and posaconazole in a murine model of invasive aspergillosis, it has no protective activity without an antifungal drug (Sionov *et al.*, 2005). Furthermore, it could not reverse the immunosuppression due to steroid treatment. By contrast, administration of GM-CSF could reverse the immunosuppressive effects of corticosteroid treatment of mice and restore the fungicidal activity of alveolar macrophages against *Aspergillus* conidia (Brummer *et al.*, 2001). Immunosuppressed mice with pulmonary aspergillosis displayed a reduced fungal burden after inhalation of this growth factor. Recombinant human M-CSF was reported to protect neutropenic rabbits against challenge with *A. fumigatus* (Gonzalez *et al.*, 2001). This protection was independent of neutrophils but rather involved pulmonary alveolar macrophages whose number and function were increased with the treatment. Growth factors were also evaluated in clinical trials in which they showed some good antifungal potential. Prophylactic use of haematopoietic CSFs reduced the periods of neutropenia and number of febrile neutropenic episodes in high-risk cancer patients under cytotoxic chemotherapy (Ozer *et al.*, 2000). This treatment helps to reduce bacterial and fungal infections. No evidence of a pro-leukaemic effect was observed when CSFs were used (Giles, 1998). However, the use of G-CSFs was shown to induce graft versus host disease (GVHD), which is the result of the host cells recognized by the graft as foreign ones, and to then impair immune recovery in allogenic stem-cell transplant recipients (Ringden *et al.*, 2004; Volpi *et al.*, 2001). Furthermore, by contrast with GM-CSF, prophylactic use of G-CSF in patients undergoing induction chemotherapy for acute myeloid leukaemia resulted in no significant reduction in the frequency of IFIs, associated with a high mortality (Giles, 1998). A long term follow-up study highlighted the beneficial combination of M-CSF with antifungal treatment in stem-cell transplant recipients with progressive fungal infections (Nemunaitis *et al.*, 1993). A better overall survival could be observed. Animal and clinical studies showed the potential of haematopoietic CSFs to prevent IFIs in high risk populations and to potentiate antifungal treatment in severe non-responding infections. Depending on the nature of the factor, different antifungal activity can be observed. GM-CSF gave better results than G-CSF, suggesting, as for leucocyte transfusions, that restoring the neutrophil count is not sufficient to limit fungal infections and that the monocyte–macrophage lineage is essential for successful clearance of invasive fungi.

Cytokines

As described under 'Adaptive immunity', several cytokines could be used as adjuvant therapy of IFIs. The Th1 response is very efficient against such infections. Recovery of this response in predisposed patients could overcome IFIs. Although FN-γ is a key cytokine for induction and maintenance of Th1 response, it was not reproducibly and significantly efficient against murine candidiasis (Balish *et al.*, 1998; Kaposzta *et al.*, 1998; Lavigne *et al.*, 1998). Studies with animal models of *C. neoformans* and *A. fumigatus* infections report that this cytokine is effective at preventing fungal infections and potentiating the activities of amphotericin B and a novel polyene macrolide antibiotic (SJA-95) with antifungal activity (Desai and Naik, 2008), respectively (Clemons *et al.*, 2001; Naik *et al.*, 2011). Clinical trials of adjunctive immunotherapy in HIV-infected patients with cryptococcal meningitis gave promising results for the use of recombinant IFN-γ in addition to antifungal therapy to prevent fungal infection (Jarvis *et al.*, 2012; Pappas *et al.*, 2004). Recombinant IFN-γ was safe, well tolerated and may be used in addition to antifungal treatment in refractory cryptococcal cases as stated in the clinical practice guidelines for the management of cryptococcosis

(Perfect *et al.*, 2010). Combination of IFN-γ with standard antifungal treatment did not affect overall mortality but promoted more rapid clearance of *C. neoformans* from the cerebrospinal fluid than standard antifungal treatment alone. Clearance of the fungus is considered as a marker of a positive treatment response (Bicanic *et al.*, 2009).

IFN-γ use for the treatment of invasive aspergillosis is less well documented as studies are restricted to preclinical assays and clinical case reports (Al-Tawfiq and Al-Abdely, 2010; Kelleher *et al.*, 2006; Saulsbury, 2001). These reports describe the successful treatment of invasive aspergillosis with combinations of IFN-γ and antifungal drugs, but randomized, controlled clinical trials are needed for any relevant conclusion as for *Cryptococcus* infection. Patients with chronic granulomatous disease (CGD) are at high risk of invasive aspergillosis. Neutrophils from these patients exhibit poor fungicidal activity against *A. fumigatus* but treatment of the neutrophils with recombinant IFN-γ (Rex *et al.*, 1991). Although recombinant IFN-γ is licensed as prophylactic treatment in patients with CGD, invasive aspergillosis remains a persistent problem in patients receiving IFN-γ (Lee *et al.*, 2010; Segal and Walsh, 2006). Although it is not supported by clinical trials, IFN-γ may aggravate GVHD in allogeneic haematopoietic stem-cell transplant recipients (Lange, 2008; Wang *et al.*, 2009).

Other pro-inflammatory cytokines are involved in up-regulation of the antifungal Th1 response. These cytokines have been less well studied than IFNγ as possible adjuvants to prevent fungal infections. Animal studies have been performed to investigate whether IL-12, IL-15, and TNF-α could overcome fungal infections, but some limitations of their use have been highlighted as they may induce an efficient antifungal inflammatory response which may become overactive with detrimental effects on the host. IL-12 is a good example of the dual role of a pro-inflammatory cytokine. This cytokine can protect neutropenic mice from disseminated candidiasis but can also induce death from septic shock in non-neutropenic mice (Romani *et al.*, 1995). IL-12 can furthermore increase susceptibility to fungal infections as observed in patients receiving autologous stem-cell transplants who developed IFIs following IL-12 treatment (Toren *et al.*, 1997). However, *in vitro* and *in vivo* assays have highlighted the protective role of IL-12 against fungi: it can enhance antifungal activity of monocytes against *A. fumigatus* hyphae (Roilides *et al.*, 1999) and can prevent pulmonary and invasive cryptococcal disease (Antachopoulos and Walsh, 2012), respectively. IL-18, which also promotes the Th1 response, could be a good candidate for immunotherapy even though preclinical and clinical trials need to be performed (Kawakami *et al.*, 2000). IL-15, which enhances the antifungal activities of granulocytes and NK cytotoxicity (Ma *et al.*, 2002; Mody *et al.*, 1998; Musso *et al.*, 1998; Tran *et al.*, 2003; Vazquez *et al.*, 1998; Winn *et al.*, 2003), should also be considered as part of antifungal treatment even if it is not effective alone (Parayath *et al.*, 2000). Further investigations are required to confirm its antifungal potential. TNF-α is a key cytokine of the Th1 response and is crucial for protection against fungal invasion. Therapies using antibodies to block this cytokine increase the susceptibility of patients to IFIs (Marie and Guglielmino, 2010; Tsiodras *et al.*, 2008). *In vitro* and *in vivo* assays have shown that TNF-α and its agonists enhance resistance to *A. fumigatus* and *C. neoformans* (Kawakami *et al.*, 1996; Mehrad *et al.*, 1999). As TNF-α is linked to different inflammatory diseases (Bradley, 2008), its use in immunotherapy needs further experimental assays to identify putative secondary effects linked to this cytokine.

Specific therapy

This approach is reviewed in Chapter 12 and will not be developed further here. Briefly, adaptive T-cell therapy is another possible approach to overcome fungal infection. It is a T-cell-independent vaccination, which does not induce immunologic memory. For example, antifungal Th1-type immune reconstitution can be induced during transplantation by transferring fungus-specific Th clones together with the graft or by transfusing DCs already primed to trigger fungus-specific Th1 immunity (Bacci *et al.*, 2002; Bozza *et al.*, 2003; Cenci *et al.*, 2000; Perruccio *et al.*, 2004, 2005). The point, of course, is not to develop immunity that could trigger rejection of the graft but to specifically target the fungus.

Conclusion

Control of IFIs mainly relies on a functional immune system. Innate immune cells play a key role in defence against invading fungi, most of which are opportunistic pathogens whose main risk factor for invasion is disruption of the immune system. The severity of diseases caused by primary fungal pathogens is also linked to immunosuppression. Restoration of a normal immune system, or at least some of its parts, is crucial to limit fungal colonization and invasion, and is a promising alternative to standard antifungal treatments. It may then partially overcome problems encountered with antifungal resistance and early diagnosis of IFIs. Thanks to our knowledge on risk factors for invasive fungal infections, immunosuppressed patients may undergo prophylactic immunotherapy. However, special awareness should be taken not to enhance the primary clinical outcome. The main questions are then to decide when the patients are at risk of IFIs and which immunotherapies may prevent them. Pre-clinical *in vitro* studies are very helpful to target protective factors that can be afterwards tested in animal models. These studies are crucial to decipher protective versus fostering fungal spread depending on the immune activation and associated immune cells. They may also enable analysis of the relationship between structure and activity of fungal cell wall components and its outcome on modulation of immune response. Only clinical trials can determine the usefulness of selected non-specific immunotherapies. Some of these studies have given promising results and highlighted antifungal activity of immunotherapy against different fungal species, opening up the door to possible universal antifungal immunotherapy. Wider clinical trials, preferentially randomized trials, with relevant numbers of patients are required to confirm these data. Furthermore, other immunotherapies such as adoptive transfer of NK cells or challenge with non-tested cytokines (IL-25, etc.) that promote the Th1 response should be investigated.

References

Al-Fattani, M.A., and Douglas, L.J. (2006). Biofilm matrix of *Candida albicans* and *Candida tropicalis*: chemical composition and role in drug resistance. J. Med. Microbiol. *55*, 999–1008.

Alspaugh, J.A., and Granger, D.L. (1991). Inhibition of *Cryptococcus neoformans* replication by nitrogen oxides supports the role of these molecules as effectors of macrophage-mediated cytostasis. Infect. Immun. *59*, 2291–2296.

Al-Tawfiq, J.A., and Al-Abdely, H.M. (2010). Vertebral osteomyelitis due to *Aspergillus fumigatus* in a patient with chronic granulomatous disease successfully treated with antifungal agents and interferon-gamma. Med. Mycol. *48*, 537–541.

Alvarez, M., Saylor, C., and Casadevall, A. (2008). Antibody action after phagocytosis promotes *Cryptococcus neoformans* and *Cryptococcus gattii* macrophage exocytosis with biofilm-like microcolony formation. Cell. Microbiol. *10*, 1622–1633.

Amitani, R., Murayama, T., Nawada, R., Lee, W.J., Niimi, A., Suzuki, K., Tanaka, E., and Kuze, F. (1995). Aspergillus culture filtrates and sputum sols from patients with pulmonary aspergillosis cause damage to human respiratory ciliated epithelium *in vitro*. Eur. Respir. J. *8*, 1681–1687.

Andes, D., Nett, J., Oschel, P., Albrecht, R., Marchillo, K., and Pitula, A. (2004). Development and characterization of an *in vivo* central venous catheter *Candida albicans* biofilm model. Infect. Immun. 72, 6023–6031.

Angebault, C., Djossou, F., Abelanet, S., Permal, E., Ben Soltana, M., Diancourt, L., Bouchier, C., Woerther, P.L., Catzeflis, F., Andremont, A., *et al.* (2013). *Candida albicans* is not always the preferential yeast colonizing humans: a study in Wayampi Amerindians. J. Infect. Dis. *208*, 1705–1716.

Antachopoulos, C., and Walsh, T.J. (2012). Immunotherapy of Cryptococcus infections. Clin. Microbiol. Infect. *18*, 126–133.

Antachopoulos, C., Katragkou, A., and Roilides, E. (2012). Immunotherapy against invasive mold infections. Immunotherapy *4*, 107–120.

Artis, D. (2008). Epithelial-cell recognition of commensal bacteria and maintenance of immune homeostasis in the gut. Nat. Rev. Immunol. *8*, 411–420.

van Asbeck, E.C., Hoepelman, A.I., Scharringa, J., Herpers, B.L., and Verhoef, J. (2008). Mannose binding lectin plays a crucial role in innate immunity against yeast by enhanced complement activation and enhanced uptake of polymorphonuclear cells. BMC Microbiol. *8*, 229.

Ascioglu, S., Rex, J.H., de Pauw, B., Bennett, J.E., Bille, J., Crokaert, F., Denning, D.W., Donnelly, J.P., Edwards, J.E., Erjavec, Z., *et al.* (2002). Defining opportunistic invasive fungal infections in immunocompromised patients with cancer and hematopoietic stem cell transplants: an international consensus. Clin. Infect. Dis. *34*, 7–14.

Bacci, A., Montagnoli, C., Perruccio, K., Bozza, S., Gaziano, R., Pitzurra, L., Velardi, A., d'Ostiani, C.F., Cutler, J.E., and Romani, L. (2002). Dendritic cells pulsed with fungal RNA induce protective immunity to *Candida albicans* in hematopoietic transplantation. J Immunol. *168*, 2904–2913.

Baddley, J.W. (2011). Clinical risk factors for invasive aspergillosis. Med. Mycol. *49* (Suppl. 1), S7-S12.

Baddley, J.W., Perfect, J.R., Oster, R.A., Larsen, R.A., Pankey, G.A., Henderson, H., Haas, D.W., Kauffman, C.A., Patel, R., Zaas, A.K., *et al.* (2008). Pulmonary cryptococcosis in patients without HIV infection: factors associated with disseminated disease. Eur. J. Clin. Microbiol. Infect. Dis. *27*, 937–943.

Balish, E., Wagner, R.D., Vazquez-Torres, A., Pierson, C., and Warner, T. (1998). Candidiasis in interferon-gamma knockout (IFN-gamma-/-) mice. J. Infect. Dis. *178*, 478–487.

Balish, E., Wagner, R.D., Vazquez-Torres, A., Jones-Carson, J., Pierson, C., and Warner, T. (1999). Mucosal and systemic candidiasis in IL-8Rh-/- BALB/c mice. J. Leukoc. Biol. *66*, 144–150.

Balkus, J.E., Richardson, B.A., Mandaliya, K., Kiarie, J., Jaoko, W., Ndinya-Achola, J.O., Marrazzo, J., Farquhar, C., and McClelland, R.S. (2011). Establishing and sustaining a healthy vaginal environment: analysis of data from a randomized trial of periodic presumptive treatment for vaginal infections. J. Infect. Dis. *204*, 323–326.

Baquir, B., Lemaire, S., Van Bambeke, F., Tulkens, P.M., Lin, L., and Spellberg, B. (2012). Macrophage killing of bacterial and fungal pathogens is not inhibited by intense intracellular accumulation of the lipoglycopeptide antibiotic oritavancin. Clin. Infect. Dis. *54* (Suppl. 3), S229–232.

Basu, S., Quilici, C., Zhang, H.H., Grail, D., and Dunn, A.R. (2008). Mice lacking both G-CSF and IL-6 are more susceptible to *Candida albicans* infection: critical role of neutrophils in defense against *Candida albicans*. Growth Factors *26*, 23–34.

Belkaid, Y., and Artis, D. (2013). Immunity at the Barriers. Eur. J. Immunol. *43*, 3096–3097.

Bellanger, A.P., Millon, L., Khoufache, K., Rivollet, D., Bieche, I., Laurendeau, I., Vidaud, M., Botterel, F., and Bretagne, S. (2009). *Aspergillus fumigatus* germ tube growth and not conidia ingestion induces expression of inflammatory mediator genes in the human lung epithelial cell line A549. J. Med. Microbiol. *58*, 174–179.

Bendel, C.M., Hess, D.J., Garni, R.M., Henry-Stanley, M., and Wells, C.L. (2003). Comparative virulence of *Candida albicans* yeast and filamentous forms in orally and intravenously inoculated mice. Crit. Care Med. *31*, 501–507.

Bennett, J.E., Friedman, M.M., and Dupont, B. (1987). Receptor-mediated clearance of *Aspergillus galactomannan*. J. Infect. Dis. *155*, 1005–1010.

Bhatia, S., McCullough, J., Perry, E.H., Clay, M., Ramsay, N.K., and Neglia, J.P. (1994). Granulocyte transfusions: efficacy in treating fungal infections in neutropenic patients following bone marrow transplantation. Transfusion *34*, 226–232.

Bianchi, M., Niemiec, M.J., Siler, U., Urban, C.F., and Reichenbach, J. (2011). Restoration of anti-Aspergillus defense by neutrophil extracellular traps in human chronic granulomatous disease after gene therapy is calprotectin-dependent. J. Allergy Clin. Immunol. *127*, 1243–1252 e1247.

Bicanic, T., Muzoora, C., Brouwer, A.E., Meintjes, G., Longley, N., Taseera, K., Rebe, K., Loyse, A., Jarvis, J., Bekker, L.G., *et al.* (2009). Independent association between rate of clearance of infection and clinical outcome of HIV-associated cryptococcal meningitis: analysis of a combined cohort of 262 patients. Clin. Infect. Dis. *49*, 702–709.

Bidula, S., Kenawy, H., Ali, Y.M., Sexton, D., Schwaeble, W.J., and Schelenz, S. (2013). Role of ficolin-A and lectin complement pathway in the innate defense against pathogenic Aspergillus species. Infect. Immun. *81*, 1730–1740.

Bittencourt, V.C., Figueiredo, R.T., da Silva, R.B., Mourao-Sa, D.S., Fernandez, P.L., Sassaki, G.L., Mulloy, B., Bozza, M.T., and Barreto-Bergter, E. (2006). An alpha-glucan of *Pseudallescheria boydii* is involved in fungal phagocytosis and Toll-like receptor activation. J. Biol. Chem. *281*, 22614–22623.

Blot, F. (2003). [Infections of intravascular perfusion sets]. La Revue du Praticien *53*, 2119–2127.

Bonnin, A. (2012). [The commensal-pathogen transition in invasive *Candida albicans* infection: molecular and cellular approaches]. Bulletin de l'Academie Nationale de Medecine *196*, 139–149.

Bourgeois, C., Majer, O., Frohner, I.E., Tierney, L., and Kuchler, K. (2010). Fungal attacks on mammalian hosts: pathogen elimination requires sensing and tasting. Curr. Opin. Microbiol. *13*, 401–408.

Bouza, E., San Juan, R., Munoz, P., Pascau, J., Voss, A., and Desco, M.; Cooperative Group of the European Study Group on Nosocomial Infections (ESGNI) (2004). A European perspective on intravascular catheter-related infections: report on the microbiology workload, aetiology and antimicrobial susceptibility (ESGNI-005 Study). Clin. Microbiol. Infect. *10*, 838–842.

Bouzani, M., Ok, M., McCormick, A., Ebel, F., Kurzai, O., Morton, C.O., Einsele, H., and Loeffler, J. (2011). Human NK cells display important antifungal activity against *Aspergillus fumigatus*, which is directly mediated by IFN-gamma release. J. Immunol. *187*, 1369–1376.

Boxer, L.A., Ingraham, L.M., Allen, J., Oseas, R.S., and Baehner, R.L. (1981). Amphotericin-B promotes leukocyte aggregation of nylon-wool-fiber-treated polymorphonuclear leukocytes. Blood *58*, 518–523.

Boyle, K.B., Gyori, D., Sindrilaru, A., Scharffetter-Kochanek, K., Taylor, P.R., Mocsai, A., Stephens, L.R., and Hawkins, P.T. (2011). Class IA phosphoinositide 3-kinase beta and delta regulate neutrophil oxidase activation in response to *Aspergillus fumigatus* hyphae. J. Immunol. *186*, 2978–2989.

Bozza, S., Perruccio, K., Montagnoli, C., Gaziano, R., Bellocchio, S., Burchielli, E., Nkwanyuo, G., Pitzurra, L., Velardi, A., and Romani, L. (2003). A dendritic cell vaccine against invasive aspergillosis in allogeneic hematopoietic transplantation. Blood *102*, 3807–3814.

Bozza, S., Clavaud, C., Giovannini, G., Fontaine, T., Beauvais, A., Sarfati, J., D'Angelo, C., Perruccio, K., Bonifazi, P., Zagarella, S., *et al.* (2009). Immune sensing of *Aspergillus fumigatus* proteins, glycolipids, and polysaccharides and the impact on Th immunity and vaccination. J. Immunol. *183*, 2407–2414.

Bradley, J.R. (2008). TNF-mediated inflammatory disease. J. Pathol. *214*, 149–160.

Bradsher, R.W., Chapman, S.W., and Pappas, P.G. (2003). Blastomycosis. Infect. Dis. Clin. North Am. *17*, 21–40, vii.

Brakhage, A.A., Bruns, S., Thywissen, A., Zipfel, P.F., and Behnsen, J. (2010). Interaction of phagocytes with filamentous fungi. Curr. Opin. Microbiol. *13*, 409–415.

Brown, G.D., Taylor, P.R., Reid, D.M., Willment, J.A., Williams, D.L., Martinez-Pomares, L., Wong, S.Y., and Gordon, S. (2002). Dectin-1 is a major beta-glucan receptor on macrophages. J. Exp. Med. *196*, 407–412.

Brummer, E., and Stevens, D.A. (1994). Anticryptococcal activity of macrophages: role of mouse strain, C5, contact, phagocytosis, and L-arginine. Cell. Immunol. *157*, 1–10.

Brummer, E., Nassar, F., and Stevens, D.A. (1994). Effect of macrophage colony-stimulating factor on anticryptococcal activity of bronchoalveolar macrophages: synergy with fluconazole for killing. Antimicrob. Agents Chemother. *38*, 2158–2161.

Brummer, E., Maqbool, A., and Stevens, D.A. (2001). *In vivo* GM-CSF prevents dexamethasone suppression of killing of *Aspergillus fumigatus* conidia by bronchoalveolar macrophages. J. Leukoc. Biol. *70*, 868–872.

Brummer, E., Maqbool, A., and Stevens, D.A. (2002). Protection of peritoneal macrophages by granulocyte/macrophage colony-stimulating factor (GM-CSF) against dexamethasone suppression of killing of Aspergillus, and the effect of human GM-CSF. Microbes Infect. *4*, 133–138.

Bruns, S., Kniemeyer, O., Hasenberg, M., Aimanianda, V., Nietzsche, S., Thywissen, A., Jeron, A., Latge, J.P., Brakhage, A.A., and Gunzer, M. (2010). Production of extracellular traps against *Aspergillus fumigatus in vitro* and in infected lung tissue is dependent on invading neutrophils and influenced by hydrophobin RodA. PLoS Pathog. *6*, e1000873.

Cambi, A., Netea, M.G., Mora-Montes, H.M., Gow, N.A., Hato, S.V., Lowman, D.W., Kullberg, B.J., Torensma, R., Williams, D.L., and Figdor, C.G. (2008). Dendritic cell interaction with *Candida albicans* critically depends on N-linked mannan. J. Biol. Chem. *283*, 20590–20599.

Cameron, M.L., Granger, D.L., Weinberg, J.B., Kozumbo, W.J., and Koren, H.S. (1990). Human alveolar and peritoneal macrophages mediate fungistasis independently of L-arginine oxidation to nitrite or nitrate. Am. Rev. Respir. Dis. *142*, 1313–1319.

Carroll, S.F., Guillot, L., and Qureshi, S.T. (2007). Mammalian model hosts of cryptococcal infection. Comp. Med. *57*, 9–17.

Carroll, S.F., Lafferty, E.I., Flaczyk, A., Fujiwara, T.M., Homer, R., Morgan, K., Loredo-Osti, J.C., and Qureshi, S.T. (2012). Susceptibility to progressive *Cryptococcus neoformans* pulmonary infection is regulated by loci on mouse chromosomes 1 and 9. Infect. Immun. *80*, 4167–4176.

Casadevall, A., and Pirofski, L.A. (2001). Adjunctive immune therapy for fungal infections. Clin. Infect. Dis. *33*, 1048–1056.

Cenci, E., Mencacci, A., Del Sero, G., Bistoni, F., and Romani, L. (1997a). Induction of protective Th1 responses to *Candida albicans* by antifungal therapy alone or in combination with an interleukin-4 antagonist. J. Infect. Dis. *176*, 217–226.

Cenci, E., Perito, S., Enssle, K.H., Mosci, P., Latge, J.P., Romani, L., and Bistoni, F. (1997b). Th1 and Th2 cytokines in mice with invasive aspergillosis. Infect. Immun. *65*, 564–570.

Cenci, E., Mencacci, A., Del Sero, G., d'Ostiani, C.F., Mosci, P., Bacci, A., Montagnoli, C., Kopf, M., and Romani, L. (1998). IFN-gamma is required for IL-12 responsiveness in mice with *Candida albicans* infection. J. Immunol. *161*, 3543–3550.

Cenci, E., Mencacci, A., Bacci, A., Bistoni, F., Kurup, V.P., and Romani, L. (2000). T cell vaccination in mice with invasive pulmonary aspergillosis. J. Immunol. *165*, 381–388.

Chanock, S.J., and Gorlin, J.B. (1996). Granulocyte transfusions. Time for a second look. Infect. Dis. Clin. North Am. *10*, 327–343.

Chaturvedi, V., Wong, B., and Newman, S.L. (1996). Oxidative killing of *Cryptococcus neoformans* by human neutrophils. Evidence that fungal mannitol protects by scavenging reactive oxygen intermediates. J. Immunol. *156*, 3836–3840.

Cheng, S.C., Joosten, L.A., Kullberg, B.J., and Netea, M.G. (2012). Interplay between *Candida albicans* and the mammalian innate host defense. Infect. Immun. *80*, 1304–1313.

Clemons, K.V., and Stevens, D.A. (2005). The contribution of animal models of aspergillosis to understanding pathogenesis, therapy and virulence. Med. Mycol. *43* (Suppl. 1), S101–110.

Clemons, K.V., and Stevens, D.A. (2012). Orogastrointestinal model of mucosal and disseminated candidiasis. Methods Mol. Biol. *845*, 557–567.

Clemons, K.V., Grunig, G., Sobel, R.A., Mirels, L.F., Rennick, D.M., and Stevens, D.A. (2000). Role of IL-10 in invasive aspergillosis: increased resistance of IL-10 gene knockout mice to lethal systemic aspergillosis. Clin. Exp. Immunol. *122*, 186–191.

Clemons, K.V., Lutz, J.E., and Stevens, D.A. (2001). Efficacy of recombinant gamma interferon for treatment of systemic cryptococcosis in SCID mice. Antimicrob. Agents Chemother. *45*, 686–689.

Collin, B., Clancy, C.J., and Nguyen, M.H. (1999). Antifungal resistance in non- albicans *Candida* species. Drug Resist. Updat. *2*, 9–14.

Cornistein, W., Mora, A., Orellana, N., Capparelli, F.J., and del Castillo, M. (2013). [Candida: epidemiology and risk factors for non-albicans species]. Enfermedades Infecciosas y Microbiologia Clinica *31*, 380–384.

Cuenca-Estrella, M., Bassetti, M., Lass-Florl, C., Racil, Z., Richardson, M., and Rogers, T.R. (2011). Detection and investigation of invasive mould disease. J. Antimicrob. Chemother. *66* (Suppl. 1), i15–24.

Cunha, C., Carvalho, A., Esposito, A., Bistoni, F., and Romani, L. (2012). DAMP signaling in fungal infections and diseases. Front. Immunol. *3*, 286.

Cupps, T.R., and Fauci, A.S. (1982). Corticosteroid-mediated immunoregulation in man. Immunol. Rev. *65*, 133–155.

Da Silva, C.A., Hartl, D., Liu, W., Lee, C.G., and Elias, J.A. (2008). TLR-2 and IL-17A in chitin-induced macrophage activation and acute inflammation. J. Immunol. *181*, 4279–4286.

Da Silva, C.A., Pochard, P., Lee, C.G., and Elias, J.A. (2010). Chitin particles are multifaceted immune adjuvants. Am. J. Respir. Crit. Care Med. *182*, 1482–1491.

Dale, D.C., Liles, W.C., Llewellyn, C., and Price, T.H. (1998). Effects of granulocyte–macrophage colony-stimulating factor (GM-CSF) on neutrophil kinetics and function in normal human volunteers. Am. J. Hematol. *57*, 7–15.

Dalle, F., Wachtler, B., L'Ollivier, C., Holland, G., Bannert, N., Wilson, D., Labruere, C., Bonnin, A., and Hube, B. (2010). Cellular interactions of *Candida albicans* with human oral epithelial cells and enterocytes. Cell. Microbiol. *12*, 248–271.

Deak, E., Nelson, M., Hernandez-Rodriguez, Y., Gade, L., Baddley, J., Momany, M., Steele, C., and Balajee, S.A. (2011). *Aspergillus terreus* accessory conidia are multinucleated, hyperpolarizing structures that display differential dectin staining and can induce heightened inflammatory responses in a pulmonary model of aspergillosis. Virulence 2, 200–207.

Del Rosso, J.Q., and Kircik, L.H. (2013). Optimizing topical antifungal therapy for superficial cutaneous fungal infections: focus on topical naftifine for cutaneous dermatophytosis. J. Drugs Dermatol. *12*, s165–171.

Desai, S.K., and Naik, S.R. (2008). Probable mechanism(s) of antifungal activity of SJA-95, a heptaene polyene antibiotic. Indian J. Pharm. Sci. *70*, 165–169.

Domer, J.E., and Kobayashi, G.S. (2004). Human fungal pathogens, Vol XII (Springer-Verlag, Berlin).

Dongari-Bagtzoglou, A., and Kashleva, H. (2003). *Candida albicans* triggers interleukin-8 secretion by oral epithelial cells. Microb. Pathog. *34*, 169–177.

Echeverria, P.M., Kett, D.H., and Azoulay, E. (2011). *Candida* prophylaxis and therapy in the ICU. Semin. Respir. Crit. Care Med. 32, 159–173.

Edwards, L., Williams, A.E., Krieg, A.M., Rae, A.J., Snelgrove, R.J., and Hussell, T. (2005). Stimulation via Toll-like receptor 9 reduces *Cryptococcus neoformans*-induced pulmonary inflammation in an IL-12-dependent manner. Eur. J. Immunol. 35, 273–281.

Eggimann, P., and Pittet, D. (2006). Postoperative fungal infections. Surg. Infect. 7 (Suppl. 2), S53–56.

Enoch, D.A., Ludlam, H.A., and Brown, N.M. (2006). Invasive fungal infections: a review of epidemiology and management options. J. Med. Microbiol. 55, 809–818.

Evron, R. (1980). *In vitro* phagocytosis of *Candida albicans* by peritoneal mouse macrophages. Infect. Immun. *28*, 963–971.

Faro-Trindade, I., Willment, J.A., Kerrigan, A.M., Redelinghuys, P., Hadebe, S., Reid, D.M., Srinivasan, N., Wainwright, H., Lang, D.M., Steele, C., *et al.* (2012). Characterisation of innate fungal recognition in the lung. PLoS One 7, e35675.

Fauci, A.S., Dale, D.C., and Balow, J.E. (1976). Glucocorticosteroid therapy: mechanisms of action and clinical considerations. Ann. Intern. Med. *84*, 304–315.

Fedor, M.E., and Rubinstein, A. (2006). Effects of long-term low-dose corticosteroid therapy on humoral immunity. Annals Allergy Asthma Immunol. *97*, 113–116.

Ferwerda, G., Meyer-Wentrup, F., Kullberg, B.J., Netea, M.G., and Adema, G.J. (2008). Dectin-1 synergizes with TLR2 and TLR4 for cytokine production in human primary monocytes and macrophages. Cell. Microbiol. *10*, 2058–2066.

Filler, S.G. (2006). Candida-host cell receptor–ligand interactions. Curr. Opin. Microbiol. *9*, 333–339.

Fortis, A.A., Lianou, P.E., and Papavassilliou, J.T. (1998). Adherence of *Staphylococcus aureus*, *Klebsiella pneumoniae* and *Candida albicans* to human buccal epithelial cells, from healthy persons and HIV carriers, under the influence of Broncho Vaxom *in vitro* and ascorbic acid *in vivo*. APMIS *106*, 441–448.

Fradin, C., Jouault, T., Mallet, A., Mallet, J.M., Camus, D., Sinay, P., and Poulain, D. (1996). Beta-1,2-linked oligomannosides inhibit *Candida albicans* binding to murine macrophage. J. Leukoc. Biol. *60*, 81–87.

Fradin, C., Poulain, D., and Jouault, T. (2000). beta-1,2-linked oligomannosides from *Candida albicans* bind to a 32-kilodalton macrophage membrane protein homologous to the mammalian lectin galectin-3. Infect. Immun. *68*, 4391–4398.

Fradin, C., De Groot, P., MacCallum, D., Schaller, M., Klis, F., Odds, F.C., and Hube, B. (2005). Granulocytes govern the transcriptional response, morphology and proliferation of *Candida albicans* in human blood. Mol. Microbiol. *56*, 397–415.

Fradin, C., Mavor, A.L., Weindl, G., Schaller, M., Hanke, K., Kaufmann, S.H., Mollenkopf, H., and Hube, B. (2007). The early transcriptional response of human granulocytes to infection with *Candida albicans* is not essential for killing but reflects cellular communications. Infect. Immun. *75*, 1493–1501.

Gallo, R.L., and Hooper, L.V. (2012). Epithelial antimicrobial defence of the skin and intestine. Nat. Rev. Immunol. *12*, 503–516.

Gangneux, J.P., Camus, C., and Philippe, B. (2010). Epidemiology of invasive aspergillosis and risk factors in non neutropaenic patients. Rev. Mal. Respir. *27*, e34–46.

Ganz, T. (2003). The role of antimicrobial peptides in innate immunity. Integr. Comp. Biol. 43, 300–304.

Garber, G. (2001). An overview of fungal infections. Drugs *61* (Suppl. 1), 1–12.

Gaziri, G., Gaziri, L.C., Kikuchi, R., Scanavacca, J., and Felipe, I. (1999). Phagocytosis of *Candida albicans* by concanavalin-A activated peritoneal macrophages. Med. Mycol. 37, 195–200.

Gersuk, G.M., Underhill, D.M., Zhu, L., and Marr, K.A. (2006). Dectin-1 and TLRs permit macrophages to distinguish between different *Aspergillus fumigatus* cellular states. J. Immunol. *176*, 3717–3724.

Gessner, M.A., Werner, J.L., Lilly, L.M., Nelson, M.P., Metz, A.E., Dunaway, C.W., Chan, Y.R., Ouyang, W., Brown, G.D., Weaver, C.T., *et al.* (2012). Dectin-1-dependent interleukin-22 contributes to early innate lung defense against *Aspergillus fumigatus*. Infect. Immun. *80*, 410–417.

Ghannoum, M.A., and Elteen, K.A. (1987). Effect of growth of *Candida* spp. in the presence of various glucocorticoids on the adherence to human buccal epithelial cells. Mycopathologia *98*, 171–178.

Gil, M.L., and Gozalbo, D. (2009). Role of Toll-like receptors in systemic *Candida albicans* infections. Front. Biosci. (Landmark Ed) *14*, 570–582.

Giles, F.J. (1998). Monocyte-macrophages, granulocyte–macrophage colony-stimulating factor, and prolonged survival among patients with acute myeloid leukemia and stem cell transplants. Clin. Infect. Dis. *26*, 1282–1289.

Giraldo, P., von Nowaskonski, A., Gomes, F.A., Linhares, I., Neves, N.A., and Witkin, S.S. (2000). Vaginal colonization by *Candida* in asymptomatic women with and without a history of recurrent vulvovaginal candidiasis. Obstet. Gynecol. *95*, 413–416.

Giri, S., and Kindo, A.J. (2012). A review of *Candida* species causing blood stream infection. Indian J. Med. Microbiol. *30*, 270–278.

Gonzalez, C.E., Lyman, C.A., Lee, S., Del Guercio, C., Roilides, E., Bacher, J., Gehrt, A., Feuerstein, E., Tsokos, M., and Walsh, T.J. (2001). Recombinant human macrophage colony-stimulating factor augments pulmonary host defences against *Aspergillus fumigatus*. Cytokine *15*, 87–95.

Gould, D. (2011). Diagnosis, prevention and treatment of fungal infections. Nurs. Stand. *25*, 38–47; quiz 48.

Gresnigt, M.S., Netea, M.G., and van de Veerdonk, F.L. (2012). Pattern recognition receptors and their role in invasive aspergillosis. Ann. N.Y. Acad. Sci. *1273*, 60–67.

Guarner, F. (2005). The intestinal flora in inflammatory bowel disease: normal or abnormal? Curr. Opin. Gastroenterol. *21*, 414–418.

Guillot, L., Carroll, S.F., Homer, R., and Qureshi, S.T. (2008). Enhanced innate immune responsiveness to pulmonary *Cryptococcus neoformans* infection is associated with resistance to progressive infection. Infect. Immun. *76*, 4745–4756.

Gullo, A. (2009). Invasive fungal infections: the challenge continues. Drugs *69* (Suppl. 1), 65–73.

Gupta, P.N., Ichhpujani, R.L., Bhatia, R., Arora, D.R., and Chugh, T.D. (1982). Effect of dexamethasone on *in vitro* growth of *Candida albicans*. J. Commun. Dis. *14*, 259–262.

Hamad, M. (2008). Antifungal immunotherapy and immunomodulation: a double-hitter approach to deal with invasive fungal infections. Scand. J. Immunol. *67*, 533–543.

Harler, M.B., and Reichner, J. (2001). Increased neutrophil motility by beta-glucan in the absence of chemoattractant. Shock *16*, 419–424.

Hawser, S.P., and Douglas, L.J. (1994). Biofilm formation by *Candida* species on the surface of catheter materials *in vitro*. Infect. Immun. *62*, 915–921.

Hawser, S.P., and Douglas, L.J. (1995). Resistance of *Candida albicans* biofilms to antifungal agents *in vitro*. Antimicrob. Agents Chemother. *39*, 2128–2131.

Hernandez-Santos, N., and Gaffen, S.L. (2012). Th17 cells in immunity to *Candida albicans*. Cell Host Microbe *11*, 425–435.

Hidore, M.R., and Murphy, J.W. (1986). Natural cellular resistance of beige mice against *Cryptococcus neoformans*. J. Immunol. *137*, 3624–3631.

Hohl, T.M., Van Epps, H.L., Rivera, A., Morgan, L.A., Chen, P.L., Feldmesser, M., and Pamer, E.G. (2005). *Aspergillus fumigatus* triggers inflammatory responses by stage-specific beta-glucan display. PLoS Pathog. *1*, e30.

Holmes, A.R., Bandara, B.M., and Cannon, R.D. (2002). Saliva promotes *Candida albicans* adherence to human epithelial cells. J. Dent. Res. *81*, 28–32.

Ibrahim-Granet, O., Philippe, B., Boleti, H., Boisvieux-Ulrich, E., Grenet, D., Stern, M., and Latge, J.P. (2003). Phagocytosis and intracellular fate of *Aspergillus fumigatus* conidia in alveolar macrophages. Infect. Immun. *71*, 891–903.

Ip, W.K., and Lau, Y.L. (2004). Role of mannose-binding lectin in the innate defense against *Candida albicans*: enhancement of complement activation, but lack of opsonic function, in phagocytosis by human dendritic cells. J. Infect. Dis. *190*, 632–640.

Jaffe, M.L., Arai, H., and Nabel, G.J. (1996). Mechanisms of tumor-induced immunosuppression: evidence for contact-dependent T cell suppression by monocytes. Mol. Med. *2*, 692–701.

Jarvis, J.N., Meintjes, G., Rebe, K., Williams, G.N., Bicanic, T., Williams, A., Schutz, C., Bekker, L.G., Wood, R., and Harrison, T.S. (2012). Adjunctive interferon-gamma immunotherapy for the treatment of HIV-associated cryptococcal meningitis: a randomized controlled trial. AIDS *26*, 1105–1113.

Jawhara, S., and Poulain, D. (2007). *Saccharomyces boulardii* decreases inflammation and intestinal colonization by *Candida albicans* in a mouse model of chemically-induced colitis. Med. Mycol. *45*, 691–700.

Jawhara, S., Thuru, X., Standaert-Vitse, A., Jouault, T., Mordon, S., Sendid, B., Desreumaux, P., and Poulain, D. (2008). Colonization of mice by *Candida albicans* is promoted by chemically induced colitis and augments inflammatory responses through galectin-3. J. Infect. Dis. *197*, 972–980.

Jawhara, S., Habib, K., Maggiotto, F., Pignede, G., Vandekerckove, P., Maes, E., Dubuquoy, L., Fontaine, T., Guerardel, Y., and Poulain, D. (2012). Modulation of intestinal inflammation by yeasts and cell wall extracts: strain dependence and unexpected anti-inflammatory role of glucan fractions. PLoS One *7*, e40648.

Jayatilake, J.A., Samaranayake, L.P., Lu, Q., and Jin, L.J. (2007). IL-1alpha, IL-1ra and IL-8 are differentially induced by *Candida* in experimental oral candidiasis. Oral Dis. *13*, 426–433.

Jouault, T., Bernigaud, A., Lepage, G., Trinel, P.A., and Poulain, D. (1994). The *Candida albicans* phospholipomannan induces *in vitro* production of tumour necrosis factor-alpha from human and murine macrophages. Immunology *83*, 268–273.

Jouault, T., Ibata-Ombetta, S., Takeuchi, O., Trinel, P.A., Sacchetti, P., Lefebvre, P., Akira, S., and Poulain, D. (2003). *Candida albicans* phospholipomannan is sensed through toll-like receptors. J. Infect. Dis. *188*, 165–172.

Jouault, T., El Abed-El Behi, M., Martinez-Esparza, M., Breuilh, L., Trinel, P.A., Chamaillard, M., Trottein, F., and Poulain, D. (2006). Specific recognition of *Candida albicans* by macrophages requires galectin-3 to discriminate *Saccharomyces cerevisiae* and needs association with TLR2 for signaling. J. Immunol. *177*, 4679–4687.

Jouault, T., Sarazin, A., Martinez-Esparza, M., Fradin, C., Sendid, B., and Poulain, D. (2009). Host responses to a versatile commensal: PAMPs and PRRs interplay leading to tolerance or infection by *Candida albicans*. Cell. Microbiol. *11*, 1007–1015.

Kamada, N., Chen, G.Y., Inohara, N., and Nunez, G. (2013). Control of pathogens and pathobionts by the gut microbiota. Nat. Immunol. *14*, 685–690.

Kamberi, M., Brummer, E., and Stevens, D.A. (2002). Regulation of bronchoalveolar macrophage proinflammatory cytokine production by dexamethasone and granulocyte–macrophage colony-stimulating factor after stimulation by *Aspergillus conidia* or lipopolysaccharide. Cytokine *19*, 14–20.

Kaposzta, R., Tree, P., Marodi, L., and Gordon, S. (1998). Characteristics of invasive candidiasis in gamma interferon- and interleukin-4-deficient mice: role of macrophages in host defense against *Candida albicans*. Infect. Immun. *66*, 1708–1717.

Karabinis, A., Hill, C., Leclercq, B., Tancrede, C., Baume, D., and Andremont, A. (1988). Risk factors for candidemia in cancer patients: a case–control study. J. Clin. Microbiol. *26*, 429–432.

Kasperkovitz, P.V., Cardenas, M.L., and Vyas, J.M. (2010). TLR9 is actively recruited to *Aspergillus fumigatus* phagosomes and requires the N-terminal proteolytic cleavage domain for proper intracellular trafficking. J. Immunol. *185*, 7614–7622.

Kasperkovitz, P.V., Khan, N.S., Tam, J.M., Mansour, M.K., Davids, P.J., and Vyas, J.M. (2011). Toll-like receptor 9 modulates macrophage antifungal effector function during innate recognition of *Candida albicans* and *Saccharomyces cerevisiae*. Infect. Immun. *79*, 4858–4867.

Katragkou, A., and Roilides, E. (2012). Immunotherapy of infections caused by rare filamentous fungi. Clin. Microbiol. Infect. *18*, 134–139.

Kawakami, K., Qifeng, X., Tohyama, M., Qureshi, M.H., and Saito, A. (1996). Contribution of tumour necrosis factor-alpha (TNF-alpha) in host defence mechanism against *Cryptococcus neoformans*. Clin. Exp. Immunol. *106*, 468–474.

Kawakami, K., Qureshi, M.H., Zhang, T., Koguchi, Y., Yara, S., Takeda, K., Akira, S., Kurimoto, M., and Saito, A. (2000). Involvement of endogenously synthesized interleukin (IL)–18 in the protective effects of IL-12 against pulmonary infection with *Cryptococcus neoformans* in mice. FEMS Immunol. Med. Microbiol. *27*, 191–200.

Kelleher, P., Goodsall, A., Mulgirigama, A., Kunst, H., Henderson, D.C., Wilson, R., Newman-Taylor, A., and Levin, M. (2006). Interferon-gamma therapy in two patients with progressive chronic pulmonary aspergillosis. Eur. Respir. J. *27*, 1307–1310.

Kelly, R.M., Chen, J., Yauch, L.E., and Levitz, S.M. (2005). Opsonic requirements for dendritic cell-mediated responses to *Cryptococcus neoformans*. Infect. Immun. *73*, 592–598.

Kennedy, A.D., Willment, J.A., Dorward, D.W., Williams, D.L., Brown, G.D., and DeLeo, F.R. (2007). Dectin-1 promotes fungicidal activity of human neutrophils. Eur. J. Immunol. *37*, 467–478.

Kennedy, M.J., and Volz, P.A. (1985). Ecology of *Candida albicans* gut colonization: inhibition of *Candida* adhesion, colonization, and dissemination from the gastrointestinal tract by bacterial antagonism. Infect. Immun. *49*, 654–663.

Koguchi, Y., and Kawakami, K. (2002). Cryptococcal infection and Th1-Th2 cytokine balance. Int. Rev. Immunol. *21*, 423–438.

Koh, A.Y., Kohler, J.R., Coggshall, K.T., Van Rooijen, N., and Pier, G.B. (2008). Mucosal damage and neutropenia are required for *Candida albicans* dissemination. PLoS Pathog. *4*, e35.

Kohatsu, L., Hsu, D.K., Jegalian, A.G., Liu, F.T., and Baum, L.G. (2006). Galectin-3 induces death of *Candida* species expressing specific beta-1,2-linked mannans. J. Immunol. *177*, 4718–4726.

Kozel, T.R., Pfrommer, G.S., and Redelman, D. (1987). Activated neutrophils exhibit enhanced phagocytosis of *Cryptococcus neoformans* opsonized with normal human serum. Clin. Exp. Immunol. *70*, 238–246.

Krasowska, A., Murzyn, A., Dyjankiewicz, A., Lukaszewicz, M., and Dziadkowiec, D. (2009). The antagonistic effect of *Saccharomyces boulardii* on *Candida albicans* filamentation, adhesion and biofilm formation. FEMS Yeast Res. *9*, 1312–1321.

Krcmery, V., and Barnes, A.J. (2002). Non-albicans *Candida* spp. causing fungaemia: pathogenicity and antifungal resistance. J. Hosp. Infect. *50*, 243–260.

Kriengkauykiat, J., Ito, J.I., and Dadwal, S.S. (2011). Epidemiology and treatment approaches in management of invasive fungal infections. Clin. Epidemiol. *3*, 175–191.

Krishnan, S., Manavathu, E.K., and Chandrasekar, P.H. (2009). *Aspergillus flavus*: an emerging non-fumigatus *Aspergillus* species of significance. Mycoses *52*, 206–222.

Kuhara, T., Uchida, K., and Yamaguchi, H. (2000). Therapeutic efficacy of human macrophage colony-stimulating factor, used alone and in combination with antifungal agents, in mice with systemic *Candida albicans* infection. Antimicrob. Agents Chemother. *44*, 19–23.

Kuhn, D.M., and Ghannoum, M.A. (2004). *Candida* biofilms: antifungal resistance and emerging therapeutic options. Curr. Opin. Investig. Drugs *5*, 186–197.

Kullberg, B.J., Netea, M.G., Vonk, A.G., and van der Meer, J.W. (1999). Modulation of neutrophil function in host defense against disseminated *Candida albicans* infection in mice. FEMS Immunol. Med. Microbiol. *26*, 299–307.

Lamouse-Smith, E.S., Tzeng, A., and Starnbach, M.N. (2011). The intestinal flora is required to support antibody responses to systemic immunization in infant and germ free mice. PLoS One *6*, e27662.

Lange, A. (2008). Genetic factors predicting IFN-gamma generation potential in patients with sarcoidosis and after haematopoietic stem cell transplantation. Int. J. Immunogenet. *35*, 385–388.

Lavigne, L.M., Schopf, L.R., Chung, C.L., Maylor, R., and Sypek, J.P. (1998). The role of recombinant murine IL-12 and IFN-gamma in the pathogenesis of a murine systemic *Candida albicans* infection. J. Immunol. *160*, 284–292.

Leal, S.M., Jr., Cowden, S., Hsia, Y.C., Ghannoum, M.A., Momany, M., and Pearlman, E. (2010). Distinct roles for Dectin-1 and TLR4 in the pathogenesis of *Aspergillus fumigatus* keratitis. PLoS Pathog. *6*, e1000976.

Lee, C.G., Da Silva, C.A., Lee, J.Y., Hartl, D., and Elias, J.A. (2008). Chitin regulation of immune responses: an old molecule with new roles. Curr. Opin. Immunol. *20*, 684–689.

Lee, E., Oh, S.H., Kwon, J.W., Kim, B.J., Yu, J., Park, C.J., and Hong, S.J. (2010). A case report of chronic granulomatous disease presenting with aspergillus pneumonia in a 2-month old girl. Korean J. Pediatr. *53*, 722–726.

Lee, J.J., Chung, I.J., Park, M.R., Kook, H., Hwang, T.J., Ryang, D.W., and Kim, H.J. (2001). Clinical efficacy of granulocyte transfusion therapy in patients with neutropenia-related infections. Leukemia *15*, 203–207.

Lee, R.T., Hsu, T.L., Huang, S.K., Hsieh, S.L., Wong, C.H., and Lee, Y.C. (2011). Survey of immune-related, mannose/fucose-binding C-type lectin receptors reveals widely divergent sugar-binding specificities. Glycobiology *21*, 512–520.

Lehrnbecher, T., Schmidt, S., Koehl, U., Schuster, F.R., Uharek, L., Klingebiel, T., and Tramsen, L. (2011). Adoptive antifungal T cell immunotherapy--into the clinic? Med. Mycol. *49* (Suppl. 1), S164–169.

Li, D., Dong, B., Tong, Z., Wang, Q., Liu, W., Wang, Y., Liu, W., Chen, J., Xu, L., Chen, L., *et al.* (2012a). MBL-mediated opsonophagocytosis of *Candida albicans* by human neutrophils is coupled with intracellular Dectin-1-triggered ROS production. PLoS One *7*, e50589.

Li, L., Wang, J., Zhang, W., Yang, J., Chen, L., and Lv, S. (2012b). Risk factors for invasive mold infections following allogeneic hematopoietic stem cell transplantation: a single center study of 190 recipients. Scand. J. Infect. Dis. *44*, 100–107.

Li, X., Utomo, A., Cullere, X., Choi, M.M., Milner, D.A., Jr., Venkatesh, D., Yun, S.H., and Mayadas, T.N. (2011). The beta-glucan receptor Dectin-1 activates the integrin Mac-1 in neutrophils via Vav protein signaling to promote *Candida albicans* clearance. Cell Host Microbe *10*, 603–615.

Lin, L., Ibrahim, A.S., Xu, X., Farber, J.M., Avanesian, V., Baquir, B., Fu, Y., French, S.W., Edwards, J.E., Jr., and Spellberg, B. (2009). Th1-Th17 cells mediate protective adaptive immunity against Staphylococcus aureus and *Candida albicans* infection in mice. PLoS Pathog. *5*, e1000703.

Linden, J.R., Kunkel, D., Laforce-Nesbitt, S.S., and Bliss, J.M. (2013). The role of galectin-3 in phagocytosis of *Candida albicans* and *Candida* parapsilosis by human neutrophils. Cell. Microbiol. *15*, 1127–1142.

Lipovsky, M.M., Gekker, G., Anderson, W.R., Molitor, T.W., Peterson, P.K., and Hoepelman, A.I. (1997). Phagocytosis of nonopsonized *Cryptococcus neoformans* by swine microglia involves CD14 receptors. Clin. Immunol. Immunopathol. *84*, 208–211.

Lo, H.J., Kohler, J.R., DiDomenico, B., Loebenberg, D., Cacciapuoti, A., and Fink, G.R. (1997). Nonfilamentous C. albicans mutants are avirulent. Cell *90*, 939–949.

Loose, D.S., Stevens, D.A., Schurman, D.J., and Feldman, D. (1983). Distribution of a corticosteroid-binding protein in *Candida* and other fungal genera. J. Gen. Microbiol. *129*, 2379–2385.

Lu, Q., Jayatilake, J.A., Samaranayake, L.P., and Jin, L. (2006). Hyphal invasion of *Candida albicans* inhibits the expression of human beta-defensins in experimental oral candidiasis. J. Invest. Dermatol. *126*, 2049–2056.

Luther, K., Torosantucci, A., Brakhage, A.A., Heesemann, J., and Ebel, F. (2007). Phagocytosis of *Aspergillus fumigatus* conidia by murine macrophages involves recognition by the dectin-1 beta-glucan receptor and Toll-like receptor 2. Cell. Microbiol. *9*, 368–381.

Ma, L.L., Spurrell, J.C., Wang, J.F., Neely, G.G., Epelman, S., Krensky, A.M., and Mody, C.H. (2002). CD8 T cell-mediated killing of *Cryptococcus neoformans* requires granulysin and is dependent on CD4 T cells and IL-15. J. Immunol. *169*, 5787–5795.

Ma, L.L., Wang, C.L., Neely, G.G., Epelman, S., Krensky, A.M., and Mody, C.H. (2004). NK cells use perforin rather than granulysin for anticryptococcal activity. J. Immunol. *173*, 3357–3365.

Maccallum, D.M. (2012). Hosting infection: experimental models to assay *Candida* virulence. Int. J. Microbiol. *2012*, 363764.

McCormick, A., Heesemann, L., Wagener, J., Marcos, V., Hartl, D., Loeffler, J., Heesemann, J., and Ebel, F. (2010a). NETs formed by human neutrophils inhibit growth of the pathogenic mold *Aspergillus fumigatus*. Microbes Infect. *12*, 928–936.

McCormick, A., Loeffler, J., and Ebel, F. (2010b). *Aspergillus fumigatus*: contours of an opportunistic human pathogen. Cell. Microbiol. *12*, 1535–1543.

McGreal, E.P., Rosas, M., Brown, G.D., Zamze, S., Wong, S.Y., Gordon, S., Martinez-Pomares, L., and Taylor, P.R. (2006). The carbohydrate-recognition domain of Dectin-2 is a C-type lectin with specificity for high mannose. Glycobiology *16*, 422–430.

Mahmud, N., Klipa, D., and Ahsan, N. (2010). Antibody immunosuppressive therapy in solid-organ transplant: Part I. MAbs *2*, 148–156.

Mambula, S.S., Simons, E.R., Hastey, R., Selsted, M.E., and Levitz, S.M. (2000). Human neutrophil-mediated nonoxidative antifungal activity against *Cryptococcus neoformans*. Infect. Immun. *68*, 6257–6264.

Maneu, V., Yanez, A., Murciano, C., Molina, A., Gil, M.L., and Gozalbo, D. (2011). Dectin-1 mediates *in vitro* phagocytosis of *Candida albicans* yeast cells by retinal microglia. FEMS Immunol. Med. Microbiol. *63*, 148–150.

Marie, I., and Guglielmino, E. (2010). [Non tuberculous anti-TNF associated opportunistic infections]. La Revue de Medecine Interne *31*, 353–360.

Mariscalco, M.M. (2006). Innate immunity in critical care. Semin. Pediatr. Infect. Dis. *17*, 25–35.

Marr, K.A., Balajee, S.A., Hawn, T.R., Ozinsky, A., Pham, U., Akira, S., Aderem, A., and Liles, W.C. (2003). Differential role of MyD88 in macrophage-mediated responses to opportunistic fungal pathogens. Infect. Immun. *71*, 5280–5286.

Martinez-Solano, L., Reales-Calderon, J.A., Nombela, C., Molero, G., and Gil, C. (2009). Proteomics of RAW 264.7 macrophages upon interaction with heat-inactivated *Candida albicans* cells unravel an anti-inflammatory response. Proteomics 9, 2995–3010.

Marwaha, A.K., Leung, N.J., McMurchy, A.N., and Levings, M.K. (2012). TH17 Cells in Autoimmunity and Immunodeficiency: Protective or Pathogenic? Front. Immunol. *3*, 129.

Maschmeyer, G., Haas, A., and Cornely, O.A. (2007). Invasive aspergillosis: epidemiology, diagnosis and management in immunocompromised patients. Drugs *67*, 1567–1601.

Mehrad, B., Strieter, R.M., and Standiford, T.J. (1999). Role of TNF-alpha in pulmonary host defense in murine invasive aspergillosis. J. Immunol. *162*, 1633–1640.

Merkel, G.J., and Scofield, B.A. (2000). The effects of *Cryptococcus neoformans*-secreted antigens on tumor necrosis factor-alpha-induced intercellular adhesion molecule-1 expression on human lung epithelial cells. FEMS Immunol. Med. Microbiol. *29*, 329–332.

Meyer, D.K., and Schuster, M. (2002). Fungal Infections. In HIV infection in primary care, R.M. Buckley, and S.J. Gluckman, eds. (W.B. Saunders Company, Philadelphia, PA), pp. 79–84.

Miele, P.S., Levy, C.S., Smith, M.A., Dugan, E.M., Cooke, R.H., Light, J.A., and Lucey, D.R. (2002). Primary cutaneous fungal infections in solid organ transplantation: a case series. Am. J. Transplant. 2, 678–683.

Miller, M.F., and Mitchell, T.G. (1991). Killing of *Cryptococcus neoformans* strains by human neutrophils and monocytes. Infect. Immun. *59*, 24–28.

Miyazato, A., Nakamura, K., Yamamoto, N., Mora-Montes, H.M., Tanaka, M., Abe, Y., Tanno, D., Inden, K., Gang, X., Ishii, K., *et al.* (2009). Toll-like receptor 9-dependent activation of myeloid dendritic cells by Deoxynucleic acids from *Candida albicans*. Infect. Immun. 77, 3056–3064.

Mody, C.H., Spurrell, J.C., and Wood, C.J. (1998). Interleukin-15 induces antimicrobial activity after release by *Cryptococcus neoformans*-stimulated monocytes. J. Infect. Dis. *178*, 803–814.

Montagnoli, C., Fallarino, F., Gaziano, R., Bozza, S., Bellocchio, S., Zelante, T., Kurup, W.P., Pitzurra, L., Puccetti, P., and Romani, L. (2006). Immunity and tolerance to Aspergillus involve functionally distinct regulatory T cells and tryptophan catabolism. J. Immunol. *176*, 1712–1723.

Morace, G., and Borghi, E. (2010). Fungal infections in ICU patients: epidemiology and the role of diagnostics. Minerva Anestesiol. *76*, 950–956.

Morrison, B.E., Park, S.J., Mooney, J.M., and Mehrad, B. (2003). Chemokine-mediated recruitment of NK cells is a critical host defense mechanism in invasive aspergillosis. J. Clin. Invest. *112*, 1862–1870.

Mousset, S., Hermann, S., Klein, S.A., Bialleck, H., Duchscherer, M., Bomke, B., Wassmann, B., Bohme, A., Hoelzer, D., and Martin, H. (2005). Prophylactic and interventional granulocyte transfusions in patients with haematological malignancies and life-threatening infections during neutropenia. Ann. Hematol. *84*, 734–741.

Moyes, D.L., and Naglik, J.R. (2011). Mucosal immunity and *Candida albicans* infection. Clin. Dev. Immunol. *2011*, 346307.

Moyes, D.L., Runglall, M., Murciano, C., Shen, C., Nayar, D., Thavaraj, S., Kohli, A., Islam, A., Mora-Montes, H., Challacombe, S.J., *et al.* (2010). A biphasic innate immune MAPK response discriminates between the yeast and hyphal forms of *Candida albicans* in epithelial cells. Cell Host Microbe *8*, 225–235.

Mukherjee, S., Feldmesser, M., and Casadevall, A. (1996). J774 murine macrophage-like cell interactions with *Cryptococcus neoformans* in the presence and absence of opsonins. J. Infect. Dis. *173*, 1222–1231.

Mullick, A., Elias, M., Harakidas, P., Marcil, A., Whiteway, M., Ge, B., Hudson, T.J., Caron, A.W., Bourget, L., Picard, S., *et al.* (2004). Gene expression in HL60 granulocytoids and human polymorphonuclear leukocytes exposed to *Candida albicans*. Infect. Immun. *72*, 414–429.

Murphy, J.W., and McDaniel, D.O. (1982). In vitro reactivity of natural killer (NK) cells against *Cryptococcus neoformans*. J. Immunol. *128*, 1577–1583.

Murzyn, A., Krasowska, A., Stefanowicz, P., Dziadkowiec, D., and Lukaszewicz, M. (2010). Capric acid secreted by *S. boulardii* inhibits *C. albicans* filamentous growth, adhesion and biofilm formation. PLoS One *5*, e12050.

Muskett, H., Shahin, J., Eyres, G., Harvey, S., Rowan, K., and Harrison, D. (2011). Risk factors for invasive fungal disease in critically ill adult patients: a systematic review. Crit. Care *15*, R287.

Musso, T., Calosso, L., Zucca, M., Millesimo, M., Puliti, M., Bulfone-Paus, S., Merlino, C., Savoia, D., Cavallo, R., Ponzi, A.N., *et al.* (1998). Interleukin-15 activates proinflammatory and antimicrobial functions in polymorphonuclear cells. Infect. Immun. *66*, 2640–2647.

Naglik, J.R., Fidel, P.L., Jr., and Odds, F.C. (2008). Animal models of mucosal *Candida* infection. FEMS Microbiol. Lett. *283*, 129–139.

Naik, S.R., Thakare, V.N., Desai, S.K., and Rahalkar, P.R. (2011). Study of immunological aspects of aspergillosis in mice and effect of polyene macrolide antibiotic (SJA-95) and IFN-gamma: a possible role of IFN-gamma as an adjunct in antifungal therapy. Immunol. Lett. *141*, 68–73.

Nakamura, K., Miyazato, A., Xiao, G., Hatta, M., Inden, K., Aoyagi, T., Shiratori, K., Takeda, K., Akira, S., Saijo, S., *et al.* (2008). Deoxynucleic acids from *Cryptococcus neoformans* activate myeloid dendritic cells via a TLR9-dependent pathway. J. Immunol. *180*, 4067–4074.

Nemunaitis, J., Shannon-Dorcy, K., Appelbaum, F.R., Meyers, J., Owens, A., Day, R., Ando, D., O'Neill, C., Buckner, D., and Singer, J. (1993). Long-term follow-up of patients with invasive fungal disease who received adjunctive therapy with recombinant human macrophage colony-stimulating factor. Blood *82*, 1422–1427.

Netea, M.G., Van Der Graaf, C.A., Vonk, A.G., Verschueren, I., Van Der Meer, J.W., and Kullberg, B.J. (2002). The role of toll-like receptor (TLR) 2 and TLR4 in the host defense against disseminated candidiasis. J. Infect. Dis. *185*, 1483–1489.

Netea, M.G., Van der Graaf, C., Van der Meer, J.W., and Kullberg, B.J. (2004). Recognition of fungal pathogens by Toll-like receptors. Eur. J. Clin. Microbiol. Infect. Dis. 23, 672–676.

Netea, M.G., Ferwerda, G., van der Graaf, C.A., Van der Meer, J.W., and Kullberg, B.J. (2006a). Recognition of fungal pathogens by toll-like receptors. Curr. Pharm. Des. *12*, 4195–4201.

Netea, M.G., Gow, N.A., Munro, C.A., Bates, S., Collins, C., Ferwerda, G., Hobson, R.P., Bertram, G., Hughes, H.B., Jansen, T., *et al.* (2006b). Immune sensing of *Candida albicans* requires cooperative recognition of mannans and glucans by lectin and Toll-like receptors. J. Clin. Invest. *116*, 1642–1650.

Netea, M.G., van de Veerdonk, F., Verschueren, I., van der Meer, J.W., and Kullberg, B.J. (2008). Role of TLR1 and TLR6 in the host defense against disseminated candidiasis. FEMS Immunol. Med. Microbiol. *52*, 118–123.

Neth, O., Jack, D.L., Dodds, A.W., Holzel, H., Klein, N.J., and Turner, M.W. (2000). Mannose-binding lectin binds to a range of clinically relevant microorganisms and promotes complement deposition. Infect. Immun. *68*, 688–693.

Nosanchuk, J.D., and Taborda, C.P. (2013). Vaccines and immunotherapy against fungi: the new frontier. Front. Microbiol. *4*, 6.

Nucci, M., and Colombo, A.L. (2002). Risk factors for breakthrough candidemia. Eur. J. Clin. Microbiol. Infect. Dis. *21*, 209–211.

Oliveira Fde, M., Severo, C.B., Guazzelli, L.S., and Severo, L.C. (2007). *Cryptococcus gattii* fungemia: report of a case with lung and brain lesions mimicking radiological features of malignancy. Rev. Inst. Med. Trop. Sao Paulo *49*, 263–265.

Olivere, J.W., Meier, P.A., Fraser, S.L., Morrison, W.B., Parsons, T.W., and Drehner, D.M. (1999). Coccidioidomycosis--the airborne assault continues: an unusual presentation with a review of the history, epidemiology, and military relevance. Aviat. Space Environ. Med. *70*, 790–796.

Osterholzer, J.J., Milam, J.E., Chen, G.H., Toews, G.B., Huffnagle, G.B., and Olszewski, M.A. (2009). Role of dendritic cells and alveolar macrophages in regulating early host defense against pulmonary infection with *Cryptococcus neoformans*. Infect. Immun. *77*, 3749–3758.

Osterholzer, J.J., Chen, G.H., Olszewski, M.A., Zhang, Y.M., Curtis, J.L., Huffnagle, G.B., and Toews, G.B. (2011). Chemokine receptor 2-mediated accumulation of fungicidal exudate macrophages in mice that clear cryptococcal lung infection. Am. J. Pathol. *178*, 198–211.

d'Ostiani, C.F., Del Sero, G., Bacci, A., Montagnoli, C., Spreca, A., Mencacci, A., Ricciardi-Castagnoli, P., and Romani, L. (2000). Dendritic cells discriminate between yeasts and hyphae of the fungus *Candida albicans*. Implications for initiation of T helper cell immunity *in vitro* and *in vivo*. J. Exp. Med. *191*, 1661–1674.

Owen, J.L., and Mohamadzadeh, M. (2013). Microbial activation of gut dendritic cells and the control of mucosal immunity. J. Interferon Cytokine Res. *33*, 619–631.

Ozer, H., Armitage, J.O., Bennett, C.L., Crawford, J., Demetri, G.D., Pizzo, P.A., Schiffer, C.A., Smith, T.J., Somlo, G., Wade, J.C., *et al.* (2000). 2000 update of recommendations for the use of hematopoietic colony-stimulating factors: evidence-based, clinical practice guidelines. American Society of Clinical Oncology Growth Factors Expert Panel. J. Clin. Oncol. *18*, 3558–3585.

Pagano, L., Akova, M., Dimopoulos, G., Herbrecht, R., Drgona, L., and Blijlevens, N. (2011). Risk assessment and prognostic factors for mould-related diseases in immunocompromised patients. J. Antimicrob. Chemother. *66* (Suppl. 1), i5–14.

Panepinto, J.C., Komperda, K.W., Hacham, M., Shin, S., Liu, X., and Williamson, P.R. (2007). Binding of serum mannan binding lectin to a cell integrity-defective *Cryptococcus neoformans* ccr4Delta mutant. Infect. Immun. 75, 4769–4779.

Pappas, P.G., Bustamante, B., Ticona, E., Hamill, R.J., Johnson, P.C., Reboli, A., Aberg, J., Hasbun, R., and Hsu, H.H. (2004). Recombinant interferon- gamma 1b as adjunctive therapy for AIDS-related acute cryptococcal meningitis. J. Infect. Dis. *189*, 2185–2191.

Parayath, K.E., Harrison, T.S., and Levitz, S.M. (2000). Effect of interleukin (IL)–15 priming on IL-12 and interferon-gamma production by pathogen-stimulated peripheral blood mononuclear cells from human immunodeficiency virus-seropositive and -seronegative donors. J. Infect. Dis. *181*, 733–736.

Park, S.J., Hughes, M.A., Burdick, M., Strieter, R.M., and Mehrad, B. (2009). Early NK cell-derived IFN-{gamma} is essential to host defense in neutropenic invasive aspergillosis. J Immunol. *182*, 4306–4312.

Peltroche-Llacsahuanga, H., Schnitzler, N., Schmidt, S., Tintelnot, K., Lutticken, R., and Haase, G. (2000). Phagocytosis, oxidative burst, and killing of *Candida* dubliniensis and *Candida albicans* by human neutrophils. FEMS Microbiol. Lett. *191*, 151–155.

Perfect, J.R. (2013). Fungal diagnosis: how do we do it and can we do better? Curr. Med. Res. Opin. *29* (Suppl. 4), 3–11.

Perfect, J.R., and Casadevall, A. (2002). Cryptococcosis. Infect. Dis. Clin. North Am. *16*, 837–874, v-vi.

Perfect, J.R., Dismukes, W.E., Dromer, F., Goldman, D.L., Graybill, J.R., Hamill, R.J., Harrison, T.S., Larsen, R.A., Lortholary, O., Nguyen, M.H., *et al.* (2010). Clinical practice guidelines for the management of cryptococcal disease: 2010 update by the infectious diseases society of america. Clin. Infect. Dis. *50*, 291–322.

Perruccio, K., Bozza, S., Montagnoli, C., Bellocchio, S., Aversa, F., Martelli, M., Bistoni, F., Velardi, A., and Romani, L. (2004). Prospects for dendritic cell vaccination against fungal infections in hematopoietic transplantation. Blood Cells Mol. Dis. *33*, 248–255.

Perruccio, K., Tosti, A., Burchielli, E., Topini, F., Ruggeri, L., Carotti, A., Capanni, M., Urbani, E., Mancusi, A., Aversa, F., *et al.* (2005). Transferring functional immune responses to pathogens after haploidentical hematopoietic transplantation. Blood *106*, 4397–4406.

Person, A.K., Chudgar, S.M., Norton, B.L., Tong, B.C., and Stout, J.E. (2010). *Aspergillus niger*: an unusual cause of invasive pulmonary aspergillosis. J. Med. Microbiol. *59*, 834–838.

Peters, C., Minkov, M., Matthes-Martin, S., Potschger, U., Witt, V., Mann, G., Hocker, P., Worel, N., Stary, J., Klingebiel, T., *et al.* (1999). Leucocyte transfusions from rhG-CSF or prednisolone stimulated donors for treatment of severe infections in immunocompromised neutropenic patients. Br. J. Haematol. *106*, 689–696.

Petrikkos, G., and Skiada, A. (2007). Recent advances in antifungal chemotherapy. Int. J. Antimicrob. Agents *30*, 108–117.

Pfaller, M.A., and Diekema, D.J. (2010). Epidemiology of invasive mycoses in North America. Crit. Rev. Microbiol. *36*, 1–53.

Pfaller, M.A., Messer, S.A., Moet, G.J., Jones, R.N., and Castanheira, M. (2011). *Candida* bloodstream infections: comparison of species distribution and resistance to echinocandin and azole antifungal agents in Intensive Care Unit (ICU) and non-ICU settings in the SENTRY Antimicrobial Surveillance Program (2008–2009). Int. J. Antimicrob. Agents *38*, 65–69.

Pikman, R., and Ben-Ami, R. (2012). Immune modulators as adjuncts for the prevention and treatment of invasive fungal infections. Immunotherapy *4*, 1869–1882.

Pippard, M.J., Dalgleish, A., Gibson, P., Malkovsky, M., and Webster, A.D. (1986). Acquired immunodeficiency with disseminated cryptococcosis. Arch. Dis. Child *61*, 289–291.

Porcaro, I., Vidal, M., Jouvert, S., Stahl, P.D., and Giaimis, J. (2003). Mannose receptor contribution to *Candida albicans* phagocytosis by murine E-clone J774 macrophages. J. Leukoc. Biol. *74*, 206–215.

Poulain, D. (2013). *Candida albicans*, plasticity and pathogenesis. Crit. Rev. Microbiol. [Epub ahead of print]

Pullen, G.R., and Hosking, C.S. (1985). Differentiated HL60 promyelocytic leukaemia cells have a deficient myeloperoxidase/halide killing system. Clin. Exp. Immunol. *62*, 304–309.

Qiu, Y., Davis, M.J., Dayrit, J.K., Hadd, Z., Meister, D.L., Osterholzer, J.J., Williamson, P.R., and Olszewski, M.A. (2012). Immune modulation mediated by cryptococcal laccase promotes pulmonary growth and brain dissemination of virulent *Cryptococcus neoformans* in mice. PLoS One *7*, e47853.

Rahman, D., Mistry, M., Thavaraj, S., Challacombe, S.J., and Naglik, J.R. (2007). Murine model of concurrent oral and vaginal *Candida albicans* colonization to study epithelial host–pathogen interactions. Microbes Infect. *9*, 615–622.

Ramirez-Ortiz, Z.G., Specht, C.A., Wang, J.P., Lee, C.K., Bartholomeu, D.C., Gazzinelli, R.T., and Levitz, S.M. (2008). Toll-like receptor 9-dependent immune activation by unmethylated CpG motifs in *Aspergillus fumigatus* DNA. Infect. Immun. *76*, 2123–2129.

Rasmussen, L., and Arvin, A. (1982). Chemotherapy-induced immunosuppression. Environ. Health Perspect. *43*, 21–25.

Rex, J.H., Bennett, J.E., Gallin, J.I., Malech, H.L., DeCarlo, E.S., and Melnick, D.A. (1991). In vivo interferon-gamma therapy augments the *in vitro* ability of chronic granulomatous disease neutrophils to damage Aspergillus hyphae. J. Infect. Dis. *163*, 849–852.

Richardson, M., and Rautemaa, R. (2009). How the host fights against *Candida* infections. Front. Biosci. *1*, 246–257.

Richie, D.L., Feng, X., Hartl, L., Aimanianda, V., Krishnan, K., Powers-Fletcher, M.V., Watson, D.S., Galande, A.K., White, S.M., Willett, T., *et al.* (2011). The virulence of the opportunistic fungal pathogen *Aspergillus fumigatus* requires cooperation between the endoplasmic reticulum-associated degradation pathway (ERAD) and the unfolded protein response (UPR). Virulence *2*, 12–21.

Ricicova, M., Kucharikova, S., Tournu, H., Hendrix, J., Bujdakova, H., Van Eldere, J., Lagrou, K., and Van Dijck, P. (2010). *Candida albicans* biofilm formation in a new *in vivo* rat model. Microbiology *156*, 909–919.

Ringden, O., Labopin, M., Gorin, N.C., Le Blanc, K., Rocha, V., Gluckman, E., Reiffers, J., Arcese, W., Vossen, J.M., Jouet, J.P., *et al.* (2004). Treatment with granulocyte colony-stimulating factor after allogeneic bone marrow transplantation for acute leukemia increases the risk of graft-versus-host disease and death: a study from the Acute Leukemia Working Party of the European Group for Blood and Marrow Transplantation. J. Clin. Oncol. *22*, 416–423.

Roeder, A., Kirschning, C.J., Rupec, R.A., Schaller, M., Weindl, G., and Korting, H.C. (2004). Toll-like receptors as key mediators in innate antifungal immunity. Med. Mycol. *42*, 485–498.

Roilides, E., Tsaparidou, S., Kadiltsoglou, I., Sein, T., and Walsh, T.J. (1999). Interleukin-12 enhances antifungal activity of human mononuclear phagocytes against *Aspergillus fumigatus*: implications for a gamma interferon-independent pathway. Infect. Immun. *67*, 3047–3050.

Rolston, K.V. (2013). Cryptococcosis due to *Cryptococcus gattii*. Clin. Infect. Dis. *57*, 552–554.

Romani, L. (2011). Immunity to fungal infections. Nat. Rev. Immunol. *11*, 275–288.

Romani, L., Bistoni, F., Mencacci, A., Cenci, E., Spaccapelo, R., and Puccetti, P. (1995). IL12 in *Candida albicans* infections. Res. Immunol. *146*, 532–538.

Romani, L., Mencacci, A., Cenci, E., Spaccapelo, R., Toniatti, C., Puccetti, P., Bistoni, F., and Poli, V. (1996). Impaired neutrophil response and CD4+ T helper cell 1 development in interleukin 6-deficient mice infected with *Candida albicans*. J. Exp. Med. *183*, 1345–1355.

Rosenberg, Z.F., and Fauci, A.S. (1989). Immunology of AIDS: approaches to understanding the immunopathogenesis of HIV infection. Ric. Clin. Lab. *19*, 189–209.

Ruping, M.J., Vehreschild, J.J., and Cornely, O.A. (2008). Patients at high risk of invasive fungal infections: when and how to treat. Drugs *68*, 1941–1962.

Ruzicka, F., Hola, V., Mahelova, M., and Prochazkova, A. (2012). [Yeast colonization of urinary catheters and the significance of biofilm formation]. Klin. Mikrobiol. Infekc. Lek. *18*, 115–119.

Safdar, A. (2013). Immunotherapy for invasive mold disease in severely immunosuppressed patients. Clin. Infect. Dis. *57*, 94–100.

Safdar, A., Hanna, H.A., Boktour, M., Kontoyiannis, D.P., Hachem, R., Lichtiger, B., Freireich, E.J., and Raad, II (2004). Impact of high-dose granulocyte transfusions in patients with cancer with candidemia: retrospective case–control analysis of 491 episodes of *Candida* species bloodstream infections. Cancer *101*, 2859–2865.

Sales-Campos, H., Tonani, L., Cardoso, C.R., and Kress, M.R. (2013). The immune interplay between the host and the pathogen in *Aspergillus fumigatus* lung infection. BioMed Res. Int. *2013*, 693023.

Salkowski, C.A., and Balish, E. (1991). Role of natural killer cells in resistance to systemic cryptococcosis. J. Leukoc. Biol. *50*, 151–159.

Santos, W.R., Meyer, W., Wanke, B., Costa, S.P., Trilles, L., Nascimento, J.L., Medeiros, R., Morales, B.P., Bezerra Cde, C., Macedo, R.C., *et al.* (2008). Primary endemic *Cryptococcosis gattii* by molecular type VGII in the state of Para, Brazil. Mem. Inst. Oswaldo Cruz *103*, 813–818.

Sasada, M., and Johnston, R.B., Jr. (1980). Macrophage microbicidal activity. Correlation between phagocytosis-associated oxidative metabolism and the killing of *Candida* by macrophages. J. Exp. Med. *152*, 85–98.

Sato, K., Yang, X.L., Yudate, T., Chung, J.S., Wu, J., Luby-Phelps, K., Kimberly, R.P., Underhill, D., Cruz, P.D., Jr., and Ariizumi, K. (2006). Dectin-2 is a pattern recognition receptor for fungi that couples with the Fc receptor gamma chain to induce innate immune responses. J. Biol. Chem. *281*, 38854–38866.

Saulsbury, F.T. (2001). Successful treatment of aspergillus brain abscess with itraconazole and interferon-gamma in a patient with chronic granulomatous disease. Clin. Infect. Dis. *32*, E137–139.

Saunus, J.M., Wagner, S.A., Matias, M.A., Hu, Y., Zaini, Z.M., and Farah, C.S. (2010). Early activation of the interleukin-23-17 axis in a murine model of oropharyngeal candidiasis. Mol. Oral Microbiol. *25*, 343–356.

Schaller, M., Boeld, U., Oberbauer, S., Hamm, G., Hube, B., and Korting, H.C. (2004). Polymorphonuclear leukocytes (PMNs) induce protective Th1-type cytokine epithelial responses in an *in vitro* model of oral candidosis. Microbiology *150*, 2807–2813.

Schaller, M., Zakikhany, K., Weindl, G., Borelli, C., Korting, H.C., and Hube, B. (2008). [Characterization of the interaction between *Candida albicans* and host cells: In vitro model using reconstituted human skin and mucosa]. Hautarzt *59*, 964–970.

Schneider, J.J., Unholzer, A., Schaller, M., Schafer-Korting, M., and Korting, H.C. (2005). Human defensins. J. Mol. Med. *83*, 587–595.

Schop, J. (2007). Protective immunity against *Cryptococcus neoformans* infection. Mcgill J. Med. *10*, 35–43.

Schwartz, R.A. (2004). Superficial fungal infections. Lancet *364*, 1173–1182.

Segal, B.H., and Walsh, T.J. (2006). Current approaches to diagnosis and treatment of invasive aspergillosis. Am. J. Respir. Crit. Care Med. *173*, 707–717.

Segal, B.H., Kwon-Chung, J., Walsh, T.J., Klein, B.S., Battiwalla, M., Almyroudis, N.G., Holland, S.M., and Romani, L. (2006). Immunotherapy for fungal infections. Clin. Infect. Dis. *42*, 507–515.

Seidel, M.G., Peters, C., Wacker, A., Northoff, H., Moog, R., Boehme, A., Silling, G., Grimminger, W., and Einsele, H. (2008). Randomized phase III study of granulocyte transfusions in neutropenic patients. Bone Marrow Transplant. *42*, 679–684.

Serrano-Gomez, D., Leal, J.A., and Corbi, A.L. (2005). DC-SIGN mediates the binding of *Aspergillus fumigatus* and keratinophylic fungi by human dendritic cells. Immunobiology *210*, 175–183.

Shao, P.L., Huang, L.M., and Hsueh, P.R. (2007). Recent advances and challenges in the treatment of invasive fungal infections. Int. J. Antimicrob. Agents *30*, 487–495.

Singla, N., Gulati, N., Kaistha, N., and Chander, J. (2012). *Candida* colonization in urine samples of ICU patients: determination of etiology, antifungal susceptibility testing and evaluation of associated risk factors. Mycopathologia *174*, 149–155.

Sionov, E., Mendlovic, S., and Segal, E. (2005). Experimental systemic murine aspergillosis: treatment with polyene and caspofungin combination and G-CSF. J. Antimicrob. Chemother. *56*, 594–597.

Sorrell, T.C., and Chen, S.C. (2009). Fungal-derived immune modulating molecules. Adv. Exp. Med. Biol. *666*, 108–120.

Springer, D.J., Phadke, S., Billmyre, B., and Heitman, J. (2012). *Cryptococcus gattii*, no longer an accidental pathogen? Curr. Fungal Infect. Rep. *6*, 245–256.

Stadnyk, A.W. (2002). Intestinal epithelial cells as a source of inflammatory cytokines and chemokines. Can. J. Gastroenterol. *16*, 241–246.

Steele, C., and Fidel, P.L., Jr. (2002). Cytokine and chemokine production by human oral and vaginal epithelial cells in response to *Candida albicans*. Infect. Immun. *70*, 577–583.

Stuehler, C., Khanna, N., Bozza, S., Zelante, T., Moretti, S., Kruhm, M., Lurati, S., Conrad, B., Worschech, E., Stevanovic, S., *et al.* (2011). Cross-protective TH1 immunity against *Aspergillus fumigatus* and *Candida albicans*. Blood *117*, 5881–5891.

Svajger, U., Anderluh, M., Jeras, M., and Obermajer, N. (2010). C-type lectin DC-SIGN: an adhesion, signalling and antigen-uptake molecule that guides dendritic cells in immunity. Cell. Signal. *22*, 1397–1405.

Szabo, E.K., and MacCallum, D.M. (2011). The contribution of mouse models to our understanding of systemic candidiasis. FEMS Microbiol. Lett. *320*, 1–8.

Takahashi, K., Kita, E., Konishi, M., Yoshimoto, E., Mikasa, K., Narita, N., and Kimura, H. (2003). Translocation model of *Candida albicans* in DBA-2/J mice with protein calorie malnutrition mimics hematogenous candidiasis in humans. Microb. Pathog. 35, 179–187.

Talaat, M., Hafez, S., Saied, T., Elfeky, R., El-Shoubary, W., and Pimentel, G. (2010). Surveillance of catheter-associated urinary tract infection in 4 intensive care units at Alexandria university hospitals in Egypt. Am. J. Infect. Control 38, 222–228.

Thywissen, A., Heinekamp, T., Dahse, H.M., Schmaler-Ripcke, J., Nietzsche, S., Zipfel, P.F., and Brakhage, A.A. (2011). Conidial dihydroxynaphthalene melanin of the human pathogenic fungus *Aspergillus fumigatus* interferes with the host endocytosis pathway. Front. Microbiol. 2, 96.

Tomee, J.F., Wierenga, A.T., Hiemstra, P.S., and Kauffman, H.K. (1997). Proteases from *Aspergillus fumigatus* induce release of proinflammatory cytokines and cell detachment in airway epithelial cell lines. J. Infect. Dis. *176*, 300–303.

Toren, A., Or, R., Ackerstein, A., and Nagler, A. (1997). Invasive fungal infections in lymphoma patients receiving immunotherapy following autologous bone marrow transplantation (ABMT). Bone Marrow Transplant. *20*, 67–69.

Tran, P., Ahmad, R., Xu, J., Ahmad, A., and Menezes, J. (2003). Host's innate immune response to fungal and bacterial agents *in vitro:* up-regulation of interleukin-15 gene expression resulting in enhanced natural killer cell activity. Immunology *109*, 263–270.

Tsiodras, S., Samonis, G., Boumpas, D.T., and Kontoyiannis, D.P. (2008). Fungal infections complicating tumor necrosis factor alpha blockade therapy. Mayo Clin. Proc. *83*, 181–194.

Tsoni, S.V., and Brown, G.D. (2008). β-Glucans and dectin-1. Ann. N. Y. Acad. Sci. *1143*, 45–60.

Tuite, N.L., and Lacey, K. (2013). Overview of invasive fungal infections. Methods Mol. Biol. *968*, 1–23.

Uematsu, S., and Fujimoto, K. (2010). The innate immune system in the intestine. Microbiol. Immunol. *54*, 645–657.

Urban, C.F., Reichard, U., Brinkmann, V., and Zychlinsky, A. (2006). Neutrophil extracellular traps capture and kill *Candida albicans* yeast and hyphal forms. Cell. Microbiol. *8*, 668–676.

Urban, C.F., Ermert, D., Schmid, M., Abu-Abed, U., Goosmann, C., Nacken, W., Brinkmann, V., Jungblut, P.R., and Zychlinsky, A. (2009). Neutrophil extracellular traps contain calprotectin, a cytosolic protein complex involved in host defense against *Candida albicans*. PLoS Pathog. *5*, e1000639.

Van Cutsem, J. (1993). Therapy of experimental meningeal and disseminated cryptococcosis. Mycoses *36*, 357–367.

Van der Waaij, D., and Van der Waaij, B.D. (1990). The colonization resistance of the digestive tract in different animal species and in man; a comparative study. Epidemiol. Infect. *105*, 237–243.

van de Veerdonk, F.L., Marijnissen, R.J., Kullberg, B.J., Koenen, H.J., Cheng, S.C., Joosten, I., van den Berg, W.B., Williams, D.L., van der Meer, J.W., Joosten, L.A., *et al.* (2009). The macrophage mannose receptor induces IL-17 in response to *Candida albicans*. Cell Host Microbe 5, 329–340.

van de Veerdonk, F.L., Kullberg, B.J., and Netea, M.G. (2010). Pathogenesis of invasive candidiasis. Curr. Opin. Crit. Care *16*, 453–459.

Vautier, S., Sousa Mda, G., and Brown, G.D. (2010). C-type lectins, fungi and Th17 responses. Cytokine Growth Factor Rev. *21*, 405–412.

Vazquez, N., Walsh, T.J., Friedman, D., Chanock, S.J., and Lyman, C.A. (1998). Interleukin-15 augments superoxide production and microbicidal activity of human monocytes against *Candida albicans*. Infect. Immun. *66*, 145–150.

Viriyakosol, S., Fierer, J., Brown, G.D., and Kirkland, T.N. (2005). Innate immunity to the pathogenic fungus Coccidioides posadasii is dependent on Toll-like receptor 2 and Dectin-1. Infect. Immun. 73, 1553–1560.

Voigt, J., Hunniger, K., Bouzani, M., Jacobsen, I.D., Barz, D., Hube, B., Loffler, J., and Kurzai, O. (2013). Human NK cells act as phagocytes against *Candida albicans* and mount an inflammatory response which modulates neutrophil antifungal activity. J. Infect. Dis. *209*, 616–626.

Volpi, I., Perruccio, K., Tosti, A., Capanni, M., Ruggeri, L., Posati, S., Aversa, F., Tabilio, A., Romani, L., Martelli, M.F., *et al.* (2001). Postgrafting administration of granulocyte colony-stimulating factor impairs functional immune recovery in recipients of human leukocyte antigen haplotype-mismatched hematopoietic transplants. Blood *97*, 2514–2521.

Wagener, J., Weindl, G., de Groot, P.W., de Boer, A.D., Kaesler, S., Thavaraj, S., Bader, O., Mailander-Sanchez, D., Borelli, C., Weig, M., *et al.* (2012). Glycosylation of *Candida albicans* cell wall proteins is critical for induction of innate immune responses and apoptosis of epithelial cells. PLoS One 7, e50518.

Walsh, T.J., and Dixon, D.M. (1996). Spectrum of Mycoses. In Medical Microbiology, S. Baron, ed. (University of Texas Medical Branch at Galveston, Galveston, TX).

Wang, H., Asavaroengchai, W., Yeap, B.Y., Wang, M.G., Wang, S., Sykes, M., and Yang, Y.G. (2009). Paradoxical effects of IFN-gamma in graft-versus-host disease reflect promotion of lymphohematopoietic graft-versus-host reactions and inhibition of epithelial tissue injury. Blood *113*, 3612–3619.

Wasylnka, J.A., and Moore, M.M. (2002). Uptake of *Aspergillus fumigatus* Conidia by phagocytic and non-phagocytic cells *in vitro:* quantitation using strains expressing green fluorescent protein. Infect. Immun. *70*, 3156–3163.

Weindl, G., Naglik, J.R., Kaesler, S., Biedermann, T., Hube, B., Korting, H.C., and Schaller, M. (2007). Human epithelial cells establish direct antifungal defense through TLR4-mediated signaling. J. Clin. Invest. *117*, 3664–3672.

Weindl, G., Wagener, J., and Schaller, M. (2010). Epithelial cells and innate antifungal defense. J. Dent. Res. *89*, 666–675.

Weindl, G., Wagener, J., and Schaller, M. (2011). Interaction of the mucosal barrier with accessory immune cells during fungal infection. Int. J. Med. Microbiol. *301*, 431–435.

Wells, C.A., Salvage-Jones, J.A., Li, X., Hitchens, K., Butcher, S., Murray, R.Z., Beckhouse, A.G., Lo, Y.L., Manzanero, S., Cobbold, C., *et al.* (2008). The macrophage-inducible C-type lectin, mincle, is an essential component of the innate immune response to *Candida albicans*. J. Immunol. *180*, 7404–7413.

Werner, J.L., Metz, A.E., Horn, D., Schoeb, T.R., Hewitt, M.M., Schwiebert, L.M., Faro-Trindade, I., Brown, G.D., and Steele, C. (2009). Requisite role for the dectin-1 beta-glucan receptor in pulmonary defense against *Aspergillus fumigatus*. J. Immunol. *182*, 4938–4946.

Wheat, L.J., and Kauffman, C.A. (2003). Histoplasmosis. Infect. Dis. Clin. North Am. *17*, 1–19, vii.

Wiechula, B., Cholewa, K., Ekiel, A., Romanik, M., Dolezych, H., and Martirosian, G. (2010). [HBD-1 and hBD-2 are expressed in cervico-vaginal lavage in female genital tract due to microbial infections]. Ginekol. Pol. *81*, 268–271.

Winn, R.M., Gil-Lamaignere, C., Roilides, E., Simitsopoulou, M., Lyman, C.A., Maloukou, A., and Walsh, T.J. (2003). Selective effects of interleukin (IL)-15 on antifungal activity and IL-8 release by polymorphonuclear leukocytes in response to hyphae of Aspergillus species. J. Infect. Dis. *188*, 585–590.

Wiseman, J.C., Ma, L.L., Marr, K.J., Jones, G.J., and Mody, C.H. (2007). Perforin-dependent cryptococcal microbicidal activity in NK cells requires PI3K-dependent ERK1/2 signaling. J. Immunol. *178*, 6456–6464.

Wormley, F.L., Jr., Perfect, J.R., Steele, C., and Cox, G.M. (2007). Protection against cryptococcosis by using a murine gamma interferon-producing *Cryptococcus neoformans* strain. Infect. Immun. *75*, 1453–1462.

Wright, D.G. (1984). Symposium on infectious complications of neoplastic disease (Part II). Leukocyte transfusions: thinking twice. Am. J. Med. *76*, 637–644.

Wright, D.G., Robichaud, K.J., Pizzo, P.A., and Deisseroth, A.B. (1981). Lethal pulmonary reactions associated with the combined use of amphotericin B and leukocyte transfusions. New Engl. J. Med. *304*, 1185–1189.

Xiao, G., Miyazato, A., Inden, K., Nakamura, K., Shiratori, K., Nakagawa, K., Miyazawa, T., Suzuki, K., Kaku, M., and Kawakami, K. (2008). *Cryptococcus neoformans* inhibits nitric oxide synthesis caused by CpG-oligodeoxynucleotide-stimulated macrophages in a fashion independent of capsular polysaccharides. Microbiol. Immunol. *52*, 171–179.

Yamamoto, Y., Klein, T.W., and Friedman, H. (1997). Involvement of mannose receptor in cytokine interleukin-1beta (IL-1beta), IL-6, and granulocyte-macrophage colony-stimulating factor responses, but not in chemokine macrophage inflammatory protein 1beta (MIP-1beta), MIP-2, and KC responses, caused by attachment of *Candida albicans* to macrophages. Infect. Immun. *65*, 1077–1082.

Yamasaki, S., Ishikawa, E., Sakuma, M., Hara, H., Ogata, K., and Saito, T. (2008). Mincle is an ITAM-coupled activating receptor that senses damaged cells. Nat. Immunol. *9*, 1179–1188.

Yan, L., Yang, C., and Tang, J. (2013). Disruption of the intestinal mucosal barrier in *Candida albicans* infections. Microbiol. Res. *168*, 389–395.

Yano, J., Lilly, E.A., Steele, C., Fortenberry, D., and Fidel, P.L., Jr. (2005). Oral and vaginal epithelial cell anti-*Candida* activity is acid labile and does not require live epithelial cells. Oral Microbiol. Immunol. *20*, 199–205.

Zelante, T., De Luca, A., Bonifazi, P., Montagnoli, C., Bozza, S., Moretti, S., Belladonna, M.L., Vacca, C., Conte, C., Mosci, P., *et al.* (2007). IL-23 and the Th17 pathway promote inflammation and impair antifungal immune resistance. Eur. J. Immunol. *37*, 2695–2706.

Zelante, T., Bozza, S., De Luca, A., D'Angelo, C., Bonifazi, P., Moretti, S., Giovannini, G., Bistoni, F., and Romani, L. (2009). Th17 cells in the setting of Aspergillus infection and pathology. Med. Mycol. *47* (Suppl. 1), S162–169.

Zhou, Q., Gault, R.A., Kozel, T.R., and Murphy, W.J. (2007). Protection from direct cerebral cryptococcus infection by interferon-gamma-dependent activation of microglial cells. J. Immunol. *178*, 5753–5761.

Zwolinska-Wcislo, M., Brzozowski, T., Budak, A., Sliwowski, Z., Drozdowicz, D., Kwiecien, S., Trojanowska, D., Rudnicka-Sosin, L., Mach, T., Konturek, S.J., *et al.* (2007). [Studies on the influence of *Candida* fungal colonization on the healing process of inflammatory lesions in the colon in rat animal model]. Przegl. Lek. *64*, 124–129.

Antifungal Vaccines and Immunotherapeutics

12

Agostinho Carvalho, Cristina Cunha, Claudia Galosi and Luigina Romani

Abstract

Fungal diseases represent an important immunological paradigm as they can result from either immune deficiency or hyper-inflammatory responses. The expanding number of patients suffering from life-threatening fungal diseases as a consequence of advances in medical care, alongside the frequently limited success of antifungal therapy, has resulted in unacceptable rates of mortality, morbidity and associated costs. Consequently, there is a pressing need for new treatments and preventative measures to combat fungal diseases and, towards this goal, safe and efficacious vaccines would constitute a major step forward. Thus, an improved understanding of the dynamics of the host–fungus interaction, together with the dissection of the mechanisms underlying heterogeneous immune responses to vaccines, set up the foundations to devise new successful vaccination strategies. This chapter reviews recent advances in innate and adaptive immunity to fungi, and highlights how antifungal mechanisms of resistance and tolerance might be exploited for the design of antifungal vaccines.

Introduction: the host–fungus interaction

An unprecedented number of fungal and fungal-like diseases has been documented over the past two decades (Fisher *et al.*, 2012). Of the 1.5 million fungal species estimated to exist, roughly 150–400 have been associated with human and/or animal disease. While only a few species are actually true pathogens for healthy individuals, opportunistic and otherwise commensal fungi can trigger severe, life-threatening infections in immunosuppressed individuals, ranging from superficial mucocutaneous disease to invasive deep-seated infections (Brown *et al.*, 2012). The constantly growing incidence in fungal diseases witnessed in the last decades is correlated with advances in medical care, long-term hospitalization and the increasing numbers among immunocompromised populations due to acquired (e.g. HIV infection) or treatment-induced immunodeficiency in patients receiving haematopoietic stem cell or solid organ transplants, or undergoing anticancer therapy (Pfaller and Diekema, 2010). Consequently, fungi are nowadays positioned as the fourth main cause of hospital-acquired infection in 'at-risk' populations (Pfaller and Diekema, 2010).

The main fungal pathogens affecting humans include *Aspergillus fumigatus*, *Cryptococcus neoformans*, *Histoplasma capsulatum*, *Paracoccidioides brasiliensis*, *Coccidioides immitis* and *posadasii*, *Blastomyces dermatitidis*, *Sporothrix schenckii*, and the commensals *Candida albicans*, *Pneumocystis jirovecii* and *Malassezia* spp. Of interest, new and improved sequencing technologies have recently allowed the identification of a new species of dimorphic fungus (*Emmonsia* spp.) pathogenic to HIV-infected humans (Kenyon *et al.*, 2013; Latge, 2013). Among fungi, pathogenicity strategies are highly variable and often define unique species-specific signatures, as these involve particular mechanisms for gaining access to the host, adhering to and colonizing a niche, evading and subverting immune defences, and multiplying (Rappleye and Goldman, 2008). For example, establishment of disease is often critically dependent on the ability of fungi to reversibly switch morphotypes (Gow *et al.*, 2012). This trait has forced the host immune system to constantly evolve its repertoire of cross-regulatory and overlapping antifungal host responses at different body sites (Romani, 2011). Thus, in the context of a dynamic host–fungus interaction, the strategies used by the host to limit fungal infectivity are necessarily varied in order to cope with the multitude of fungal survival strategies; in retaliation, fungi have perfected their own elaborate tactics to evade or sabotage host defences and to survive.

The ability of fungi to survive and thrive within the human host highlights particular adaptation mechanisms that allow them to persist relatively unnoticed by the immune system (Cooney and Klein, 2008). For example, *A. fumigatus* expresses surface proteins in its conidial form that mask immunogenic cell wall components, rendering it immunologically inert (Aimanianda *et al.*, 2009). This 'peaceful' coexistence may deviate into overt disease under conditions of immune deregulation which may modify the environmental conditions perceived by the fungus. Fungi are exceptionally proficient at sensing their surroundings and responding to cues that benefit their survival in changing environments. In this regard, sensing of mammalian cytokines appears to be one major regulator of fungal virulence and resistance to local antifungal defences (Zelante *et al.*, 2012). Besides an immunologically active milieu, fungi have often to sustain extreme environmental stress conditions, such as hypoxia (Grahl *et al.*, 2011) and limited nutrient availability (Amich *et al.*, 2013). Adaptation to these circumstances relies on profound structural and metabolic changes that, by alerting host immunity, may paradoxically contribute to disease.

Currently available antifungal drugs have been proved efficient *in vitro* and in animal models of fungal diseases. However, the limited accuracy in the clinical diagnosis of fungal infection often delaying a proper antifungal therapy supports the development of novel treatments and in particular of antifungal vaccination strategies (Cassone and Casadevall, 2012; Iannitti *et al.*, 2012). For this purpose, a clear understanding of the mechanisms of antifungal immunity is ultimately required to foster the development of vaccines or strategies aiming at modulating the host's immune response. The continuous interaction between the host and fungi and the commensal relationship with some of them may pose significant challenges in eliciting durable protective immunity, owing to repeated exposure or sensitivity to fungal antigens. A balanced host–fungus relationship and the concomitant fine tuning of pro- and anti-inflammatory signalling allowing for host survival irrespective of pathogen elimination is a prerequisite for coexistence and requires the concerted actions of both innate and adaptive immune systems (Romani, 2011). Therefore, generation of antifungal immunity through vaccination approaches represents a difficult challenge which relies on the establishment and preservation of a precarious equilibrium between pathogen clearance and restriction of tissue damage, while preserving the host microbiota ecology.

This chapter focuses on the current understanding of resistance and tolerance mechanisms of antifungal immune protection and how these might be exploited for the design of effective antifungal vaccines.

Immune response to fungi

Resistance and tolerance mechanisms of antifungal protection

The primary function of the immune system is to sense and remove invading pathogens through a variety of host resistance effector mechanisms. Fig. 12.1 provides an overview of the major pattern recognition receptors (PRRs) involved in innate sensing of common fungal pathogens. While resistance is defined as the capacity to reduce pathogen burden during infection, tolerance is meant to alleviate the substantial cost of resistance systems to host fitness (Medzhitov *et al.*, 2012). Even in the absence of overt tissue damage, resistance mechanisms commonly occur at a cost to normal tissue homeostasis, thus causing immunopathology. This means that an optimal immune response is defined by the balance between efficient pathogen clearance and an acceptable level of immunopathology. Inflammation is an indispensable process required for immune resistance,

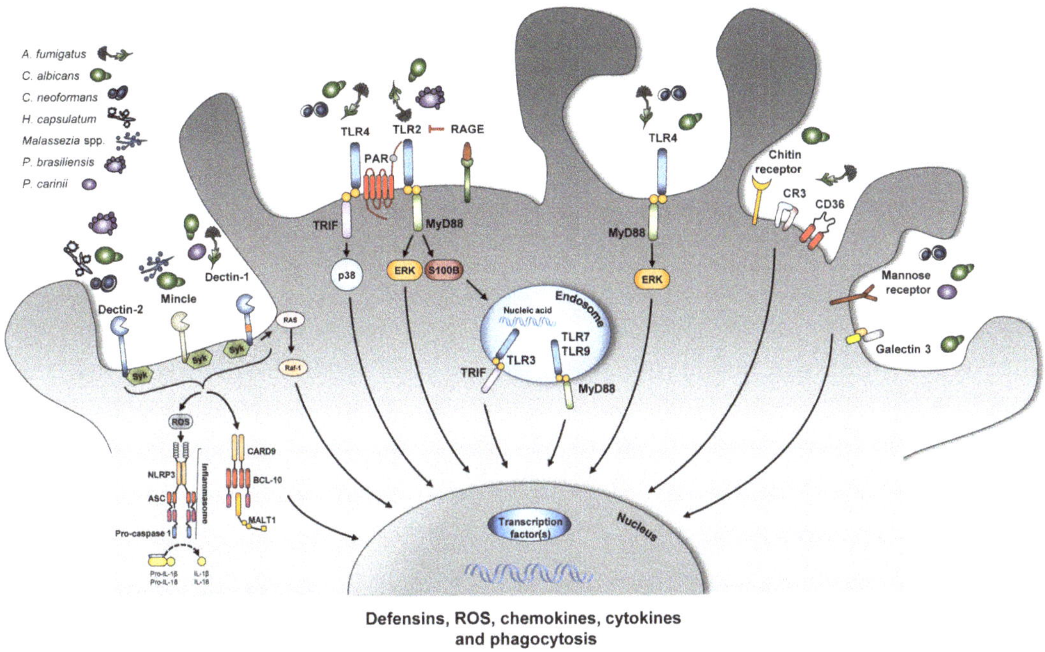

Figure 12.1 Innate recognition of fungi. The major PRRs involved in fungal sensing and the species known to be recognized by each of these PRRs are shown. Upon engagement, PRRs activate multiple intracellular signalling pathways resulting in the nuclear translocation of transcription factors that induce gene expression culminating in specific antifungal immune responses. Abbreviations: ASC; apoptosis-associated speck-like protein containing a CARD; BCL-10, B cell lymphoma 10; CD36, cluster of differentiation 36; CR3, complement receptor 3; ERK, extracellular signal-regulated kinase; MALT1, mucosa-associated lymphoid tissue lymphoma translocation protein 1; MAPK, mitogen-associated protein kinase; NLRP3, nuclear oligomerization domain-like receptor family, pyrin domain containing 3; PAR, protease-activated receptor; RAGE, receptor for advanced glycation endproducts; ROS, reactive oxygen species.

particularly at mucosal tissues, during the shift from the rapid innate to the slower adaptive response. However, the downside of this powerful mechanism of protection against fungi is the collateral damage to the host, which may be more devastating than infection itself. Thus, the ability to tolerate a pathogen's presence is a distinct host defence strategy that may have evolved to favour protective mechanisms without necessarily resorting to pathogen elimination. A vast number of tolerance mechanisms, despite relatively less known than resistance mechanisms, protect the host from immune- or pathogen-induced damage (Cobbold *et al.*, 2010; Saraiva and O'Garra, 2010). Therefore, the term tolerance refers to the multitude of anti-inflammatory mechanisms, including immunological tolerance or unresponsiveness to self-antigens. Whether unnecessary immune responses against 'self' or environmental antigens and commensal microorganisms occur is not clearly defined, although evidence has demonstrated that fungal sensitization contributes to autoreactivity against self-antigens due to shared epitopes with homologous fungal allergens (Zeller *et al.*, 2008).

Innate antifungal immunity is equipped with sensing mechanisms able to discriminate fungal morphotypes and that are hard-wired to activate distinct adaptive immune responses with protective and non-protective functions against different fungal pathogens. Stimulation of antigen-presenting cells such as macrophages, dendritic cells (DCs), and as described more recently, epithelial cells, leads to activation and recruitment of lymphocytes, and the development of adaptive antifungal immune responses. Fig. 12.2 illustrates the distinct T cell responses elicited during fungal infection. There is extensive plasticity in T cell

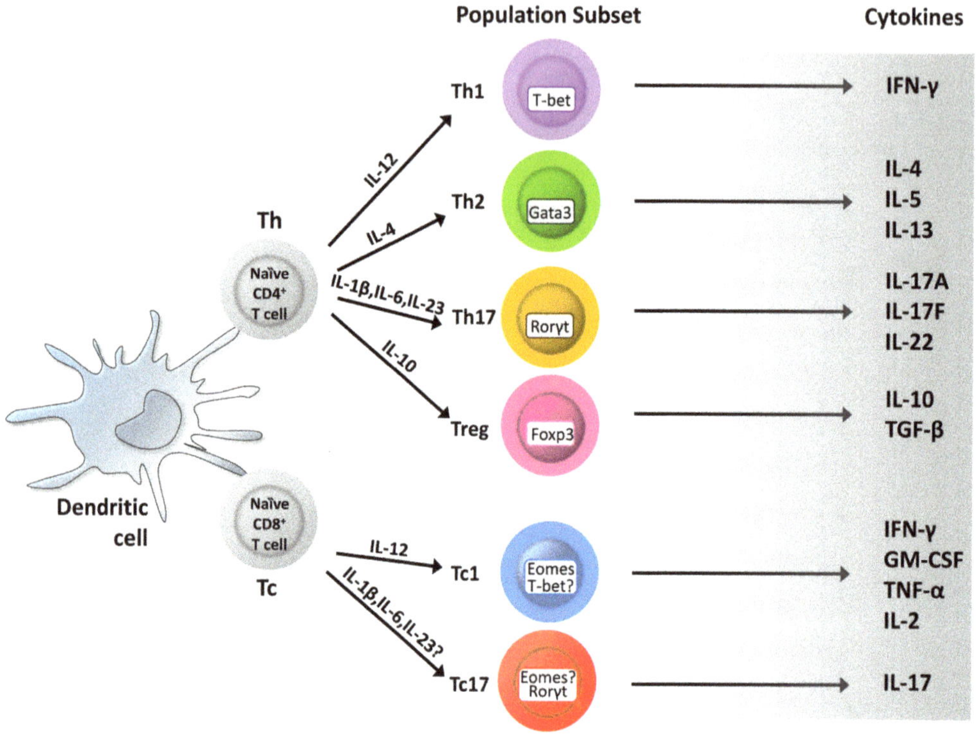

Figure 12.2 T cell responses to fungal infection. During infection, antigen-presenting cells (most notably DCs) prime naïve CD4+ or CD8+ T cells by the presentation of pathogen-associated antigens. Depending on the expression pattern of lineage-specific transcription factors and of cytokine production, primed CD4+ T and CD8+ T cells can differentiate into Th1, Th2, Th17 or Treg, and Tc1 or Tc17 cells, respectively. These T cell subsets are then able to produce different cytokines with specific roles in response to fungal infection.

responses to fungi and the heterogeneity of the CD4+ and CD8+ T cell repertoire may account for the multiplicity and redundancy of effector mechanisms through which these cells participate in the control of fungal infections. Once committed, T cells monitor the host for infection and mobilize appropriate effector functions by inducing cytokines and cytolytic molecules, which are instrumental in rallying and activating professional phagocytes to the site of microbial deposition, thus providing a prompt and effective control of infection (Romani, 2011).

T helper (Th)1 and Th17 cells, the most prominent providers of antifungal resistance, and T regulatory (Treg) cells, which limit the inflammation-associated deleterious effects, crucially contribute to the activation and preservation of these two disparate antifungal mechanisms. The fact that individuals displaying a combined deficiency of Th1 and Th17 pathways are extremely susceptible to fungal diseases (van de Veerdonk *et al.*, 2011) further highlights the crucial role played by both pathways in protection against fungi. Thus, the Th1/Th17 pathways and Treg cells, capable of fine tuning protective antimicrobial immunity to lessen harmful immune pathology, are key components of the current view of immunity to fungi. In this regard, indoleamine 2,3-dioxygenase 1 (IDO1), a haem enzyme that catalyses the first and rate-limiting step in tryptophan catabolism is of central importance (Fallarino *et al.*, 2012). Its downstream catabolites (collectively referred to as kynurenines) provide the host with adequate protective immune mechanisms without necessarily eliminating the pathogen or causing undesirable tissue damage (Zelante *et al.*, 2009). In particular, IDO1 activity induces differentiation of Treg cells and inhibits Th17 cells, thus playing a central role in cell lineage commitment in experimental fungal infections. Under these circumstances, the Th17 pathway, by restraining tryptophan catabolism, may instead favour pathology (Romani *et al.*, 2008b). Altogether, these findings provide evidence supporting the seemingly paradoxical association of chronic inflammation with fungal disease.

Activation of antifungal immunity

Dendritic cells

Immunity to fungi is a dynamic interplay between every arm of the immune system. Innate immune mechanisms are used by the host to respond to a broad range of fungal pathogens in an acute and conserved fashion. Multiple cell populations and cell-signalling pathways are involved in antigen-independent recognition of the fungus by the innate immune system. Applying systems biology approaches to this complex process has resulted in a better appreciation of the intricate crosstalk provided by temporal changes in mediators, metabolites, and cell phenotypes underlining these coordinated processes (Santamaria *et al.*, 2011).

The induction of innate immunity via PRRs shapes the development of the adaptive immune response (Fig. 12.1). As sentinels of the immune system, DCs are responsible for sampling antigenic material in the environment, shaping T cell responses through secretion of cytokines and priming T cells via antigen presentation (Steinman, 2012) (Fig. 12.2). Priming of T cells by DCs is mediated by the presentation of pathogen-associated antigens on major histocompatibility complex class I or class II molecules for priming of $CD8^+$ or $CD4^+$ T cells, respectively. After priming of naïve $CD4^+$ T cells, the response is generally described as Th1, Th2, Th17 or Treg based on the expression pattern of lineage-specific transcription factors and of cytokine production. Priming of naïve $CD8^+$ T cells instead leads to differentiation of cytotoxic T (Tc) cells. Thus, the unique talent of DCs to decode fungal antigens, translating them into qualitatively different adaptive T cell immune responses, positions them as logical cellular targets for the development of antifungal vaccines (Romani, 2011; Roy and Klein, 2012; Wuthrich *et al.*, 2012a).

DCs have the exceptional competence to initiate distinct adaptive antifungal immune responses as a result of cooperation between subsets (Romani *et al.*, 2008a; Romani and Puccetti, 2006) and activation of distinct intracellular signalling pathways (Bonifazi *et al.*, 2009, 2010). The functional plasticity of DCs is mostly defined by their ability to discriminate between different fungal species and morphotypes, as a consequence of their vast repertoire of PRRs (Fig. 12.1). Indeed, whole-genome transcriptional analysis of fungus-pulsed DCs revealed the presence of a specific transcriptional program governing fungal recognition (Rizzetto and Cavalieri, 2010). Among DC subsets, plasmacytoid (p)DCs have an outstanding role in fungal infections. Not only are they rapidly recruited and activated in response to fungi at mucosal sites, such as the lung (Bonifazi *et al.*, 2010; Carvalho *et al.*, 2012b; De Luca *et al.*, 2012; Ramirez-Ortiz *et al.*, 2011) and gut (Bonifazi *et al.*, 2009; De Luca *et al.*, 2007), but infusion of pDCs in bone marrow-transplanted mice has been found to trigger Th1/Treg cell priming leading to decreased fungal growth and inflammatory pathology, and interestingly, transplantation tolerance (Romani *et al.*, 2006).

DCs are now being exploited to improve vaccine efficacy (Steinman, 2008). Fungus-pulsed DCs or RNA-transfected DCs acted, indeed, as potent antifungal vaccines in experimental haematopoietic stem-cell transplantation (Bozza *et al.*, 2003). Protection was associated with myeloid and T cell recovery, the activation of Th1 cells, and the concomitant interleukin (IL)-10-driven Treg cells. Thus, tolerogenic DCs proved to be pivotal in the generation of a dominant form of regulation controlling inflammation, pathogen immunity and tolerance in transplant recipients, eventually contributing to prevention of graft-versus-host reactions and decreased incidence of aspergillosis. These results, along with the finding that fungus-pulsed DCs could reverse T cell anergy of patients with fungal diseases, further supports the usefulness of targeting DCs for antifungal vaccination strategies (Bozza *et al.*, 2004). The screening of signalling pathways in DCs has also been recently exploited for the development of therapeutics targeting inflammation in experimental fungal infections and diseases. *In vivo* modulation of inflammatory or anti-inflammatory pathways in DCs by intranasally delivered small interfering RNA (siRNA) was found to modulate resistance and tolerance to infection (Bonifazi *et al.*, 2010), thus putting forward a potential application of siRNA therapeutics in attenuating inflammation in respiratory fungal infections and diseases.

T-cell responses to fungal infection

Generation of qualitatively different adaptive T cell immune responses is crucial for antifungal immunity (Fig. 12.2). A prevailing Th1 response driven by IL-12 is essentially required for the expression of protective immunity to fungi and is mandatory for the design of effective antifungal vaccines (Spellberg *et al.*, 2008). Through the release of the signature cytokine interferon (IFN)-γ and by helping the production of opsonizing antibodies, induction of Th1 cells is instrumental for the optimal activation of phagocytes at sites of infection. Of interest, Th1 cell-based cross-protection for different fungi has been recently accomplished with a universal immunogenic epitope from the cell wall of *A. fumigatus*, thus providing attractive opportunities for immunotherapeutic strategies (Stuehler *et al.*, 2011). In this regard, it is also worth mentioning that specific innate receptor activation (e.g. dectin-1) can alter the Th1 profile of *A. fumigatus*-specific CD4$^+$ T cells, enabling Th17 differentiation (Rivera *et al.*, 2011). This implies that the design of any new antifungal vaccine should take into account the PRRs and downstream signalling events activated in order to predict the outcome of vaccine-induced T cell-mediated immunity. It is also noteworthy that correction of the Th17/Treg imbalance in murine cystic fibrosis by IDO1 activity promoted lung immune homeostasis and control of *A. fumigatus*-associated lung inflammation (Iannitti *et al.*, 2013). Further supporting the pivotal role of Th1 cell activation, patients with asymptomatic forms of paracoccidioidomycosis demonstrate a Th1-biased reactivity, as opposed to patients with severe disease and poor prognosis, which most commonly display a dominant Th2 response. Thus, the finding that oestradiol favours Th1 type of immunity may explain why paracoccidioidomycosis is much more frequent in men than in women (Pinzan *et al.*, 2010).

A shift in dominance from Th1 to Th2 is a common feature of progressive fungal diseases, with IL-4, IL-5 and IL-13 providing the signals for commitment to the Th2 cell lineage. Despite occurring often concomitantly with vigorous Th1 responses, Th2 reactivity results in a net effect of poor control of the fungal burden, as it induces alternatively activated macrophages that are permissive for intracellular fungal growth and may be impervious to signalling by IFN-γ (Voelz *et al.*, 2009). In addition, Th2 reactivity alters pulmonary physiology and increases airway resistance, thus adding to the severity of infection and contributing to fungus-associated allergic responses (Jain *et al.*, 2009). Accordingly, disrupting Th2 responses to allergic bronchopulmonary aspergillosis in patients with cystic fibrosis (Kreindler *et al.*, 2010) or restricting IL-4 production in experimental models of histoplasmosis (Szymczak and Deepe, 2009) restored antifungal resistance. In contrast to these observations, pneumocystosis is the only known example of fungal disease in which a Th2 cell-dependent humoral immunity may afford some degree of protection (Rapaka *et al.*, 2010).

Th17 cells are a separate lineage of effector Th cells (Kaufmann and Kuchroo, 2009). It is now well accepted that Th17 cell activation occurs in fungal infections (Hernandez-Santos and Gaffen, 2012), mainly through the spleen tyrosine kinase (Syk)/caspase-associated recruitment domain 9 (CARD9), myeloid differentiation primary response gene 88 (MyD88) and mannose receptor signalling pathways in DCs and macrophages leading to the production of unique cytokines such as IL-17, IL-17F and IL-22. This signalling is inhibited by Raf-1 and Toll/IL-1 receptor domain-containing adapter-inducing IFN-β (TRIF)/type I IFN pathways, indicating that the molecular pathways defining activation or inhibition of Th17 cells are present downstream of both C-type lectin receptors (CLRs) and Toll-like receptors (TLRs). Indeed, the central role of Th17 cells in antifungal immunity is supported by studies reporting Mendelian susceptibility to fungal infections of individuals with inborn errors of dectin-1, CARD9, signal transducer and activator of transcription (STAT)1, STAT3 and specifically, IL-17 immunity (Ferwerda *et al.*, 2009; Glocker *et al.*, 2009; Puel *et al.*, 2011; van de Veerdonk *et al.*, 2011). In experimental fungal infections, however, IL-17-dependent immunity has been reported to be either essential (Conti *et al.*, 2009; Huang *et al.*, 2004; Saijo *et al.*, 2007; Wuthrich *et al.*, 2011) or not (De Luca *et al.*, 2010; Hardison *et al.*, 2010; Lin *et al.*, 2009). This suggests that the protective or unfavourable effects of this pathway may rely on the stage and site of infection, as well as the fungal pathogen involved. In addition, Th17 cells are present in the fungus-specific T cell memory repertoire in humans (Acosta-Rodriguez *et al.*, 2007; Bozza *et al.*, 2009a) and mediate some (Wuthrich *et al.*, 2011), but not all (De Luca *et al.*, 2012) vaccine-induced protection in mice. This could be explained by the notion that Th17 cells, although found early during the initiation of an immune response, are involved in a broad range of both Th1 and Th2 type of responses. Indeed, a role for Th17 cells in supporting Th1 cell responses has been shown in experimental mucosal candidiasis (Conti *et al.*, 2009; De Luca *et al.*, 2010). Thus, these findings point to an important regulatory function of the Th17 cell pathway in promoting Th1 and restraining Th2 type of immunity.

As mentioned above, Th17 cells can concomitantly synthesize IL-22, a member of the IL-10 family of cytokines, which has been shown to play a more significant role than IL-17 in host defence in the lung and gut (Zenewicz and Flavell, 2008). Recent findings suggest that the IL-23/IL-22/defensins pathway is crucially involved in the control of fungal growth at mucosal and non-mucosal sites in both candidiasis and aspergillosis, particularly in conditions of Th1 deficiency. Interestingly, memory IL-22$^+$ CD4$^+$ cells specific for *C. albicans* are present in humans (Liu *et al.*, 2009), but are defective in patients with chronic mucocutaneous candidiasis (Eyerich *et al.*, 2008). Indeed, IL-22 has recently been demonstrated to be required for the control of *C. albicans* growth at mucosal sites in the absence of Th1 and Th17 cells (De Luca *et al.*, 2010) and for the balance between resistance and tolerance mechanisms in human and murine vaginal candidiasis (De Luca *et al.*, 2013). Consistently, patients with autosomal dominant hyper-IgE syndrome owing to dominant negative mutations of STAT3 have a defective Th17 cell response to *C. albicans* (Milner *et al.*, 2008). Of interest, dectin-1-mediated production of IL-22 in the lung has also been demonstrated to contribute to early innate immune resistance to *A. fumigatus* (Gessner *et al.*, 2012), although it paradoxically promoted lung inflammation and immunopathology during persistent fungal exposure in an fungal allergy model (Lilly *et al.*, 2012). These seemingly discrepant findings further add to the complexity of IL-22 function in antifungal mucosal immunity and point to the existence of regulatory events leading to its production that depend on the stage and site of infection.

In experimental fungal infections, inflammatory immunity and immune tolerance in the respiratory or gastrointestinal mucosa have been shown to be controlled by the coordinated activation of different Treg cell subsets. Treg cell activity contributes to host tissue damage restriction during infection, but may conversely handicap the efficacy of protective immunity by promoting fungal persistence (Romani and Puccetti, 2006) and eventually, immunosuppression (Ferreira *et al.*, 2010). Altogether, by controlling the quality and magnitude of effector innate and adaptive responses, the spectrum of Treg cell skills may go from protective tolerance and immune homeostasis to dominant effector activities. This scenario is exemplified in experimental aspergillosis, in which inflammation is controlled early by natural Treg cells suppressing neutrophils whereas later tolerogenic Treg cells inhibit Th2 cells and prevent fungal allergy (Montagnoli *et al.*, 2006).

Antifungal vaccination

Challenges in antifungal vaccine design

Vaccines represent the most useful immunological application for human health. They are the only medical tool that, when used prophylactically allow disease elimination and even pathogen eradication. However, the truth is that since the first safe and efficacious vaccine has eradicated smallpox, and after more than a dozen lethal bacterial and viral infections have been kept at bay by vaccination, there is still no vaccine against any fungal disease (Cassone and Casadevall, 2012). However, the increasing medical need and the attention of both researchers and pharmaceutical industry representatives has been changing the perspective regarding the development of antifungal vaccines. The combination of biotechnological progress leading to the generation of well-characterized, highly immunogenic, stable and standardized fungal antigens on one hand, and the recognition of our inability to adequately deal and treat most fungal diseases at acceptable cost on the other hand, are the major determinants of such developments (Cassone and Casadevall, 2012). In recent years, a number of antifungal vaccine candidates has been reported, refined and tested in animal models for their safety, immunogenicity and efficacy and at least two of them are now under study in human clinical trials (Cassone, 2013). A list with the major fungal vaccine candidates is shown in Table 12.1.

With the possible exception of vaccines against endemic fungal diseases, such as coccidioidomycosis, paracoccidioidomycosis and histoplasmosis, and chronic fungal diseases such as recurrent vulvovaginal candidiasis, medical need for antifungal vaccines is more pressing in clinical settings consisting of immunocompromised or otherwise debilitated hosts. There are however some remarkable obstacles for the implementation of such strategies in these patients. Many immunocompromised subjects who are prone to develop aggressive fungal infections such as invasive aspergillosis and deep-seated candidiasis are those with haematological malignancies and cancer, and that need to be subjected to transplant and/or immunosuppressive, anticancer therapy immediately after diagnosis. Consequently, vaccine development for this group poses a problem of leaving too narrow a window to consider vaccination and/or initiate an appropriate schedule of vaccination. In addition, despite claims that residual, effective immune responses could effectively deal with the pathogen even after removal or loss of major groups of antifungal defensive cells such as phagocytes and $CD4^+$ T cells (Koh *et al.*, 2008; Wuthrich *et al.*, 2003), it remains uncertain whether long-lasting protection could be obtained in these subjects because of their underlying immune defect. One such example regards haematological patients undergoing haploidentical haematopoietic stem-cell transplantation, which display delayed immune reconstitution and impaired immune function rendering them more susceptible to invasive aspergillosis and other post-transplant infectious complications (Oevermann *et al.*, 2012). One potential approach to ensure that molecular mechanisms involved in regulating vaccine-related immune responses do not breakdown after immunization is the manipulation of IDO1 activity and its immunoregulatory properties (Fallarino and Grohmann, 2011).

A large set of individuals that may nonetheless benefit from these approaches includes those with no obvious immune deficits, but that, as the result of specific medical interventions, are placed at risk of life-threatening fungal diseases. For example, patients with breaches in their cutaneous or mucosal defences, bearing permanent central venous catheters, or simply having a long hospital stay, particularly in the intensive care setting, may be suitable candidates for antifungal vaccination. Hence, and given the public health problem that fungal diseases represent in these patients, it is conceivable that vaccines against opportunistic fungi will eventually be generated and primarily used in one or more of these subject categories. Vaccine design in these patients is however not free of impediments or potential side-effects. Ideally, it should consider the nature of the relationship between medically significant fungi and their human hosts (Hamad, 2012). For example, vaccination against commensals may disrupt the asymptomatic and possibly beneficial relationship between the fungus and its host. In the case of fungal pathogens with the ability to establish latency or to cause chronic or recurrent infections, vaccination may carry the risk of reactivating the target pathogen. Finally, vaccination-related exacerbations of allergic responses may arise when vaccinating against moulds causing allergic disease.

Deciphering antifungal immunity to vaccination strategies

An efficacious vaccine relies on the elicitation of pathogen-specific immune memory that mediates long-lasting protection from infection or disease. Fig. 12.3 illustrates the major cellular- and humoral-mediated correlates of immunity/protection elicited in response to fungi. Given the

Table 12.1 Major antifungal vaccine candidates

Target pathogen(s)	Vaccine candidate(s)	Type of antigen	Mechanism(s) of protection	Reference(s)
Aspergillus	Asp f16, Asp f3, Pep1, Gel1, glucans	Recombinant/ subunit	Th1	Bozza *et al*. (2002, 2009a), Ito *et al*. (2006)
Aspergillus, *Candida*	Crf1	Recombinant	Th1	Stuehler *et al*. (2011)
Aspergillus, *Coccidioides*, *Candida*, *Cryptococcus*	Heat-killed *Saccharomyces cerevisiae*	Heat-killed whole	Th1, Th2, Th17, antibodies?	Liu *et al*. (2011)
Blastomyces	Live attenuated	Whole	Th1, Th17	Wuthrich *et al*. (2011)
Candida	Als1, Als3, rAls3p-N (NDV-3)[a]	Recombinant/ subunit	Th1, Th17, antibodies	Ibrahim *et al*. (2013), Lin *et al*. (2009)
Candida	Fba peptide	Subunit	Antibodies	Cutler *et al*. (2011)
Candida	Hyr1	Recombinant/ subunit	Antibodies	Luo *et al*. (2011)
Candida	Live attenuated	Whole	T cells, antibodies	Martinez-Lopez *et al*. (2008), Saville *et al*. (2009)
Candida	Mannan	Subunit/ recombinant	Th1, antibodies	Paulovicova *et al*. (2007), Xin *et al*. (2008)
Candida	Eno1	Recombinant	Antibodies	Li *et al*. (2011)
Candida	Sap2p PEV-7[a]	Recombinant truncated	Antibodies	Sandini *et al*. (2011)
Candida, *Cryptococcus*	Laminarin	Subunit	Antibodies	Bromuro *et al*. (2010), Rachini *et al*. (2007), Torosantucci *et al*. (2005)
Coccidioides	Live attenuated	Whole	Th1, Th17, Th2?	Hung *et al*. (2011), Wuthrich *et al*. (2011), Xue *et al*. (2009)
Coccidioides	T-cell epitopes	Recombinant	Th1, Th17, Th2?	Hurtgen *et al*. (2012), Tarcha *et al*. (2006b)
Cryptococcus	Glucuronoxylomannan (GXM), galactoxylomannan	Subunit	Antibodies?	Chow and Casadevall (2011)
Cryptococcus	GXM mimotopes	Subunit/ recombinant	Antibodies	Datta *et al*. (2008)
Cryptococcus	Mannoprotein	Subunit/ recombinant	Th1, antibodies	Specht *et al*. (2007)
Histoplasma	HSP60	Subunit	Th1	Scheckelhoff and Deepe (2002)
Histoplasma	Live, heat-killed	Whole	Th1, Th17	Wuthrich *et al*. (2011)
Paracoccidioides	HSP60	Recombinant	Th1	de Bastos Ascenco Soares *et al*. (2008)
Paracoccidioides	Mycobacterial HSP65	Recombinant	Th1	Ribeiro *et al*. (2009)
Paracoccidioides	P10 peptide	Subunit	Th1, antibodies	Mayorga *et al*. (2012)
Paracoccidioides	rPb27	Recombinant	Antibodies	Fernandes *et al*. (2011)
Pneumocystis	Kexin	Recombinant/ subunit	Antibodies, $CD8^+$ T cells	Zheng *et al*. (2005)
Pneumocystis	p55	Recombinant	Partial, antibodies?	Duan *et al*. (2011), Feng *et al*. (2011)

[a]rAls3p-N (NDV-3) and Sap2p PEV-7 are under study in human clinical trials.

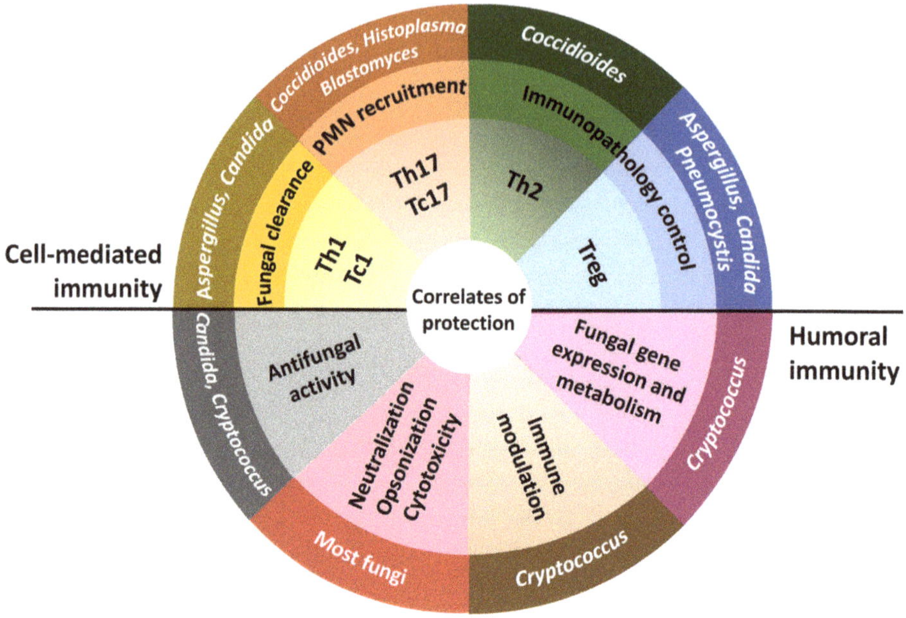

Figure 12.3 Cellular- and humoral-mediated protective immune responses against fungi. Shown are the different correlates of protection to the various fungal pathogens elicited either through the activity of T cells (cell-mediated immunity) or secreted antibodies (humoral immunity).

array of fungal motifs present at the cell surface, as well as those that become available to immune sensing upon processing of the fungus by phagocytic cells, it is obvious that vaccine-induced protection to attenuated fungal strains occurs through distinct PRRs and downstream signalling adapters (De Luca *et al.*, 2012; Wuthrich *et al.*, 2011). Indeed, analyses of naïve to effector to memory transition has revealed changes in expression of innate immune receptors as one major early molecular signature upon vaccination (Pulendran *et al.*, 2010).

Acquisition of vaccine immunity to live attenuated strains of *B. dermatitidis*, *H. capsulatum* and *C. posadasii* was found to require signalling through MyD88 leading to induction of Th17 cells (Wuthrich *et al.*, 2011). Of note, IL-1 was found to promote the expansion of naïve $CD4^+$ T cells, allowing their entry into the lungs, and inducing their differentiation to protective Th17 cells, thereby enhancing the efficacy of safe, but otherwise ineffective heat-inactivated formulations and subunit vaccines against blastomycosis (Wuthrich *et al.*, 2013). CLRs and downstream signalling through CARD9 have also been recently reported to play a key role in driving Th17-dependent vaccine immunity (Wang *et al.*, 2014). Specifically, acquired immunity to *B. dermatitidis* was found to require dectin-2, whereas vaccination against *H. capsulatum* and *C. posadasii* infection instead relied on innate sensing by dectin-1 and dectin-2, but not macrophage-inducible C-type lectin (Mincle). In contrast to these observations, protection to *A. fumigatus* relied on TRIF-dependent differentiation of Th1 cells (De Luca *et al.*, 2012), even though vaccination with purified *A. fumigatus* antigens was dependent on the MyD88 pathway in the presence of the appropriate adjuvant (Carvalho *et al.*, 2012b), a finding pointing to the crucial role of adjuvants in promoting T cell differentiation along specific effector pathways. Thus, fungal innate sensing is one critical step in mounting immune responses eventually defining appropriate effector responses maximizing protection (Levitz and Golenbock, 2012). Fungi have evolved their own intricate mechanisms to prevent activation of effector T cell responses. For example, mucosal vaccination was found to subvert T cell priming by impairing chemokine signals on egress of inflammatory monocytes from the bone marrow and their recruitment to the lung (Wuthrich *et al.*, 2012b). This finding is even more significant considering the role of inflammatory monocytes in orchestrating antifungal immunity in the lung by regulating neutrophil conidiocidal activity and their own differentiation to dendritic cells (Espinosa *et al.*, 2014).

Given the complexity of innate immune signalling networks activated in response to fungal antigens and the strategies devised by fungi to subvert these at the different body sites, engagement of individual PRRs or in combinations will have to be weighed in order to achieve the best vaccine-specific responses appropriate for each fungal pathogen. The systematic elucidation of mechanisms regulating the extent, quality, and persistence of vaccine-induced humoral and T cell-dependent immunity will contribute to a more balanced design of potentially useful vaccines (Pulendran and Ahmed, 2011). Examples of such approaches include the novel vaccine platform consisting of hollow yeast-derived β-glucan particles that combine adjuvancity and high load antigen delivery to induce strong humoral and Th1- and Th17-biased T cell responses (Huang *et al.*, 2010) and the glycoconjugate vaccines capable of eliciting remarkably potent B cell responses by providing immunogenic epitopes to CD4$^+$ T cells (Bromuro *et al.*, 2010; Rachini *et al.*, 2007; Torosantucci *et al.*, 2005; Xin *et al.*, 2008).

In any case, and despite the unequivocal role other cells may play, Th1-mediated protection has been reported across nearly all clinically relevant fungal diseases (Table 12.1). For example, crude antigen preparations from *A. fumigatus* or recombinant fungal antigens alone (Diaz-Arevalo *et al.*, 2011) or in conjunction with CpG oligonucleotides as adjuvants (Bozza *et al.*, 2009b; Cenci *et al.*, 2000; Ito *et al.*, 2006; Stuehler *et al.*, 2011) and the recombinant Als3p-N protein of *C. albicans* formulated with alum adjuvant (designated as NDV-3 and currently under clinical trial) (Ibrahim *et al.*, 2013) are among the most important fungal vaccine candidates leading to induction of strong Th1 responses. Th1 differentiation has also been shown for mannosylated cryptococcal antigens (Lam *et al.*, 2005), *B. dermatitidis* adhesin antigen (Wuthrich *et al.*, 2003), heat shock protein 60 from *P. brasiliensis* (de Bastos Ascenco Soares *et al.*, 2008) and *H. capsulatum* (Deepe and Gibbons, 2002), and the multivalent vaccines comprising complexes of protein antigens of *Coccidioides* spp., administered in combination with adjuvants (Shubitz *et al.*, 2006; Tarcha *et al.*, 2006a).

Exploiting CD8$^+$ T cell immunity in antifungal vaccination

Fungal infections in CD4$^+$ T cell lymphopenic patients have risen at alarming rates due to the HIV pandemic, and despite antifungal therapy, they carry a mortality rate that can reach 70% depending on the type of infection (Miller *et al.*, 2013; Nacher *et al.*, 2013; Park *et al.*, 2009). Even though opportunistic fungi reemerge in HIV patients, primary systemic fungal pathogens also cause life-threatening disease when effector CD4$^+$ T cells are absent or functionally impaired (Kenyon *et al.*, 2013). Hence, understanding the role of CD8$^+$ T cell immunity against fungal infections may aid in the development of antifungal vaccines tailored to such patients. Indeed, and despite CD4$^+$ Th1 cells have been historically considered the cornerstone of cell-mediated defence against intracellular fungi (Cassone and Casadevall, 2012; Iannitti *et al.*, 2012), CD8$^+$ T cells have also recently earned a place in this category (Nanjappa and Klein, 2014) (Fig. 12.2). Antifungal immunity in the absence of CD4$^+$ T cell immunity is mediated by CD8$^+$ Tc1 cells that secrete proinflammatory cytokines and cytotoxic factors including perforin and granulysin. However, like the prominent role played by Th17 cells in intact hosts, recent studies have shown that antifungal immunity can also be mediated by a distinct lineage of CD8$^+$ T cells able to produce IL-17A (Tc17 cells) (Hernandez-Santos *et al.*, 2013; Nanjappa *et al.*, 2012a). For example, in a mouse model of vaccination against blastomycosis, both the numbers and function of protective antifungal memory CD8$^+$ T cells were maintained even in the absence of CD4$^+$ T cell help (Nanjappa *et al.*, 2012b). Under these circumstances, Tc17 cells have been found to be non-redundant for vaccine immunity to fungal infection by mediating protection in a neutrophil-dependent manner (Nanjappa *et al.*, 2012a) and to compensate for the lack of classic Th17 cells in CD4-deficient hosts with oral mucosal candidiasis (Hernandez-Santos *et al.*, 2013). Indeed, $CD4^+$ T cells may not be required for generating and maintaining normal memory homeostasis of $CD8^+$ T cells during fungal infection. These findings suggest a unique priming environment that facilitates their long-term survival and a distinct effector cell profile that is different from that following viral or bacterial immunization. The causes for these functional differences could be many (Nanjappa and Klein, 2014). In these models, live attenuated fungal strains were used, which provide a unique inflammatory milieu distinct from that using subunit/recombinant/DNA vaccines. In addition, whole cell-based vaccines, along with an appropriate inflammatory microenvironment, can give rise to a broad repertoire of $CD8^+$ T cells, which may be less prone to exhaustion or terminal differentiation. Finally, $CD8^+$ T cell responses arise in a setting that gives rises to multifunctional cytokine profiles, including the production of IL-2, which promotes and helps memory cell homeostasis.

Immunological memory is the hallmark of vaccination. Pathogen-specific memory T and B cells can persist in the absence of immunogen and are able to overcome a pathogen by 'remembering' the strategy used to defeat previous infections (Sallusto *et al.*, 2010). In addition to central memory T cells present in secondary lymphoid organs, which scrutinize the presence of remote pathogens via DCs, effector memory T cells reside in peripheral non-lymphoid tissues such as the skin and mucosa. As stated, since many fungal infections occur in the setting of immune deficiency where $CD4^+$ T cells may be absent, reduced or dysfunctional, it becomes crucial to understand the limits of memory maintenance under these unusual circumstances. Memory $CD8^+$ cytotoxic T cells are induced in fungal infections (Nanjappa *et al.*, 2012b) and exhibit a pleiotropic activity by mediating protection via production of IFN-γ and cytolytic activity against fungus-laden cells or the fungus itself (Carvalho *et al.*, 2012b; De Luca *et al.*, 2012). As such, $CD8^+$ T cells, especially if long-lasting, are regarded as ideal candidates for expansion at mucosal surfaces by vaccination strategies. Of note, antifungal effector Tc17 cells were found to persist over two months in a CD4-deficient model of blastomycosis and portrayed a molecular signature that portended their long-term survival (Nanjappa *et al.*, 2012a). The long-term fate and plasticity of Tc17 cells remain to be determined, although it is likely that the requirements for Th17/Tc17 memory homeostasis are regulated differently from that for Th1 and Tc1 cells.

The recent evidence proposing a role for metabolism (Pearce *et al.*, 2009) and bioenergetics (van der Windt *et al.*, 2013) in harnessing T cell memory and promoting rapid recall rates opens up new perspectives on how epigenetic and environmental mechanisms modulate memory differentiation and quality, thus opening new avenues for vaccine development. Finally, additional subsets of T cells may also become important targets for new vaccines, such as the invariant natural killer T cells that activate antifungal responses and promote airway hyperreactivity through the recognition of β-1,3-glucans (Cohen *et al.*, 2011) and the glycosphingolipid asperamide B (Albacker *et al.*, 2013) present in the fungal cell wall. Ultimately, the origin of the antifungal vaccine, the route of antigen delivery and the mode of antigen routing and presentation are critical for the successful application of a fungal vaccine. Indeed, recent evidence has highlighted striking differences in antigen presentation pathways in DCs leading to the activation of $CD4^+$ or $CD8^+$ T cells, with either cell type being able to provide antifungal memory in the absence of the other under specific conditions (De Luca *et al.*, 2012). These findings further stress the need to better understand immunological memory at a basic level, including information obtained from suitable animal models, in order to personalize vaccination strategies against fungal diseases. Refinement of these approaches could lead to antifungal vaccines and adjuvants tailored to the different target populations, administered either alone or in combination with immunomodulators modulating antigen trafficking and presentation pathways. It is conceivable that, as recently proposed (Cassone, 2013), the development of a multivalent vaccine inducing an immune response against multiple, unrelated fungal virulence traits may be the best approach towards the first antifungal vaccine approved for human use.

Correlates of protection: passive and active immunotherapy

Like natural infection, antifungal vaccines exploit the redundancy in the immune system to afford protection though a wide variety of mechanisms (Iannitti *et al.*, 2012). Antifungal vaccines can be defined as active immunotherapies since they boost the immune system to specifically attack fungal pathogens on the basis of one or more specific antigens (Carvalho *et al.*, 2012a). On the other hand, passive immunotherapy strategies are comprised of laboratory-synthesized antibodies or other immune system components that can be administered to patients and do not stimulate the immune system nor 'actively' respond to infection in the way a vaccine does. In this regard, a number of monoclonal human recombinant antibodies and their fragments have already been tested in experimental fungal diseases (Casadevall and Pirofski, 2012). Most antifungal vaccines induce antibody responses and their titre threshold may therefore predict vaccine efficacy and may serve as vaccine surrogate markers even when the mechanism of protection is cell-mediated (Spellberg *et al.*, 2008). It is worth mentioning that one of the antifungal vaccines presently under clinical trial is a formulation based on recombinant secreted aspartic proteinase 2 proteins that induce protection through elicitation of neutralizing antibodies at the vaginal level and protects against recurrent vulvovaginal candidiasis (Sandini *et al.*, 2011). Indeed, and even though the current view is that protection against fungi is prevalently provided by $CD4^+$ or $CD8^+$ T cells, antibodies are known to globally contribute to the optimal function of T cell-mediated immunity in several ways (Casadevall and Pirofski, 2012). This may suggest that the concomitant administration of antibodies and vaccination may be potentially useful to further enhance or modulate the immune response.

Several immunotherapeutic strategies for fungal infections have been attempted in the past or are currently under study. Colony-stimulating factors (CSFs) and cytokines, mainly IFN-γ, have been utilized in the clinical management of fungal diseases (Roilides *et al.*, 2002). However, the use of CSFs is still controversial; granulocyte–macrophage CSF has been shown to reprogram human DCs for activation of inflammatory Th1 and Th17 cells, and to serve a non-redundant function in the initiation of autoimmune inflammation (Codarri *et al.*, 2011). Most antifungal immunotherapies have however tried to limit inflammation while stimulating protective immune responses to fungi. In this regard, the long pentraxin 3 (PTX3) is receiving great attention as a potential therapeutic agent with anti-inflammatory properties in aspergillosis. PTX3 deficiency renders immunocompetent mice as well as human patients undergoing stem-cell transplantation highly susceptible to invasive aspergillosis (Cunha *et al.*, 2014; Garlanda *et al.*, 2002). Of interest, exogenous administration of PTX3 reverts the susceptibility phenotype in mice (Garlanda *et al.*, 2002) and *in vitro* using human PTX3-deficient neutrophils (Cunha *et al.*, 2014). More recently, it has been shown that PTX3 alone (Lo Giudice *et al.*, 2010) or in combination with antifungals (Lo Giudice *et al.*, 2012) can be used to confer protection in animal models of invasive aspergillosis. This beneficial effect has also been demonstrated in nicotinamide adenine dinucleotide phosphate oxidase-deficient mice, in which early administration of PTX3 enhanced the conidiocidal activity of neutrophils and limited the inflammatory pathology characteristic of these animals (D'Angelo *et al.*, 2009).

As mentioned above, innate signalling defines the induction of specific adaptive immune responses. This implies that selected PRR ligands might be useful candidate adjuvants to dampen inflammatory responses associated with infections. Thymosin α1 (Tα1), a naturally occurring thymic peptide, which is approved in 30 countries for treatment of a number of viral infections and as an immune adjuvant, has been found to promote maturation of and cytokine production by DCs via TLR signalling (Pierluigi *et al.*, 2010). By inducing IDO1 activity and controlling the balance between the activity of immunogenic and tolerogenic DCs, Tα1 acted as an endogenous immune regulator capable of inducing protective immunity to *A. fumigatus*. The potential usefulness of PRR agonists to treat fungal infections was also recently addressed in a clinical trial assessing the efficacy of topical applications of imiquimod (a TLR7 ligand) to treat patients afflicted with the chronic infection chromoblastomycosis (da Gloria Teixeira de Sousa *et al.*, 2014). In this study, all four patients tested displayed a significant improvement of fungus-associated lesions, even in the absence of concurrent oral antifungal therapy.

Altogether, it seems that correlates of protection may differ quantitatively and qualitatively depending on the type and route of vaccine and, perhaps, on whether the purpose is to prevent systemic or mucosal infection. The ideal antifungal vaccine would be expected to induce both mucosal and systemic immunity as a way to provide both the first and second line of defence against infection and to compensate for the apparently fading memory responses at mucosal sites.

Vaccinomics and personalized vaccinology

The deciphering of the intricacies in the immune response to vaccines demands for the integration of advanced immunology approaches, systems biology, immunogenetic profiling, and bioinformatics in the areas of pathogen and host biology and their reciprocal interaction. Vaccinomics is a recently generated term that encompasses the field of immunogenomics applied to the elucidation of the mechanisms underlying heterogeneity in immune responses to vaccines (Poland and Oberg, 2010). Common genetic variability in immune genes has been disclosed as a major determinant of the quality of the immune response to fungi, particularly in the relative absence of immune redundancy, as observed for invasive aspergillosis (Cunha *et al.*, 2013) and candidaemia (Smeekens *et al.*, 2013). Of note, the first step towards uncovering the genetic susceptibility to *Cryptococcus* infections in HIV-infected individuals has also been recently reported (Rohatgi *et al.*, 2013). For these reasons, immunogenetic profiling is nowadays regarded as a promising methodology to exploit in the future towards improved diagnosis and therapy of fungal diseases (Cunha and Carvalho, 2012). Not only is this relevant but also functional polymorphisms in certain immune genes, including that coding for IFN-γ (Ovsyannikova *et al.*, 2013), have been found to underlie variable immune responses to a given vaccine. Thus, it is not surprising that measurement of *A. fumigatus*-specific immune responses in haematological patients has also been recently proposed as a promising immunodiagnostic approach (Potenza *et al.*, 2013). These data suggest that, by interpreting immunogenetic signatures and identifying even subtle differences in immune profiles, it may be possible to discriminate response efficiencies to antifungal vaccines. The finding that genetic deficiency of TLR3 is associated with susceptibility to invasive aspergillosis among haematopoietic stem-cell transplant recipients due to the inability of DCs to activate protective memory CD8$^+$ T cells provides one such example (Carvalho *et al.*, 2012b).

The complexity of innate and adaptive immune responses to fungi implies many challenges, and data from functional genomics, transcriptomics, proteomics and metabolomics will probably be needed to identify the best predictors of vaccine efficacy (Kennedy and Poland, 2011). Along this line, host microbiomics should also be given special consideration in view of their protective function against colonization by endogenous and exogenous microorganisms and the modulation of host immune responses (Kamada *et al.*, 2013). As a matter of fact, it was recently shown that catabolites originating from the breakdown of tryptophan by intestinal microbiota activated the aryl hydrocarbon receptor, thereby fine tuning mucosal reactivity to *C. albicans* (Zelante *et al.*, 2013). Further supporting this concept, dectin-1 has also been found to act as a major intestinal immune homeostatic element that by interacting with the rich fungal community contributes to the 'peaceful' coexistence with bacteria in the mammalian gut, thus affording protection from colitis (Iliev *et al.*, 2012). Whether the endogenous lung microbiota also plays a role in susceptibility to and progression of fungal diseases is currently a hot topic of research. In the end, all this information can be used in the directed and rational development of new vaccines – predictive vaccinology – which will predict the likelihood of a vaccine response or an adverse response to a vaccine in each target population. By bringing together the fields of basic immunology and vaccine research, creation of individualized immune profiles that have the potential for customizing adjuvants and vaccines for fungal diseases may be within our grasp. In summary, the paradigm shifts in vaccinology that have been witnessed in the past decade of vaccine research may help us to move from an empiric approach towards the optimization of vaccine design and development. It is expected that our increased understanding of the fundamental mechanisms by which the innate immune system orchestrates protective immune responses to fungi will be integrated into clinical practices to guide vaccine development.

Concluding remarks

The control of inflammation leading to tolerance, the molecular bases underlying immune regulation and deregulation, and the way in which commensal but opportunistic fungal pathogens can switch from a 'friendly' affinity with the host to a pathological relationship, by evading or subverting host inflammation, are challenging issues in the field of medical mycology and infection-related immunological disorders. As mentioned above, a related question is how and whether the bacterial and fungal biota contributes to the regulation of inflammation in health and disease. By the use of multidisciplinary approaches based on whole-genome immunogenetics, cutting-edge 'omics' techniques, advanced bioinformatics and systems biology applied to immune profiling, it will be possible to challenge existing paradigms in the fields of fungal immunopathology. This will probably lead to the discovery of specific immunogenetic signatures and the development of patient-tailored therapeutic vaccination approaches for mucosal and systemic fungal diseases.

Acknowledgements

We thank Cristina Massi-Benedetti for editorial assistance. Cristina Cunha was supported by the Fundação para a Ciência e Tecnologia, Portugal (contract SFRH/BPD/96176/2013).

References

Acosta-Rodriguez, E.V., Rivino, L., Geginat, J., Jarrossay, D., Gattorno, M., Lanzavecchia, A., Sallusto, F., and Napolitani, G. (2007). Surface phenotype and antigenic specificity of human interleukin 17-producing T helper memory cells. Nat. Immunol. *8*, 639–646.

Aimanianda, V., Bayry, J., Bozza, S., Kniemeyer, O., Perruccio, K., Elluru, S.R., Clavaud, C., Paris, S., Brakhage, A.A., Kaveri, S.V., *et al.* (2009). Surface hydrophobin prevents immune recognition of airborne fungal spores. Nature *460*, 1117–1121.

Albacker, L.A., Chaudhary, V., Chang, Y.J., Kim, H.Y., Chuang, Y.T., Pichavant, M., Dekruyff, R.H., Savage, P.B., and Umetsu, D.T. (2013). Invariant natural killer T cells recognize a fungal glycosphingolipid that can induce airway hyperreactivity. Nat. Med. *19*, 1297–1304.

Amich, J., Schafferer, L., Haas, H., and Krappmann, S. (2013). Regulation of sulphur assimilation is essential for virulence and affects iron homeostasis of the human-pathogenic mould *Aspergillus fumigatus*. PLoS Pathog. *9*, e1003573.

Bonifazi, P., D'Angelo, C., Zagarella, S., Zelante, T., Bozza, S., De Luca, A., Giovannini, G., Moretti, S., Iannitti, R.G., Fallarino, F., *et al.* (2010). Intranasally delivered siRNA targeting PI3K/Akt/mTOR inflammatory pathways protects from aspergillosis. Mucosal. Immunol. *3*, 193–205.

Bonifazi, P., Zelante, T., D'Angelo, C., De Luca, A., Moretti, S., Bozza, S., Perruccio, K., Iannitti, R.G., Giovannini, G., Volpi, C., *et al.* (2009). Balancing inflammation and tolerance *in vivo* through dendritic cells by the commensal *Candida albicans*. Mucosal. Immunol. *2*, 362–374.

Bozza, S., Clavaud, C., Giovannini, G., Fontaine, T., Beauvais, A., Sarfati, J., D'Angelo, C., Perruccio, K., Bonifazi, P., Zagarella, S., *et al.* (2009a). Immune sensing of *Aspergillus fumigatus* proteins, glycolipids, and polysaccharides and the impact on Th immunity and vaccination. J. Immunol. *183*, 2407–2414.

Bozza, S., Clavaud, C., Giovannini, G., Fontaine, T., Beauvais, A., Sarfati, J., D'Angelo, C., Perruccio, K., Bonifazi, P., Zagarella, S., *et al.* (2009b). Immune sensing of *Aspergillus fumigatus* proteins, glycolipids, and polysaccharides and the impact on Th immunity and vaccination. J. Immunol. *183*, 2407–2414.

Bozza, S., Gaziano, R., Lipford, G.B., Montagnoli, C., Bacci, A., Di Francesco, P., Kurup, V.P., Wagner, H., and Romani, L. (2002). Vaccination of mice against invasive aspergillosis with recombinant *Aspergillus* proteins and CpG oligodeoxynucleotides as adjuvants. Microbes Infect. *4*, 1281–1290.

Bozza, S., Montagnoli, C., Gaziano, R., Rossi, G., Nkwanyuo, G., Bellocchio, S., and Romani, L. (2004). Dendritic cell-based vaccination against opportunistic fungi. Vaccine *22*, 857–864.

Bozza, S., Perruccio, K., Montagnoli, C., Gaziano, R., Bellocchio, S., Burchielli, E., Nkwanyuo, G., Pitzurra, L., Velardi, A., and Romani, L. (2003). A dendritic cell vaccine against invasive aspergillosis in allogeneic hematopoietic transplantation. Blood *102*, 3807–3814.

Bromuro, C., Romano, M., Chiani, P., Berti, F., Tontini, M., Proietti, D., Mori, E., Torosantucci, A., Costantino, P., Rappuoli, R., *et al.* (2010). β-glucan-CRM197 conjugates as candidates antifungal vaccines. Vaccine *28*, 2615–2623.

Brown, G.D., Denning, D.W., Gow, N.A., Levitz, S.M., Netea, M.G., and White, T.C. (2012). Hidden killers: human fungal infections. Sci. Transl. Med. *4*, 165rv113.

Carvalho, A., Cunha, C., Bistoni, F., and Romani, L. (2012a). Immunotherapy of aspergillosis. Clin. Microbiol. Infect. *18*, 120–125.

Carvalho, A., De Luca, A., Bozza, S., Cunha, C., D'Angelo, C., Moretti, S., Perruccio, K., Iannitti, R.G., Fallarino, F., Pierini, A., *et al.* (2012b). TLR3 essentially promotes protective class I-restricted memory CD8(+) T-cell responses to *Aspergillus fumigatus* in hematopoietic transplanted patients. Blood *119*, 967–977.

Casadevall, A., and Pirofski, L.A. (2012). Immunoglobulins in defense, pathogenesis, and therapy of fungal diseases. Cell Host Microbe *11*, 447–456.

Cassone, A. (2013). Development of vaccines for *Candida albicans*: fighting a skilled transformer. Nat. Rev. Microbiol. *11*, 884–891.

Cassone, A., and Casadevall, A. (2012). Recent progress in vaccines against fungal diseases. Curr. Opin. Microbiol. *15*, 427–433.

Cenci, E., Mencacci, A., Bacci, A., Bistoni, F., Kurup, V.P., and Romani, L. (2000). T cell vaccination in mice with invasive pulmonary aspergillosis. J. Immunol. *165*, 381–388.

Chow, S.K., and Casadevall, A. (2011). Evaluation of *Cryptococcus neoformans* galactoxylomannan–protein conjugate as vaccine candidate against murine cryptococcosis. Vaccine *29*, 1891–1898.

Cobbold, S.P., Adams, E., Nolan, K.F., Regateiro, F.S., and Waldmann, H. (2010). Connecting the mechanisms of T-cell regulation: dendritic cells as the missing link. Immunol. Rev. *236*, 203–218.

Codarri, L., Gyulveszi, G., Tosevski, V., Hesske, L., Fontana, A., Magnenat, L., Suter, T., and Becher, B. (2011). RORgammat drives production of the cytokine GM-CSF in helper T cells, which is essential for the effector phase of autoimmune neuroinflammation. Nat. Immunol. *12*, 560–567.

Cohen, N.R., Tatituri, R.V., Rivera, A., Watts, G.F., Kim, E.Y., Chiba, A., Fuchs, B.B., Mylonakis, E., Besra, G.S., Levitz, S.M., *et al.* (2011). Innate recognition of cell wall beta-glucans drives invariant natural killer T cell responses against fungi. Cell Host Microbe *10*, 437–450.

Conti, H.R., Shen, F., Nayyar, N., Stocum, E., Sun, J.N., Lindemann, M.J., Ho, A.W., Hai, J.H., Yu, J.J., Jung, J.W., *et al.* (2009). Th17 cells and IL-17 receptor signaling are essential for mucosal host defense against oral candidiasis. J. Exp. Med. *206*, 299–311.

Cooney, N.M., and Klein, B.S. (2008). Fungal adaptation to the mammalian host: it is a new world, after all. Curr. Opin. Microbiol. *11*, 511–516.

Cunha, C., Aversa, F., Lacerda, J.F., Busca, A., Kurzai, O., Grube, M., Loffler, J., Maertens, J.A., Bell, A.S., Inforzato, A., *et al.* (2014). Genetic PTX3 deficiency and aspergillosis in stem-cell transplantation. New Engl. J. Med. *370*, 421–432.

Cunha, C., Aversa, F., Romani, L., and Carvalho, A. (2013). Human genetic susceptibility to invasive aspergillosis. PLoS Pathog. *9*, e1003434.

Cunha, C., and Carvalho, A. (2012). Host genetics and invasive fungal diseases: towards improved diagnosis and therapy? Expert Rev. Anti Infect. Ther. *10*, 257–259.

Cutler, J.E., Corti, M., Lambert, P., Ferris, M., and Xin, H. (2011). Horizontal transmission of *Candida albicans* and evidence of a vaccine response in mice colonized with the fungus. PLoS One *6*, e22030.

D'Angelo, C., De Luca, A., Zelante, T., Bonifazi, P., Moretti, S., Giovannini, G., Iannitti, R.G., Zagarella, S., Bozza, S., Campo, S., *et al.* (2009). Exogenous pentraxin 3 restores antifungal resistance and restrains inflammation in murine chronic granulomatous disease. J. Immunol. *183*, 4609–4618.

da Gloria Teixeira de Sousa, M., Junior, W.B., Spina, R., Lota, P.R., Valente, N.S., Brown, G.D., Criado, P.R., and Benard, G. (2014). Topical application of Imiquimod as a treatment for Chromoblastomycosis. Clin. Infect. Dis. *58*, 1734–1737.

Datta, K., Lees, A., and Pirofski, L.A. (2008). Therapeutic efficacy of a conjugate vaccine containing a peptide mimotope of cryptococcal capsular polysaccharide glucuronoxylomannan. Clin. Vaccine Immunol. *15*, 1176–1187.

de Bastos Ascenco Soares, R., Gomez, F.J., de Almeida Soares, C.M., and Deepe, G.S., Jr. (2008). Vaccination with heat shock protein 60 induces a protective immune response against experimental *Paracoccidioides brasiliensis* pulmonary infection. Infect. Immun. *76*, 4214–4221.

De Luca, A., Carvalho, A., Cunha, C., Iannitti, R.G., Pitzurra, L., Giovannini, G., Mencacci, A., Bartolommei, L., Moretti, S., Massi-Benedetti, C., *et al.* (2013). IL-22 and IDO1 affect immunity and tolerance to murine and human vaginal candidiasis. PLoS Pathog. *9*, e1003486.

De Luca, A., Iannitti, R.G., Bozza, S., Beau, R., Casagrande, A., D'Angelo, C., Moretti, S., Cunha, C., Giovannini, G., Massi-Benedetti, C., *et al.* (2012). CD4(+) T cell vaccination overcomes defective cross-presentation of fungal antigens in a mouse model of chronic granulomatous disease. J. Clin. Invest. *122*, 1816–1831.

De Luca, A., Montagnoli, C., Zelante, T., Bonifazi, P., Bozza, S., Moretti, S., D'Angelo, C., Vacca, C., Boon, L., Bistoni, F., *et al.* (2007). Functional yet balanced reactivity to Candida albicans requires TRIF, MyD88, and IDO-dependent inhibition of Rorc. J. Immunol. *179*, 5999–6008.

De Luca, A., Zelante, T., D'Angelo, C., Zagarella, S., Fallarino, F., Spreca, A., Iannitti, R.G., Bonifazi, P., Renauld, J.C., Bistoni, F., *et al.* (2010). IL-22 defines a novel immune pathway of antifungal resistance. Mucosal Immunol. *3*, 361–373.

Deepe, G.S., Jr., and Gibbons, R.S. (2002). Cellular and molecular regulation of vaccination with heat shock protein 60 from *Histoplasma capsulatum*. Infect. Immun. *70*, 3759–3767.

Diaz-Arevalo, D., Bagramyan, K., Hong, T.B., Ito, J.I., and Kalkum, M. (2011). CD4+ T cells mediate the protective effect of the recombinant Asp f3-based anti-aspergillosis vaccine. Infect. Immun. *79*, 2257–2266.

Duan, Y.N., Yi, L.H., Chen, J.L., Zhu, D.D., Wang, J.X., Feng, J.R., Qin, Y.W., and Zhu, Y. (2011). Protective effect of DNA vaccine with the gene encoding 55kDa antigen fragment against Pneumocystis carinii in mice. Asian Pac. J. Trop. Med. *4*, 353–356.

Espinosa, V., Jhingran, A., Dutta, O., Kasahara, S., Donnelly, R., Du, P., Rosenfeld, J., Leiner, I., Chen, C.C., Ron, Y., *et al.* (2014). Inflammatory monocytes orchestrate innate antifungal immunity in the lung. PLoS Pathog. *10*, e1003940.

Eyerich, K., Foerster, S., Rombold, S., Seidl, H.P., Behrendt, H., Hofmann, H., Ring, J., and Traidl-Hoffmann, C. (2008). Patients with chronic mucocutaneous candidiasis exhibit reduced production of Th17-associated cytokines IL-17 and IL-22. J. Invest. Dermatol. *128*, 2640–2645.

Fallarino, F., and Grohmann, U. (2011). Using an ancient tool for igniting and propagating immune tolerance: IDO as an inducer and amplifier of regulatory T cell functions. Curr. Med. Chem. *18*, 2215–2221.

Fallarino, F., Grohmann, U., and Puccetti, P. (2012). Indoleamine 2,3-dioxygenase: from catalyst to signaling function. Eur. J. Immunol. *42*, 1932–1937.

Feng, Y., Guo, S., Jiang, T., Han, X., Liu, P., Wu, T., and Luo, Y. (2011). Active immunization against *Pneumocystis carinii* with p55-v3 DNA vaccine in rats. Can. J. Microbiol. *57*, 375–381.

Fernandes, V.C., Martins, E.M., Boeloni, J.N., Coitinho, J.B., Serakides, R., and Goes, A.M. (2011). Additive effect of rPb27 immunization and chemotherapy in experimental paracoccidioidomycosis. PLoS One *6*, e17885.

Ferreira, M.C., de Oliveira, R.T., da Silva, R.M., Blotta, M.H., and Mamoni, R.L. (2010). Involvement of regulatory T cells in the immunosuppression characteristic of patients with paracoccidioidomycosis. Infect. Immun. *78*, 4392–4401.

Ferwerda, B., Ferwerda, G., Plantinga, T.S., Willment, J.A., van Spriel, A.B., Venselaar, H., Elbers, C.C., Johnson, M.D., Cambi, A., Huysamen, C., *et al.* (2009). Human dectin-1 deficiency and mucocutaneous fungal infections. New Engl. J. Med. *361*, 1760–1767.

Fisher, M.C., Henk, D.A., Briggs, C.J., Brownstein, J.S., Madoff, L.C., McCraw, S.L., and Gurr, S.J. (2012). Emerging fungal threats to animal, plant and ecosystem health. Nature *484*, 186–194.

Garlanda, C., Hirsch, E., Bozza, S., Salustri, A., De Acetis, M., Nota, R., Maccagno, A., Riva, F., Bottazzi, B., Peri, G., *et al.* (2002). Non-redundant role of the long pentraxin PTX3 in anti-fungal innate immune response. Nature *420*, 182–186.

Gessner, M.A., Werner, J.L., Lilly, L.M., Nelson, M.P., Metz, A.E., Dunaway, C.W., Chan, Y.R., Ouyang, W., Brown, G.D., Weaver, C.T., *et al.* (2012). Dectin-1-dependent interleukin-22 contributes to early innate lung defense against *Aspergillus fumigatus*. Infect. Immun. *80*, 410–417.

Glocker, E.O., Hennigs, A., Nabavi, M., Schaffer, A.A., Woellner, C., Salzer, U., Pfeifer, D., Veelken, H., Warnatz, K., Tahami, F., *et al.* (2009). A homozygous CARD9 mutation in a family with susceptibility to fungal infections. New Engl. J. Med. *361*, 1727–1735.

Gow, N.A., van de Veerdonk, F.L., Brown, A.J., and Netea, M.G. (2012). *Candida albicans* morphogenesis and host defence: discriminating invasion from colonization. Nat. Rev. Microbiol. *10*, 112–122.

Grahl, N., Puttikamonkul, S., Macdonald, J.M., Gamcsik, M.P., Ngo, L.Y., Hohl, T.M., and Cramer, R.A. (2011). *In vivo* hypoxia and a fungal alcohol dehydrogenase influence the pathogenesis of invasive pulmonary aspergillosis. PLoS Pathog. *7*, e1002145.

Hamad, M. (2012). Universal fungal vaccines: could there be light at the end of the tunnel? Hum. Vaccin. Immunother. *8*, 1758–1763.

Hardison, S.E., Wozniak, K.L., Kolls, J.K., and Wormley, F.L., Jr. (2010). Interleukin-17 is not required for classical macrophage activation in a pulmonary mouse model of *Cryptococcus neoformans* infection. Infect. Immun. *78*, 5341–5351.

Hernandez-Santos, N., and Gaffen, S.L. (2012). Th17 cells in immunity to *Candida albicans*. Cell Host Microbe *11*, 425–435.

Hernandez-Santos, N., Huppler, A.R., Peterson, A.C., Khader, S.A., McKenna, K.C., and Gaffen, S.L. (2013). Th17 cells confer long-term adaptive immunity to oral mucosal *Candida albicans* infections. Mucosal. Immunol. *6*, 900–910.

Huang, H., Ostroff, G.R., Lee, C.K., Specht, C.A., and Levitz, S.M. (2010). Robust stimulation of humoral and cellular immune responses following vaccination with antigen-loaded beta-glucan particles. MBio *1*, e00164–10.

Huang, W., Na, L., Fidel, P.L., and Schwarzenberger, P. (2004). Requirement of interleukin-17A for systemic anti-*Candida albicans* host defense in mice. J. Infect. Dis. *190*, 624–631.

Hung, C.Y., Gonzalez, A., Wuthrich, M., Klein, B.S., and Cole, G.T. (2011). Vaccine immunity to coccidioidomycosis occurs by early activation of three signal pathways of T helper cell response (Th1, Th2, and Th17). Infect. Immun. *79*, 4511–4522.

Hurtgen, B.J., Hung, C.Y., Ostroff, G.R., Levitz, S.M., and Cole, G.T. (2012). Construction and evaluation of a novel recombinant T cell epitope-based vaccine against Coccidioidomycosis. Infect. Immun. *80*, 3960–3974.

Iannitti, R.G., Carvalho, A., Cunha, C., De Luca, A., Giovannini, G., Casagrande, A., Zelante, T., Vacca, C., Fallarino, F., Puccetti, P., *et al.* (2013). Th17/Treg imbalance in murine cystic fibrosis is linked to indoleamine 2,3-dioxygenase deficiency but corrected by kynurenines. Am. J. Respir. Crit. Care Med. *187*, 609–620.

Iannitti, R.G., Carvalho, A., and Romani, L. (2012). From memory to antifungal vaccine design. Trends Immunol. *33*, 467–474.

Ibrahim, A.S., Luo, G., Gebremariam, T., Lee, H., Schmidt, C.S., Hennessey, J.P., Jr., French, S.W., Yeaman, M.R., Filler, S.G., and Edwards, J.E., Jr. (2013). NDV-3 protects mice from vulvovaginal candidiasis through T- and B-cell immune response. Vaccine *31*, 5549–5556.

Iliev, I.D., Funari, V.A., Taylor, K.D., Nguyen, Q., Reyes, C.N., Strom, S.P., Brown, J., Becker, C.A., Fleshner, P.R., Dubinsky, M., *et al.* (2012). Interactions between commensal fungi and the C-type lectin receptor Dectin-1 influence colitis. Science *336*, 1314–1317.

Ito, J.I., Lyons, J.M., Hong, T.B., Tamae, D., Liu, Y.K., Wilczynski, S.P., and Kalkum, M. (2006). Vaccinations with recombinant variants of *Aspergillus fumigatus* allergen

Asp f 3 protect mice against invasive aspergillosis. Infect. Immun. *74*, 5075–5084.

Jain, A.V., Zhang, Y., Fields, W.B., McNamara, D.A., Choe, M.Y., Chen, G.H., Erb-Downward, J., Osterholzer, J.J., Toews, G.B., Huffnagle, G.B., *et al.* (2009). Th2 but not Th1 immune bias results in altered lung functions in a murine model of pulmonary *Cryptococcus* neoformans infection. Infect. Immun. *77*, 5389–5399.

Kamada, N., Seo, S.U., Chen, G.Y., and Nunez, G. (2013). Role of the gut microbiota in immunity and inflammatory disease. Nat. Rev. Immunol. *13*, 321–335.

Kaufmann, S.H., and Kuchroo, V.K. (2009). Th17 cells. Microbes Infect. *11*, 579–583.

Kennedy, R.B., and Poland, G.A. (2011). The top five "game changers" in vaccinology: toward rational and directed vaccine development. OMICS *15*, 533–537.

Kenyon, C., Bonorchis, K., Corcoran, C., Meintjes, G., Locketz, M., Lehloenya, R., Vismer, H.F., Naicker, P., Prozesky, H., van Wyk, M., *et al.* (2013). A Dimorphic Fungus Causing Disseminated Infection in South Africa. New Engl. J. Med. *369*, 1416–1424.

Koh, A.Y., Kohler, J.R., Coggshall, K.T., Van Rooijen, N., and Pier, G.B. (2008). Mucosal damage and neutropenia are required for *Candida albicans* dissemination. PLoS Pathog. *4*, e35.

Kreindler, J.L., Steele, C., Nguyen, N., Chan, Y.R., Pilewski, J.M., Alcorn, J.F., Vyas, Y.M., Aujla, S.J., Finelli, P., Blanchard, M., *et al.* (2010). Vitamin D3 attenuates Th2 responses to Aspergillus fumigatus mounted by CD4+ T cells from cystic fibrosis patients with allergic bronchopulmonary aspergillosis. J. Clin. Invest. *120*, 3242–3254.

Latge, J.P. (2013). Oh, to Be New. New Engl. J. Med. *369*, 1464–1466.

Levitz, S.M., and Golenbock, D.T. (2012). Beyond empiricism: informing vaccine development through innate immunity research. Cell *148*, 1284–1292.

Li, W., Hu, X., Zhang, X., Ge, Y., Zhao, S., Hu, Y., and Ashman, R.B. (2011). Immunisation with the glycolytic enzyme enolase confers effective protection against *Candida albicans* infection in mice. Vaccine *29*, 5526–5533.

Lilly, L.M., Gessner, M.A., Dunaway, C.W., Metz, A.E., Schwiebert, L., Weaver, C.T., Brown, G.D., and Steele, C. (2012). The beta-glucan receptor dectin-1 promotes lung immunopathology during fungal allergy via IL-22. J. Immunol. *189*, 3653–3660.

Lin, L., Ibrahim, A.S., Xu, X., Farber, J.M., Avanesian, V., Baquir, B., Fu, Y., French, S.W., Edwards, J.E., Jr., and Spellberg, B. (2009). Th1-Th17 cells mediate protective adaptive immunity against *Staphylococcus aureus* and *Candida albicans* infection in mice. PLoS Pathog. *5*, e1000703.

Liu, M., Clemons, K.V., Bigos, M., Medovarska, I., Brummer, E., and Stevens, D.A. (2011). Immune responses induced by heat killed *Saccharomyces cerevisiae*: a vaccine against fungal infection. Vaccine *29*, 1745–1753.

Liu, Y., Yang, B., Zhou, M., Li, L., Zhou, H., Zhang, J., Chen, H., and Wu, C. (2009). Memory IL-22-producing CD4+ T cells specific for Candida albicans are present in humans. Eur. J. Immunol. *39*, 1472–1479.

Lo Giudice, P., Campo, S., De Santis, R., and Salvatori, G. (2012). Effect of PTX3 and voriconazole combination in a rat model of invasive pulmonary aspergillosis. Antimicrob. Agents Chemother. *56*, 6400–6402.

Lo Giudice, P., Campo, S., Verdoliva, A., Rivieccio, V., Borsini, F., De Santis, R., and Salvatori, G. (2010). Efficacy of PTX3 in a rat model of invasive aspergillosis. Antimicrob. Agents Chemother. *54*, 4513–4515.

Luo, G., Ibrahim, A.S., French, S.W., Edwards, J.E., Jr., and Fu, Y. (2011). Active and passive immunization with rHyr1p-N protects mice against hematogenously disseminated candidiasis. PLoS One *6*, e25909.

Martinez-Lopez, R., Nombela, C., Diez-Orejas, R., Monteoliva, L., and Gil, C. (2008). Immunoproteomic analysis of the protective response obtained from vaccination with *Candida albicans* ecm33 cell wall mutant in mice. Proteomics *8*, 2651–2664.

Mayorga, O., Munoz, J.E., Lincopan, N., Teixeira, A.F., Ferreira, L.C., Travassos, L.R., and Taborda, C.P. (2012). The role of adjuvants in therapeutic protection against paracoccidioidomycosis after immunization with the P10 peptide. Front. Microbiol. *3*, 154.

Medzhitov, R., Schneider, D.S., and Soares, M.P. (2012). Disease tolerance as a defense strategy. Science *335*, 936–941.

Miller, R.F., Huang, L., and Walzer, P.D. (2013). Pneumocystis pneumonia associated with human immunodeficiency Virus. Clin. Chest Med. *34*, 229–241.

Milner, J.D., Brenchley, J.M., Laurence, A., Freeman, A.F., Hill, B.J., Elias, K.M., Kanno, Y., Spalding, C., Elloumi, H.Z., Paulson, M.L., *et al.* (2008). Impaired T(H)17 cell differentiation in subjects with autosomal dominant hyper-IgE syndrome. Nature *452*, 773–776.

Montagnoli, C., Fallarino, F., Gaziano, R., Bozza, S., Bellocchio, S., Zelante, T., Kurup, W.P., Pitzurra, L., Puccetti, P., and Romani, L. (2006). Immunity and tolerance to *Aspergillus* involve functionally distinct regulatory T cells and tryptophan catabolism. J. Immunol. *176*, 1712–1723.

Nacher, M., Adenis, A., Mc Donald, S., Gomes, M.D.M., Singh, S., Lima, I.L., Leite, R.M., Hermelijn, S., Wongsokarijo, M., Van Eer, M., *et al.* (2013). Disseminated histoplasmosis in HIV-infected patients in South America: a neglected killer continues on its rampage. PLoS Neglect. Trop. Dis. *7*, e2319.

Nanjappa, S.G., Heninger, E., Wuthrich, M., Gasper, D.J., and Klein, B.S. (2012a). Tc17 cells mediate vaccine immunity against lethal fungal pneumonia in immune deficient hosts lacking CD4+ T cells. PLoS Pathog. *8*, e1002771.

Nanjappa, S.G., Heninger, E., Wuthrich, M., Sullivan, T., and Klein, B. (2012b). Protective antifungal memory CD8(+) T cells are maintained in the absence of CD4(+) T cell help and cognate antigen in mice. J. Clin. Invest. *122*, 987–999.

Nanjappa, S.G., and Klein, B.S. (2014). Vaccine immunity against fungal infections. Curr. Opin. Immunol. *28C*, 27–33.

Oevermann, L., Lang, P., Feuchtinger, T., Schumm, M., Teltschik, H.M., Schlegel, P., and Handgretinger, R. (2012). Immune reconstitution and strategies for rebuilding the immune system after haploidentical stem cell transplantation. Ann. N. Y. Acad. Sci. *1266*, 161–170.

Ovsyannikova, I.G., Haralambieva, I.H., Kennedy, R.B., O'Byrne, M.M., Pankratz, V.S., and Poland, G.A. (2013). Genetic variation in IL18R1 and IL18 genes and inteferon gamma ELISPOT response to smallpox vaccination: an unexpected relationship. J. Infect. Dis. *208*, 1422–1430.

Park, B.J., Wannemuehler, K.A., Marston, B.J., Govender, N., Pappas, P.G., and Chiller, T.A. (2009). Estimation of the current global burden of cryptococcal meningitis among persons living with HIV/AIDS. AIDS *23*, 525–530.

Paulovicova, E., Machova, E., Tulinska, J., and Bystricky, S. (2007). Cell and antibody mediated immunity induced by vaccination with novel Candida dubliniensis mannan immunogenic conjugate. Int. Immunopharmacol. *7*, 1325–1333.

Pearce, E.L., Walsh, M.C., Cejas, P.J., Harms, G.M., Shen, H., Wang, L.S., Jones, R.G., and Choi, Y. (2009). Enhancing CD8 T-cell memory by modulating fatty acid metabolism. Nature *460*, 103–107.

Pfaller, M.A., and Diekema, D.J. (2010). Epidemiology of invasive mycoses in North America. Crit. Rev. Microbiol. *36*, 1–53.

Pierluigi, B., D'Angelo, C., Fallarino, F., Moretti, S., Zelante, T., Bozza, S., De Luca, A., Bistoni, F., Garaci, E., and Romani, L. (2010) Thymosin alpha1: the regulator of regulators? Ann. N. Y. Acad. Sci. *1194*, 1–5.

Pinzan, C.F., Ruas, L.P., Casabona-Fortunato, A.S., Carvalho, F.C., and Roque-Barreira, M.C. (2010). Immunological basis for the gender differences in murine *Paracoccidioides brasiliensis* infection. PLoS One *5*, e10757.

Poland, G.A., and Oberg, A.L. (2010). Vaccinomics and bioinformatics: accelerants for the next golden age of vaccinology. Vaccine *28*, 3509–3510.

Potenza, L., Vallerini, D., Barozzi, P., Riva, G., Forghieri, F., Beauvais, A., Beau, R., Candoni, A., Maertens, J., Rossi, G., *et al.* (2013). Characterization of specific immune responses to different *Aspergillus* antigens during the course of invasive aspergillosis in hematologic patients. PLoS One *8*, e74326.

Puel, A., Cypowyj, S., Bustamante, J., Wright, J.F., Liu, L., Lim, H.K., Migaud, M., Israel, L., Chrabieh, M., Audry, M., *et al.* (2011). Chronic mucocutaneous candidiasis in humans with inborn errors of interleukin-17 immunity. Science *332*, 65–68.

Pulendran, B., and Ahmed, R. (2011). Immunological mechanisms of vaccination. Nat. Immunol. *12*, 509–517.

Pulendran, B., Li, S., and Nakaya, H.I. (2010). Systems vaccinology. Immunity *33*, 516–529.

Rachini, A., Pietrella, D., Lupo, P., Torosantucci, A., Chiani, P., Bromuro, C., Proietti, C., Bistoni, F., Cassone, A., and Vecchiarelli, A. (2007). An anti-beta-glucan monoclonal antibody inhibits growth and capsule formation of *Cryptococcus neoformans in vitro* and exerts therapeutic, anticryptococcal activity *in vivo*. Infect. Immun. *75*, 5085–5094.

Ramirez-Ortiz, Z.G., Lee, C.K., Wang, J.P., Boon, L., Specht, C.A., and Levitz, S.M. (2011). A nonredundant role for plasmacytoid dendritic cells in host defense against the human fungal pathogen *Aspergillus fumigatus*. Cell Host Microbe *9*, 415–424.

Rapaka, R.R., Ricks, D.M., Alcorn, J.F., Chen, K., Khader, S.A., Zheng, M., Plevy, S., Bengten, E., and Kolls, J.K. (2010). Conserved natural IgM antibodies mediate innate and adaptive immunity against the opportunistic fungus *Pneumocystis murina*. J. Exp. Med. *207*, 2907–2919.

Rappleye, C.A., and Goldman, W.E. (2008). Fungal stealth technology. Trends Immunol. *29*, 18–24.

Ribeiro, A.M., Bocca, A.L., Amaral, A.C., Faccioli, L.H., Galetti, F.C., Zarate-Blades, C.R., Figueiredo, F., Silva, C.L., and Felipe, M.S. (2009). DNAhsp65 vaccination induces protection in mice against *Paracoccidioides brasiliensis* infection. Vaccine *27*, 606–613.

Rivera, A., Hohl, T.M., Collins, N., Leiner, I., Gallegos, A., Saijo, S., Coward, J.W., Iwakura, Y., and Pamer, E.G. (2011). Dectin-1 diversifies *Aspergillus fumigatus*-specific T cell responses by inhibiting T helper type 1 CD4 T cell differentiation. J. Exp. Med. *208*, 369–381.

Rizzetto, L., and Cavalieri, D. (2010). A systems biology approach to the mutual interaction between yeast and the immune system. Immunobiology *215*, 762–769.

Rohatgi, S., Gohil, S., Kuniholm, M.H., Schultz, H., Dufaud, C., Armour, K.L., Badri, S., Mailliard, R.B., and Pirofski, L.A. (2013). Fc gamma receptor 3A polymorphism and risk for HIV-associated cryptococcal disease. MBio *4*, e00573–00513.

Roilides, E., Lamaignere, C.G., and Farmaki, E. (2002). Cytokines in immunodeficient patients with invasive fungal infections: an emerging therapy. Int. J. Infect. Dis. *6*, 154–163.

Romani, L. (2011). Immunity to fungal infections. Nat. Rev. Immunol. *11*, 275–288.

Romani, L., Bistoni, F., Perruccio, K., Montagnoli, C., Gaziano, R., Bozza, S., Bonifazi, P., Bistoni, G., Rasi, G., Velardi, A., *et al.* (2006). Thymosin alpha1 activates dendritic cell tryptophan catabolism and establishes a regulatory environment for balance of inflammation and tolerance. Blood *108*, 2265–2274.

Romani, L., Fallarino, F., De Luca, A., Montagnoli, C., D'Angelo, C., Zelante, T., Vacca, C., Bistoni, F., Fioretti, M.C., Grohmann, U., *et al.* (2008a). Defective tryptophan catabolism underlies inflammation in mouse chronic granulomatous disease. Nature *451*, 211–215.

Romani, L., and Puccetti, P. (2006). Protective tolerance to fungi: the role of IL-10 and tryptophan catabolism. Trends Microbiol. *14*, 183–189.

Romani, L., Zelante, T., De Luca, A., Fallarino, F., and Puccetti, P. (2008b). IL-17 and therapeutic kynurenines in pathogenic inflammation to fungi. J. Immunol. *180*, 5157–5162.

Roy, R.M., and Klein, B.S. (2012). Dendritic cells in antifungal immunity and vaccine design. Cell Host Microbe *11*, 436–446.

Saijo, S., Fujikado, N., Furuta, T., Chung, S.H., Kotaki, H., Seki, K., Sudo, K., Akira, S., Adachi, Y., Ohno, N., *et al.* (2007). Dectin-1 is required for host defense against

Pneumocystis carinii but not against *Candida albicans*. Nat. Immunol. *8*, 39–46.

Sallusto, F., Lanzavecchia, A., Araki, K., and Ahmed, R. (2010). From vaccines to memory and back. Immunity 33, 451–463.

Sandini, S., La Valle, R., Deaglio, S., Malavasi, F., Cassone, A., and De Bernardis, F. (2011). A highly immunogenic recombinant and truncated protein of the secreted aspartic proteases family (rSap2t) of *Candida albicans* as a mucosal anticandidal vaccine. FEMS Immunol. Med. Microbiol. *62*, 215–224.

Santamaria, R., Rizzetto, L., Bromley, M., Zelante, T., Lee, W., Cavalieri, D., Romani, L., Miller, B., Gut, I., Santos, M., *et al.* (2011). Systems biology of infectious diseases: a focus on fungal infections. Immunobiology *216*, 1212–1227.

Saraiva, M., and O'Garra, A. (2010). The regulation of IL-10 production by immune cells. Nat. Rev. Immunol. *10*, 170–181.

Saville, S.P., Lazzell, A.L., Chaturvedi, A.K., Monteagudo, C., and Lopez-Ribot, J.L. (2009). Efficacy of a genetically engineered *Candida albicans* tet-NRG1 strain as an experimental live attenuated vaccine against hematogenously disseminated candidiasis. Clin. Vacc. Immunol. *16*, 430–432.

Scheckelhoff, M., and Deepe, G.S., Jr. (2002). The protective immune response to heat shock protein 60 of *Histoplasma capsulatum* is mediated by a subset of V beta 8.1/8.2+ T cells. J. Immunol. *169*, 5818–5826.

Shubitz, L.F., Yu, J.J., Hung, C.Y., Kirkland, T.N., Peng, T., Perrill, R., Simons, J., Xue, J., Herr, R.A., Cole, G.T., *et al.* (2006). Improved protection of mice against lethal respiratory infection with *Coccidioides posadasii* using two recombinant antigens expressed as a single protein. Vaccine *24*, 5904–5911.

Smeekens, S.P., van de Veerdonk, F.L., Kullberg, B.J., and Netea, M.G. (2013). Genetic susceptibility to *Candida* infections. EMBO Mol. Med. *5*, 805–813.

Specht, C.A., Nong, S., Dan, J.M., Lee, C.K., and Levitz, S.M. (2007). Contribution of glycosylation to T cell responses stimulated by recombinant *Cryptococcus neoformans* mannoprotein. J. Infect. Dis. *196*, 796–800.

Spellberg, B., Ibrahim, A.S., Lin, L., Avanesian, V., Fu, Y., Lipke, P., Otoo, H., Ho, T., and Edwards, J.E., Jr. (2008). Antibody titer threshold predicts anti-candidal vaccine efficacy even though the mechanism of protection is induction of cell-mediated immunity. J. Infect. Dis. *197*, 967–971.

Steinman, R.M. (2012). Decisions about dendritic cells: past, present, and future. Annu. Rev. Immunol. *30*, 1–22.

Stuehler, C., Khanna, N., Bozza, S., Zelante, T., Moretti, S., Kruhm, M., Lurati, S., Conrad, B., Worschech, E., Stevanovic, S., *et al.* (2011). Cross-protective TH1 immunity against *Aspergillus fumigatus* and *Candida albicans*. Blood *117*, 5881–5891.

Szymczak, W.A., and Deepe, G.S., Jr. (2009). The CCL7–CCL2-CCR2 axis regulates IL-4 production in lungs and fungal immunity. J. Immunol. *183*, 1964–1974.

Tarcha, E.J., Basrur, V., Hung, C.Y., Gardner, M.J., and Cole, G.T. (2006a). Multivalent recombinant protein vaccine against coccidioidomycosis. Infect. Immun. *74*, 5802–5813.

Tarcha, E.J., Basrur, V., Hung, C.Y., Gardner, M.J., and Cole, G.T. (2006b). A recombinant aspartyl protease of *Coccidioides posadasii* induces protection against pulmonary coccidioidomycosis in mice. Infect. Immun. *74*, 516–527.

Torosantucci, A., Bromuro, C., Chiani, P., De Bernardis, F., Berti, F., Galli, C., Norelli, F., Bellucci, C., Polonelli, L., Costantino, P., *et al.* (2005). A novel glyco-conjugate vaccine against fungal pathogens. J. Exp. Med. *202*, 597–606.

van de Veerdonk, F.L., Plantinga, T.S., Hoischen, A., Smeekens, S.P., Joosten, L.A., Gilissen, C., Arts, P., Rosentul, D.C., Carmichael, A.J., Smits-van der Graaf, C.A., *et al.* (2011). STAT1 mutations in autosomal dominant chronic mucocutaneous candidiasis. New Engl. J. Med. *365*, 54–61.

van der Windt, G.J., O'Sullivan, D., Everts, B., Huang, S.C., Buck, M.D., Curtis, J.D., Chang, C.H., Smith, A.M., Ai, T., Faubert, B., *et al.* (2013). CD8 memory T cells have a bioenergetic advantage that underlies their rapid recall ability. Proc. Natl. Acad. Sci. U.S.A. *110*, 14336–14341.

Voelz, K., Lammas, D.A., and May, R.C. (2009). Cytokine signaling regulates the outcome of intracellular macrophage parasitism by *Cryptococcus neoformans*. Infect. Immun. 77, 3450–3457.

Wang, H., LeBert, V., Hung, C.Y., Galles, K., Saijo, S., Lin, X., Cole, G.T., Klein, B.S., and Wuthrich, M. (2014). C-type lectin receptors differentially induce th17 cells and vaccine immunity to the endemic mycosis of North America. J. Immunol. *192*, 1107–1119.

Wuthrich, M., Deepe, G.S., Jr., and Klein, B. (2012a). Adaptive immunity to fungi. Annu. Rev. Immunol. *30*, 115–148.

Wuthrich, M., Ersland, K., Sullivan, T., Galles, K., and Klein, B.S. (2012b). Fungi subvert vaccine T cell priming at the respiratory mucosa by preventing chemokine-induced influx of inflammatory monocytes. Immunity 36, 680–692.

Wuthrich, M., Filutowicz, H.I., Warner, T., Deepe, G.S., Jr., and Klein, B.S. (2003). Vaccine immunity to pathogenic fungi overcomes the requirement for CD4 help in exogenous antigen presentation to CD8+ T cells: implications for vaccine development in immune-deficient hosts. J. Exp. Med. *197*, 1405–1416.

Wuthrich, M., Gern, B., Hung, C.Y., Ersland, K., Rocco, N., Pick-Jacobs, J., Galles, K., Filutowicz, H., Warner, T., Evans, M., *et al.* (2011). Vaccine-induced protection against 3 systemic mycoses endemic to North America requires Th17 cells in mice. J. Clin. Invest. *121*, 554–568.

Wuthrich, M., Lebert, V., Galles, K., Hu-Li, J., Ben-Sasson, S.Z., Paul, W.E., and Klein, B.S. (2013). Interleukin 1 enhances vaccine-induced antifungal T-helper 17 cells and resistance against *Blastomyces dermatitidis* infection. J. Infect. Dis. *208*, 1175–1182.

Xin, H., Dziadek, S., Bundle, D.R., and Cutler, J.E. (2008). Synthetic glycopeptide vaccines combining beta-mannan and peptide epitopes induce protection against candidiasis. Proc. Natl. Acad. Sci. U.S.A. *105*, 13526–13531.

Xue, J., Chen, X., Selby, D., Hung, C.Y., Yu, J.J., and Cole, G.T. (2009). A genetically engineered live attenuated vaccine of *Coccidioides posadasii* protects BALB/c mice against coccidioidomycosis. Infect. Immun. *77*, 3196–3208.

Zelante, T., Fallarino, F., Bistoni, F., Puccetti, P., and Romani, L. (2009). Indoleamine 2,3-dioxygenase in infection: the paradox of an evasive strategy that benefits the host. Microbes Infect. *11*, 133–141.

Zelante, T., Iannitti, R.G., Cunha, C., De Luca, A., Giovannini, G., Pieraccini, G., Zecchi, R., D'Angelo, C., Massi-Benedetti, C., Fallarino, F., *et al.* (2013). Tryptophan catabolites from microbiota engage aryl hydrocarbon receptor and balance mucosal reactivity via interleukin-22. Immunity *39*, 372–385.

Zelante, T., Iannitti, R.G., De Luca, A., Arroyo, J., Blanco, N., Servillo, G., Sanglard, D., Reichard, U., Palmer, G.E., Latge, J.P., *et al.* (2012). Sensing of mammalian IL-17A regulates fungal adaptation and virulence. Na.t Commun. *3*, 683.

Zeller, S., Glaser, A.G., Vilhelmsson, M., Rhyner, C., and Crameri, R. (2008). Immunoglobulin-E-mediated reactivity to self antigens: a controversial issue. Int. Arch. Allergy Immunol. *145*, 87–93.

Zenewicz, L.A., and Flavell, R.A. (2008). IL-22 and inflammation: leukin' through a glass onion. Eur. J. Immunol. *38*, 3265–3268.

Zheng, M., Ramsay, A.J., Robichaux, M.B., Norris, K.A., Kliment, C., Crowe, C., Rapaka, R.R., Steele, C., McAllister, F., Shellito, J.E., *et al.* (2005). CD4+ T cell-independent DNA vaccination against opportunistic infections. J. Clin. Invest. *115*, 3536–3544.

Animal Models to Study Fungal Virulence and Antifungal Drugs

13

Alix T. Coste and Sara Amorim-Vaz

Abstract

Animal models are crucial tools and a prerequisite for any clinical trials, since both the anatomo-pathology of an infection and a drug's efficacy/toxicity can only be accurately assessed *in vivo* before being used in humans. Ideally good animal models are able to closely mimic human diseases and allow analysis of the pathology, pathogen virulence traits, host response patterns, and efficacy of potential treatments. In the world of drug design, animal models are used to determine effective dose, pharmacokinetics, pharmacodynamics, and toxicity.

Unfortunately, the majority of pathogenic fungi have essentially humans as natural hosts, meaning that the perfect animal model does not exist. Optimized models have to be developed depending on the scientific question being addressed, and several animal models often have to be used in order to answer all the questions raised by a biological problem. In addition, several ethical and financial considerations could also affect the choice of the model. In this review we will focus on animal models. They are useful to assess antifungal drug efficacy and/or resistance and to analyse host–pathogen interaction that aims to determine new drug targets.

In this review, we will discuss mammalian models of infection, with mice as being the gold standard as well as alternative non-mammalian models, also called mini-host models. These models range from insects, with *Galleria mellonella* as gold standard, to single cell hosts such as eggs or amoeba, through vertebrate models such as zebrafish.

Mouse: the model of choice

Rodents and especially mice have been the most used animal models to study human pathogens. Mice or rats have advantages compared with primates. For example they have a short generation time, do not need a lot of space, and are relatively cheap. Moreover, several different strains adapted to each defined biological question have been made available. Almost all the models that have been set up in mice can also be adapted to rats, with the exception of genetically engineered animals that do not necessarily exist for rats (reviewed in Naglik *et al.*, 2008).

Different mouse models of infection exist to study the pathogenicity of the fungi or the efficacy of a particular drug against a particular pathogenic fungus. These models can be distinguished by the route of infection (Fig. 13.1 and Table 13.1) such as systemic infection through the tail or penile veins, gastrointestinal infection by gavage, mucosal infection via the tongue, vagina, and skin eye. Until now the systemic route has been utilized as the gold standard among mouse models to assess efficacy of antifungal drugs.

Mouse systemic infection

The mouse systemic infection model mimics human systemic infections. Even though this type of infection is rare compared to mucosal infections, it can be lethal in up to 50% of the cases (Miceli *et al.*, 2011; Wisplinghoff *et al.*, 2004). Therefore, efforts have been made to develop new therapies. Systemic infections are difficult to treat partly due to the fact that they are difficult to diagnose. Indeed, fungal infections often cause

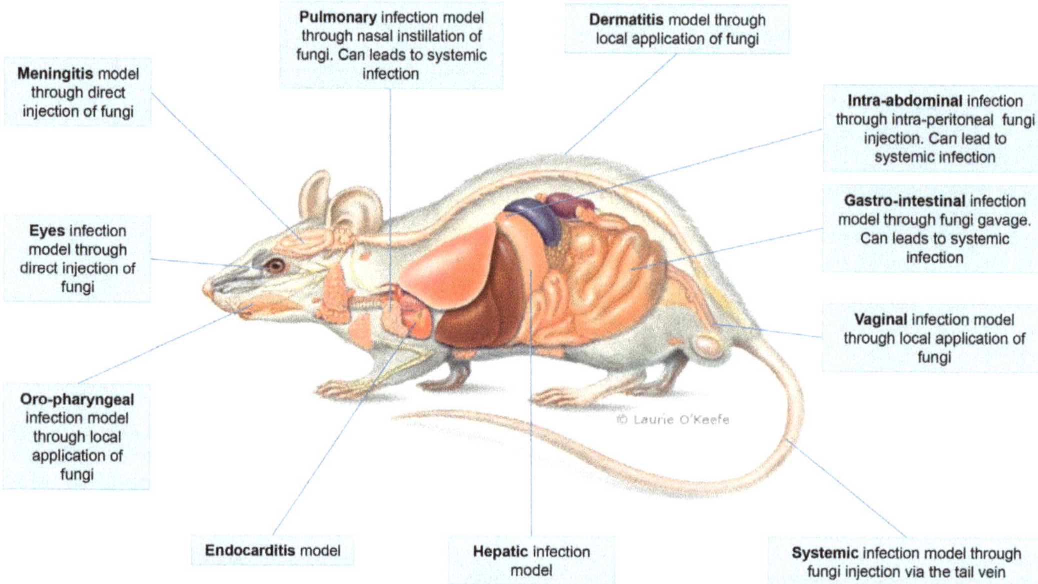

Figure 13.1 Major fungal infection models implemented in rodents. Adapted from O'Keefe, 2013.

non-specific symptoms, especially in the case of candidiasis (Szabo and MacCallum, 2011). The mouse model of systemic infection recapitulates some of the aspects of human systemic infection and especially this 'non-specific symptoms' feature.

The mouse systemic model of infection consists simply of injecting a fungal cell suspension into the tail vein of mice. This model was previously well described by MacCallum (MacCallum, 2012b). For other animal models of systemic infection, fungal cells can be injected into the tail vein of rats or in the ear vein of rabbits.

Systemic infection has been previously used to study almost all pathogenic fungi, including *Candida* spp. (Papadimitriou and Ashman, 1986), *Cryptococcus* spp. (Capilla *et al.*, 2006; Carroll *et al.*, 2007; Louria *et al.*, 1963), *Aspergillus fumigatus* (Hanson *et al.*, 1995; Mavridou *et al.*, 2010b; Schmidt, 2002), *Fusarium* spp. (Gonzalez *et al.*, 2013; Lozano-Chiu *et al.*, 1999; Spellberg *et al.*, 2006) *Histoplasma* (Clemons *et al.*, 2002) *Trichosporon* (Yang *et al.*, 2008) and others (Capilla *et al.*, 2007; MacCallum, 2012b). This model is often used to assess drug efficacy and toxicity (Andes and Safdar, 2005; Graybill *et al.*, 1997, 1998a,b; Graybill, 2000; Herrera and Guentzel, 1982; Pasko *et al.*, 1990; Restrepo *et al.*, 2000; Zarif *et al.*, 2000), pharmacodynamics or pharmacokinetics (Andes, 2003a,b; Andes *et al.*, 2003a,b,c, 2008a,b, 2010; Mavridou *et al.*, 2010b), fungal virulence (Navarro-García *et al.*, 2001; Odds *et al.*, 2001), as well as vaccine efficacy (Szabo and MacCallum, 2011). It offers different kinds of read-outs, such as post-infection mouse survival, organ colonization, histopathology analyses, cytokines measurement and recently host and fungus genome-wide gene expression assay (Mitchell, 2014). The choice of the read-out depends not only on the question being addressed, but also on the pathogen itself.

End-point survival assay

In this assay, the ability of a given pathogenic fungal strain to kill its host is evaluated in the presence or absence of a drug. After infection, animal survival is scored daily. Survival assays are performed using Log-Rank statistical survival analyses, which is to human studies. This read-out is extensively used for *C. albicans, C. neoformans, Fusarium, Aspergillus and Histoplasma* in immunocompetent mice (Clemons *et al.*, 2002; MacCallum, 2012b; Mirkov *et al.*, 2012). For some other species, immunosuppressed mice must be used in order to allow infection to occur, as is the case for *Mucor,* and other pathogenic yeast such as *Cryptococci* or *C. glabrata* (Guarro, 2011).

In order to have mice dying in a time frame that is large enough to allow measurable effects of a therapy or differences in virulence between strains, pilot experiments determining dose response relationships have to be performed to decide which fungal cell dose to inject. The chosen dose not only depends on the mouse strain, or the fungal strain but also on the fungal cells preparation method (reviewed in (Maccallum, 2012b).

This model of infection was extensively used to assess drug efficacy and drug resistance mechanism (Clemons *et al.*, 2002; Hata *et al.*, 2011; Lohberger *et al.*, 2013; MacCallum *et al.*, 2010). The tested drugs can be injected before or after the infection to test their protective abilities versus curative effect. All the currently available drugs used to treat systemic fungal infections, including azoles, echinocandins, polyenes and flucytosine have been tested for their pharmacological properties in this model of infection (Abruzzo *et al.*, 2000; Andes, 2003a; Groll and Walsh, 2006; Klepser *et al.*, 1997; Maki and Kaneko, 2013; Stiller *et al.*, 1983). While polyenes and echinocandins are administered almost exclusively intravenously or intraperitoneally, azoles could be given intraperitoneally, intravenously or orally.

For some pathogens, such as *Aspergillus fumigatus*, intravenous infection only induces limited pulmonary disease (Mavridou *et al.*, 2010a; Mirkov *et al.*, 2011, 2012). As a result, this infection model does not reflect the human pathogenesis and is not optimal for drug efficacy testing. Therefore, immunosuppression of the animal (Barchiesi *et al.*, 2005; Sionov *et al.*, 2005) or other routes of infection have been developed to better reflect the human illness (see below).

Cryptococcus neoformans virulence or drug susceptibility was also assessed through this model (Barchiesi *et al.*, 2000; Capilla *et al.*, 2006). After injection of encapsulated *C. neoformans* in the animals' bloodstream, fungal cells were found in the brain suggesting that the fungi crossed the blood–brain barrier and induced mouse death. In contrast to the human disease, *C. gattii* was shown to be unable to reach the brain of immunocompetent mice (Capilla *et al.*, 2006). To observe such brain infection with *C. gattii*, the utilization of immunodeficient mice is required (Clemons *et al.*, 1996). In this model of systemic infection, mouse strains should be cautiously chosen as a strong variability in *Cryptococcus* susceptibility was shown between immunocompetent inbred mice strains (Rhodes *et al.*, 1980) and immunodeficient mice strains (Cantorna and Balish, 1990).

End-point survival assays are difficult because the choice of the dose to inject is critical for the outcome, and also because of ethical issues. According to predefined criteria the researcher must not allow the animals to die naturally due to the infection and instead, must euthanize the animal when a certain level of severity is reached. Consequently, the time of euthanasia is critical and may significantly influence the results of the experiment. Also, double-blind scoring of the animal and a strict scoring protocol are crucial. Still some degree of subjectivity will always be present. Some studies try to find reliable criteria to define the day of euthanasia (Spellberg *et al.*, 2005; Trammell and Toth, 2011).

Moreover, some fungal pathogens poorly colonize the mice and are not able to kill them. As an example, *C. glabrata* are able to colonize the animals but do not kill them in absence of immunosuppression (Arendrup *et al.*, 2002; MacCallum, 2012a). As a result, end-point survival assays cannot be used to evaluate the virulence of the majority of the *Candida* species, at least with immunocompetent mice. In addition, some authors demonstrated that only a specific mode of animal scoring that includes weight loss and fungal burden could perfectly reflect results obtained with end-point survival assay, especially in drug efficacy testing (MacCallum *et al.*, 2009a, 2010).

Fungal burden

Fungal burden reflects the ability of the fungi to colonize the host. The fungal burden of a given organ can be evaluated by homogenization of the targeted organ and spreading it on solid media at different time-points post-infection. This readout demonstrates the ability of a fungal strain to survive in the host by infiltrating and proliferating in target organs. Evaluation of target organ colonization can also be monitored by real-time quantitative PCR using fungal genes (reviewed in

Table 13.1 Main mammal models used to study major fungal infections

	Model										
	Yeast				*Blastomycetes*		***Moulds***				
Infection	*C. albicans*	*C. tropicalis*	*C. glabrata*	*Other Candida*	*C. neoformans*	*C. gattii*	*A. fumigatus*	*Histoplasma*	*Fusarium*	*Zygomycetes*	*Dermatophytes*
Systemic infection											
AoC	Ms	Ms	Ms	Ms	Ms	Ms	Ms	Ms	Ms	Ms	Ms
Route	i.v.	Gavage	i.v.	i.v.	i.v.	i.v.	Anaesth. + i.n. instil.	i.v.	i.v.	i.v.	i.v.
Treatment		IS + Antibio.	IS	IS	IS	IS	IS	IS	IS	IS	IS
Gastro-intestinal infection											
AoC	Ms	Ms	Ms	Ms							
Route	Gavage	Gavage	Drinking water	Drinking water							
Treatment	Antibio. + IS	Antibio. + IS	Antibio. + IS	Antibio. + IS							
Oral thrush/stomatitis											
AoC	Rat/Ms	Rat/Ms	Rat/Ms	Rat/Ms							
Route	Anaesth. + cotton swab	Anaesth. + cotton swab	Anaesth. + cotton swab	Anaesth. + cotton swab							
Treatment	Antibio. + IS	Antibio. + IS	Antibio. + IS	Antibio. + IS							
Pulmonary infection											
AoC					Ms/Rab./GP	Ms/Rab./GP	Ms[b]/Rab./GP	Ms[b]/Rab./GP	Ms[b]/Rab./GP	Ms[b]/Rab./GP	
Route					Anaesth. + i.n. instil.	Anaesth. + i.n. instil.	Anaesth. + i.n. instil.	Anaesth. + i.n. instil.	Anaesth. + i.n. instil.	Anaesth. + i.n. instil.	
Treatment					IS Rab. and GP	IS Rab. and GP	IS	IS	IS	IS	
Vulvovaginal candidiasis											
AoC	Ms	Ms	Rat	Rat							
Route	Vaginal lavage	Vaginal lavage	Vaginal lavage	Vaginal lavage							
Treatment	Oestrogen + IS	Oestrogen + IS	Oestrogen + IS	Oestrogen + IS							

Dermatitis											
AoC	GP/Rab./Ms	GP/Rab./Ms	GP/Rab./Ms	GP/Rab./Ms	GP/Rab./ Ms	GP/Rab./ Ms	GP[c]/Rab./ Ms	GP[c]/Rab./ Ms	GP[c]/Rab./ Ms	GP[c]/Rab./Ms	GP[c]/Rab./Ms
Route	Topical	Topical	Topical	Topical	Topical	Topical	Topical	Topical	Topical	Topical	Topical
Treatment	IS	IS	IS	IS	IS	IS	IS	IS	IS	IS	IS
Keratitis/ophthalmitis											
AoC	Rab.	Rab.	Ms	Ms			Rab. Ms		1, Rab.; 2, Ms		
Route	i.o.	i.o.	i.o.	i.o.			i.o.		1, i.o.; 2, i.v.		
Treatment	IS	IS	IS	IS			IS		IS not required but increases infection		
Meningitis											
AoC	Ms	Ms	Rab.	Rab.	Rab.	Ms[a]	Ms Rab.			1, Ms; 2, Rab.	
Route	i.c.	i.c.	i.v.	i.v.	i.c.	i.v.	i.c			i.c	
Treatment	IS Rab.	IS Rab.	IS Rab.	IS Rab.	IS Rab.	i.v: ID Ms/ IS Rab.	IS			IS	

[a]Model of choice for *C. neoformans*. [b]Model of choice for moulds except Dermatophytes. [c]Model of choice for dermatophytes + ID mice. AoC, animal of choice; Ms, mouse; i.v., intraveinous; antibio., antibiotic treatment; IS, immunosuppressed or immunosuppression; anaesth., anaesthesia; GP, guinea pig; Rab., rabbit; i.o., intraocular; i.c., intracerebral; i.n., intranasal; instil., instillation; ID, immunodeficient.

(Brown *et al.*, 2007; Gonzalez *et al.*, 2013)) or a tag inserted in the tested fungal strain in the case of screening (Vandeputte *et al.*, 2011).

C. albicans has also been shown to preferentially infiltrate the kidneys when injected intravenously in mice, although transient infection has been demonstrated in the lung, spleen and liver. At higher doses, *C. albicans* can also reach the brain (MacCallum and Odds, 2005).

As mentioned above, *C. neoformans* has the same organ tropism in mice as in humans, reaching the central nervous system (CNS), and therefore reflects the human disease to a good extent (Barchiesi *et al.*, 2000; Capilla *et al.*, 2006). *C. gattii*, on the other hand, has a very different organ tropism in mice when compared to humans, mainly because it does not reach the CNS but infects mainly the liver and lungs and causes skin lesions in severe combined immunodeficient (SCID) mice (Capilla *et al.*, 2006).

Histoplasma preferentially infects liver and spleen when injected intravenously (Clemons *et al.*, 2002). This model was extensively used to assess drug efficacy, even if it does not reflect the natural course of infection in humans where this pathogen reaches preferentially the lung. Therefore intranasal instillation seems to be more adapted to study anti-histoplasma treatment (Smulian, 2012).

Intravenous injections of moulds have been extensively used to assess drug efficacy for both fusariosis (reviewed in Gonzalez *et al.*, 2013) and invasive aspergillosis (Barchiesi *et al.*, 2005; Shadkchan *et al.*, 2004; Sionov *et al.*, 2005). *Fusarium* colony-forming-unit (CFU) are counted in spleen and kidney (Gonzalez *et al.*, 2013) and *Aspergillus* CFU are preferentially counted in lungs and brain (Barchiesi *et al.*, 2005; Mirkov *et al.*, 2011; Shadkchan *et al.*, 2004). However, CFU counting of filamentous fungi does not precisely reproduce the amount of filamentous cells present in the original sample. For that reason, alternative methods for quantification of fungal burden for these pathogens were proposed, namely by qPCR or Chitin assay (Gonzalez *et al.*, 2013; Hummel *et al.*, 2004). Measuring fungal burden by qPCR appeared to give higher valuable results than by CFU counting.

Histopathology and host cytokine measurement

A global approach to assess the toxicity/efficacy of a treatment and/or the fungal pathogenesis is to evaluate the tissue damage through histological analyses (Asmundsdottir *et al.*, 2009; Soo *et al.*, 2010). In addition, with the help of immunocytochemistry, the host immune response can be assessed by labelling host immune cells *in situ*. This approach is quite laborious and relatively expensive due to the necessity of a histopathology platform.

Fungal pathogenesis can also be assessed by measuring the host`s cytokines production in the serum or in fungal target organs. MacCallum *et al.* (2009b) showed that early measurement of chemokines in mice infected with *C. albicans* can predict kidney immunopathology. Such analyses can be performed by using either ELISA assay or cytokine mRNA quantitation by RT-PCR. Cytokines profile have been studied in fungal infections with several species such as *Candida*, *Aspergillus*, *Cryptococcus*, *Fusarium* or *Pneumocystis* (Antachopoulos *et al.*, 2012; Antachopoulos and Walsh, 2012; Leal and Pearlman, 2012; MacCallum *et al.*, 2009b) for review see (Capilla *et al.*, 2007; Centeno-Lima *et al.*, 2002)).

This kind of read-out is more frequently applied to study host–pathogen interactions as cytokines are one of the primary mediators responsible for activation and regulation of the innate, and subsequently adaptive, immune systems of the host. In this case, mice are the perfect animal model as several biological tools exist such as genetically engineered mouse strains, antibodies or cytokines themselves.

The majority of the studies analysing the development of the innate immune response aim essentially to develop immunotherapies or therapies eliciting a protective host immune response (for review see Antachopoulos *et al.*, 2012; Antachopoulos and Walsh, 2012; Capilla *et al.*, 2007) by their own or in combination with antifungal drugs. Other studies analysing the development of a protective adaptive response are more related to vaccine development (see Chapter 12).

Special case of systemic infection through biofilm

It is important to keep in mind that often the entry portal of systemic infections are medical devices such as catheters infected by fungal biofilms. Even if biofilms are extensively studied *in vitro*, these systems cannot reproduce the complexity encountered within the host. To circumvent this problem, *in vivo* biofilm models were developed in rodents (Andes *et al.*, 2004; Vande Velde *et al.*, 2013, 2014) and rabbits (Chandra *et al.*, 2011), which consisted of establishing a fungal biofilm on a catheter before or after its implantation in veins of anaesthetized animals. After a few days of treatment with antifungal drugs, biofilm development was assessed by biofilm biomass determination through CFU counting or bioluminescence measurement (Vande Velde *et al.*, 2013, 2014).

Fungal biofilms on denture materials play a major role in oral thrush and denture stomatitis in healthy patients. Therefore, biofilm models on fixed or mobile acrylic denture material were developed in rats to study the development of infection and host-response in both immunocompetent and immunosuppressed animals. Antibiotherapy seems to play a crucial role in the establishment of the fungal infection (reviewed in Kirkpatrick *et al.*, 2012).

Mucosal model of infection

Even though the systemic model of infection is the gold standard in pathogenic fungal studies it is not always optimal. In fact, intravenous infection does not reflect the most frequent types of fungal infections (oral and vaginal mucosal infections) or the supposed major entry route of *Candida* strains, the gut (reviewed in MacCallum, 2012a). Several studies showed that every human being carries at least one species of *Candida* in their gut, with up to 70% being *C. albicans*. Only 40% of the population carries *Candida* in the oral cavity and only 20% of the women have been found to have vaginal colonization. These sites of commensal colonization are thought to be the origin of almost all cases of systemic infection (Bougnoux *et al.*, 2006; Miranda *et al.*, 2009; Odds *et al.*, 2006). For this reason, several alternative models of infection have been developed in mice to mimic localized infections such as gastrointestinal, oral, vaginal, skin and hair and to assess drug efficacy in such models (Fig. 13.1 and Table 13.1).

Gastro-intestinal model

One of the most used alternative mouse model of infection is the gastrointestinal model. This model is used essentially to mimic a natural route of *Candida* entry leading to systemic infection (Szabo and MacCallum, 2011). It is also widely used to study the properties distinguishing the commensal and virulent states of *Candida* species (Chen and Noble, 2012; Chen *et al.*, 2011; Noble, 2013).

Gastrointestinal infection can be performed in immunocompetent infant mice or in antibiotic treated and immunosuppressed adults. This model was recently well described by Clemons and Stevens (2012). After antibiotherapy using gentamicin, clindamycin, vancomycin or imipenen, *Candida* cells are administered in drinking water or by gavage. In this model of commensal colonization, fungal burden can be assessed by counting CFU in the faeces. Systemic dissemination can be obtained by subsequent immunosuppression and then, survival and fungal burden can be assessed. Immunosuppression can be achieved with chemical compounds such as prednisolone, cyclophosphamide, cortisone acetate (Clemons and Stevens, 2012; Cole *et al.*, 1996), or using immune-deficient mice (Cantorna and Balish, 1990; Clemons and Stevens, 2000, 2001; Flattery *et al.*, 1996). Also, the type of immunosuppression or the strain of immune-deficient mice that is used can have a great influence on the results.

Using this model of infection, *C. albicans* and *C. tropicalis* are able to disseminate and cause death of the mice, colonizing mainly the liver and spleen and, to a lesser extent, the kidney (Pope *et al.*, 1979). *C. parapsilosis* is not able to disseminate and can only colonize the gut (Mellado *et al.*, 2000). In contrast to other *Candida* species *C. glabrata* is unable to colonize the oesophageal tissue (Westwater *et al.*, 2007).

This model was also extensively used to assess antifungal efficacy. Pertaining to the systemic model, drugs can be administered intraperitoneally, orally, subcutaneously or intravenously

(Clemons *et al.*, 2006; Clemons and Stevens, 2000, 2001; Cole *et al.*, 1991; Flattery *et al.*, 1996; Gordee *et al.*, 1984; Kinsman and Pitblado, 1989; Ponnuvel *et al.*, 1993; Samonis *et al.*, 2001; Yordanov *et al.*, 2005). This model can be modulated by the choices of the antibiotherapy or immunosuppression used, as well as by the route of drug administration (Kinsman and Pitblado, 1989).

Some models have been used that involve both oral and gastrointestinal infections, starting with infection of the mouth which then disseminates to the gut. In this case, *C. albicans* is given in the drinking water to mice under antibiotherapy for several days. Similarly to the previous model, the liver is the most colonized organ, however mortality is rare (Clemons *et al.*, 2006).

This gastrointestinal and disseminated murine model is considered to better reflect the natural course of systemic infection that occurs in humans. This model also better reflects the type of treatment or immune status for patients who are prone to systemic candidiasis. Finally, the disseminated gastrointestinal model has the advantage that drug efficacy can be evaluated for localized and disseminated infections. Nevertheless, since this model involves antibiotherapy and immunosuppression of the animal, results have to be interpreted with caution taking into account all these manipulations.

Intra-abdominal model

Intra-abdominal candidiasis is the second most frequent form of invasive candidiasis with 25% of positive blood cultures (Carneiro *et al.*, 2011). A model of intra-abdominal candidiasis was established to mimic such kind of infection in mice (Cheng *et al.*, 2013; Felk *et al.*, 2002; Sawyer *et al.*, 1995; Vonk *et al.*, 2003). *C. albicans* or *C. glabrata* cells are injected intra-peritoneally in immunocompetent mice concomitantly with 5% w/v of sterile stools which helps to prolong the *Candida* infection. Infection can be monitored by CFU counting over time recovering peritoneal fluid. To assess the dissemination, fungal burden is determined in the eyes and the brain. In this model, all the mice display on ocular infection 24 hours post infection. Severity of the infection is also evaluated measuring the number of *Candida* abscesses in the peritoneal cavity. To date it is the only known model in which *C. glabrata* is able to kill the host (Cheng *et al.*, 2014).

Few drugs were tested in this model via intraperitoneal injection (Vonk *et al.*, 2003). Several studies evaluated the role of adhesines in the severity of this infection (Felk *et al.*, 2002; Staib *et al.*, 2002). Recently, this model was used to assess the transcriptional rewiring of *C. albicans* in response to the host environment using the nanostring technology (Cheng *et al.*, 2013).

Oral thrush model

Oral thrush is one of the most common types of fungal infections, and *Candida* infection is especially prevalent in AIDS patients or patients who undergo chemotherapy (Lopez-Ribot *et al.*, 1999; Sangeorzan *et al.*, 1994). As a result, testing drug efficacy in murine models mimicking oral thrush is essential, and several have been developed so far (for review see Costa *et al.*, 2013). To obtain a sustained infection, mice are usually pre-immunosuppressed and/or treated with antibiotics to remove normal flora that would normally compete with yeasts. *Candida* cells are then applied to the oral cavity of anaesthetized mice using a cotton swab under the tongue for 2 hours (Kamai *et al.*, 2001; Solis and Filler, 2012) or directly on the teeth, tongue, gums, and rest of oral cavity (Samaranayake and Samaranayake, 2001; Takakura *et al.*, 2003). Immunosuppressive or antibiotic therapies have to be maintained during the experiment otherwise the fungal load decreases rapidly after cessation of treatment (reviewed in MacCallum, 2012a; Naglik *et al.*, 2008; Samaranayake and Samaranayake, 2001). In the case of immunocompetent animals, the infection occurs variably and the fungal burden declines rapidly in one week to the level of a carrier state (Gladiator *et al.*, 2012; Lacasse *et al.*, 1993). In this model, fungal virulence, treatment efficacy, or host responses are measured through assessment of CFU titres, histopathology and cytokine measurement in the tongue, palatine, oral mucosa, and oesophagus.

Several drugs were tested in the oral thrush model through oral or intragastric administration (Hayama *et al.*, 2012; Ishibashi *et al.*, 2007a; Kamai *et al.*, 2003; Matsubara *et al.*, 2012; Taguchi *et al.*, 2005; Takakura *et al.*, 2003; Walsh *et al.*, 2000).

In addition to the oral model some pharyngeal models have also been developed, using immunosuppression to assess drug efficacy (Hu *et al.*, 2007; Ishibashi *et al.*, 2007a).

Vulvovaginal candidiasis (VVC)

VVC episodes occur at least once in the lifespan of 75% of healthy women (Geiger *et al.*, 1995; Ilkit and Guzel, 2011). Recurrent VVC (RVVC) infection affects 5–8% of women and strongly impacts their quality of life. With VVC and RVVC being clear public health problems, models to study fungal vaginitis were necessary. Such models were first developed in rodents to mimic this kind of infections (Capilla *et al.*, 2007; De Bernardis *et al.*, 1999; MacCallum, 2012a; Rahman *et al.*, 2012; Yano and Fidel, 2011). In this model, the mice or rats are maintained in oestrus to obtain colonization and infection. The infection is done through a vaginal lavage with yeast cells suspension. Colonization can be increased applying immunosuppression (Foldvari *et al.*, 2000). Not all the mice strains respond to oestrogen in the same way, which might influence the infection success (Calderon *et al.*, 2003; Clemons *et al.*, 2004). VVC infection models have been used to study the virulence of *C. albicans* and *C. glabrata* (Ferrari *et al.*, 2011; Fidel *et al.*, 1996), as well as other *Candida* species, but to a lesser extent (Bertini *et al.*, 2013). Similarly to other models, the mouse vaginitis model is used to analyse drug efficacy, pathogenesis and host response, through CFU counting, cytokine production, and histopathology (Berretta *et al.*, 2013; Clemons *et al.*, 2011; De Bernardis *et al.*, 1995; Fidel *et al.*, 1993; Gonzalez *et al.*, 2009; Polak, 1998; Yano *et al.*, 2012).

Cutaneous infection model

Dermal infection by fungi is uncommon in immunocompetent people and is often a marker of immunosuppression. In AIDS patients, skin is one of the most frequently involved organs, and fungal skin infection is often an early marker of AIDS (Mendiratta *et al.*, 2010; Outerbridge, 2006; Ramos *et al.*, 2010). Fungal skin infection can be caused by *Candida* species such as *C. albicans*, *C. glabrata*, *C. tropicalis*, *C. krusei*, and *C. parapsilosis*, *Aspergillus*, *Cryptococcus*, *Fusarium*, *Mucor*, *Malassezia*, *Histoplasma* and dermatophytes, including *Trichophyton*, *Microsporum* and *Epidermophyton*. Severe skin infection can propagate to deeper tissues and be the entry portal for systemic infection and in severe cases can lead to death. For all these pathogens, models of infection were developed to mimic human disease and study fungal virulence, host immune response and drug treatment efficacy (Ben-Ami and Kontoyiannis, 2012; Cambier *et al.*, 2013; Domer *et al.*, 1983; Domer and Moser, 1978; Giger *et al.*, 1978; Kamei, 2001; Ray and Wuepper, 1976; Shimamura *et al.*, 2012; Sohnle and Hahn, 1989). In such models drugs can be administered by topical, oral or systemic treatment. Read-out is essentially performed by CFU counting or fungus detection by PCR on skin samples, as well as histopathology of skin lesions and deep tissues and chemokine measurement. As the skin surface in mice is limited, larger animals such as rabbits, rats or guinea pigs are preferable. Guinea pigs, for example, are the model of choice for dermatophyte studies (reviewed in Shimamura *et al.*, 2012).

Other types of mice infection models

Other models of infection have also been developed to study particular fungal infection features, such as brain infection with *Cryptococcus*, pulmonary infection with *Aspergillus* and eye infection with *Fusarium*.

Invasive aspergillosis

Over the last decade, the incidence of invasive aspergillosis has increased markedly (Araujo *et al.*, 2008; McNeil *et al.*, 2001) creating an urgent need for adequate animal models. Several models have been described in mice using intravenous (see above) but also intranasal, intratracheal and finally inhalation infection, which better reproduces the natural entry of *Aspergillus* in humans (reviewed in Lewis and Wiederhold, 2005; Sheppard *et al.*, 2006). Nevertheless, inhalation of conidia by anaesthetized mice using aerosol chamber appears to be the most standardized method and closely resembles the human disease (Sheppard *et al.*, 2006; Steinbach *et al.*, 2004). Since epidemiological studies have shown a strong relationship between immunosuppression and *Aspergillus* susceptibility (Marr *et al.*,

2002), invasive aspergillosis models also include immunosuppression of mice with both cyclophosphamide, to induce neutropenia, and cortisone acetate. This double immunosuppression leads to lethal infection, in contrast to single-drug treatment. Moreover, several studies demonstrated that the type of immunosuppression influences the virulence of tested strains and therefore might also affect the antifungal therapy (reviewed in Kirkpatrick *et al.*, 2012).

Other pulmonary infection models

Even if the major manifestation of cryptococcosis is meningoencephalitis, pulmonary cryptococcosis is clearly underestimated and could be a source for central nervous system infection as the inhalation is the natural route of entry for *Cryptococcus* (Brizendine *et al.*, 2011). Thus, a cryptococcosis pulmonary model mimics the natural course of infection in humans better than the systemic model of infection. Effectively, in this model as in human disease, *C. gattii* does not cause fatal meningoencephalitis but fatal lung infection which is in contrast to *C. neoformans* (Ngamskulrungroj *et al.*, 2012). Immunodeficient mice are often used, because immunocompetent animals are relatively resistant to the infection (Muhammed *et al.*, 2012b). Recent studies showed that humid nebulization results in better infection rate of *Cryptococcus* and better reproducibility when compared to simple nasal instillation or intratracheal inoculation (Springer *et al.*, 2013). CFU assessment in the lungs and brain and histopathology are the main read-outs of the infection. Immunosuppressed rabbits and guinea pig were also shown to be good models to study pulmonary cryptococcosis (Carroll *et al.*, 2007).

Pulmonary infection models have also been developed in mice to mimic histoplasmosis, blastomycosis, zygomicosis and fusariosis. In all these cases, the most common and standardized methods to inoculate fungi are nasal instillation or inhalation. For almost all models, mice strains, anaesthesia and immunosuppression greatly influence the outcome of the infections and, therefore, the drug efficacy analyses (reviewed in (Capilla *et al.*, 2007; Muhammed *et al.*, 2012b; Smulian, 2012).

Intracerebral infection models

As mentioned before, the major cause of death by *C. neoformans* is meningoencephalitis. Several models have been developed to reproduce such infection (Blasi *et al.*, 1992) including direct injection of an encapsulated *Cryptococcus* suspension intracerebrally or intracisternally (reviewed in Carroll *et al.*, 2007). Using this model, survival, fungal burden or histopathology read-outs can be used to test drug efficacy or to better understand host–fungus interaction, without the need of immunosuppression (Hata *et al.*, 1996; Troke *et al.*, 1987). In this model of infection, *C. gattii* was shown to kill and destroy brain tissues more rapidly than *C. neoformans* but with no difference in brain fungal burden or dissemination in other organs (Thompson *et al.*, 2012). However, mice have a relatively small brain and so are not model of choice for this infection route. Bigger animals, such as rabbits, rats or guinea pigs, are more suitable (Carroll *et al.*, 2007; Kirkpatrick *et al.*, 2007) because they allow the collection of bigger samples of cerebrospinal fluid from the same animal several times, which could be useful in measuring host response during different time points of infection. The larger volume of the brain also allows more accurate histopathology analyses. In the case of the guinea pig and rabbit, immunosuppression is needed as those animals seem to be intrinsically resistant to cryptococcosis, probably due to their higher body temperature (Carroll *et al.*, 2007).

Other fungal pathogens can cause brain infections such as *Aspergillus*, *Candida* and *Mucor*. Therefore, several intracranial models were developed in rodents and rabbits to study these types of infections (Anand *et al.*, 1992; Blasi *et al.*, 1993, 1994; Chiller *et al.*, 2002; Morton *et al.*, 2011).

Eye infection models

Eye fungal infections such as keratitis and endophthalmitis are an important cause of visual problems, including blindness (Thomas and Kaliamurthy, 2013). Trauma remains the principal predisposition factor, but contact lenses have also become an important risk factor for keratitis, at least in western countries. Several fungal pathogens have been shown to be the etiologic agent of keratitis and endophthalmitis, but the more frequently isolated are moulds, such as *Fusarium* and *Aspergillus* (Srinivasan, 2004). Yeasts such as *Candida* species can also be found (Leal and Pearlman, 2012; Thomas and Kaliamurthy, 2013). Several models have been developed so far in mice to mimic fungal endophthalmitis and keratitis (Mayayo *et al.*, 1998; Wu *et al.*, 2003). *Fusarium* can be injected intravenously in immunocompetent mice, leading to a systemic infection with cornea as the main target organ (Mayayo *et al.*, 1998). *Aspergillus* and *Fusarium* can also be inoculated directly into the cornea (Leal *et al.*, 2010; Leal and Pearlman, 2012; Tarabishy *et al.*, 2008). Immunosuppression is not necessary for *Fusarium* but enhances the infection rate (Wu *et al.*, 2004). In contrast, *Aspergillus* and other pathogenic fungi such as *Candida* can only cause a successful eye infection in immunosuppressed animals (Wu *et al.*, 2003). Eye infection can be assessed by CFU counting or PCR, targeting fungal genes, either in the cornea or in the entire eye, and by histopathological studies that measure corneal lesions and opacification.

Since the eyes of mice are quite small, rabbits and, to a lesser extent, rats, are the most prevalent animals used for such studies. Rabbits are of particular interest since their eye size, anatomy, and physiology are closely related to that of humans (Forster and Rebell, 1975; Ishibashi *et al.*, 1986a,b; Kiryu *et al.*, 1991; Zhu *et al.*, 2011).

Less frequent models of fungi infection

Some fungi are less frequently isolated and necessitate special conditions to give rise to an infection. As a consequence, infection models for such fungi, Zygomycetes for example, need special parameters for animal experiments. These kinds of models have been already extensively reviewed and will not be detailed here (Guarro, 2011; Kamei, 2001).

Some models are performed with frequently isolated fungi but are more difficult to implement, such as the endocarditis model in rat (Marchetti *et al.*, 2000), and to a lesser degree the hepatic candidiasis model (Sawyer *et al.*, 1976, 1990, 1992).

The results obtained in all the mouse models described above (Fig. 13.1 and Table 13.1) are dependent on the mouse strain (Calderon *et al.*, 2003; Cantorna and Balish, 1990; Capilla *et al.*, 2006; Clemons *et al.*, 2004; MacCallum and Odds, 2002; Mirkov *et al.*, 2011). Systemic infections are more reproducible than mucosal infections probably due to the fact that mucosal infections require the use of anaesthesia, immunosuppressive therapy, antibiotics, oestrogen treatment or the use of transgenic animals, which may greatly influence the results (Jacobsen *et al.*, 2010a; Kinsman and Pitblado, 1989; Kirkpatrick *et al.*, 2012; Muhammed *et al.*, 2012b; Smulian, 2012). A very recent study demonstrated that important genes such as *EFG1* are differentially expressed in immunocompetent versus immunosuppressed mice like if *C. albicans* is really able to sense the immune status of the host (Kumamoto, 2014). As a result, data must be carefully interpreted and compared with other studies. Moreover, results from one model may not necessarily correlate with another model of infection. Finally, while testing drug efficacy, several studies showed the importance of the route of antifungal administration, especially for azoles, on the outcome of the study, even when performed using the same mouse model of infection (Groll and Walsh, 2006; Ishibashi *et al.*, 2007b; Pasko *et al.*, 1990; Samonis *et al.*, 2001; Stevens and Clemons, 2007; Taguchi *et al.*, 2005; Van Cutsem *et al.*, 1987, 1989).

Mini-host models

Animal experimentation in mammals is essential for scientific research and drug development as it brings us as close as possible to the complexity of the human organism, and has undoubtedly contributed to an improvement of the quality of life for people all over the world. However, this is often a subject of controversy among the public, mainly for ethical concerns. For this reason, the last decade has seen the development of laws controlling animal use in research, applying the 3R concept – 'Reduce, Refine, Replace' (Russell and Burch, 1959). Although it is true that most drug discovery processes start with *in vitro* screenings which provide valuable information, a second step involving *in vivo* models is invariably necessary, especially when it comes to antifungal resistance.

In this context, the number of animals used should be reduced to the minimum that allows enough data to be obtained to reach a definitive conclusion. Refinement refers to the obligation of the experimenter to improve the quality of life and decrease the suffering inflicted on the animals. The last 'R' stands for replace and states that, whenever the current state of knowledge allows, animal experiments must be substituted by *in vitro* methodologies, *in silico* approaches, or the use of so-called 'inferior' animal models, such as animals with a shorter lifespan (e.g. insects). In this context, several alternative non-mammalian experimental models have been developed and are being thoroughly used, including invertebrate models such as *Drosophila melanogaster* and *Caenorhabditis elegans*, as well as the vertebrate model *Danio rerio* (zebrafish). These models have the added advantage of a significantly lower price and of providing results faster than the bigger mammalian models. The next section will be dedicated to describing these models and their applications in the study of fungal pathogens and antifungal drugs.

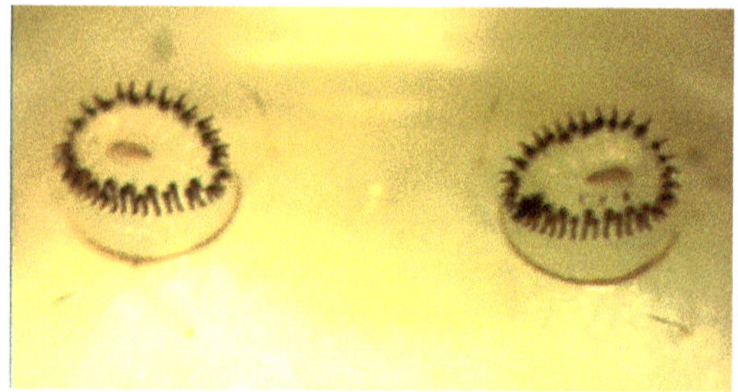

Figure 13.2 Amplified view (10×) of two *G. mellonella* prolegs.

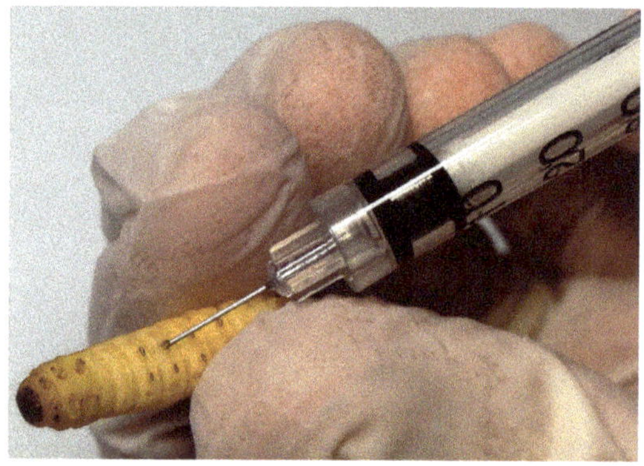

Figure 13.3 *G. mellonella* larva being injected through one of the eight abdominal prolegs.

An insect model of choice: *Galleria mellonella*

Galleria mellonella, commonly known as the greater wax moth, is used as an *in vivo* model of infection to study pathogenicity of both bacterial and fungal pathogens, as well as for determining the activity and toxicity of antimicrobial drugs.

G. mellonella is used mainly as a systemic model of infection. This model was first developed for bacterial studies (Dunphy *et al.*, 1986; Morton *et al.*, 1987; Walters and Ratcliffe, 1983) and later adapted to fungal pathogens (Cotter *et al.*, 2000). Larvae on the sixth instar are inoculated with the pathogen by injection through one of its eight prolegs (usually the last left one) (Figs. 13.2 and 13.3). Drugs can be administered in the same way, either into infected or non-infected insects, depending on whether the focus is on the antifungal effect or on its toxicity. Larvae can then be incubated at a range of temperatures and for variable periods of time, most frequently 24 or 48 hours [see Fuchs *et al.* (2010b) for detailed protocol]. The most common readouts are survival, which consists of daily scoring the number of live animals for a certain period of time (Brennan *et al.*, 2002; Cotter *et al.*, 2000; García-Rodas *et al.*, 2011; Thomaz *et al.*, 2013), fungal burden (CFU counting), which is performed by homogenizing the larval body, diluting and plating it on solid media (Bergin *et al.*, 2003; Melo *et al.*, 2013), and haemocyte counting using a microscope, since a decrease in haemocyte numbers correlates with higher fungal virulence (García-Rodas *et al.*, 2011; Melo *et al.*, 2013). Other assays include the analyses of antimicrobial peptides expression, either at the proteomic or transcriptomic levels (Bergin *et al.*, 2006; Melo *et al.*, 2013), and histological analyses to observe the distribution of fungal pathogens inside the host, particularly the formation of nodules (Fuchs *et al.*, 2010a; Mesa-Arango *et al.*, 2012; Thomaz *et al.*, 2013). The degree of melanization (Fig. 13.4), which is part of the immune response of the larvae, is another parameter that can be explored, although

Figure 13.4 Difference between healthy, PBS injected *G. mellonella* larvae (top) and infected larvae, illustrating the process of melanization (bottom).

it has been shown not to correlate with fungal virulence, but with the number of injected fungal cells (Cotter *et al.*, 2000; Thomaz *et al.*, 2013).

The *G. mellonella* model has a number of advantages among the non-mammalian animal models, such as the possibility of incubating at a large range of temperatures, inoculation by injection and the possibility of working without inducing immunosuppression. In fact, *G. mellonella* can be kept at around 15°C with no food source for up to 1 week without interfering with the results (Banville *et al.*, 2012) and, when inoculated with fungal pathogens, can be incubated between 25°C and 37°C. Therefore, and in contrast to other non-mammalian infection models, *G. mellonella* allows the study of pathogens at human body temperature and of temperature-related virulence traits (Coleman *et al.*, 2011; García-Rodas *et al.*, 2011; Mesa-Arango *et al.*, 2012). Inoculation by injection is possible thanks to the larger size of the larvae when compared to *D. melanogaster*. Furthermore, inoculation by injection permits the administration of precise amounts of both fungal cells and drugs, which is in contrast to the methods used for *D. melanogaster* or *C. elegans* (see below). Injection is done through one of the eight prolegs, without piercing the cuticle, and therefore ensures that no immune response to a wound is triggered. This eliminates the risk of the inoculation procedure interfering in the results (Cotter *et al.*, 2000).

G. mellonella has a functional immune system that is similar to the innate immune system of mammals. Six different types of haemocytes – prohaemocytes, coagulocytes, spherulocytes, oenocytoids, plasmatocytes and granulocytes – circulate within the haemocoel and compose the cellular immune response. Similar to mammalian neutrophils, the cells are able to phagocytose and kill invading pathogens through the action of reactive oxygen species (ROS) (Browne *et al.*, 2013). The haemolymph component of the insects' immune response includes antimicrobial peptides (Brown *et al.*, 2009) and enzyme cascades involved in the production of melanin, with several toxic intermediates, particularly quinones. Similarity with mammalian innate immunity is further evidenced by the discovery in *G. mellonella* of homologues to the human p47phox and p67phox proteins, which play a role in NADPH oxidase system (Fallon *et al.*, 2011; Renwick *et al.*, 2007). Moreover, the larval apolipophorin III, a pattern recognition protein that participates in the elimination of *C. albicans* cells from the haemolymph by binding to β-1,3-glucans, is closely related to the N-terminal domain of human ApoE (Dunphy *et al.*, 2003; Weers and Ryan, 2006). *G. mellonella* is equally susceptible to immunosuppressive agents (Melo *et al.*, 2013; Reeves *et al.*, 2004). The fact that innate immunity is crucial for a proper response against fungal infections in mammals (Cheng *et al.*, 2012; Romani, 1999) might explain the strong correlation between the results obtained with *G. mellonella* and those obtained with mammalian models, such as mice (Brennan *et al.*, 2002; Coleman *et al.*, 2011). Interestingly, incubation of larvae at 37°C stimulates its immune response, rendering the insects less susceptible to subsequent fungal infections (Mowlds and Kavanagh, 2008; Wojda *et al.*, 2009).

The correlation between results in *G. mellonella* and mice was demonstrated in numerous studies. Firstly, *G. mellonella* was shown to be able to differentiate between pathogenic and non pathogenic yeasts, such as *C. albicans* and *S. cerevisiae*. Additionally, the insects' survival rate was affected only by viable, and not heat killed, yeasts. Moreover, the degree of pathogenicity of different *Candida* species has been shown to correlate well with what is observed in humans (Cotter *et al.*, 2000). Analysing the survival rates after 24 and 48 hours of injection in the insects, a subsequent study showed, for a number of *C. albicans* mutants, a correlation between the virulence observed in *G. mellonella* and that reported for mice in previous works (Brennan *et al.*, 2002). *G. mellonella* was also shown to be an appropriate model to analyse the virulence of *Aspergillus fumigatus* (Reeves *et al.*, 2004), *Cryptococcus neoformans* (García-Rodas *et al.*, 2011; Mylonakis E, 2005), *Fusarium* (Coleman *et al.*, 2011), *Candida tropicalis* (Mesa-Arango *et al.*, 2012), *Candida krusei* (Scorzoni *et al.*, 2013) *Paracoccidioides lutzii* and *Histoplasma capsulatum* (Thomaz *et al.*, 2013).

Fewer studies exist concerning antifungal drug testing using *G. mellonella* as a model of infection. The results obtained with this model positively correlate with what is known from human pathogenesis or from mammalian models. Mylonakis *et al.* first demonstrated the applicability of this model to reproduce the susceptibility of *C. neoformans* to commonly used antifungal agents in a clinical setting, namely amphotericin B, fluconazole and 5-flucytosine. In this study, *G. mellonella* survival was prolonged when the drugs where used, both independently or in combination. A synergistic effect was observed between amphotericin B and 5-flucytosine, similarly to what happens in human patients (Mylonakis, 2005), which supports the use of this insect model for testing antifungal drugs both in monotherapy and in combination. *Fusarium* spp. are generally susceptible to amphotericin B, but resistant to fluconazole. When *Fusarium*-infected larvae are treated with these compounds, the same drug susceptibility profile is observed (Coleman *et al.*, 2011). *G. mellonella* infected with *C. albicans* responds to treatment with amphotericin B, fluconazole and 5-flucytosine in a similar way to that observed in human patients. Additionally, fluconazole-resistant clinical isolates are also resistant to the drug in the *G. mellonella* model. Furthermore, synergism between amphotericin B and 5-flucytosine was observed in *G. mellonella* for *C. albicans* infection as well (Li *et al.*, 2013). The response to drug treatment of other species of *Candida*, such as *C. tropicalis* (Mesa-Arango *et al.*, 2012) and *C. krusei* (Scorzoni *et al.*, 2013), is also equivalent to that observed in the clinic and with mammalian models. *G. mellonella* has already been used to investigate innovative therapeutic strategies. A good example is the use of inhibitors of Hsp90 to overcome resistance to commonly used antifungal drugs. *C. albicans* and *A. fumigatus* strains that presented resistance to azoles or echinocandins became susceptible when Hsp90 inhibitors were used in combination with those drugs. Comparable results were found *in vitro*, in the *G. mellonella* model and in a murine model of infection, underscoring the utility of this insect (Cowen *et al.*, 2009). Another study investigated the activity of astemizole and a closely related analogue (A2) against *C. neoformans* and *Cryptococcus gattii*. The fungicidal effect of the compounds was observed *in vitro*, and subsequently in the *G. mellonella* infection model in combination with fluconazole (Vu and Gelli, 2010).

However, this model has its limitations. Firstly, and in contrast with other non-mammalian infection models, the full genome of *G. mellonella* is currently not available, limiting the analyses of host reactions to infection, although a transcriptome for immunity-related genes is now available (Vogel *et al.*, 2011). The lack of a complete *G. mellonella* genome makes developing tools for genetic manipulation and generation of mutant strains difficult. Secondly, antifungal drug testing is limited to parenteral administration, since oral administration would be difficult to quantify accurately. Finally, no inbred strains of *G. mellonella* are available to the general scientific community, as opposed the standard mouse models or fly models. This is probably one of the major causes of the high variability observed between individual insects. High variability between laboratories is also predicted by the influence of environmental factors on susceptibility to infection, particularly incubation temperature (Mowlds and Kavanagh,

2008; Wojda *et al.*, 2009) and nutrition (Banville *et al.*, 2012). Therefore, the standardization of such conditions would be highly beneficial. Due to the relatively higher variability when compared to murine models, a larger number of individuals are needed to reach statistically significant results with *G. mellonella*. In spite of that, this model still constitutes a faster, cheaper and easier way of evaluating fungal virulence and antifungal drug activity and toxicity than conventional mammalian models. This is therefore a promising alternative for initial *in vivo* screenings which can narrow down the list of antifungal drugs, or fungal strains, that should subsequently be validated in mammals.

Other mini-host models

Invertebrates

Drosophila melanogaster

Drosophila melanogaster and *C. elegans* are among the most studied model systems in terms of immunity and interaction with pathogens. Even though the wild-type fly is usually resistant to fungal infections, several mutated fly lines have been constructed which are susceptible to fungal pathogens (Chamilos *et al.*, 2007a, 2009; Glittenberg *et al.*, 2011b; Lionakis and Kontoyiannis, 2012). A large number of these lines are mutated for genes involved in immune pathways, particularly the toll pathway. For that reason using different mutant strains of the fly allows the study of innate immune defence mechanisms against pathogenic fungi (Lionakis, 2011). A large panel of pathogenic fungi including *Candida, Aspergillus, Fusarium, Cryptococcus* or *Mucor* was studied in the fly model (Chamilos *et al.*, 2006; Kirkpatrick *et al.*, 2012; Lamaris *et al.*, 2007; Lionakis and Kontoyiannis, 2012). Several methods of inoculation exist, including injection of fungal cells trough the dorsal side of the thorax directly in the body cavity, ingestion of fungi necessitating an incubation of 2 days in vials containing *Candida* culture mixed with classical food plates or YEDP agar plates (Chamilos *et al.*, 2006; Glittenberg *et al.*, 2011a), or rolling the anaesthetized insect over lawns of fungal cells yeast, moulds or spores (reviewed in Chamilos *et al.*, 2007b). After infection, flies are usually incubated at 29°C which is the optimal temperature for susceptibility to fungal challenge (Chamilos *et al.*, 2006), and are transferred into new vials every 2 days. The most common read-out is survival over a one week period.

Direct injection in the body cavity completely bypasses the natural route of infection, which is similar to systemic infection in mice, and does not reflect a physiological event. Other routes of infection, such as ingestion or rolling, are the preferred methods to study the initial steps of infection and to elucidate the host–fungi interaction. Indeed, Chamilos *et al.* (2006) reported the importance of *alb1* in *A. fumigatus* virulence using the rolling or ingestion methods, but were unable to demonstrate it using the injection methods. Glittenberg *et al.* (2011a) developed an intestinal model of colonization to study the importance of the normal gut flora on candidiasis development. In general, direct injection in flies offers more reproducible results than the ingestion or rolling models, probably due to the fact that standardizing the inoculum is difficult.

As in mice, the genetic backgrounds of the flies or the methods employed for immune suppression greatly influences the results of the virulence experiments (Glittenberg *et al.*, 2011a,b; Quintin *et al.*, 2013). Namely, toll pathways seem to be very important in the control of infection, resulting in some phenotypes being visible only with Toll-deficient flies, and not with other immune-deficient lines such as spz (set for Spätzle) (Glittenberg *et al.*, 2011b; Quintin *et al.*, 2013).

To study stage-specific importance of some fungal genes across the infection course, conditional expression of fungal genes can be implemented in drosophila. Like in mice, a tet-inducible promoter can be used to control gene expression through addition of doxycycline in the flies' food (Chamilos *et al.*, 2006). This is a powerful tool, especially in the rolling or ingestion models, in which the fungi have to cross several barriers, using different virulence factors at different stages.

Drosophila has also been used in pharmacology (Chamilos *et al.*, 2006; Lionakis, 2011; Lionakis *et al.*, 2005). Drugs are usually injected dorsally. However, precise pharmacological analyses are difficult to carry out since the exact amount of drug administered is hard to monitor. Moreover, repeated injections of drugs are deleterious for the animals, limiting the possible assays.

Caenorhabditis elegans

The 1-mm-long round worm *Caenorhabditis elegans* is used as an alternative model to study fungal virulence and antifungal properties. It has been used essentially to study *Cryptococcus* virulence (Muhammed *et al.*, 2012a; Mylonakis *et al.*, 2002), but also to a lesser extent to study *C. albicans*, *C. glabrata* and *H. capsulatum* (Breger *et al.*, 2007; Mylonakis *et al.*, 2002; Pukkila-Worley *et al.*, 2011; Thakur *et al.*, 2008).

Concerning *Cryptococcus* infection, *C. elegans* is inoculated by being laid on a lawn of fungal cells which it will eat. *C. neoformans* is then able to accumulate in the worm intestine and kill it. Two different read-outs can be used: survival or progeny permissive assay. In both cases the worms have to be synchronized in the L4 stage of development. Survival assay is usually performed using worms that are unable to produce progeny (*glp4;sek1* strain). Viability of the worm is verified under a dissecting microscope after stimulation with a pike every 12 to 24 hours. In the case of progeny permissive assay, wild-type worms are infected and transferred every two days to fresh media containing a lawn of *E. coli* as food source. This passage also results in the separation of infected worms from their growing progeny. Interestingly, less virulent strains of *C. neoformans* induces the worm to produce more progeny than those infected with a wild-type strain of the fungus. This allows a rapid screening of fungal strains mutated in putative virulence factors (Muhammed *et al.*, 2012a).

C. albicans survival assay is usually performed in liquid media. After infecting the worm on a lawn of *Candida* cells, they are transferred to a 96 wells plate and incubated for 5 days, during which survival is scored (Muhammed *et al.*, 2012a). Antifungal activity assays are performed in a similar way, adding antifungal compounds to the 96-well plate before transfer of the infected worms enabling the rapid screening of compound libraries (Anastassopoulou *et al.*, 2011; Pukkila-Worley *et al.*, 2009). The toxicity of the drugs can be assessed in parallel by simply using non-infected worms and scoring their survival rate in the presence or absence of drugs.

Zebrafish (*Danio rerio*)

Zebrafishes are the only vertebrate mini-host model used for fungal infections. It was developed to mainly study *Candida* virulence (Brothers and Wheeler, 2012; Chao *et al.*, 2010; Chen *et al.*, 2013; Kuo *et al.*, 2013). Zebrafishes have an innate and adaptive immune system (Warr, 1997) and genetic tools allow its manipulation to obtain immune deficient strains which represents an advantage because it mimics human infections. On the other hand, zebrafish grow at 28–29°C, which is significantly different from human conditions.

The adults and the embryos of zebrafishes constitute two distinct models, and embryos up to 120 hours post-fertilization (hpf) are exempted from the European animal experiment regulation (The European Parliament and the Council of the European Union, 2010). In addition, adult zebrafishes can be infected systemically by intraperitoneal injection (Chao *et al.*, 2010). Read-out of virulence can be determined by survival assay, scoring mortality every 3 hours during 1 week, fungal burden determination, or histopathology. Other models are performed in embryos (Fig. 13.5). Systemic infection is performed through direct injection of *Candida* cells in the hindbrain ventricle of the larvae aged of around 30 hpf, leading to death (Brothers *et al.*, 2011; Chao *et al.*, 2010). A mucosal infection model was also developed through infection of the swim bladder by immersion of 72 hpf larvae in a *C. albicans* solution. After 1 day, the swim-bladder of the 96 hpf larvae naturally inflates and is infected by *C. albicans* (Gratacap *et al.*, 2013). Owing to the transparency of the embryo, Candida infection can be followed in real time, often using fluorescent fungi cells (Brothers *et al.*, 2013; Brothers *et al.*, 2011; Gratacap *et al.*, 2013).

Zebrafish are also used for toxicity assays with antifungal drugs. For such tests, zebrafish embryos aged from fertilization to hatching (around 72 hpf) are used (Hermsen *et al.*, 2012a,b). In

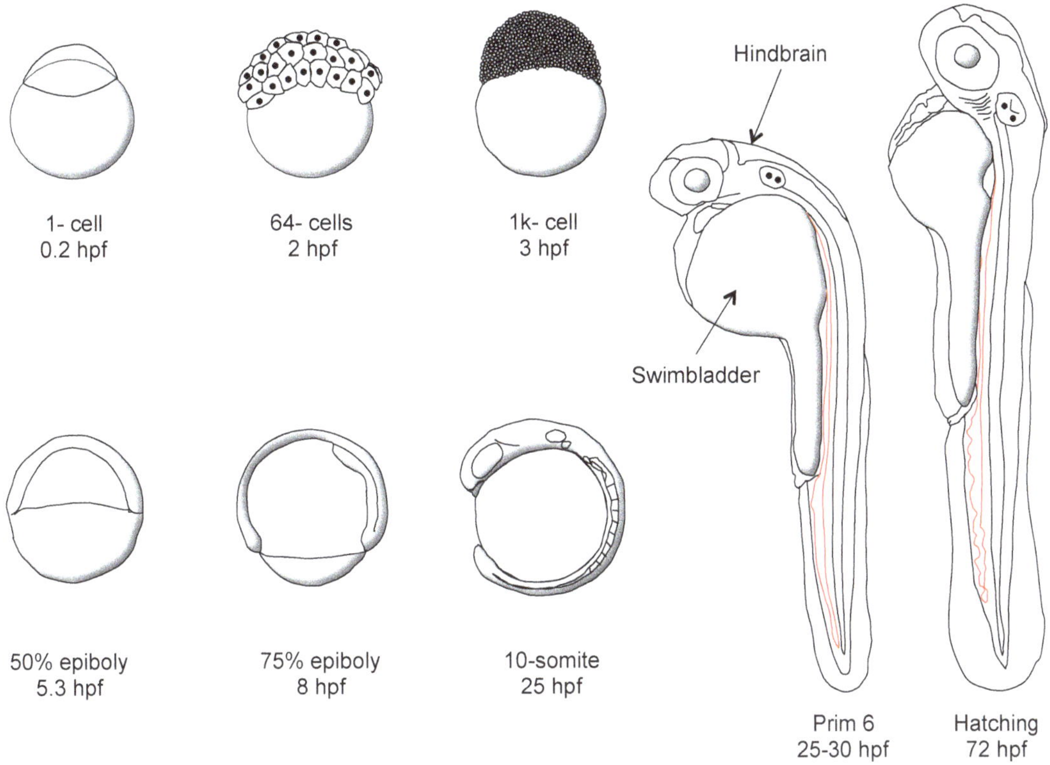

Figure 13.5 Different stages of zebrafish development. Zygote period (0–0.75 hours); cleavage period (0.75–2.25 hours); blastula period (2.25–5.25 hours); gastrula period (5.25–10 hours); segmentation period (10–24 hours); pharyngula period (24–48 hours); hatching period (48–72 hours). hpf, hours post-fertilization. For an extensive description of zebrafish development stages see (Kimmel *et al.*, 1995).

these assays, fertilized eggs between the 4- and 64-cell stages are incubated in presence of drugs for several hours. Development of the fish is then scored until 72 hpf.

Single cell models

Some models have also been implemented in very simple single cell organisms such as amoeba or the chick egg.

Eggs

For a long time, the chicken egg model of infection was used to study the virulence properties of viruses and bacteria (Adam *et al.*, 2002; Crespo *et al.*, 2009; Guy, 2008). Ten years ago, the chicken egg model was adapted to *Candida* infection (Gow *et al.*, 2003). Recently this model was successfully further developed for pathogenic fungal infection including *Candida* species and *Aspergillus fumigatus* (Jacobsen *et al.*, 2010b, 2011, 2012).

Amoeba

Amoeba has been used to study phagocytic process as they were shown to be very similar to mammalian macrophages (Cosson and Soldati, 2008). At least two species of amoeba, *Acanthamoeba castellanii* and *Dictyostelium discoideum*, have been used to study their interaction with pathogenic fungi including *C. neoformans*, *Blastomyces dermatitidis*, *Sporothrix schenckii*, and *Histoplasma capsulatum* (Fuchs and Mylonakis, 2006; Levitz, 2001; Steenbergen and Casadevall, 2003). In our laboratory, we have tried to develop a model to perform *C. albicans* virulence assay using amoeba as a host. Our investigations demonstrate that *A. castellanii* but not *D. discoideum* allow discriminating *C. albicans* mutant strains having a decrease in host colonization properties in mice. Our assay was based on the measurement of the phagocytic plaque in the feeding-bacteria layer, as described previously for bacteria (Vandeputte *et al.*, 2011), and growth of the fungus (Cosson and Soldati, 2008, and personal communication).

Conclusions

When using animal models to determine the virulence of a given strain of a fungal species, it is important to keep in mind that different phenotypes can be observed depending on the model that is used. As a result, the concept of virulence of a strain is difficult to define because it depends on the environment where it is analysed. The same problems apply to drug activity investigations. This highlights the importance of choosing the most appropriate animal model for each situation, since it will define the conditions with which the microorganism will interact and, therefore, its reaction (virulence and response to drug treatment). A consequence of using different animal models is that we can also expect differences between the behaviour of fungal cells in animals and in humans emphasizing the limitations of these models. Nevertheless, animal models have contributed greatly to our understanding of most fungal diseases. They allowed us to develop antifungal drug therapies presently being used. Finally animal models will continue to be an essential part of scientific research to improve and further develop antifungal therapies.

References

Abruzzo, G.K., Gill, C.J., Flattery, A.M., Kong, L., Leighton, C., Smith, J.G., Pikounis, V.B., Bartizal, K., and Rosen, H. (2000). Efficacy of the echinocandin caspofungin against disseminated aspergillosis and candidiasis in cyclophosphamide-induced immunosuppressed mice. Antimicrob. Agent Chemother. *44*, 2310–2318.

Adam, R., Mussa, S., Lindemann, D., Oelschlaeger, T.A., Deadman, M., Ferguson, D.J., Moxon, R., and Schroten, H. (2002). The avian chorioallantoic membrane in ovo--a useful model for bacterial invasion assays. Int. J. Med. Microbiol. *292*, 267–275.

Anand, V.K., Alemar, G., and Griswold, J.A., Jr. (1992). Intracranial complications of mucormycosis: an experimental model and clinical review. Laryngoscope *102*, 656–662.

Anastassopoulou, C.G., Fuchs, B.B., and Mylonakis, E. (2011). *Caenorhabditis elegans*-based model systems for antifungal drug discovery. Curr. Pharm. Des. *17*, 1225–1233

Andes, D. (2003a). *In vivo* pharmacodynamics of antifungal drugs in treatment of candidiasis. Antimicrob. Agent Chemother. *47*, 1179–1186.

Andes, D. (2003b). Pharmacokinetics and pharmacodynamics in the development of antifungal compounds. Curr. Opin. Investig. Drugs *4*, 991–998.

Andes, D., and Safdar, N. (2005). Efficacy of micafungin for the treatment of candidemia. Eur. J. Clin. Microbiol. Infect. Dis. *24*, 662–664.

Andes, D., Marchillo, K., Lowther, J., Bryskier, A., Stamstad, T., and Conklin, R. (2003a). *In vivo* pharmacodynamics of HMR 3270, a glucan synthase inhibitor, in a murine candidiasis model. Antimicrob. Agent Chemother. *47*, 1187–1192.

Andes, D., Marchillo, K., Stamstad, T., and Conklin, R. (2003b). *In vivo* pharmacodynamics of a new triazole, ravuconazole, in a murine candidiasis model. Antimicrob. Agent Chemother. *47*, 1193–1199.

Andes, D., Marchillo, K., Stamstad, T., and Conklin, R. (2003c). *In vivo* pharmacokinetics and pharmacodynamics of a new triazole, voriconazole, in a murine candidiasis model. Antimicrob. Agent Chemother. *47*, 3165–3169.

Andes, D., Nett, J., Oschel, P., Albrecht, R., Marchillo, K., and Pitula, A. (2004). Development and characterization of an *in vivo* central venous catheter *Candida albicans* biofilm model. Infect. Immun. *72*, 6023–6031.

Andes, D., Diekema, D., Pfaller, M., Marchillo, K., and Bohrmueller, J. (2008a). *In vivo* pharmacodynamic target investigation for micafungin against *Candida albicans* and *C. glabrata* in a neutropenic murine candidiasis model. Antimicrob. Agent Chemother. *52*, 3497–3503.

Andes, D., Diekema, D., Pfaller, M., Prince, R., Marchillo, K., Ashbeck, J., and Hou, J. (2008b). *In vivo* pharmacodynamic characterization of anidulafungin in a neutropenic murine candidiasis model. Antimicrob. Agent Chemother. *52*, 539–550.

Andes, D., Diekema, D.J., Pfaller, M.A., Bohrmuller, J., Marchillo, K., and Lepak, A. (2010). *In vivo* comparison of the pharmacodynamic targets for echinocandin drugs against *Candida* species. Antimicrob. Agents Chemother. *54*, 2497–2506.

Antachopoulos, C., and Walsh, T.J. (2012). Immunotherapy of *Cryptococcus* infections. Clin. Microbiol. Infect. *18*, 126–133.

Antachopoulos, C., Katragkou, A., and Roilides, E. (2012). Immunotherapy against invasive mold infections. Immunotherapy *4*, 107–120.

Araujo, R., Carneiro, A., Costa-Oliveira, S., Pina-Vaz, C., Rodrigues, A.G., and Guimaraes, J.E. (2008). Fungal infections after haematology unit renovation: evidence of clinical, environmental and economical impact. Eur. J. Haematol. *80*, 436–443.

Arendrup, M., Horn, T., and Frimodt-Moller, N. (2002). *In vivo* pathogenicity of eight medically relevant *Candida* species in an animal model. Infection *30*, 286–291.

Asmundsdottir, L., Erlendsdottir, H., Agnarsson, B., and Gottfredsson, M. (2009). The importance of strain variation in virulence of *Candida dubliniensis* and *Candida albicans*: results of a blinded histopathological study of invasive candidiasis. Clin. Microbiol. Infect. *15*, 576–585.

Banville, N., Browne, N., and Kavanagh, K. (2012). Effect of nutrient deprivation on the susceptibility of *Galleria mellonella* larvae to infection. Virulence 3.

Barchiesi, F., Schimizzi, A.M., Caselli, F., Novelli, A., Fallani, S., Giannini, D., Arzeni, D., Di Cesare, S., Di Francesco, L.F., Fortuna, M., *et al.* (2000). Interactions between triazoles and amphotericin B against *Cryptococcus neoformans*. Antimicrob. Agent Chemother. *44*, 2435–2441.

Barchiesi, F., Spreghini, E., Santinelli, A., Fothergill, A.W., Fallani, S., Manso, E., Pisa, E., Giannini, D., Novelli, A., Cassetta, M.I., *et al.* (2005). Efficacy of caspofungin against *Aspergillus terreus*. Antimicrob. Agent Chemother. *49*, 5133–5135.

Ben-Ami, R., and Kontoyiannis, D.P. (2012). A nonlethal murine cutaneous model of invasive aspergillosis. Methods Mol. Biol. *845*, 569–582.

Bergin, D., Brennan, M., and Kavanagh, K. (2003). Fluctuations in haemocyte density and microbial load may be used as indicators of fungal pathogenicity in larvae of *Galleria mellonella*. Microbes Infect. *5*, 1389–1395.

Bergin, D., Murphy, L., Keenan, J., Clynes, M., and Kavanagh, K. (2006). Pre-exposure to yeast protects larvae of *Galleria mellonella* from a subsequent lethal infection by *Candida albicans* and is mediated by the increased expression of antimicrobial peptides. Microbes Infect. *8*, 2105–2112.

Berretta, A.A., de Castro, P.A., Cavalheiro, A.H., Fortes, V.S., Bom, V.P., Nascimento, A.P., Marquele-Oliveira, F., Pedrazzi, V., Ramalho, L.N., and Goldman, G.H. (2013). Evaluation of mucoadhesive gels with propolis (EPP-AF) in preclinical treatment of candidiasis vulvovaginal infection. Evid. Based Complement. Alternat. Med. *2013*, 641480.

Bertini, A., De Bernardis, F., Hensgens, L.A., Sandini, S., Senesi, S., and Tavanti, A. (2013). Comparison of *Candida parapsilosis*, *Candida orthopsilosis*, and *Candida metapsilosis* adhesive properties and pathogenicity. Int. J. Med. Microbiol. *303*, 98–103.

Blasi, E., Barluzzi, R., Mazzolla, R., Mosci, P., and Bistoni, F. (1992). Experimental model of intracerebral infection with *Cryptococcus neoformans*: roles of phagocytes and opsonization. Infect. Immun. *60*, 3682–3688.

Blasi, E., Mazzolla, R., Pitzurra, L., Barluzzi, R., and Bistoni, F. (1993). Protective effect of picolinic acid on mice intracerebrally infected with lethal doses of *Candida albicans*. Antimicrob. Agent Chemother. *37*, 2422–2426.

Blasi, E., Bartoli, A., Barluzzi, R., Mazzolla, R., and Bistoni, F. (1994). Pattern of cytokine gene expression in brains of mice protected by picolinic acid against lethal intracerebral infection with *Candida albicans*. J. Neuroimmunol. *52*, 205–213.

Bougnoux, M.E., Diogo, D., Francois, N., Sendid, B., Veirmeire, S., Colombel, J.F., Bouchier, C., Van Kruiningen, H., d'Enfert, C., and Poulain, D. (2006). Multilocus sequence typing reveals intrafamilial transmission and microevolutions of *Candida albicans* isolates from the human digestive tract. J. Clin. Microbiol. *44*, 1810–1820.

Breger, J., Fuchs, B.B., Aperis, G., Moy, T.I., Ausubel, F.M., and Mylonakis, E. (2007). Antifungal chemical compounds identified using a C. elegans pathogenicity assay. PLoS Pathog. *3*, e18.

Brennan, M., Thomas, D.Y., Whiteway, M., and Kavanagh, K. (2002). Correlation between virulence of *Candida albicans* mutants in mice and *Galleria mellonella* larvae. FEMS Immunol. Med. Microbiol. *34*, 153–157.

Brizendine, K.D., Baddley, J.W., and Pappas, P.G. (2011). Pulmonary cryptococcosis. Semin. Respir. Crit. Care Med. 32, 727–734.

Brothers, K.M., and Wheeler, R.T. (2012). Non-invasive imaging of disseminated candidiasis in zebrafish larvae. J. Vis. Exp. *July 30* (65), 4051.

Brothers, K.M., Newman, Z.R., and Wheeler, R.T. (2011). Live imaging of disseminated candidiasis in Zebrafish reveals role of phagocyte oxidase in limiting filamentous growth. Eukaryot. Cell *10*, 932.

Brothers, K.M., Gratacap, R.L., Barker, S.E., Newman, Z.R., Norum, A., and Wheeler, R.T. (2013). NADPH oxidase-driven phagocyte recruitment controls *Candida albicans* filamentous growth and prevents mortality. PLoS Pathog. *9*, e1003634.

Brown, A.J.P., Odds, F.C., and Gow, N.A.R. (2007). Infection-related gene expression in *Candida albicans*. Curr. Opin. Microbiol. *10*, 307–313.

Brown, S.E., Howard, A., Kasprzak, A.B., Gordon, K.H., and East, P.D. (2009). A peptidomics study reveals the impressive antimicrobial peptide arsenal of the wax moth *Galleria mellonella*. Insect Biochem. Mol. Biol. *39*, 792–800.

Browne, N., Heelan, M., and Kavanagh, K. (2013). An analysis of the structural and functional similarities of insect hemocytes and mammalian phagocytes. Virulence *4*, 597–603.

Calderon, L., Williams, R., Martinez, M., Clemons, K., and Stevens, D. (2003). Genetic susceptibility to vaginal candidiasis. Med. Mycol. *41*, 143–147.

Cambier, L., Weatherspoon, A., Defaweux, V., Bagut, E.T., Heinen, M.P., Antoine, N., and Mignon, B. (2013). Assessment of the cutaneous immune response during *Arthroderma benhamiae* and *Arthroderma vanbreuseghemii* infection using an experimental mouse model. Br. J. Dermatol. *170*, 625–633.

Cantorna, M.T., and Balish, E. (1990). Mucosal and systemic candidiasis in congenitally immunodeficient mice. Infect. Immun. *58*, 1093–1100.

Capilla, J., Maffei, C.M., Clemons, K.V., Sobel, R.A., and Stevens, D.A. (2006). Experimental systemic infection with *Cryptococcus neoformans* var. *grubii* and *Cryptococcus gattii* in normal and immunodeficient mice. Med. Mycol. *44*, 601–610.

Capilla, J., Clemons, K.V., and Stevens, D.A. (2007). Animal models: an important tool in mycology. Med. Mycol. *45*, 657–684.

Carneiro, H.A., Mavrakis, A., and Mylonakis, E. (2011). *Candida peritonitis*: an update on the latest research and treatments. World J. Surg. *35*, 2650–2659.

Carroll, S.F., Guillot, L., and Qureshi, S.T. (2007). Mammalian model hosts of cryptococcal infection. Comp. Med. *57*, 9–17.

Centeno-Lima, S., Silveira, H., Casimiro, C., Aguiar, P., and do Rosario, V.E. (2002). Kinetics of cytokine expression in mice with invasive aspergillosis: lethal infection and protection. FEMS Immunol. Med. Microbiol. *32*, 167–173.

Chamilos, G., Lionakis, M.S., Lewis, R.E., López-Ribot, J.L., Saville, S.P., Albert, N.D., Halder, G., and Kontoyiannis, D.P. (2006). *Drosophila melanogaster* as a facile model for large-scale studies of virulence mechanisms and antifungal drug efficacy in Candida species. J. Infect. Dis. *193*, 1014–1022.

Chamilos, G., Lewis, R.E., Albert, N., and Kontoyiannis, D.P. (2007a). Paradoxical effect of Echinocandins across Candida species *in vitro:* evidence for echinocandin-specific and candida species-related differences. Antimicrob. Agents Chemother. *51*, 2257–2259.

Chamilos, G., Lionakis, M.S., Lewis, R.E., and Kontoyiannis, D.P. (2007b). Role of mini-host models in the study of medically important fungi. Lancet Infect. Dis. *7*, 42–55.

Chamilos, G., Nobile, C.J., Bruno, V.M., Lewis, R.E., Mitchell, A.P., and Kontoyiannis, D.P. (2009). *Candida albicans* Cas5, a regulator of cell wall integrity, is required for virulence in murine and toll mutant fly models. J. Infect. Dis. *200*, 152–157.

Chandra, J., Long, L., Ghannoum, M.A., and Mukherjee, P.K. (2011). A rabbit model for evaluation of catheter-associated fungal biofilms. Virulence *2*, 466–474.

Chao, C., Hsu, P., Jen, C., Chen, I., Wang, C., Chan, H., Tsai, P., Tung, K., Lan, C., and Chuang, Y. (2010). Zebrafish as a model host for *Candida albicans* infection. Infect Immun. *78*, 2512–2521.

Chen, C., Pande, K., French, S.D., Tuch, B.B., and Noble, S.M. (2011). An iron homeostasis regulatory circuit with reciprocal roles in *Candida albicans* commensalism and pathogenesis. Cell Host Microbes *10*, 118–135.

Chen, C., and Noble, S.M. (2012). Post-transcriptional regulation of the Sef1 transcription factor controls the virulence of *Candida albicans* in its mammalian host. PLoS Pathog. *8*, e1002956.

Chen, Y.Y., Chao, C.C., Liu, F.C., Hsu, P.C., Chen, H.F., Peng, S.C., Chuang, Y.J., Lan, C.Y., Hsieh, W.P., and Wong, D.S. (2013). Dynamic transcript profiling of *Candida albicans* infection in zebrafish: a pathogen–host interaction study. PLoS One *8*, e72483.

Cheng, S., Clancy, C.J., Xu, W., Schneider, F., Hao, B., Mitchell, A.P., and Nguyen, M.H. (2013). Profiling of *Candida albicans* gene expression during intra-abdominal candidiasis identifies biologic processes involved in pathogenesis. J. Infect. Dis. *208*, 1529–1537.

Cheng, S., Clancy, C.J., Hao, B., Filler, S., Solis, N., and Nguyen, M. (2014). Insight into the pathogenesis of *Candida glabrata* using murine intra-abdominal candidiasis (IAC) model. Presentation at: 12th American Society for Microbiology (ASM) Conference on Candida and Candiasis, 26–30th March (New Orleans, LA, USA).

Cheng, S.C., Joosten, L.A., Kullberg, B.J., and Netea, M.G. (2012). Interplay between *Candida albicans* and the mammalian innate host defense. Infect. Immun. *80*, 1304–1313.

Chiller, T.M., Luque, J.C., Sobel, R.A., Farrokhshad, K., Clemons, K.V., and Stevens, D.A. (2002). Development of a murine model of cerebral aspergillosis. J. Infect. Dis. *186*, 574–577.

Clemons, K.V., and Stevens, D.A. (2000). Treatment of orogastrointestinal candidosis in SCID mice with fluconazole alone or in combination with recombinant granulocyte colony-stimulating factor or interferon-gamma. Med. Mycol. *38*, 213–219.

Clemons, K.V., and Stevens, D.A. (2001). Efficacy of ravuconazole in treatment of mucosal candidosis in SCID mice. Antimicrob. Agent Chemother. *45*, 3433–3436.

Clemons, K.V., and Stevens, D.A. (2012). Orogastrointestinal model of mucosal and disseminated candidiasis. Methods Mol. Biol. *845*, 557–567.

Clemons, K.V., Azzi, R., and Stevens, D.A. (1996). Experimental systemic cryptococcosis in SCID mice. J. Med. Vet. Mycol. *34*, 331–335.

Clemons, K.V., Martinez, M., Calderon, L., and Stevens, D.A. (2002). Efficacy of ravuconazole in treatment of systemic murine histoplasmosis. Antimicrob. Agent Chemother. *46*, 922–924.

Clemons, K.V., Spearow, J.L., Parmar, R., Espiritu, M., and Stevens, D.A. (2004). Genetic susceptibility of mice to *Candida albicans* vaginitis correlates with host estrogen sensitivity. Infect. Immun. *72*, 4878–4880.

Clemons, K.V., Gonzalez, G.M., Singh, G., Imai, J., Espiritu, M., Parmar, R., and Stevens, D.A. (2006). Development of an orogastrointestinal mucosal model of candidiasis with dissemination to visceral organs. Antimicrob. Agent Chemother. *50*, 2650–2657.

Clemons, K.V., Martinez, M., Axelsen, M., Thiel, S., and Stevens, D.A. (2011). Efficacy of recombinant human mannose binding lectin alone and in combination with itraconazole against murine *Candida albicans* vaginitis. Immunol. Invest. *40*, 553–568.

Cole, G.T., Seshan, K.R., Phaneuf, M., and Lynn, K.T. (1991). Chlamydospore-like cells of *Candida albicans* in the gastrointestinal tract of infected, immunocompromised mice. Can. J. Microbiol. *37*, 637–646.

Cole, G.T., Halawa, A.A., and Anaissie, E.J. (1996). The role of the gastrointestinal tract in hematogenous candidiasis: from the laboratory to the bedside. Clin. Infect. Dis. *22* (Suppl. 2), S73–88.

Coleman, J.J., Muhammed, M., Kasperkovitz, P.V., Vyas, J.M., and Mylonakis, E. (2011). Fusarium pathogenesis investigated using *Galleria mellonella* as a heterologous host. Fungal Biol. *115*, 1279–1289.

Cosson, P., and Soldati, T. (2008). Eat, kill or die: when amoeba meets bacteria. Curr. Opin. Microbiol. *11*, 271–276.

Costa, A.C., Pereira, C.A., Junqueira, J.C., and Jorge, A.O. (2013). Recent mouse and rat methods for the study of experimental oral candidiasis. Virulence *4*, 391–399.

Cotter, G., Doyle, S., and Kavanagh, K. (2000). Development of an insect model for the *in vivo* pathogenicity testing of yeasts. FEMS Immunol. Med. Microbiol. *27*, 163–169.

Cowen, L.E., Singh, S.D., Köhler, J.R., Collins, C., Zaas, A.K., Schell, W.A., Aziz, H., Mylonakis, E., Perfect, J.R., Whitesell, L., *et al.* (2009). Harnessing Hsp90 function as a powerful, broadly effective therapeutic strategy for fungal infectious disease. Proc. Natl. Acad. Sci. U.S.A. *106*, 2818–2823.

Crespo, R., Shivaprasad, H.L., Franca, M., and Woolcock, P.R. (2009). Isolation and distribution of West Nile virus in embryonated chicken eggs. Avian Dis. *53*, 608–612.

De Bernardis, F., Cassone, A., Sturtevant, J., and Calderone, R. (1995). Expression of *Candida albicans* SAP1 and SAP2 in experimental vaginitis. Infect. Immun. *63*, 1887–1892.

De Bernardis, F., Lorenzini, R., Cassone, A., Zak, O., and Sande, M. (1999). Rat model of Candida vaginal infection. In Handbook of Animals Models of Infection, O. Zak, M.A. Sande, eds. (Academic Press, San Diego, CA, USA), pp. 735–740.

Domer, J.E., and Moser, S.A. (1978). Experimental murine candidiasis: cell-mediated immunity after cutaneous challenge. Infect. Immun. *20*, 88–98.

Domer, J.E., Lyon, F.L., and Murphy, J.W. (1983). Cellular immunity in a cutaneous model of cryptococcosis. Infect. Immun. *40*, 1052–1059.

Dunphy, G.B., Morton, D.B., Kropinski, A., and Chadwick, J.M. (1986). Pathogenicity of lipopolysaccharide mutants of *Pseudomonas aeruginosa* for larvae of *Galleria mellonella*: bacterial properties associated with virulence. J. Invertebr. Pathol. *47*, 48–55.

Dunphy, G.B., Oberholzer, U., Whiteway, M., Zakarian, R., and Boomer, I. (2003). Virulence of *Candida albicans* mutants toward larval *Galleria mellonella* (Insecta, Lepidoptera, Galleridae). Can. J. Microbiol. *49*, 514–524.

Fallon, J.P., Reeves, E.P., and Kavanagh, K. (2011). The *Aspergillus fumigatus* toxin fumagillin suppresses the immune response of *Galleria mellonella* larvae by inhibiting the action of haemocytes. Microbiology *157*, 1481–1488.

Felk, A., Kretschmar, M., Albrecht, A., Schaller, M., Beinhauer, S., Nichterlein, T., Sanglard, D., Korting, H., Schafer, W., and Hube, B. (2002). *Candida albicans* hyphal formation and the expression of the Efg1-regulated proteinases Sap4 to Sap6 are required for the invasion of parenchymal organs. Infect. Immun. *70*, 3689–3700.

Ferrari, S., Sanguinetti, M., De Bernardis, F., Torelli, R., Posteraro, B., Vandeputte, P., and Sanglard, D. (2011). Loss of mitochondrial functions associated with azole resistance in *Candida glabrata* also results in enhanced virulence in mice. Antimicrob. Agents Chemother. *55*, 1852–1860.

Fidel, P.L., Jr., Lynch, M.E., and Sobel, J.D. (1993). Candida-specific Th1-type responsiveness in mice with experimental vaginal candidiasis. Infect. Immun. *61*, 4202–4207.

Fidel, P.L., Jr., Cutright, J.L., Tait, L., and Sobel, J.D. (1996). A murine model of *Candida glabrata* vaginitis. J. Infect. Dis. *173*, 425–431.

Flattery, A.M., Abruzzo, G.K., Gill, C.J., Smith, J.G., and Bartizal, K. (1996). New model of oropharyngeal and gastrointestinal colonization by *Candida albicans* in CD4+ T-cell-deficient mice for evaluation of antifungal agents. Antimicrob. Agent Chemother. *40*, 1604–1609.

Foldvari, M., Radhi, J., Yang, G., He, Z., Rennie, R., and Wearley, L. (2000). Acute vaginal candidosis model in the immunocompromized rat to evaluate delivery systems for antimycotics. Mycoses *43*, 393–401.

Forster, R.K., and Rebell, G. (1975). Animal model of *Fusarium solani* keratitis. Am. J. Ophthalmol. *79*, 510–515.

Fuchs, B.B., and Mylonakis, E. (2006). Using non-mammalian hosts to study fungal virulence and host defense. Curr. Opin. Microbiol. *9*, 346–351.

Fuchs, B., Eby, J., Nobile, C., El Khoury, J., Mitchell, A., and Mylonakis, E. (2010a). Role of filamentation in *Galleria mellonella* killing by *Candida albicans*. Microbes Infect. *12*, 488–496.

Fuchs, B.B., O'Brien, E., Khoury, J.B., and Mylonakis, E. (2010b). Methods for using *Galleria mellonella* as a model host to study fungal pathogenesis. Virulence *1*, 475–482.

García-Rodas, R., Casadevall, A., Rodriguez-Tudela, J.L., Cuenca-Estrella, M., and Zaragoza, O. (2011). *Cryptococcus neoformans* capsular enlargement and cellular gigantism during *Galleria mellonella* infection. PLoS One *6*, e24485.

Geiger, A.M., Foxman, B., and Sobel, J.D. (1995). Chronic vulvovaginal candidiasis: characteristics of women with *Candida albicans*, *C. glabrata* and no candida. Genitourin Med. *71*, 304–307.

Giger, D.K., Domer, J.E., and McQuitty, J.T., Jr. (1978). Experimental murine candidiasis: pathological and immune responses to cutaneous inoculation with *Candida albicans*. Infect. Immun. *19*, 499–509.

Gladiator, A., Wangler, N., Trautwein-Weidner, K., and LeibundGut-Landmann, S. (2012). Cutting edge: IL-17-secreting innate lymphoid cells are essential for host defense against fungal infection. J. Immunol. *190*, 521–525.

Glittenberg, M.T., Kounatidis, I., Christensen, D., Kostov, M., Kimber, S., Roberts, I., and Ligoxygakis, P. (2011a). Pathogen and host factors are needed to provoke a systemic host response to gastrointestinal infection of *Drosophila* larvae by *Candida albicans*. Disease Models & mechanisms *4*, 515–525.

Glittenberg, M.T., Silas, S., MacCallum, D.M., Gow, N.A.R., and Ligoxygakis, P. (2011b). Wild-type *Drosophila melanogaster* as an alternative model system for investigating the pathogenicity of *Candida albicans*. Dis. Model Mech. *4*, 504–514.

Gonzalez, G.M., Portillo, O.J., Uscanga, G.I., Andrade, S.E., Robledo, M., Rodriguez, C., and Elizondo, M. (2009). Therapeutic efficacy of voriconazole against a fluconazole-resistant *Candida albicans* isolate in a vaginal model. J. Antimicrob. Chemother. *64*, 571–573.

Gonzalez, G.M., Marquez, J., Trevino-Rangel Rde, J., Palma-Nicolas, J.P., Garza-Gonzalez, E., Cecenas, L.A., and Gerardo Gonzalez, J. (2013). Murine model of disseminated fusariosis: evaluation of the fungal burden by traditional CFU and quantitative PCR. Mycopathologia *176*, 219–224.

Gordee, R.S., Zeckner, D.J., Ellis, L.F., Thakkar, A.L., and Howard, L.C. (1984). In vitro and *in vivo* anti-Candida activity and toxicology of LY121019. J. Antibiot. (Tokyo) *37*, 1054–1065.

Gow, N., Knox, Y., Munro, C., and Thompson, W. (2003). Infection of chick chorioallantoic membrane (CAM) as a model for invasive hyphal growth and pathogenesis of *Candida albicans*. Med. Mycol. *41*, 331–338.

Gratacap, R.L., Rawls, J.F., and Wheeler, R.T. (2013). Mucosal candidiasis elicits NF-B activation, proinflammatory gene expression and localized neutrophilia in zebrafish. Dis. Model Mech. *6*, 1260–1270.

Graybill, J.R. (2000). The role of murine models in the development of antifungal therapy for systemic mycoses. Drug Resist. Updat. *3*, 364–383.

Graybill, J., Bocanegra, R., Luther, M., Fothergill, A., and Rinaldi, M. (1997). Treatment of murine *Candida krusei* or *Candida glabrata* infection with L- 743,872. Antimicrob. Agent Chemother. *41*, 1937–1939.

Graybill, J., Najvar, L., Bocanegra, R., Hector, R., and Luther, M. (1998a). Efficacy of Nikkomycin Z in the treatment of murine histoplasmosis. Antimicrob. Agent Chemother. *42*, 2371–2374.

Graybill, J.R., Montalbo, E., Kirkpatrick, W.R., Luther, M.F., Revankar, S.G., and Patterson, T.F. (1998b). Fluconazole versus *Candida albicans*: a complex relationship. Antimicrob. Agent Chemother. *42*, 2938–2942.

Groll, A.H., and Walsh, T.J. (2006). Antifungal efficacy and pharmacodynamics of posaconazole in experimental models of invasive fungal infections. Mycoses *49* (Suppl. 1), 7–16.

Guarro, J. (2011). Lessons from animal studies for the treatment of invasive human infections due to uncommon fungi. J. Antimicrob. Chemother. *66*, 1447–1466.

Guy, J.S. (2008). Isolation and propagation of coronaviruses in embryonated eggs. Methods Mol. Biol. *454*, 109–117.

Hanson, L.H., Clemons, K.V., Denning, D.W., and Stevens, D.A. (1995). Efficacy of oral saperconazole in systemic murine aspergillosis. J. Med. Vet. Mycol. *33*, 311–317.

Hata, K., Kimura, J., Miki, H., Toyosawa, T., Moriyama, M., and Katsu, K. (1996). Efficacy of ER-30346, a novel oral triazole antifungal agent, in experimental models of aspergillosis, candidiasis, and cryptococcosis. Antimicrob. Agent Chemother. *40*, 2243–2247.

Hata, K., Horii, T., Miyazaki, M., Watanabe, N.A., Okubo, M., Sonoda, J., Nakamoto, K., Tanaka, K., Shirotori, S., Murai, N., *et al.* (2011). Efficacy of oral E1210, a new broad-spectrum antifungal with a novel mechanism of action, in murine models of candidiasis, aspergillosis, and fusariosis. Antimicrob. Agent Chemother. *55*, 4543–4551.

Hayama, K., Ishibashi, H., Ishijima, S.A., Niimi, K., Tansho, S., Ono, Y., Monk, B.C., Holmes, A.R., Harding, D.R.K., Cannon, R.D., *et al.* (2012). A D-octapeptide drug efflux pump inhibitor acts synergistically with azoles in a murine oral candidiasis infection model. FEMS Microbiol. Lett. *328*, 130–137.

Hermsen, S.A., Pronk, T.E., van den Brandhof, E.J., van der Ven, L.T., and Piersma, A.H. (2012a). Concentration-response analysis of differential gene expression in the zebrafish embryotoxicity test following flusilazole exposure. Toxicol. Sci. *127*, 303–312.

Hermsen, S.A., Pronk, T.E., van den Brandhof, E.J., van der Ven, L.T., and Piersma, A.H. (2012b). Triazole-induced gene expression changes in the zebrafish embryo. Reprod. Toxicol. *34*, 216–224.

Herrera, C., and Guentzel, M.N. (1982). Mice with persistent gastrointestinal *Candida albicans* as a model for antifungal therapy. Antimicrob. Agent Chemother. *21*, 51–53.

Hu, W., Ninomiya, K., Ishibashi, H., Maruyama, N., Oshima, H., Yamaguchi, H., and Abe, S. (2007). A novel murine model of pharyngeal candidiasis with local symptoms characteristic of pharyngeal thrush produced by using an inhaled corticosteroid. Med. Mycol. *45*, 143–148.

Hummel, M., Baust, C., Kretschmar, M., Nichterlein, T., Schleiermacher, D., Spiess, B., Skladny, H., Morz, H., Hehlmann, R., and Buchheidt, D. (2004). Detection of Aspergillus DNA by a nested PCR assay is superior to blood culture in an experimental murine model of invasive aspergillosis. J. Med. Microbiol. *53*, 803–806.

Ilkit, M., and Guzel, A.B. (2011). The epidemiology, pathogenesis, and diagnosis of vulvovaginal candidosis: a mycological perspective. Crit. Rev. Microbiol. *37*, 250–261.

Ishibashi, H., Hisajima, T., Hu, W., Yamaguchi, H., Nishiyama, Y., and Abe, S. (2007a). A murine model of esophageal candidiasis with local characteristic symptoms. Microbiol. Immunol. *51*, 501–506.

Ishibashi, H., Uchida, K., Nishiyama, Y., Yamaguchi, H., and Abe, S. (2007b). Oral administration of itraconazole solution has superior efficacy in experimental oral and oesophageal candidiasis in mice than its intragastric administration. J. Antimicrob. Chemother. *59*, 317–320.

Ishibashi, Y., Kaufman, H.E., and Kagawa, S. (1986a). Comparison of the pathogenicities of *Fusarium solani* and *Candida albicans* in the rabbit cornea. J. Med. Vet. Mycol. *24*, 369–376.

Ishibashi, Y., Kaufman, H.E., and Kagawa, S. (1986b). Differences in development of keratomycosis after Candida inoculation in the central and peripheral cornea. J. Med. Vet. Mycol. *24*, 437–444.

Jacobsen, I.D., Brunke, S., Seider, K., Schwarzmüller, T., Firon, A., d'Enfert, C., Kuchler, K., and Hube, B. (2010a). *Candida glabrata* persistence in mice does not depend on host immunosuppression and is unaffected by fungal amino acid auxotrophy. Infect. Immun. *78*, 1066–1077.

Jacobsen, I.D., Grosse, K., Slesiona, S., Hube, B., Berndt, A., and Brock, M. (2010b). Embryonated eggs as an alternative infection model to investigate *Aspergillus fumigatus* virulence. Infect. Immun. *78*, 2995–3006.

Jacobsen, I.D., Grosse, K., Berndt, A., and Hube, B. (2011). Pathogenesis of *Candida albicans* infections in the alternative chorio-allantoic membrane chicken embryo model resembles systemic murine infections. PLoS One *6*, e19741.

Jacobsen, I.D., Grosse, K., and Hube, B. (2012). Embryonated chicken eggs as alternative infection model for pathogenic fungi. Methods Mol. Biol. *845*, 487–496.

Kamai, Y., Kubota, M., Hosokawa, T., Fukuoka, T., and Filler, S.G. (2001). New model of oropharyngeal candidiasis in mice. Antimicrob. Agents Chemother. *45*, 3195–3197.

Kamai, Y., Kubota, M., Fukuoka, T., Maeda, N., Hosokawa, T., Shibayama, T., Uchida, K., Yamaguchi, H., and

Kuwahara, S. (2003). Efficacy of CS-758, a novel triazole, against experimental fluconazole-resistant oropharyngeal candidiasis in mice. Antimicrob. Agent Chemother. *47*, 601–606.

Kamei, K. (2001). Animal models of zygomycosis--Absidia, Rhizopus, Rhizomucor, and Cunninghamella. Mycopathologia *152*, 5–13.

Kimmel, C.B., Ballard, W.W., Kimmel, S.R., Ullmann, B., and Schilling, T.F. (1995). Stages of embryonic development of the zebrafish. Dev. Dyn. *203*, 253–310.

Kinsman, O.S., and Pitblado, K. (1989). *Candida albicans* gastrointestinal colonization and invasion in the mouse: effect of antibacterial dosing, antifungal therapy and immunosuppression. Mycoses *32*, 664–674.

Kirkpatrick, W.R., Najvar, L.K., Bocanegra, R., Patterson, T.F., and Graybill, J.R. (2007). New guinea pig model of Cryptococcal meningitis. Antimicrob. Agent Chemother. *51*, 3011–3013.

Kirkpatrick, W.R., Wiederhold, N.P., Najvar, L.K., and Patterson, T.F. (2012). Animal models in mycology: what have we learned over the past 30 years. Curr. Fungal Infect. Rep. *7*, 68–78.

Kiryu, H., Yoshida, S., Suenaga, Y., and Asahi, M. (1991). Invasion and survival of Fusarium solani in the dexamethasone-treated cornea of rabbits. J. Med. Vet. Mycol. *29*, 395–406.

Klepser, M.E., Wolfe, E.J., Jones, R.N., Nightingale, C.H., and Pfaller, M.A. (1997). Antifungal pharmacodynamic characteristics of fluconazole and amphotericin B tested against *Candida albicans*. Antimicrob. Agent Chemother. *41*, 1392–1395.

Kumamoto, C.A. (2014). Candida/host interactions: responding to the host environment. Presentation at: 12th American Society for Microbiology (ASM) Conference Candida and Candidiasis, 26–30th March (New Orleans, LA, USA).

Kuo, Z.Y., Chuang, Y.J., Chao, C.C., Liu, F.C., Lan, C.Y., and Chen, B.S. (2013). Identification of infection- and defense-related genes via a dynamic host–pathogen interaction network using a *Candida albicans*–zebrafish infection model. J. Innate Immun. *5*, 137–152.

Lacasse, M., Fortier, C., Chakir, J., Cote, L., and Deslauriers, N. (1993). Acquired resistance and persistence of *Candida albicans* following oral candidiasis in the mouse: a model of the carrier state in humans. Oral Microbiol. Immunol. *8*, 313–318.

Lamaris, G.A., Chamilos, G., Lewis, R.E., and Kontoyiannis, D.P. (2007). Virulence studies of *Scedosporium* and *Fusarium* species in *Drosophila melanogaster*. J. Infect. Dis. *196*, 1860–1864.

Leal, S.M., Jr., and Pearlman, E. (2012). The role of cytokines and pathogen recognition molecules in fungal keratitis – Insights from human disease and animal models. Cytokine *58*, 107–111.

Leal, S.M., Jr., Cowden, S., Hsia, Y.C., Ghannoum, M.A., Momany, M., and Pearlman, E. (2010). Distinct roles for Dectin-1 and TLR4 in the pathogenesis of *Aspergillus fumigatus* keratitis. PLoS Pathog. *6*, e1000976.

Levitz, S.M. (2001). Does amoeboid reasoning explain the evolution and maintenance of virulence factors in *Cryptococcus neoformans*? Proc. Natl. Acad. Sci. U.S.A. *98*, 14760–14762.

Lewis, R.E., and Wiederhold, N.P. (2005). Murine model of invasive aspergillosis. Methods Mol. Med. *118*, 129–142.

Li, D.-D., Deng, L., Hu, G.-H., Zhao, L.-X., Hu, D.-D., Jiang, Y.-Y., and Wang, Y. (2013). Using *Galleria mellonella–Candida albicans* infection model to evaluate antifungal agents. Biol. Pharm. Bull. *36*, 1482–1487.

Lionakis, M.S. (2011). *Drosophila* and *Galleria* insect model hosts: New tools for the study of fungal virulence, pharmacology and immunology. Virulence *2*, 521–527.

Lionakis, M.S., and Kontoyiannis, D.P. (2012). *Drosophila melanogaster* as a model organism for invasive aspergillosis. Methods Mol. Biol. *845*, 455–468.

Lionakis, M.S., Lewis, R.E., May, G.S., Wiederhold, N.P., Albert, N.D., Halder, G., and Kontoyiannis, D.P. (2005). Toll-deficient *Drosophila* flies as a fast, high-throughput model for the study of antifungal drug efficacy against invasive aspergillosis and *Aspergillus* virulence. J. Infect. Dis. *191*, 1188–1195.

Lohberger, A., Coste, A., and Sanglard, D. (2013). Distinct roles of the drug resistance transcription factors TAC1, MRR1 and UPC2 from *Candida albicans* in virulence. Eukaryot. Cell. *13*, 127–142.

Lopez-Ribot, J.L., McAtee, R.K., Perea, S., Kirkpatrick, W.R., Rinaldi, M.G., and Patterson, T.F. (1999). Multiple resistant phenotypes of *Candida albicans* coexist during episodes of oropharyngeal candidiasis in human immunodeficiency virus-infected patients. Antimicrob. Agent Chemother. *43*, 1621–1630.

Louria, D.B., Kaminski, T., and Finkel, G. (1963). Further studies on immunity in experimental cryptococcosis. J. Exp. Med. *117*, 509–520.

Lozano-Chiu, M., Arikan, S., Paetznick, V.L., Anaissie, E.J., Loebenberg, D., and Rex, J.H. (1999). Treatment of murine fusariosis with SCH 56592. Antimicrob. Agent Chemother. *43*, 589–591.

MacCallum, D.M. (2012a). Hosting infection: experimental models to assay *Candida* virulence. Int. J. Microbiol. *2012*, 1–12.

MacCallum, D.M. (2012b). Mouse intravenous challenge models and applications. Methods Mol. Biol. *845*, 499–509.

MacCallum, D.M., and Odds, F.C. (2002). Efficacy of parenteral itraconazole against disseminated *Candida albicans* infection in two mouse strains. J. Antimicrob. Chemother. *50*, 225–229.

MacCallum, D.M., and Odds, F.C. (2005). Temporal events in the intravenous challenge model for experimental *Candida albicans* infections in female mice. Mycoses *48*, 151–161.

MacCallum, D.M., Castillo, L., Nather, K., Munro, C., Brown, A., Gow, N., and Odds, F. (2009a). Property differences among the four major *Candida albicans* strain clades. Eukaryot Cell. *8*, 373–387.

MacCallum, D.M., Castillo, L., Brown, A.J.P., Gow, N.A.R., and Odds, F.C. (2009b). Early-expressed chemokines predict kidney immunopathology in experimental disseminated *Candida albicans* infections. PLoS One *4*, e6420.

MacCallum, D.M., Coste, A., Ischer, F., Jacobsen, M.D., Odds, F.C., and Sanglard, D. (2010). Genetic dissection of azole resistance mechanisms in *Candida albicans* and their validation in a mouse model of disseminated infection. Antimicrob. Agent Chemother. *54*, 1476–1483.

McNeil, M.M., Nash, S.L., Hajjeh, R.A., Phelan, M.A., Conn, L.A., Plikaytis, B.D., and Warnock, D.W. (2001). Trends in mortality due to invasive mycotic diseases in the United States, 1980–1997. Clin. Infect. Dis. 33, 641–647.

Maki, K., and Kaneko, S. (2013). Effective concentration-based serum pharmacodynamics for antifungal azoles in a murine model of disseminated *Candida albicans* infection. Eur. J. Drug Metab. Pharmacokinet. 38, 261–268.

Marchetti, O., Entenza, J.M., Sanglard, D., Bille, J., Glauser, M.P., and Moreillon, P. (2000). Fluconazole plus cyclosporine: a fungicidal combination effective against experimental endocarditis due to *Candida albicans*. Antimicrob. Agents Chemother. *44*, 2932–2938.

Marr, K.A., Carter, R.A., Crippa, F., Wald, A., and Corey, L. (2002). Epidemiology and outcome of mould infections in hematopoietic stem cell transplant recipients. Clin. Infect. Dis. *34*, 909–917.

Matsubara, V.H., Silva, E.G., Paula, C.R., Ishikawa, K.H., and Nakamae, A.E. (2012). Treatment with probiotics in experimental oral colonization by *Candida albicans* in murine model (DBA/2). Oral Dis. *18*, 260–264.

Mavridou, E., Bruggemann, R.J., Melchers, W.J., Mouton, J.W., and Verweij, P.E. (2010a). Efficacy of posaconazole against three clinical *Aspergillus fumigatus* isolates with mutations in the cyp51A gene. Antimicrob. Agent Chemother. *54*, 860–865.

Mavridou, E., Bruggemann, R.J., Melchers, W.J., Verweij, P.E., and Mouton, J.W. (2010b). Impact of cyp51A mutations on the pharmacokinetic and pharmacodynamic properties of voriconazole in a murine model of disseminated aspergillosis. Antimicrob. Agent Chemother. *54*, 4758–4764.

Mayayo, E., Guarro, J., and Pujol, I. (1998). Endogenous endophthalmitis by *Fusarium solani*: an animal experimental model. Med. Mycol. *36*, 249–253.

Mellado, E., Cuenca-Estrella, M., Regadera, J., Gonzalez, M., Diaz-Guerra, T.M., and Rodriguez-Tudela, J.L. (2000). Sustained gastrointestinal colonization and systemic dissemination by *Candida albicans, Candida tropicalis* and *Candida parapsilosis* in adult mice. Diagn. Microbiol. Infect. Dis. *38*, 21–28.

Melo, N.R.d., Abdrahman, A., Greig, C., Mukherjee, K., Thornton, C., Ratcliffe, N.A., Vilcinskas, A., and Butt, T.M. (2013). Myriocin significantly increases the mortality of a non-mammalian model host during *Candida* pathogenesis. PLoS One *8*, e78905.

Mendiratta, V., Mittal, S., Jain, A., and Chander, R. (2010). Mucocutaneous manifestations in children with human immunodeficiency virus infection. Indian J. Dermatol. Venereol. Leprol. *76*, 458–466.

Mesa-Arango, A.C., Forastiero, A., Bernal-Martínez, L., Cuenca-Estrella, M., Mellado, E., and Zaragoza, O. (2012). The non-mammalian host *Galleria mellonella* can be used to study the virulence of the fungal pathogen *Candida tropicalis* and the efficacy of antifungal drugs during infection by this pathogenic yeast. Med. Mycol. *51*, 461–472.

Miceli, M.H., Díaz, J.A., and Lee, S.A. (2011). Emerging opportunistic yeast infections. Lancet Infect. Dis. *11*, 142–151.

Miranda, L.N., van der Heijden, I.M., Costa, S.F., Sousa, A.P., Sienra, R.A., Gobara, S., Santos, C.R., Lobo, R.D., Pessoa, V.P., Jr., and Levin, A.S. (2009). Candida colonisation as a source for candidaemia. J. Hosp. Infect. *72*, 9–16.

Mirkov, I., Stojanovic, I., Glamoclija, J., Stosic-Grujicic, S., Zolotarevski, L., Kataranovski, D., and Kataranovski, M. (2011). Differential mechanisms of resistance to sublethal systemic *Aspergillus fumigatus* infection in immunocompetent BALB/c and C57BL/6 mice. Immunobiology *216*, 234–242.

Mirkov, I., Stosic-Grujicic, S., and Kataranovski, M. (2012). Host immune defense against *Aspergillus fumigatus*: insight from experimental systemic (disseminated) infection. Immunol. Res. *52*, 120–126.

Mitchell, A.P. (2014). Pathogen response to the environment of invasive infection. Presentation at: 12th American Society for Microbiology (ASM) Conference Candida and Candidiasis, 26–30th March (New Orleans, LA, USA).

Morton, C.O., Clemons, K.V., Springer, J., Mueller, J.G., Rogers, T.R., Stevens, D.A., Kurzai, O., Einsele, H., and Loeffler, J. (2011). Real-time PCR and quantitative culture for monitoring of experimental *Aspergillus fumigatus* intracranial infection in neutropenic mice. J. Med. Microbiol. *60*, 913–919.

Morton, D.B., Dunphy, G.B., and Chadwick, J.S. (1987). Reactions of hemocytes of immune and non-immune *Galleria mellonella* larvae to *Proteus mirabilis*. Dev. Comp. Immunol. *11*, 47–55.

Mowlds, P., and Kavanagh, K. (2008). Effect of pre-incubation temperature on susceptibility of *Galleria mellonella* larvae to infection by *Candida albicans*. Mycopathologia *165*, 5–12.

Muhammed, M., Coleman, J.J., and Mylonakis, E. (2012a). *Caenorhabditis elegans*: a nematode infection model for pathogenic fungi. Methods Mol. Biol. *845*, 447–454.

Muhammed, M., Feldmesser, M., Shubitz, L.F., Lionakis, M.S., Sil, A., Wang, Y., Glavis-Bloom, J., Lewis, R.E., Galgiani, J.N., Casadevall, A., *et al.* (2012b). Mouse models for the study of fungal pneumonia: a collection of detailed experimental protocols for the study of Coccidioides, Cryptococcus, Fusarium, Histoplasma and combined infection due to Aspergillus-Rhizopus. Virulence 3, 329–338.

Mylonakis, E., Ausubel, F.M., Perfect, J.R., Heitman, J., and Calderwood, S.B. (2002). Killing of *Caenorhabditis elegans* by *Cryptococcus neoformans* as a model of yeast pathogenesis. Proc. Natl. Acad. Sci. U.S.A. *99*, 15675–15680.

Mylonakis E., Moreno, R., El Khoury J.B., Idnurm, A., Heitman, J., Calderwood, S.B., Ausubel, F.M., and Diener, A. (2005). *Galleria mellonella* as a model system to study *Cryptococcus neoformans* pathogenesis. Infect. Immun. 73, 3842–3850.

Naglik, J.R., Fidel, P.L., Jr., and Odds, F.C. (2008). Animal models of mucosal Candida infection. FEMS Microbiol. Lett. *283*, 129–139.

Navarro-García, F., Sánchez, M., Nombela, C., and Pla, J. (2001). Virulence genes in the pathogenic yeast *Candida albicans*. FEMS Microbiol. Rev. *25*, 245–268.

Ngamskulrungroj, P., Chang, Y., Sionov, E., and Kwon-Chung, K.J. (2012). The primary target organ of *Cryptococcus gattii* is different from that of *Cryptococcus neoformans* in a murine model. MBio *3*, e00103–12.

Noble, S.M. (2013). *Candida albicans* specializations for iron homeostasis: from commensalism to virulence. Curr. Opin. Microbiol. *16*, 708–715.

Odds, F.C., Gow, N.A., and Brown, A.J. (2001). Fungal virulence studies come of age. Genome Biol. *2*, REVIEWS1009.

Odds, F.C., Davidson, A., Jacobsen, M., Tavanti, A., Whyte, J., Kibbler, C., Ellis, D., Maiden, M., Shaw, D., and Gow, N. (2006). *Candida albicans* strain maintenance, replacement, and microvariation demonstrated by multilocus sequence typing. J. Clin. Microbiol. *44*, 3647–3658.

O'Keefe, L. (2013). Carcinogen Selector – Integumentary System (Sigma-Aldrich, St Louis, MO). Available at: http://www.sigmaaldrich.com/life-science/cell-biology/cancer-research/learning-center/carcinogen-selector/integumentary-system.html

Outerbridge, C.A. (2006). Mycologic disorders of the skin. Clin. Tech. Small Anim. Pract. *21*, 128–134.

Papadimitriou, J.M., and Ashman, R.B. (1986). The pathogenesis of acute systemic candidiasis in a susceptible inbred mouse strain. J. Pathol. *150*, 257–265.

Pasko, M.T., Piscitelli, S.C., and Van Slooten, A.D. (1990). Fluconazole: a new triazole antifungal agent. DICP *24*, 860–867.

Polak, A. (1998). Experimental models in antifungal chemotherapy. Mycoses *41*, 1–30.

Ponnuvel, K.M., Rama, C.P., and Menon, T. (1993). Systemic and gastrointestinal candidiasis of infant mice as model for antifungal therapy. Indian J. Exp. Biol. *31*, 450–452.

Pope, L.M., Cole, G.T., Guentzel, M.N., and Berry, L.J. (1979). Systemic and gastrointestinal candidiasis of infant mice after intragastric challenge. Infect. Immun. *25*, 702–707.

Pukkila-Worley, R., Holson, E., Wagner, F., and Mylonakis, E. (2009). Antifungal drug discovery through the study of invertebrate model hosts. Curr. Med. Chem. *16*, 1588–1595.

Pukkila-Worley, R., Ausubel, F.M., and Mylonakis, E. (2011). *Candida albicans* infection of *Caenorhabditis elegans* induces antifungal immune defenses. PLoS Pathog. *7*, e1002074.

Quintin, J., Asmar, J., Matskevich, A.A., Lafarge, M.C., and Ferrandon, D. (2013). The *Drosophila* toll pathway controls but does not clear *Candida glabrata* infections. J. Immunol. *190*, 2818–2827.

Rahman, D., Mistry, M., Thavaraj, S., Naglik, J.R., and Challacombe, S.J. (2012). Murine model of concurrent oral and vaginal *Candida albicans* colonisation. Methods Mol. Biol. *845*, 527–535.

Ramos, E.S.M., Lima, C.M., Schechtman, R.C., Trope, B.M., and Carneiro, S. (2010). Superficial mycoses in immunodepressed patients (AIDS). Clin. Dermatol. *28*, 217–225.

Ray, T.L., and Wuepper, K.D. (1976). Experimental cutaneous candidiasis in rodents. J. Invest. Dermatol. *66*, 29–33.

Reeves, E.P., Messina, C.G., Doyle, S., and Kavanagh, K. (2004). Correlation between gliotoxin production and virulence of *Aspergillus fumigatus* in *Galleria mellonella*. Mycopathologia *158*, 73–79.

Renwick, J., Reeves, E.P., Wientjes, F.B., and Kavanagh, K. (2007). Translocation of proteins homologous to human neutrophil p47phox and p67phox to the cell membrane in activated hemocytes of *Galleria mellonella*. Dev. Comp. Immunol. *31*, 347–359.

Restrepo, M.I., Najvar, L.K., Barchiesi, F., Luther, M.F., and Graybill, J.R. (2000). Treatment of murine cryptococcal meningitis with UR-9751 and UR-9746. Med. Mycol. *38*, 221–224.

Rhodes, J.C., Wicker, L.S., and Urba, W.J. (1980). Genetic control of susceptibility to *Cryptococcus neoformans* in mice. Infect. Immun. *29*, 494–499.

Romani, L. (1999). Immunity to *Candida albicans*: Th1, Th2 cells and beyond. Curr. Opin. Microbiol. *2*, 363–367.

Russell, W.M.S., and Burch, R.L. (1959). The Principles of Humane Experimental Technique (Methuen & Co. London, UK).

Samaranayake, Y.H., and Samaranayake, L.P. (2001). Experimental oral candidiasis in animal models. Clin. Microbiol. Rev. *14*, 398–429.

Samonis, G., Maraki, S., Hajiioannou, I., Chatzinikolaou, I., Rolston, K.V., Bodey, G.P., and Kontoyiannis, D.P. (2001). Effect of ketoconazole, itraconazole and fluconazole on the gastrointestinal colonization of mice by *Candida albicans*. J. Chemother. *13*, 66–69.

Sangeorzan, J.A., Bradley, S.F., He, X., Zarins, L.T., Ridenour, G.L., Tiballi, R.N., and Kauffman, C.A. (1994). Epidemiology of oral candidiasis in HIV-infected patients: colonization, infection, treatment, and emergence of fluconazole resistance. Am. J. Med. *97*, 339–346.

Sawyer, R.G., Adams, R.B., Rosenlof, L.K., May, A.K., and Pruett, T.L. (1995). The role of *Candida albicans* in the pathogenesis of experimental fungal/bacterial peritonitis and abscess formation. Am. Surg. *61*, 726–731.

Sawyer, R.T., Moon, R.J., and Beneke, E.S. (1976). Hepatic clearance of *Candida albicans* in rats. Infect. Immun. *14*, 1348–1355.

Sawyer, R.T., Horst, M.N., Garner, R.E., Hudson, J., Jenkins, P.R., and Richardson, A.L. (1990). Altered hepatic clearance and killing of *Candida albicans* in the isolated perfused mouse liver model. Infect. Immun. *58*, 2869–2874.

Sawyer, R.T., Garner, R.E., and Hudson, J.A. (1992). Effect of lectins on hepatic clearance and killing of *Candida albicans* by the isolated perfused mouse liver. Infect. Immun. *60*, 1041–1046.

Schmidt, A. (2002). Animal models of aspergillosis – also useful for vaccination strategies? Mycoses *45*, 38–40.

Scorzoni, L., de Lucas, M.P., Mesa-Arango, A.C., Fusco-Almeida, A.M., Lozano, E., Cuenca-Estrella, M., Mendes-Giannini, M.J., and Zaragoza, O. (2013). Antifungal efficacy during *Candida krusei* infection in non-conventional models correlates with the yeast *in vitro* susceptibility profile. PLoS One *8*, e60047.

Shadkchan, Y., Shemesh, E., Mirelman, D., Miron, T., Rabinkov, A., Wilchek, M., and Osherov, N. (2004). Efficacy of allicin, the reactive molecule of garlic, in inhibiting *Aspergillus* spp. *in vitro*, and in a murine model of disseminated aspergillosis. J. Antimicrob. Chemother. *53*, 832–836.

Sheppard, D.C., Graybill, J.R., Najvar, L.K., Chiang, L.Y., Doedt, T., Kirkpatrick, W.R., Bocanegra, R., Vallor, A.C., Patterson, T.F., and Filler, S.G. (2006). Standardization of an experimental murine model of invasive pulmonary aspergillosis. Antimicrob. Agent Chemother. *50*, 3501–3503.

Shimamura, T., Kubota, N., and Shibuya, K. (2012). Animal model of dermatophytosis. J. Biomed. Biotechnol. *2012*, 125384.

Sionov, E., Mendlovic, S., and Segal, E. (2005). Experimental systemic murine aspergillosis: treatment with polyene and caspofungin combination and G-CSF. J. Antimicrob. Chemother. *56*, 594–597.

Smulian, A.G. (2012). Invasive models of histoplasmosis. Methods Mol. Biol. *845*, 519–525.

Sohnle, P.G., and Hahn, B.L. (1989). Effect of immunosuppression on epidermal defenses in a murine model of cutaneous candidiasis. J. Lab. Clin. Med. *113*, 700–707.

Solis, N.V., and Filler, S.G. (2012). Mouse model of oropharyngeal candidiasis. Nat. Protoc. *7*, 637–642.

Soo, G.W., Law, J.H., Kan, E., Tan, S.Y., Lim, W.Y., Chay, G., Bukhari, N.I., and Segarra, I. (2010). Differential effects of ketoconazole and primaquine on the pharmacokinetics and tissue distribution of imatinib in mice. Anticancer Drugs *21*, 695–703.

Spellberg, B., Ibrahim, A.S., Edwards, J.E., Jr., and Filler, S.G. (2005). Mice with disseminated candidiasis die of progressive sepsis. J. Infect. Dis. *192*, 336–343.

Spellberg, B., Schwartz, J., Fu, Y., Avanesian, V., Adler-Moore, J., Edwards, J.E., Jr., and Ibrahim, A.S. (2006). Comparison of antifungal treatments for murine fusariosis. J. Antimicrob. Chemother. *58*, 973–979.

Springer, D.J., Saini, D., Byrnes, E.J., Heitman, J., and Frothingham, R. (2013). Development of an aerosol model of *Cryptococcus* reveals humidity as an important factor affecting the viability of *Cryptococcus* during aerosolization. PLoS One *8*, e69804.

Srinivasan, M. (2004). Fungal keratitis. Curr. Opin. Ophthalmol. *15*, 321–327.

Staib, P., Kretschmar, M., Nichterlein, T., Hof, H., and Morschhauser, J. (2002). Transcriptional regulators Cph1p and Efg1p mediate activation of the *Candida albicans* virulence gene SAP5 during infection. Infect. Immun. *70*, 921–927.

Steenbergen, J., and Casadevall, A. (2003). The origin and maintenance of virulence for the human pathogenic fungus *Cryptococcus neoformans*. Microbes Infect. *5*, 667–675.

Steinbach, W.J., Benjamin, D.K., Jr., Trasi, S.A., Miller, J.L., Schell, W.A., Zaas, A.K., Foster, W.M., and Perfect, J.R. (2004). Value of an inhalational model of invasive aspergillosis. Med. Mycol. *42*, 417–425.

Stevens, D.A., and Clemons, K.V. (2007). Azole therapy of clinical and experimental coccidioidomycosis. Ann. N. Y. Acad. Sci. *1111*, 442–454.

Stiller, R., Bennett, J., Scholer, H., Wall, M., Polak, A., and Stevens, D. (1983). Correlation of *in vitro* susceptibility test results with *in vivo* response: flucytosine therapy in a systemic candidiasis model. J. Infect. Dis. *147*, 1070–1077.

Szabo, E.K., and MacCallum, D.M. (2011). The contribution of mouse models to our understanding of systemic candidiasis. FEMS Microbiol. Lett. *320*, 1–8.

Taguchi, Y., Ishibashi, H., Takizawa, T., Inoue, S., Yamaguchi, H., and Abe, S. (2005). Protection of oral or intestinal candidiasis in mice by oral or intragastric administration of herbal food, clove (*Syzygium aromaticum*). Nihon Ishinkin Gakkai Zasshi *46*, 27–33.

Takakura, N., Sato, Y., Ishibashi, H., Oshima, H., Uchida, K., Yamaguchi, H., and Abe, S. (2003). A novel murine model of oral candidiasis with local symptoms characteristic of oral thrush. Microbiol. Immunol. *47*, 321–326.

Tarabishy, A.B., Aldabagh, B., Sun, Y., Imamura, Y., Mukherjee, P.K., Lass, J.H., Ghannoum, M.A., and Pearlman, E. (2008). MyD88 regulation of Fusarium keratitis is dependent on TLR4 and IL-1R1 but not TLR2. J. Immunol. *181*, 593–600.

Thakur, J.K., Arthanari, H., Yang, F., Pan, S.J., Fan, X., Breger, J., Frueh, D.P., Gulshan, K., Li, D.K., Mylonakis, E., *et al.* (2008). A nuclear receptor-like pathway regulating multidrug resistance in fungi. Nature *452*, 604–609.

The European Parliament and the Council of the European Union (2010). Directive 2010/63/EU of the European Parliament and of the Council of 22 September 2010 on the protection of animals used for scientific purposes. Offic. J. Eur. Union *276*, 33–79.

Thomas, P.A., and Kaliamurthy, J. (2013). Mycotic keratitis: epidemiology, diagnosis and management. Clin. Microbiol. Infect. *19*, 210–220.

Thomaz, L., García-Rodas, R., Guimarães, A.J., Taborda, C.P., Zaragoza, O., and Nosanchuk, J.D. (2013). *Galleria mellonella* as a model host to study *Paracoccidioides lutzii* and *Histoplasma capsulatum*. Virulence *4*, 139–146.

Thompson, G.R., 3rd, Wiederhold, N.P., Najvar, L.K., Bocanegra, R., Kirkpatrick, W.R., Graybill, J.R., and Patterson, T.F. (2012). A murine model of *Cryptococcus gattii* meningoencephalitis. J. Antimicrob. Chemother. *67*, 1432–1438.

Trammell, R.A., and Toth, L.A. (2011). Markers for predicting death as an outcome for mice used in infectious disease research. Comp. Med. *61*, 492–498.

Troke, P.F., Andrews, R.J., Marriott, M.S., and Richardson, K. (1987). Efficacy of fluconazole (UK-49,858) against experimental aspergillosis and cryptococcosis in mice. J. Antimicrob. Chemother. *19*, 663–670.

Van Cutsem, J., Van Gerven, F., and Janssen, P.A. (1987). Activity of orally, topically, and parenterally administered itraconazole in the treatment of superficial and deep mycoses: animal models. Rev. Infect. Dis. *9* (Suppl. 1), S15–32.

Van Cutsem, J., Van Gerven, F., and Janssen, P.A. (1989). Oral and parenteral therapy with saperconazole (R 66905) of invasive aspergillosis in normal and immunocompromised animals. Antimicrob. Agent Chemother. *33*, 2063–2068.

Vandeputte, P., Ischer, F., Sanglard, D., and Coste, A.T. (2011). *In vivo* systematic analysis of *Candida albicans* Zn2-Cys6 transcription factors mutants for mice organ colonization. PLoS One *6*, e26962.

Vande Velde, G., Kucharikova, S., Schrevens, S., Himmelreich, U., and Van Dijck, P. (2013). Towards non-invasive monitoring of pathogen–host interactions during *Candida albicans* biofilm formation using *in vivo* bioluminescence. Cell. Microbiol. *16*, 115–130.

Vande Velde, G., Kucharikova, S., Van Dijck, P., and Himmelreich, U. (2014). Bioluminescence imaging of fungal biofilm development in live animals. Methods Mol. Biol. *1098*, 153–167.

Vogel, H., Altincicek, B., Glöckner, G., and Vilcinskas, A. (2011). A comprehensive transcriptome and immune-gene repertoire of the lepidopteran model host *Galleria mellonella*. BMC Genomics *12*, 308.

Vonk, A.G., Netea, M.G., van Krieken, J.H., Verweij, P.E., van der Meer, J.W., and Kullberg, B.J. (2003). Treatment of intra-abdominal abscesses caused by *Candida albicans* with antifungal agents and recombinant murine granulocyte colony-stimulating factor. Antimicrob. Agent Chemother. *47*, 3688–3693.

Vu, K., and Gelli, A. (2010). Astemizole and an analogue promote fungicidal activity of fluconazole against *Cryptococcus neoformans* var. *grubii* and *Cryptococcus gattii*. Med. Mycol. *48*, 255–262.

Walsh, T.J., Gonzalez, C.E., Piscitelli, S., Bacher, J.D., Peter, J., Torres, R., Shetti, D., Katsov, V., Kligys, K., and Lyman, C.A. (2000). Correlation between *in vitro* and *in vivo* antifungal activities in experimental fluconazole-resistant oropharyngeal and esophageal candidiasis. J. Clin. Microbiol. *38*, 2369–2373.

Walters, J.B., and Ratcliffe, N.A. (1983). Studies on the *in vivo* cellular reactions of insects: fate of pathogenic and non-pathogenic bacteria in *Galleria mellonella* nodules. J. Insect. Physiol. *29*, 417–424.

Warr, G.W. (1997). The adaptive immune system of fish. Dev. Biol. Stand. *90*, 15–21.

Weers, P.M., and Ryan, R.O. (2006). Apolipophorin III: role model apolipoprotein. Insect Biochem. Mol. Biol. *36*, 231–240.

Westwater, C., Schofield, D.A., Nicholas, P.J., Paulling, E.E., and Balish, E. (2007). *Candida glabrata* and *Candida albicans*; dissimilar tissue tropism and infectivity in a gnotobiotic model of mucosal candidiasis. FEMS Immunol. Med. Microbiol. *51*, 134–139.

Wisplinghoff, H., Bischoff, T., Tallent, S.M., Seifert, H., Wenzel, R.P., and Edmond, M.B. (2004). Nosocomial bloodstream infections in US hospitals: analysis of 24,179 cases from a prospective nationwide surveillance study. Clin. Infect. Dis. *39*, 309–317.

Wojda, I., Kowalski, P., and Jakubowicz, T. (2009). Humoral immune response of *Galleria mellonella* larvae after infection by *Beauveria bassiana* under optimal and heat-shock conditions. J. Insect Physiol. *55*, 525–531.

Wu, T.G., Wilhelmus, K.R., and Mitchell, B.M. (2003). Experimental keratomycosis in a mouse model. Invest. Ophthalmol. Vis. Sci. *44*, 210–216.

Wu, T.G., Keasler, V.V., Mitchell, B.M., and Wilhelmus, K.R. (2004). Immunosuppression affects the severity of experimental *Fusarium solani* keratitis. J. Infect. Dis. *190*, 192–198.

Yang, R.Y., Wang, W.L., Ao, J.H., Hao, Z.F., Zhang, J., and Wang, C.M. (2008). Pathogenicity of *Trichosporon asahii* in a murine model of disseminated trichosporonosis. Chin. Med. J. (Engl) *121*, 2557–2560.

Yano, J., and Fidel, P.L., Jr. (2011). Protocols for vaginal inoculation and sample collection in the experimental mouse model of *Candida vaginitis*. J. Vis. Exp. *Dec 8* (58), 3382.

Yano, J., Noverr, M.C., and Fidel, P.L., Jr. (2012). Cytokines in the host response to *Candida vaginitis*: Identifying a role for non-classical immune mediators, S100 alarmins. Cytokine *58*, 118–128.

Yordanov, M., Dimitrova, P., Patkar, S., Falcocchio, S., Xoxi, E., Saso, L., and Ivanovska, N. (2005). Ibogaine reduces organ colonization in murine systemic and gastrointestinal *Candida albicans* infections. J. Med. Microbiol. *54*, 647–653.

Zarif, L., Graybill, J.R., Perlin, D., Najvar, L., Bocanegra, R., and Mannino, R.J. (2000). Antifungal activity of amphotericin B cochleates against *Candida albicans* infection in a mouse model. Antimicrob. Agent Chemother. *44*, 1463–1469.

Zhu, J.L., Gao, X.R., Cui, H.P., Lang, L.L., Li, Q., and Liao, X. (2011). Experimental model of *Fusarium solani* keratitis in rats. Int. J. Ophthalmol. *4*, 371–376.

Current Books of Interest

Thermophilic Microorganisms	2015
Flow Cytometry in Microbiology: Technology and Applications	2015
Probiotics and Prebiotics: Current Research and Future Trends	2015
Epigenetics: Current Research and Emerging Trends	2015
Advanced Vaccine Research Methods for the Decade of Vaccines	2015
Bacteria-Plant Interactions: Advanced Research and Future Trends	2015
Aeromonas	2015
Antibiotics: Current Innovations and Future Trends	2015
Leishmania: Current Biology and Control	2015
Acanthamoeba: Biology and Pathogenesis (2nd edition)	2015
Microarrays: Current Technology, Innovations and Applications	2014
Metagenomics of the Microbial Nitrogen Cycle: Theory, Methods and Applications	2014
Pathogenic *Neisseria*: Genomics, Molecular Biology and Disease Intervention	2014
Proteomics: Targeted Technology, Innovations and Applications	2014
Biofuels: From Microbes to Molecules	2014
Human Pathogenic Fungi: Molecular Biology and Pathogenic Mechanisms	2014
Applied RNAi: From Fundamental Research to Therapeutic Applications	2014
Halophiles: Genetics and Genomes	2014
Molecular Diagnostics: Current Research and Applications	2014
Phage Therapy: Current Research and Applications	2014
Bioinformatics and Data Analysis in Microbiology	2014
The Cell Biology of Cyanobacteria	2014
Pathogenic *Escherichia coli*: Molecular and Cellular Microbiology	2014
Campylobacter Ecology and Evolution	2014
Burkholderia: From Genomes to Function	2014
Myxobacteria: Genomics, Cellular and Molecular Biology	2014
Next-generation Sequencing: Current Technologies and Applications	2014
Omics in Soil Science	2014
Applications of Molecular Microbiological Methods	2014
Mollicutes: Molecular Biology and Pathogenesis	2014
Genome Analysis: Current Procedures and Applications	2014
Bacterial Toxins: Genetics, Cellular Biology and Practical Applications	2013
Bacterial Membranes: Structural and Molecular Biology	2014
Cold-Adapted Microorganisms	2013

Full details at www.caister.com

Index

D

E

F

G

K

M

N

U

V

www.ingramcontent.com/pod-product-compliance
Lightning Source LLC
LaVergne TN
LVHW071336210926
848486LV00003B/158
9781910190012